Soil Science: Agricultural and Environmental Prospectives

Khalid Rehman Hakeem • Javaid Akhtar
Muhammad Sabir
Editors

Soil Science: Agricultural and Environmental Prospectives

Springer

Editors
Khalid Rehman Hakeem
Universiti Putra Malaysia
Selangor, Malaysia

Muhammad Sabir
Institute of Soil and Environmental Science
University of Agriculture Faisalabad
Faisalabad, Pakistan

Javaid Akhtar
Institute of Soil and Environmental Science
University of Agriculture Faisalabad
Faisalabad, Pakistan

ISBN 978-3-319-34449-2 ISBN 978-3-319-34451-5 (eBook)
DOI 10.1007/978-3-319-34451-5

Library of Congress Control Number: 2016947711

Printed on acid-free paper

This Springer imprint is published by Springer Nature
The registered company is Springer International Publishing AG Switzerland

This book is dedicated to Abdul Sattar Edhi
(Popularly known as the Angel of Mercy*)*

(1928–2016)

A prominent Pakistani philanthropist, social activist, ascetic, humanitarian and the founder of the Edhi Foundation in Pakistan

Foreword

The soil scientists currently are striving hard toward the transformation of agriculture into a more sustainable enterprise. All modern technologies are needed like geographical information systems, global positioning systems, as well as computer applications in crop production and natural resource management. During the last two decades, suggestions for a new type of soil science have come to the forefront, with more attention being paid toward a soil care approach in a closer contact with the society. There is a greater need for the young researchers to look back, note the achievements, and try to learn from the past. The authors in this book have tried to look forward by presenting the shifts in research foci. They have focused on the identity of soil science, directions for the future on a global scale, and the environmental and agricultural aspects in this field. The contributors here have also tried to actualize views on the soil science. Several well-known colleagues from different parts of the world have participated in this attempt and the book has been completed.

Our mother soil is not just a dust; it is a vital resource sustaining the miracle of life on this planet. The researchers in this field have a very important task to increase crop productivity but at the same time prevent our soils from erosion and pollution. Yes, the "green revolution" of the 1960s was a success of researchers working in the field of soils. Today, many of these workers are striving hard to find the ways of feeding the world's people in an agriculturally, environmentally, and economically sustainable way. In this sense, fundamental understanding of soil biology, chemistry, pedology, and physics has to be applied to the environmental problems caused by production.

The researchers in this field have to use geographic information systems (GIS) to analyze different aspects of soils and create soil maps for managing the soils for sustainable crop production as well as other products. There is a need for researchers in this field to get involved in creating environmental impact statements, erosion control, mine reclamation, and industrial site restoration. It all comes primarily under an applied agricultural field that is the soils.

Soil researchers have to apply their basic understanding of soils to many environmental problems, together with the concerns about soil, water, and air pollution. The book presents 18 chapters from an array of scientists.

Chapter 1 gives an appraisal of conservation tillage on the soil physical properties and highlights the information on tillage systems like *conventional tillage*, *intensive tillage*, and *conservation tillage*, the principles of conservation agriculture, comparison of tillage systems, conservation tillage effects on soil physical properties, and constraints in the adoption of conservation tillage.

Chapter 2 discusses the *degraded soil origin, types, and management and presents a detailed information on the* causes of land degradation, processes of land degradation, soil erosion, soil salinization, waterlogging, decline in soil fertility, types of land degradation, soil salinity, causes of salt-affected soils, impact of salt-affected soil on plants, reclamation of salt-affected soils, management of salt-affected soils, soil erosion, conservation technologies, soil acidity, effects of soil acidity on crop production, and finally agroforestry.

Chapter 3 summarizes the nitrogen management in rice-wheat cropping system in salt-affected soils with emphasis on the extent and nature of salt-affected soils, relationship between soil properties and salinity/sodicity, ionic and osmotic stresses, salinity stress at the cellular level, salinity impact on the N metabolism, interaction of salinity and N fertilization, the interactive effect of salinity and Ca^{2+}, and NUE in wheat and rice under saline conditions. At the end, contributors are presenting an important aspect in the soils that is nitrate leaching followed by salinity/sodicity and N management.

The management of acid sulfate soils for sustainable rice cultivation in Malaysia, constraints in acid sulfate soils, aluminum toxicity, and iron toxicity have been covered in Chap. 4.

In Chap. 5, approaches to remediate petroleum hydrocarbon-contaminated soils have been presented with emphasis on the health hazards of petroleum contamination, approaches to remediate petroleum contamination, physical approaches, chemical approaches, biological approaches, bioremediation, phytodegradation/phytotransformation, phytostabilization, phytovolatilization, and advantages and disadvantages of phytoremediation, and at the end, plant-assisted bioremediation and microbial-assisted phytoremediation have been discussed.

In Chap. 6, environmental impacts of nitrogen use in agriculture and mitigation strategies have been evaluated. The information includes nitrogen in the environment, nitrate leaching from soils, nitrate-related regulations, contribution of water and food to NO_3 ingestion, nitrate-related ecological issues in aquatic ecosystems, physical transport mechanisms of NO_3, factors involved in the NO_3 leaching environments, options to minimize NO_3 leaching, and fertilizer/soil/irrigation-based management options and strategies.

The topic on potassium for sustainable agriculture has been covered in Chap. 7, which deals with the potassium dynamics in soils, in plants, and in agriculture, environmental stresses due to K, sustainable soil fertility and K, human health and K interactions, and finally K evaluation in the soils.

Chapter 8 gives an overview of weathering and approaches to evaluation of weathering indices for soil profile studies with emphasis on physical/chemical weathering, relationship between physical/chemical weathering, quantification of weathering, the criteria applied in evaluating the utility of weathering indices, and applications of weathering indices.

The pesticide pollution in the agricultural soils of Pakistan has been discussed in Chap. 9. It covers the classification and the use of pesticides, the history of pesticides, pesticide use in the world and agricultural sector of Pakistan, major crops in Pakistan and pesticide use, pesticide occurrence in agricultural soils of Pakistan, groundwater and surface water pollution by pesticides in Pakistan, fate of pesticides in soils, toxicity of pesticides in soil, risk associated with pesticide use, and integrated pest management in Pakistan.

The problems and solutions related to the iron biofortification of cereals grown under calcareous soils have been summarized in Chap. 10. The information presented includes the status and forms of Fe in soil, iron deficiency in calcareous soils, strategies to overcome iron deficiency, significance of iron for plants, severity of iron deficiency in crops, strategies to overcome Fe deficiency in plants, organic amendments and nutrient availability, iron and human health, strategies to combat deficiency in humans, approaches for iron biofortification, nutritional factors affecting Fe bioavailability, and finally the models used for determination of iron bioavailability.

Chapter 11 discusses boron toxicity in salt-affected soils and effects on plants. Main features of this chapter are salinity, oxidative stress and plant growth, physiological responses as well as physiological and biochemical mechanisms of plants for salinity tolerance, forms of boron, sources and toxicity in soils and plants, toxicity symptoms in plants, toxicity effects on plant growth and physiology, activity of antioxidant enzymes in response to boron toxicity, photosynthetic features under boron toxicity, environment salinity and boron toxicity, and physiological and biochemical aspects.

In Chap. 12, silicon, a beneficial nutrient under salt stress, and its uptake mechanism and mode of action are presented, with details on the uptake in cereals, distribution in the mature cereal plant, silicon-mediated mechanisms improving salinity tolerance, and future prospects/missing links.

The topic of extensive research on the soil microflora has been evaluated in Chap. 13, which includes information on the effect of environment on soil microflora, advantages, and anthropogenic activities responsible for deteriorating effects on soil microflora.

Chapter 14 presents an overview of the arbuscular mycorrhizal fungi – a boon for plant nutrition and soil health. It includes detailed information on the host specificity, structural features of these mycorrhizae and their role to maintain a plant-soil nutrient balance, rhizosphere, concept and molecular signaling in the context of promoting mycorrhizal symbiosis, symbiotic relationships, their benefits in the context of sustainability of agroecosystems, sustainable soil health, biota, and soil structure and management.

An overview on the *Azotobacter chroococcum* – a potential biofertilizer in agriculture – has been discussed in Chap. 15. It covers information on the research on *A. chroococcum spp.* in crop production, its significance in plant nutrition and contribution to soil fertility and use as microbial inoculant, synthesis of growth-promoting substances, stimulation of rhizospheric microbes, protection from phytopathogens, improvement of nutrient uptake, and ultimately biological nitrogen fixation.

Sources and composition of wastewater, threats to plants and soils, industrial/domestic wastes, pesticides and insecticides, hospital/pharmaceutical wastes, nutrients, and impacts on soil and plant health are the main features discussed in Chap. 16.

Chapter 17 describes commonly used and emerging cost-effective amendments for heavy metal immobilization. There is a dire need to develop procedures to determine immobilization efficacy that could be used to assess the in situ short- and long-term environmental stability of metal immobilization.

In the last chapter, climate change and its impacts on carbon sequestration, biodiversity, and agriculture have been evaluated in the light of the difference between weather and climate, greenhouse effect/gases, global warming potential of greenhouse gases, non-greenhouse influences of climate, and global warming and its impacts in the future and on soil carbon sequestration.

The title selection of this volume is a highly challenging one, as it involves the "soils," the most complicated biomaterial present on earth. A science-oriented approach vis-a-vis an emphasis on a healthy ecosystem approach in crop production and sustainability of natural resources has been presented at length. The information gathered from this book will be helpful in the understanding of fundamental properties of and processes in soils, both of which have agricultural and environmental benefits.

I trust this book will serve its purpose, and when read carefully, it will stimulate thinking among the young researchers.

Dr. Münir Öztürk (M.Sc., Ph.D., D.Sc.)
Fellow of the Islamic World
Academy of Sciences
Professor (Emer.) of Ecology
& Environmental Sciences
Ex-Chairman Botany Department
and Founder Director Centre for Environmental Studies,
Faculty of Science, Ege University,
35100 Bornova- Izmir, Turkey
Consultant Fellow, Faculty of Forestry,
Universiti Putra Malaysia, Selangor-Malaysia
Distinguished Visiting Scientist, ICCBS,
Karachi University, Pakistan
http://ege.academia.edu/MunirOzturk
Citations: http://scholar.google.com.pk/citations?user=ooL4g4wAAAAJ&hl=en

Preface

Soil is a natural resource which supports life on earth. It provides a natural medium for plant growth, raw material for industries, and energy production. Soil is composed of mineral grains that come from weathering of the rocks which finally constitute soil particles like sand, silt, and clay. Soil formations are very slow processes which took place over thousands of years as a result of physical, chemical, and biological processes. Human interventions, climate, and living organisms are involved in this extremely slow process which ensues in soil formation. Soil is the largest reservoir of biodiversity which contains almost one-third of all living organisms. Soil performs different functions ranging from provision of livelihood and habitats of humans, animals, plants, and soil organisms to sustainability of environmental quality.

Being a natural and universal sink for the variety of the pollutants, soil occupies the pivotal position in the environment and maintaining its quality. The soil plays an important role in purification and recycling of air, water, and nutrients and thus maintains different natural cycles with ensuring the sustainability of life on earth. Soil purifies and transforms nutrients and other chemical substances and thus maintains the quality of groundwater, provides plants with nutrients, and affects the climate. Soil is the primary production factor for agriculture and forestry. Fertile soils provide the basis for the entire food chain, and thus the soil is inevitable for sustaining life on earth. However, its improper use and the underestimation of its importance are a matter of serious concern which may have dire consequences over a period of time. Environmental pollution is affecting soil productivity and thus its capacity to sustain life on earth. Different types of pollutants are added into soils like agricultural nutrients and pollutants, as well as local contamination and pollution at abandoned sites. In addition to pollutant load, soil sustainability is threatened by soil erosion caused by wind and water. Soil erosion not only depletes soil fertility but also affects environmental quality. Soil erosion is the result of intensive agriculture and unscientific management of soil resources.

In this book, we have tried to integrate literature focusing on the issue related to soil productivity, different practices to manage these issues, and then the role of the soil in environmental and agricultural sustainability. The chapters in this book highlight importance of soil as a natural resource for agricultural productivity and environmental sustainability.

We are highly grateful to all our contributors for readily accepting our invitation and for not only sharing their knowledge and research but for venerably integrating their expertise in dispersed information from diverse fields in composing the chapters and enduring editorial suggestions to finally produce this venture. We greatly appreciate their commitment. We are also thankful to Prof. Munir Ozturk for his suggestions and writing the foreword for this volume.

We thank the Springer International team for their generous cooperation at every stage of the book production.

Selangor, Malaysia Khalid Rehman Hakeem
Faisalabad, Pakistan Javaid Akhtar
Faisalabad, Pakistan Muhammad Sabir

Contents

An Appraisal of Conservation Tillage on the Soil Physical Properties

Sartaj A. Wani, Tahir Ali, M. Nayeem Sofi, M. Ramzan, and Khalid Rehman Hakeem

Contents

Abstract Farming systems today have many implications than before because of the growing concerns about agricultural sustainability and environment. Soil management is aimed at the maintenance of optimal soil physical quality for crop

S.A. Wani (✉) • T. Ali • M.N. Sofi
Division of Soil Science, Sher-e-Kashmir University of Agricultural Sciences and Technology of Kashmir, Shalimar, 190001 Srinagar, Jammu and Kashmir, India
e-mail: sartajmess@gmail.com

M. Ramzan
Department of Botany, Aligarh Muslim University, Aligarh, Uttar Pradesh, India

K.R. Hakeem (✉)
Faculty of Forestry, Universiti Putra Malaysia, 43400 Serdang, Selangor, Malaysia
e-mail: kur.hakeem@gmail.com

K.R. Hakeem et al. (eds.), *Soil Science: Agricultural and Environmental Prospectives*, DOI 10.1007/978-3-319-34451-5_1

production. The conventional tillage practices resulted in losses of soil, water and nutrients, and degraded the soil with low organic matter content and a fragile physical structure. The conservation tillage in its many and varied forms holds promise for the sustainability of agricultural productivity and environment by reducing greenhouse gas emissions, improvement in dynamic soil physical properties in general like soil aggregate stability, structure, soil strength, bulk density etc, that provides key information about the soil quality. All these aspects are reviewed with some detailed information on the benefits of conservation tillage. The aim of the present review is to analyze and discuss the conservation tillage and its impacts on physical aspects of soil health.

Keywords Conservation tillage • Soil • Sustainability • Physical Properties

1 Introduction

The greatest challenge to the world in the years to come is to provide food to burgeoning population, which would likely to rise 8909 million in 2050 and there is urgent need to increase in total food production on sustainable basis, without compromising on natural resources and environment. The growth rate in agriculture has been the major detriment in world food production. Today farming systems have more obvious and detectable social, ecological, economic and environmental implications than ever before because of the growing concerns about agricultural sustainability and the environment (Shrestha and Clements 2003). Maintenance of soil quality would reduce the problems of land degradation, decreasing soil fertility and rapidly declining production levels that occur in many parts of the world which not only lack the basic principles of good farming practices but weak technical know-how as well. The global importance of soil conservation and the control and mitigation of land degradation (Derpsch 2005) are more highly recognized now than at any time in the past. This is because rising populations and rising incomes in the middle classes, as well as increased capacity of human interventions to cause ecosystem degradation, are now of such magnitude that for the first time in history how we manage the land can impact directly on global environmental goods and services. This grave concern on environmental values is the major driving force on the geopolitical agenda for soil conservation and this is expected to increase in the future as society better understands the important linkages between soil quality and the environment (Dumanski 2015).

The importance of conserving soil resources and reducing soil erosion came first to national attention in the United States during the 'Dust Bowl' period in the early 1930s, when the combination of drought, intensive tillage practices, crop failure and wind-driven erosion of millions of acres of farmland occurred in the Great Plains of the US (Chauhan et al. 2006 and Gajri et al. 2009). In the subsequent decades, several conservation tillage production systems and other latest technical management services emerged in the mid-west and the south-east US, to address the soil loss concerns. Similarly, yet somewhat later, recognition of the importance of controlling soil erosion losses has also been a major driver for the development of conservation tillage systems (Coughenour and Chamala 2000). The introduction of Paraquat herbicide became the source of chemical foundation for no-tillage farming, which led the scientists to think that crop could be drilled into soil with minimum of tillage or no- tillage at all, when weeds are controlled chemically (Gajri et al. 2009). Conservation tillage was practiced in 1995 on about 40 × 106 ha or 35.5 % of planted area in USA. Because of the important role that surface cover or roughness has in mitigating soil erosion losses, the concept of conservation tillage during this time eventually became linked with the specific management goal of maintaining at least 30 % crop residue on the soil surface after planting.

Soil tillage one of the basic and important components of agricultural production technology, greatly influencing agricultural sustainability through its effects on soil processes, soil properties, and crop growth. Soil tillage is one of the very important factors in agriculture that would affect soil physical properties and yield of crops (Rashidi and Keshavarzpour 2008). The tillage would aim to create a soil environment favorable to the plant growth and development. Among different crop production factors, tillage contributes up to 20 % (Khurshid et al. 2006). An important effect of soil tillage on sustainability is through its impact on the environment e.g. soil degradation, water quality, emission of greenhouse gases from soil-related processes (Kassam et al. 2010). As a sub-system of a crop production system, different practices in tillage can be used to achieve many agronomic objectives, like soil conditioning, weed or pest suppression, crop residue management, incorporation or mixing (placement or redistribution of substances such as fertilizers, manures, seeds, residues, sometimes from a less favourable location to a more favourable spatial distribution), segregation (consolidation of rocks, root crops, soil crumb sizes), land forming (changing the shape of the soil surface e.g. ridging, roughening and furrowing).

Degradation of soil structure in some situations leads to a continuous soil compaction of fine particles with low levels of organic matter. Such soils are more prone to soil loss through water and wind erosion eventually resulting in desertification, as experienced in USA in the 1930s (Biswas 1984). The conventional soil tillage system practices resulted not only in loss of soil, water and nutrients in the field, but degraded the soil with low organic matter content and a fragile physical structure, which in turn led to low crop yields and low water and fertilizer use efficiency (Wang et al. 2007). The impact of tillage on soil, environment etc. depends on the combination of tillage operations and their timing in the tillage system to provide specific functions in given situations. The genetic yield potential of a crop cannot be realized even when all the other requirements are fulfilled unless the soil physical environment is maintained at its optimum level. No doubt, if these soils are managed properly for good physical health, the yield potential of different crops can be increased significantly (Indoria et al. 2016). However, the soil physical management technologies are location specific and the benefits from their adoption are greatly depend on the rainfall intensity, slope and texture of the soil besides the prevailing crop/cropping system (Indoria et al. 2016). Therefore, scientists and policy makers put emphasis on alternative form of conservation tillage systems. Conservation tillage increases the amount of crop residue left in the soil after harvest, thereby reducing soil erosion and increasing organic matter, soil aggregation, water infiltration and water holding capacity compared with conventional tillage systems (Basic 2004). Reduced tillage, mulching and crop rotation have the potential of reversing physical, chemical and biological degradation of soils (Dexter 2004) under different climatic conditions and soil types (Daraghmeh et al. 2009). Compared to conventional tillage, there are several benefits from conservation tillage such as economic benefits to labour, cost and time saved, erosion protection, soil and water conservation (Glab and Kulig 2008) and increases of soil fertility or reduce nutrient loss (Wang and Gao 2004; Limousin and Tessier 2007). One of the most successful soil management techniques in agricultural land is no-tillage management (NT), and it is being applied worldwide (Barbera et al. 2012; Lieskovský and Kenderessy 2014). Therefore, the first step in making sustainable production management decisions is to understand the practices associated with each tillage system. The different tillage systems are described below.

2 Tillage Systems

2.1 *Conventional Tillage*

Conventional tillage is a tillage system using cultivation as the major means of seedbed preparation and weed control. Conventional tillage is defined by the *Conservation Tillage Information Center* in West Lafayette, Indiana, USA (CTIC 2004) as any tillage and planting system that leaves less than 15% residue cover after planting, or less than 560 kg per hectare of small grain residue equivalent throughout the critical wind erosion period. It is based on mechanical soil manipulation involving a sequence of soil tillage, such as mouldboard ploughing followed by one or two harrowings, to produce a fine seedbed and also the removal of most of the plant residue from the previous crop.

2.2 *Intensive Tillage*

Multiple field operations or practices with implements such as a mould board, disk, and/or chisel plough are used to describe intensive tillage systems. Then a finisher with a harrow, rolling basket and cutter can be used to prepare the seed bed. Intensive tillage systems leave less than 15 % crop residue and cover less than 560 kg/ha of small grain residue on the surface. These types of tillage systems are often referred to as conventional tillage systems but as reduced and conservation tillage systems have been more widely adopted, it is often not appropriate to refer to this type of system as conventional.

2.3 Conservation Tillage

Conservation tillage (CT) is defined by the Conservation Tillage Information Center (CTIC 1993) as any tillage and planting system that covers 30% or more of the soil surface with crop residue after planting, to reduce soil erosion by water. The FAO definition of conservation tillage canters on avoiding mechanical soil disturbance, maintaining continuous soil cover, and adopting diverse cropping systems (Kassam et al. 2014). It is the collective umbrella term which is given for no-tillage, direct-drilling, minimum tillage, ridge tillage (Baker et al. 2002). Main aim of conservation tillage is to boost agricultural production by increasing the efficiency of farm resources, and facilitating to reduce land degradation through integrated management of available land, water, and natural resources combined with external inputs (SoCo 2009). However, the success or failure of conservation tillage depends on the use of herbicides, crop residue and efficiency of planting equipments to place seed in soil below the residues.

Conservation tillage systems is being practised worldwide and currently about 100 million ha has been adopted throughout the world. Six countries have more than 1 million ha area under no tillage systems. South America has the highest adoption rates, and has more area under permanent no-till and permanent soil cover. United States has the maximum area under conservation agriculture, followed by Brazil, Argentina, Canada, Australia and Paraguay. Adoption of no-tillage systems for sowing of winter-season crops including wheat planted after rice has shown tremendous increase in South Asia in the last few years (Fig. 1). The CTIC (1993) has sub-divided the conservation tillage into following four systems:

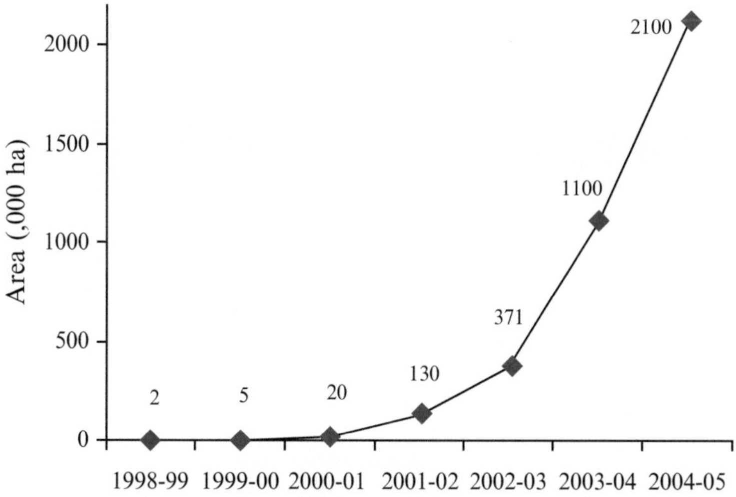

Fig. 1 Increase in area under zero-till winter season crops including wheat planted after rice in South Asia

2.3.1 No-Tillage (No-Till, Zero-Till, Slot Planting, Sod Planting, Eco-fallow, Chemical- Fallow, Direct Drilling)

Several variations of minimum tillage systems are in use globally, varying in degree from almost no tillage to nearly full conventional tillage (Unger 1984). The CTIC defines no-till as a system in which the soil is left undisturbed from harvest to planting except for nutrient injection. Tillage is essentially eliminated with no-till system. The only tillage that is used is the soil disturbance in a narrow slot created by coulters or seed openers (Conservation Tillage Systems and Management 2000). Planting or drilling is accomplished in a narrow seedbed or slot created by coulters, row cleaners, disk openers in row chisels or roto-tillers. Compared to other tillage systems, no-till also minimizes fuel and labour requirements. Pre-emergence or post-emergence surface applications of one or two herbicides properly timed is sufficient to control weeds. Recent advancements in herbicides make weed control with no-till easier than it used to be. Early pre-plant applications, longer-lasting residual herbicides and a wide variety of post-emerge products are helping assure weed control success with no-till (Lal 1998). Residue, when uniformly spread, increases water infiltration and reduces soil moisture evaporation as shown in Fig. 2 (Gajri et al. 2009). In a long-term tillage study, higher soil moisture under no-till corn production was observed throughout the growing season. Significantly less evaporation occurred under no-till early in the growing season. This conservation of soil water may carry the no-till crop through short drought periods without severe moisture stresses developing in the plants. However, the extra water conserved under no-till can occasionally be detrimental under conditions in which excessive soil water contributes to denitrification losses. Soil compaction in no-tillage soils was not found to be a problem. Saturated hydraulic conductivity measurements suggest better water movement in no-tillage compared with existing system of

Fig. 2 No-till leaves the maximum crop residue for soil protection

conventional tillage. In 2011, South America had 44 % of the total global area under no-tillage, followed by North America i.e. 32 %. Europe had 1.35 million ha under no-tillage which is about 1% of the total global area (Friedrich et al. 2012).

2.3.2 Reduced Tillage

Reduced tillage system that is less intensive than conventional systems. Under this conservation tillage system, the number of tillage operations is minimized by either the elimination of one or more tillage operations or combining together of primary and secondary tillage operations. Only those tillage operations are operated and performed that are absolutely necessary for crop production under a given set of soil, crop and climatic conditions (Gajri et al. 2009). Land preparations and seeding is completed in one operation. Ploughing is normally eliminated, but the total field surface is still worked by tillage equipment. The crop residues are retained on the soil surface for as long as possible if the objective is to conserve soil and soil moisture during rainy season. Under irrigated conditions, reduced tillage may be practiced after removing residues from the surface. This tillage practice has largely been adopted in alluvial soils of Indo-Gangetic Plains, where wheat is planted with minimum tillage operations in the lean fields the surface (Gajri et al. 2009) Figs. 4, 5, 6.

2.3.3 Ridge Tillage

Ridge tillage or Ridge-till is a reduced disturbance planting system in which crops are planted and grown on ridges formed during the previous growing season and by shallow, in-season cultivation equipment. In ridge-till, the soil is also left undisturbed from harvest to planting except for possible fertilizer injection (Gajri et al. 2009). The ridge beds are established and maintained through the use of specialized cultivators and planters designed to work in heavy crop residues (Fig. 2). Tillage is generally very shallow, disturbing only the ridge tops. Planting is completed in a seedbed prepared on ridges with sweeps, disk openers, coulters, or row cleaners. Residue is left on the surface between ridges. A band application of herbicide behind the planter provides weed control in the row (Opara-Nadi 1993). Ridge tillage is primarily intended for the production of agronomic row crops like corn, soybeans, cotton, sorghum and sunflower. Level or gently sloping fields, especially those with poorly drained soils are well suited to ridge systems. A ridge tillage system is an excellent choice for soils that are often too wet. Ridges in the ridge-till system work quite well to provide drainage on poorly drained soils. Ridge-tillage system reduces erosion by leaving the soil covered with residue until planting. After planting, 30–50 % residues may be left, but it is not uniformly distributed on the surface (Fig. 3).

Fig. 3 Ridge-till system with crops planted on ridges

2.3.4 Stubble Mulch Tillage

Mulch tillage or Mulch-till is a category that includes all conservation tillage practices other than no-till and ridge-till. Mulch tillage is described as a tillage system in which a significant portion of crop residue is left on the soil surface to cover soil surface (SCSA 1987). It is usually accomplished by substituting chisel plows, sweep cultivators, or disk harrows for the moldboard plow or disk plow in primary tillage. This change in implements is attractive because residues are not buried deep in the soil and good aerobic decomposition is thus encouraged (Gajri et al. 2009). Weed control is accomplished with herbicides and/or cultivation.

3 The Principles of Conservation Agriculture

Conservation agriculture emphasizes that the soil is a living body, essential to sustain quality of life on the surface of earth. Conservation tillage in particular recognizes the importance of the upper 0–20 cm of soil as the most active zone, but also the zone most vulnerable to erosion and degradation. Most environmental functions and services that are essential to support terrestrial life are concentrated in the micro, meso, and macro fauna and flora which live and interact in this zone.

The principles of conservation agriculture and the activities to be supported are described as follows:

- *Maintaining permanent soil cover and promoting minimal mechanical disturbance of soil through zero tillage systems, to ensure sufficient residual biomass to enhance soil and water conservation and control soil erosion.* This improves soil aggregation, soil biological activity and soil biodiversity, water quality and increases soil carbon sequestration. It greatly enhances water infiltration,

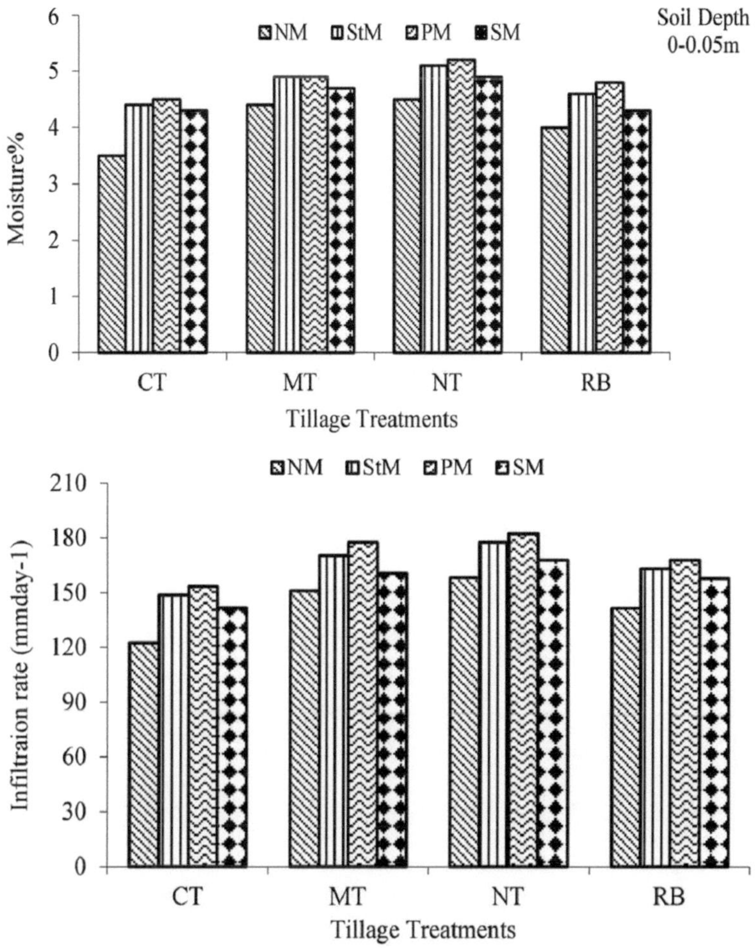

Fig. 4 Effect of tillage & water management practices on soil water content at harvesting of maize (3 years average), were, *CT* Conv. Till., *MT* Min. Till., *NT* No Till., *RB* Raised bed, *NM* No Mulch, *StM* Straw mulch, *PM* Polythene Mulch and *SM* Soil Mulch

improves efficient water use efficiency and maintains optimum temperature, moisture and increased insurance against drought.

- *Promoting a healthy, living soil through crop rotations, cover crops and involving integrated pest management technologies.* These practices reduce requirements for pesticides and herbicides, control off-site pollution. The objective is to create a healthy soil microenvironment that is naturally aerated, better able to receive, hold and supply plant available water, maintain nutrient cycling and better able to decompose and mitigate pollutants.
- *Promoting application of fertilizers, pesticides, herbicides and fungicides in balance with crop requirements.* By feeding the soil medium rather than fertilize the crop to be grown, will reduce chemical pollution, improve water quality and maintain the natural ecological integrity of the soil, while optimizing crop productivity and economic returns.

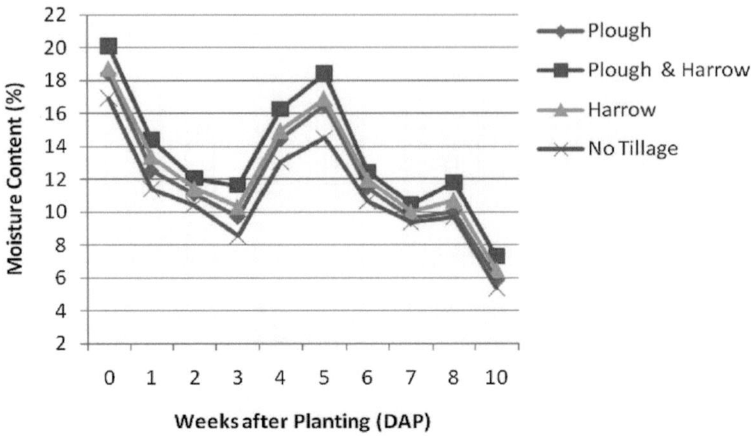

Fig. 5 Effect of tillage practice on moisture content: 0–10 cm (2009)

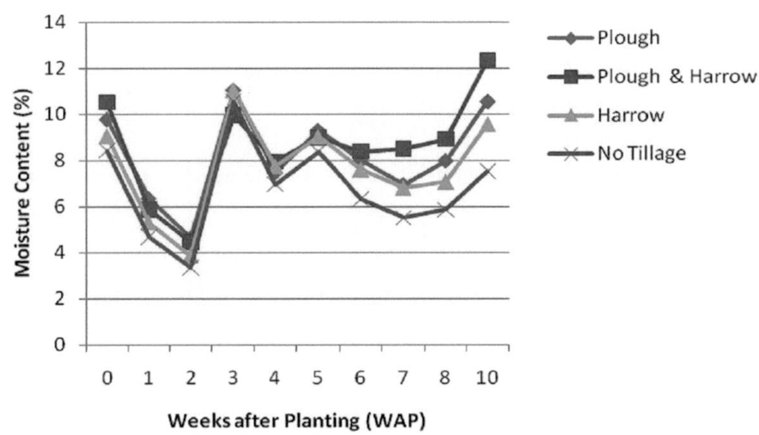

Fig. 6 Effect of tillage practice on moisture content: 0–10 cm (2010)

- *Promoting precision placement of inputs to reduce costs, optimize efficiency of operations, and prevent environmental damage.* Site specific treatment of problems, increased economic and field operation efficiencies, improved environmental protection and reduced (optimized) input costs. Precision is exercised at many steps in field operations especially during seed, fertilizer and spray placement; permanent wheel placement to stop random compaction.
- *Promoting legume fallows (including herbaceous and tree fallows where suitable), composting and the use of manures and other organic soil amendments.* This improves soil structure and biodiversity and reduces the need for inorganic fertilizers; improves soil physical properties and soil biodiversity.
- *Promoting agroforestry for fiber, fruit and medicinal purposes.* Agroforestry (trees on farms) provides many opportunities for value added production, efficient utilization of soil resources, contour hedges for erosion control, to conserve and enhance biodiversity.

4 Comparison of Tillage Systems

Typical advantages and disadvantages of different tillage systems are shown below. This information is useful in determining the suitability of tillage systems or combinations of systems for various situations. The most important advantage of conservation tillage is to control considerably soil erosion effectively and economically and the resulting sedimentation as well, a major water pollutant. Compared to the commonly used disk system, no-till saves about 1–1/2 gal/ac in fuel and 20 min of labour/ac. Labour savings allow a larger area to be farmed without additional equipment or help. Conservation tillage systems therefore represent alternatives at a time when economics require flexibility in crop production.

Comparison of selected tillage systems and typical field operations

System	Typical field operations	Major advantages	Major disadvantages
Mouldboard plough	Fall or spring plough; plough one or two spring disking	Suited for poorly drained Major soil erosion. Well-tilled seedbed.	Fuel and labour costs maximum, Soil moisture loss high, field cultivations; Timeliness considerations, very good incorporation.
Chisel plough	Fall chisel; one or two spring diskings or field cultivations; plant; cultivate. than fall plow or fall disk	Less wind erosion and fuel requirements Well adapted to poorly drained soils. Good to excellent incorporation.	Erosion controls less but significant soil moisture loss. Medium to high labour because of rough surface.
Disk	Fall or spring disk; spring disk and/or field cultivate; plant; cultivate.	Less erosion than from cleanly tilled systems. Well adapted for lighter to medium textured, well-drained soils. Good to excellent incorporation.	Little erosion control. High soil moisture loss
Ridge-till	Chop stalks (on furrow irrigation); plant on ridges; cultivate for weed control and to rebuild ridges.	Excellent erosion control if on contour. Well adapted to wide range of soils. Excellent for furrow irrigation. Ridges warm up and dry out quickly. Fuel and labour costs	No incorporation Narrow row soybeans and small grains not well suited. No forage crops. Machinery Low modifications required
Strip-till	Fall strip-till; spray; plant On cleared strips; post emergent spray as needed.	Clears residue from row area to allow pre-plant soil warming and drying. Injection of nutrients directly into row area. Well suited for poorly drained soils.	Cost of pre plant operation Strips may dry too much, crust, or erode without residue. Not suited for drilled crops. Potential for different nitrogen fertilizer losses
No-till	Spray; plant into undisturbed surface; post emergent spray	Maximum erosion control. Soil moisture conservation as needed. Minimum fuel and labour costs.	No incorporation increased dependence on herbicides. Limitations with poorly drained soils

5 Conservation Tillage Effects on Soil Physical Properties

Effects of conservation tillage on soil properties vary and these variations depend on the particular system chosen. No-tillage management affects the pedological, hydrological and geomorphological processes (García-Orenes et al. 2009; Gao et al. 2014). Conservation tillage improved soil structure thereby enhancing nutrient recycling, water availability and biodiversity while reducing water and wind erosion and improving surface and ground water quality (Lahmar 2010). Soil physical properties in general have significantly improved under different types of conservation tillage (Wang et al. 2005). According to (Lal 1997), and generally with no-till physical properties are more favourable than tillage-based systems. No-till (NT) systems, which maintain high surface soil coverage, have resulted insignificant change in soil properties especially in the upper few centimetres (Anikwe and Ubochi 2007). Macro-porosity hence soil aggregation are increased substantially with build-up of active organic matter fraction and consolidation of other macro and micro organisms. Infiltration and internal drainage are generally improved, as is soil water-holding capacity. Infiltration capacity at optimum rate in no-till managed soils is generally quite desirable for long term water retention, but under high infiltration rate it may lead to more rapid leaching of nitrates and other water-soluble nutrient related chemicals. Residue-covered soils are generally cooler and moister. This is an advantage in the hot part of the year, but may be detrimental to early crop growth in the cool spring of temperate regions.

In cool regions, adoption of conservation tillage has hinders due to restricted soil drainage leading to soil conditions as wetter and cooler than with conventional tillage and hence yield levels may be low. Reduced yields have discouraged the adoption of conservation tillage in these regions. However, limited pre-planting tillage over the crop rows or ridge tillage are conservation tillage systems that allow at least part of the soil to warm faster and largely overcome these problems.

5.1 Soil Structure and Soil Aggregation

Soil aggregation can be improved by management practices that decrease agro-ecosystem disturbances, improve soil fertility, increase organic inputs, increase plant cover, and decrease soil organic carbon decomposition rate. The stability of soil structure and soil aggregates is strongly linked with physical protection of soil organic matter within soil aggregates (Bachmann et al. 2008). The stability of aggregates depends on the strength of intra-aggregate bonds and distribution of possible failure zones related to the geometries of air filled pores, small cracks, strength of mineral-organic bonds and other cementing agents between soil particles (Bronick and Lal 2005; Kodesova et al. 2009). Better aggregation and improved pore size distribution (Bhattacharyya et al. 2006a) was observed by the adoption of zero tillage. In Gottingen, Germany, Jacobs et al. (2009) found that minimum tillage

Table 1 Effect of different tillage treatments on soil structure (mean of 2006 and 2007)

	Aggregate proportion in size class (%)				
Tillage	>2 mm	2–1 mm	1–0.25 mm	<0.25 mm	MWD
NT	56.45a	25.62b	16.57c	1.36b	2.47a
RT	46.86b	27.43ab	23.30b	2.41a	2.20b
CT	25.96c	30.46a	41.81a	1.78ab	1.63c

Letters indicate results of LSD test. *NT* no tillage, *RT* ridge tillage, *CT* conventional tillage

(MT), compared with CT, did not only improve aggregate stability but also increased the concentrations of SOC and N within the aggregates in the upper 5–8 cm soil depth after 37–40 years of tillage treatments. Zhang et al. (2012) reported the soil aggregate distribution and mean weight diameter (MWD) under different tillage systems at the depth of 0–20 cm (Table 1).

5.2 *Bulk Density, Porosity and Penetration Resistance*

Studies showed that macropore connectivity increased under NT, which allowed for increased water infiltration. There were inconsistent responses in total porosity and soil bulk density as compared with conventional tillage practices (Stubbs et al. 2004). Macropores typically are characteristic features of root channels, earthworm made holes, and other continuous non uniform channels. The population of earthworms is generally higher under NT systems than under full-width tillage systems (Blanco-Canqui and Lal 2007). Total porosity, which is directly related to bulk density, tends to be lower under NT, at least initially. Studies by Jiang et al. (2007) showed that micropores (<10 μm) will increase under NT, even in soils that have a clay pan. In areas being converting from CT to NT, bulk density will increase and typically will not decrease unless there is an increase in soil organic matter or proportion of macro-porosity. Practices such as adding residue mulch, growing winter cover crops with fibrous rooting systems, and including small grain in the rotation along with no-till increase porosity (Villamil et al. 2006; Blanco-Canqui and Lal 2007; Jemai et al. 2013). Raczkowski et al. (2012) conducted an 8-year NT study in North Carolina on sandy soils under a rotation of low-residue cotton, soybeans, and peanuts along with high-residue corn and sorghum. This study showed that NT resulted in higher bulk density, lower macroporosity, and higher microporosity compared to CT. These results were expected because low-residue crops were used and the area was in a thermic soil temperature regime under a humid climate, in which residue decomposes quickly.

Penetration resistance (kPa) of the soil can be regarded as a factor determining the quality of its structure and depends on its physical and mechanical properties. Tillage increased bulk density and penetration resistance of the soils significantly as

Table 2 Effect of different tillage treatments on soil physical properties (mean of 2006 and 2007)

Treatments	Soil bulk density (gcm^{-3})	Soil penetration resistance (k Pa)	Soil moisture content (%)
CT	1.41 c	560 c	19.6 a
RT	1.47 b	815 b	18.4 b
MT	1.50 ab	1105 a	17.1 c
NT	1.52 a	1250 a	16.8 c

CT Conv. Till., *RT* Reduced till. *MT* Minimum Till., *NT* No till

compared to zero-tillage (Carman 1997; Zhang et al. 2013). Rashidi and Keshavarzpour (2008) observed that the highest soil bulk density of 1.52 g cm^{-3} was obtained for the NT treatment and lowest (1.41 g cm^{-3}) for the CT treatment. The highest soil penetration resistance of 1250 kPa was obtained for the NT treatment and lowest (560 kPa) for the CT treatment. The highest soil moisture content of 19.6 % was obtained for the CT treatment and lowest (16.8 %) for the NT treatment shown in Table 2.

5.3 *Soil Strength and Stability*

The term 'strength' describes the level of stress or pressure that a soil can resist without undergoing irreversible deformation or change in shape while as 'stability' is used to describe the ability of soil to retain a coherent structure in the presence of free water. Both strength and stability are necessary if the soil is to retain its structure against imposed or external stresses. These imposed stresses may be natural such as raindrop impact or floods or may be anthropogenic such as those imposed by vehicular traffic. A complication with soil is that its strength must not be too great otherwise; plant roots and other organisms will not be able to penetrate. Aggregate stability improved with only 3 years of NT on sandy piedmont soils in Georgia under thermic and humid climatic conditions, which cause a more rapid breakdown of organic matter (Franzluebbers and Stuedemann 2008). Brock (1999) reported that water-stable aggregates cannot be sustained with CT since the residue cover is not sufficient to protect against surface crusting and that long-term NT is needed to increase aggregate stability. Conservation tillage encourages microbial activity through increase in organic matter thereby leading to increased soil aggregate stability (Table 3).

Table 3 Effects of tillage systems on soil porosity and aggregate stability

Soil components	Comparison of conservation tillage vs. conventional tillage	References
Aggregate stability	More stable in surface layer.	Ball et al. (1996), Arshad et al. (1999), and Stenberg et al. (2000)
	More clay decreases differences between tillage system	Tebru¨ gge & Du¨ ring (1999)
Total porosity	Greater or no difference in surface layer (0–5 cm) with no tillage	Guerif (1994) and Rasmussen (1999)
	No difference in tilled layer with shallow tillage	Ball and O'Sullivan (1987) and Kay and VandenBygaart (2002)
	Less in the untilled layer and the whole topsoil	
Soil bulk density	Smaller or no difference in surface layer (0–5 cm) with no tillage	Tebru¨ gge & Du¨ ring (1999)
	Greater in the untilled layer and the whole topsoil	Arshad et al. (1999), Rasmussen (1999) and Deen and Kataki (2003)
	No difference or higher in the subsoil	Tebru¨ gge & Du¨ ring (1999)

5.4 Hydraulic Conductivity, Infiltration Rate and Moisture Content

In general, under no-tillage system bulk density in the upper layer was increased, thereby causing a decrease in the amount of macro-pores and lowering saturated hydraulic conductivity, in comparison to when the conventional and reduced tillage soil systems. Conservation tillage provides excellent technique of *in-situ* soil moisture conservation in rainfed areas during creation of dust mulch through repeated tillage (Sharma et al. 1990). Increase in hydraulic conductivity and infiltration in zero tillage as compared to conventional tillage was reported by Mc Garry et al. (2000), attributing it to earthworm activities and termite galleries. Srivastava et al. (2000) found significantly lower hydraulic conductivity in zero tillage plots as compared to chiseling and roto-tilling, resulting from more favorable soil physical conditions created by these tilling equipments. After rice and wheat harvest, the Ksat (saturated hydraulic conductivity) values estimated in the laboratory in the upper 15 cm soil depth under zero tillage plots were higher than that of the tilled plots (Bhattacharyya et al. 2006b, 2008). The decrease of Ksat by tillage in the surface soil layer may result from the possible destruction of soil aggregates and reduction of non-capillary pores where as in zero tillage plots the pore continuity was probably maintained due to better aggregate stability and pore geometry (Bhattacharyya et al. 2006a). Sharma et al. (2011) showed that the no tillage retained the highest moisture followed by minimum tillage, raised bed and conventional tillage in inceptisols under semi arid regions of India (Fig. 4). Tillage treatments influenced the water intake and infiltration rate (IR) increased in the order of NT>MT>RB>CT and in mulching treatment the order was PM>StM>SM>NM. The maximum mean value of IR (182.4 mm/day) was obtained in case of no tillage and polythene mulch combination and minimum (122.4 mm/day) was recorded in CT and no mulch combination.

According to Aikin and Afuakwa (2012) in upper 10 cm soil layer for the 2009 major growing season, tillage treatments showed significant influence in moisture content from the planting date through the first 5 weeks after planting, and after harvest. There was no significant difference in moisture content from the sixth to the eighth week after planting in the 0–10 cm soil layer. In the 0–10 cm soil layer for the 2010 major growing season, tillage treatments significantly affected moisture content only on the planting date, the seventh and eighth week after planting, and after harvest. Tillage treatment caused significant difference in moisture content in the 10–20 cm soil layer throughout the 2009 major growing season except at 7 and 8 weeks after planting. Tillage treatment did not significantly influence soil moisture content in the 10–20 cm soil layer from the planting date through the first 5 weeks after planting during the 2010 major growing season. Significant differences in soil moisture content were observed at 6, 7 and 8 weeks after planting, and after harvest. Plots with disc ploughing followed by disc harrowing tillage had the highest soil moisture contents. The lowest soil moisture contents were located in the no tillage plots.

5.5 *Soil Aeration and Soil Temperature*

Plant roots and soil fauna require oxygen, and aerobic microbes are important decomposers. Air permeability is a measure of how easily air convection occurs through soil in response to pressure gradients. Pressure gradients can be generated naturally by air turbulence above the soil surface, and this can lead to air flows through the tilled layers of soils especially when they contain pores larger than about 5 mm (Kimball and Lemon 1971).

Temperature is one of the physical states of soil that is rarely analyzed because of its greater variability in time. Radecki (1986) stated that dark soils show greater warmth of the surface layer directly after agricultural treatment than when not treated. Under field conditions the soil temperature can be altered by mulching and vegetation by acting as buffers and preventing incoming and outgoing radiations. In arid and semiarid regions or in summers, crop residues left on the soil surface as a mulch as compared to incorporation, removal or burning are known to be beneficial for crop production (Dao 1996). If used as mulch, the residue can modify soil temperature and soil temperature is lowered by the plant residues left on the soil surface in no tillage (Rasmussen 1999).

5.6 *Soil Erosion*

Degradation of agricultural soils as a result of excessive tillage has spurred interest in no-till cropping systems. These systems help to maintain the physical conditions of a relatively undisturbed soil. Residue is left on the surface of the soil, making it less susceptible to wind and water erosion (Baker and Saxton 2007). Tillage accelerates mineralization (breakdown) of crop residue and loss of soil organic matter (Stubbs et al. 2004). Soil erosion by wind and water occurs in all environments (Hudson 1995). Low intensity tillage favours consolidation of soil through better structure thereby imparting resistance to erosion. According to Lal et al. (2007) NT technologies are very effective in reducing soil and crop residue disturbance, moderating soil evaporation and minimizing erosion losses. The presence of residues at the soil surface in different types of tillage systems has a tremendous effect on runoff and erosion (Basic 2004). The residues also have an effect on soil temperature, soil reaction, nutrient distribution and availability, population and activities of soil fauna, and, therefore, on soil organic matter content. Because of effectiveness in controlling erosion, no-tillage makes crop production possible on sloping lands that would under clean tillage result in enormous erosion problems (Lal 1999). No-tillage systems also ensure significant increases in water conservation. Soil carbon sequestration can be accomplished by management systems that add high amounts of biomass to the soil, cause minimal soil disturbance, conserve soil and water, improve soil structure, and enhance soil fauna activity.

6 Constraints in the Adoption of Conservation Tillage

Factors for non-adoption of conservation tillage include climate, soil, levels of crop residue, (mixed) cropping systems etc. Conservation tillage is a function of weather of a particular region and changes with the type of soils (Lal 1999). A humid temperate climate and political support can be the main reason for lower adoption of no-tillage (Derpsch et al. 2010; Mader and Berner 2012). Among the disadvantages are more difficult weed control, specific machinery and cropping systems requirements and soil specificity. For example each crop has specific soil preparation requirements (for example soil temperature, alleopathic response) and irrigation needs that create challenges for the universal adoption of conservation tillage. One of the main obstacles to implementing conservation tillage is furrow or surface irrigation. Crop residues that build up under conservation tillage can cause blockage impeding the progress of irrigation (Lal 1999). For this reason, maintaining furrows to supply water is seen as an impediment to the adoption of conservation tillage and cover cropping practices unless measures are taken to allow efficient irrigation, or unless water delivery systems are changed to subsurface drip (SDI) or low pressure overhead sprinklers. Moreover, the increased need for herbicides to control weeds in conventional tillage under furrow irrigation is potentially an environmental and economic issue for farmers.

7 Conclusion

Continuing soil degradation is threatening food security and the livelihood of millions of farm households throughout the world. Soil types and their various reactions to tillage are of paramount importance in determining the superiority of one practice over the other. There is a need to develop precise objective and quantitative indices of assessing soil health attributes. Conservation and recycling of nutrients is a major feature of any organic farming system (National Standard 2005). Issues of conservation have assumed importance in view of the widespread resource degradation and the need to reduce production costs, increase profitability and make agriculture more competitive. There is considerable evidence that CT can provide a wide range of benefits to the environment and wildlife, some of these being similar to that provided by set-aside. Evolution and accelerated adoption of zero tillage is a significant step paving way for more comprehensive conservation agriculture systems. The new technologies, on one hand, are exciting the farmers to take up new ways of managing their resources more productively and, on the other and throwing new challenges to the scientific community to solve emerging problems associated with new technologies. Future research and development efforts would need to take a more holistic view of sustainability concerns in developing and promoting technological, policy and institutional options for sustained resource use and profitability of production systems. CT has the potential to provide some of the benefits while also allowing farmers to continue cropping, but most will achieve it as part of an integrated approach to crop management. In addition, by preserving soil and maintaining it in optimum condition, crop yields are sustained thereby reducing the need to convert remaining natural habitats to agriculture.

References

Aikins SHM, Afuakwa JJ (2012) Effect of four different tillage practices on soil physical properties under cowpea. Agric Biol J N Am 1:17–24

Anikwe MAN, Ubochi JN (2007) Short term changes in soil properties under tillage systems and their effect on sweet potato (*Ipomea batatas* L.) grow than yield in an Ultisol in south-eastern Nigeria. Aus J Soil Res 45:351–358

Arshad MA, Gill KS, Coy GR (1999) Wheat yield and weed population as influenced by three tillage systems on a clay soil in temperate continental climate. Soil Tillage Res 28:227–238

Bachmann J, Guggenberger G, Baumgart T, Ellerbrock RH, Urbanek E, Goebel MO, Kaiser K, Horn R, Fischer WR (2008) Physical carbon-sequestration mechanisms under special consideration of soil wettability. J Plant Nutr Soil Sci 171:14–26

Baker CJ, Saxton KE, Ritchie WR (2002) No-tillage seeding: science and practice, 2nd edn. CAB International, Oxford

Baker CJ, Saxton KE (2007) No-tillage seeding in conservation griculture. Second edn. Food and Agriculture Organization of the United Nations. CAB International, Nosworthy Way Wallingford, Cambridge, MA

Ball BC, O'Sullivan MF (1987) Cultivation and nitrogen requirements for drilled and broadcast winter barley on a surface water gley (gleysol). Soil Tillage Res 9:103–112

Barbera V, Poma I, Gristina L, Novara A, Egli M (2012) Long term cropping systems and tillage management effects on soil organic carbon stock and steady state level of C sequestration rates in a semiarid environment. Land Degrad Dev 23:82–91
Basic F (2004) Tillage and crop management effects on soil erosion in central Croatia. Soil Tillage Res 78:197–206
Bhattacharyya R, Prakash V, Kundu S, Gupta HS (2006a) Effect of tillage and crop rotations on pore size distribution and soil hydraulic conductivity in sandy clay loam soil of the Indian Himalayas. Soil Tillage Res 82:129–140
Bhattacharyya R, Singh RD, Chandra S, Kundu S, Gupta HS (2006b) Effect of tillage and irrigation on yield and soil properties under rice-wheat system on a sandy clay loam soil of Uttaranchal. Ind J Agr Sci 76(7):405–409
Bhattacharyya R, Kundu S, Pandey SC, Singh KP, Gupta HS (2008) Tillage and irrigation effects on crop yields and soil properties under the rice–wheat system in the Indian Himalayas. Agric Water Manag 95:993–1002
Biswas MR (1984) Agricultural production and the environment: a review. Environ Conserv 11(3):253–259
Blanco-Canqui H, Lal R (2007) Impacts of long-term wheat straw management on soil hydraulic properties under no-tillage. Soil Sci Soc Am J 71:1166–1173
Brock BG (1999) Soil quality and soil erosion, chapter 10. Soil and Water Conservation Society Ankeny lowa. CRC Press, Boca Raton/New York/Washington, DC
Bronick CJ, Lal R (2005) Soil structure and management: a review. Geoderma 124:3–22
Carman K (1997) Effect of different tillage systems on soil properties and wheat yield in Middle Anatoliarc. Soil Tillage Res 40:201–207
Chauhan BS, Gill GS, Preston C (2006) Tillage system effects on weed ecology, herbicide activity and persistence: a review. Aus J Exp Agric 46:1557–1570
Coughenour CM, Chamala S (2000) Conservation tillage and cropping innovation: constructing the new culture of agriculture. Iowa State Uni Press, Ames
Conservation Tillage Systems and Management (2000) MWPS-45. Crop residue management with no-till, ridgetill, mulch-till and strip-till, 2nd edn. Midwest Plan Service, Iowa State University, Ames
CTIC (Conservation Tillage Information Center) (1993) Conservation tillage definitions and types of systems. Conserv impact 11(5):6
CTIC (Conservation Tillage Information Center) (2004) National Crop Residue Management Survey. CTIC, West Lafayette
Dao TH (1996) Tillage system crop residue effect on surface compaction of a paleustoll. Agron J 88:141–148
Daraghmeh OA, Jensen JR, Petersen CT (2009) Soil structure stability under conventional and reduced tillage in a sandy loam. Geoderma 150:64–71
Deen W, Kataki PK (2003) Carbon sequestration in a long-term conventional versus conservation tillage experiment. Soil Tillage Res 74:143–150
Derpsch R (2005) The extent of conservation agriculture adoption worldwide: implications and impact. In: Proceedings of the third international congress of conservation agriculture. Nairobi, Kenya
Derpsch R, Friedrich T, Kassam A, Li H (2010) Current status of adoption of no-till farming in the world and some of its main benefits. Int J Agric Biol Eng 3:1–25
Dexter AR (2004) Soil physical quality. Part 1. Theory, effects of soil texture, density and organic matter, and effects on root growth. Geoderma 120:201–214
Dumanski J (2015) Evolving concepts and opportunities in soil conservation. Int Soil Water Conserv Res 3:1–14
Franzluebbers AJ, Stuedemann JA (2008) Soil physical responses to cattle grazing cover crops under conventional and no tillage in the Southern Piedmont, USA. Soil Tillage Res 100:141–153
Friedrich T, Derpsch R, Kassam A (2012) Overview of the global spread of conservation agriculture, Field actions science reports, Special Issue 6, 2012

Gajri PR, Majumdar SP, Sharma PK (2009) Tillage. Fundamentals of soil science, 2nd edn. Indian Society of Soil Science, Pusa, New Delhi, pp 162–179

Gao X, Wu P, Zhao X, Wang J, Shi Y (2014) Effects of land use on soil moisture variation in a semi-arid catchment: implications for land and agricultural water management. Land Degrad Dev 25:163–172

Garcia-Orenes F, Cerda A, Mataix-Solera J, Guerrero C, Bodi MB, Arcenegui V, Zornoza R, Sempere JG (2009) Effects of agricultural management on surface soil properties and soil water losses in eastern Spain. Soil Tillage Res 106:117–123

Glab T, Kulig B (2008) Effect of mulch and tillage system on soil porosity under wheat (*Triticum aestivum*). Soil Tillage Res 99:169–178

Guerif J (1994) Effects of compaction on soil strength parameters. In: Soane BD, Van Ouwerkerk C (eds) Soil Compaction in Crop Production. Elsevier science Publishers, Amsterdam, pp 479–500

Hudson N (1995) Soil conservation. B.T. Batsford, London, 391 pp

Indoria AK, Sharma KL, Reddy SK, Rao CS (2016) Role of soil physical properties in soil health management and crop productivity in rainfed systems–II. Manage Technol Crop Prod Curr Sci 3(110):320–328

Jacobs A, Rauber R, Ludwig B (2009) Impact of reduced tillage on carbon and nitrogen storage of two Haplic Luvisols after 40 years. Soil Tillage Res 102:158–164

Jemai I, Aissa NB, Saida BG, Moncef BH, Tahar G (2013) Impact of three and seven years of no-tillage on the soil water storage, in the plant root zone, under a dry subhumid Tunisian climate. Soil Tillage Res 126:26–33

Jiang P, Anderson SH, Kitchen NR, Sadler EJ, Sudduth KA (2007) Landscape and conservation management effects on hydraulic properties of a claypan-soil toposequence. Soil Sci Soc Am J 71:803–811

Kassam AH, Friedrich T, Derpsch R (2010) Conservation Agriculture in the 21st century: a paradigm of sustainable agriculture. European Congress on conservation agriculture, Madrid, October 4–6 2010

Kassam A, Derpsch R, Freidrich T (2014) Global achievements in soil and water conservation the case of conservation agriculture. ISWCR 2(1):5–13

Kay BD, VandenBygaart AJ (2002) Conservation tillage and depth stratification of porosity and soil organic matter. Soil Tillage Res 66(2):107–118

Khurshidi K, Iqbal M, Arif MS, Nawaz A (2006) Effect of tillage and mulch on soil physical properties and growth of maize. Int J Agric Biol 8(5):593–596

Kimball BA, Lemon ER (1971) Air turbulence effects upon soil gas exchange. Soil Sci Soc Am Proc 35:16–21

Kodesova R, Rohoskova M, Zigova A (2009) Comparison of aggregate stability within six soil profiles under conventional tillage using various laboratory tests. Biologia 64:550–554

Lahmar R (2010) Adoption of conservation agriculture in Europe -Lessons of the KASSA project. Land Use Policy 27:4–10

Lal R (1997) Long term tillage and maize monoculture effects on a tropical Alfisol in western Nigeria. Crop yield and soil physical properties. Soil Tillage Res 42:145–160

Lal R (1998) Tillage effects on soil degradation, soil resilience, soil quality and sustainability. Soil Tillage Res 27:1–8

Lal R (1999) Residue management, conservation tillage and soil restoration for mitigating greenhouse effect by CO_2 – enrichment. Soil Tillage Res 43:81–107

Lal R, Suleimenov M, Doraiswamy P, Wall P, Hansen D (2007) Climate change and terrestrial C sequestration in Central Asia. Taylor and Francis, Amsterdam, 493pp

Lieskovsky J, Kenderessy P (2014) Modelling the effect of vegetation cover and different tillage practices on soil erosion in vineyards: a case study in VRÁBLE (SLOVAKIA) using watem/sedem. Land Degrad Dev 25:288–296

Limousin G, Tessier D (2007) Effects of no-tillage on chemical gradients and topsoil acidification. Soil Tillage Res 92:167–174

Mader P, Berner A (2012) Development of reduced tillage systems in organic farming in Europe. Renewable Agric Food Syst 27:7–11

Mc Garry D, Bridgc BJ, Radford BJ (2000) Contrasting soil physical properties after zero and traditional tillage of an alluvial soil in the semi-arid subtropics. Soil Tillage Res 53:105–115

Mc Garry D, Bridge BJ, Radford BJ (2000). Contrasting soil physical properties after zero and traditional tillage of an alluvial soil in the semi-arid subtropics. Soil and Tillage Research 53, 105-115. National Standard (2005) National standard for organic and bio-dynamic produce, edition 3.1, As amended January 2005, Organic industry export consultative committee, c/o Australian Quarantine and inspection service, GPO Box 858, Canberra, ACT, 2601

Opara-Nadi OA (1993) Conservation tillage for increased crop production. In: Soil tillage in Africa: needs and challenges, soil resources, management and conservation service land and water development division. FAO, Rome, 1993. Issue 69

Raczkowski CW, Mueller JP, Busscher WJ, Bell MC, McGraw ML (2012) Soil physical properties of agricultural systems in a large-scale study. Soil Tillage Res 119:50–59

Radecki A (1986) Studia nad moiliwobcia zastosowania siewu bezposredniego na czamych ziemiachwla8ciwych. Wydawn. SGGW-AR Warszawa 86

Rashidi M, Keshavarzpour F (2008) Effect of different tillage methods on grain yield and yield components of maize (*Zea mays* L.). Int J Agric Biol 9:274–277

Rasmussen KJ (1999) Impact of ploughless soil tillage on yield and soil quality: a Scandinavian review. Soil Tillage Res 53:3–14

SCSA (Soil Conservation Society of America) (1987) Glossary of soil science terms. Soil Science Conservation Society of America, Ankeny

Sharam PK, Khawara PC, Tewatia RK (1990) Residue soil moisture and wheat yield in relation to mulching and tillage preceding rain-fed crop. Soil Tillage Res 12:279–284

Sharma PA, Vikas A, Sharma RK (2011) Impact of tillage and mulch management on economics, energy requirement and crop performance in maize–wheat rotation in rainfed subhumid inceptisols. India Eur J Agron 34:46–51

Shrestha A, Clements DR (2003) Emerging trends in cropping systems research. J Crop Prod 8:1–32

SoCo (Sustainable agriculture and soil conservation) (2009) Soil-friendly farming systems and practices: fact sheet no. 5: conservation agriculture. Available on: http://eusoils.jrc.ec.europa.eu/projects/SOCO/FactSheets/ENFactSheet-05.pdf

Srivastava AP, Panwar JS, Garg RN (2000) Influence of tillage on soil properties and wheat productivity in rice (*Oryza sativa*)-wheat (*Triticum aestivum*) cropping system. Indian J Agric Sci 70(4):207–210

Stenberg M, Stenberg B, Rydberg T (2000) Effects of reduced tillage and liming on microbial activity and soil properties in a weakly-structured soil. Appl Soil Ecol 14:135–145

Stubbs TL, Kennedy AC, Schillinger WF (2004) Soil ecosystem changes during the transition to no-till cropping. J Crop Improv 11(1/2):21–22

Tebrugge F, During RA (1999) Reducing tillage intensity: a review of results from a long-term study in Germany. Soil Tillage Res 53:15–28

Unger PW (1984) Tillage sysyetems for soil and water conservation. FAO Soils Bull 54:278

Villamil MB, Bollero GA, Darmody RD, Simmons FW, Bullock DG (2006) No-till corn/soybean systems including winter cover crops: effects on soil properties. Soil Sci Soc Am J 70:1936–1944

Wang ZC, Gao HW (2004) Conservation tillage and sustainable farming. Agricultural Science and Technology Press, Beijing. Water Conserv Res (3):1–14

Wang XB, Cai DX, Hoogmoed WB, Oenema Perdok UD (2005) Scenario analysis of tillage, residue and fertilization management effects on soil organic carbon dynamics. Pedosphere 15(4):473–483

Wang XB, Cai DX, Perdok UD, Hoogmoed WB, Oenema O (2007) Development in conservation tillage in rainfed regions of North China. Soil Tillage Res 93(2):239–250

Zhang S, Li Q, Zhang X, Wei K, Chen L, Liang W (2012) Effects of conservation tillage on soil aggregation and aggregate binding agents in black soil of Northeast China. Soil Tillage Res 124:196–202

Zhang M, Wang F, Chen F, Malemela MP, Zhang H (2013) Comparison of three tillage systems in the wheat-maize system on carbon sequestration in the North China Plain. J Cleaner Prod 54:101–107

Degraded Soils: Origin, Types and Management

Muhammad Zia-ur-Rehman, Ghulam Murtaza, Muhammad Farooq Qayyum, Saifullah, Muhammad Rizwan, Shafaqat Ali, Fatima Akmal, and Hinnan Khalid

Contents

M. Zia-ur-Rehman (✉) • G. Murtaza • Saifullah • F. Akmal • H. Khalid
Institute of Soil and Environmental Sciences, University of Agriculture Faisalabad, Faisalabad 38040, Pakistan
e-mail: ziasindhu1399@gmail.com

M.F. Qayyum
Department of Soil Sciences, BZU, Multan, Pakistan

M. Rizwan • S. Ali
Department of Environmental Sciences, GC University, Faisalabad 38040, Pakistan

K.R. Hakeem et al. (eds.), *Soil Science: Agricultural and Environmental Prospectives*, DOI 10.1007/978-3-319-34451-5_2

Abstract The cultivated lands are continuously degrading and the extent is increasing because of different natural environmental and anthropogenic activities. Soil degradation due to salinization, erosion, water logging etc. makes environment difficult for plant growth resulting in reduced agricultural production. Soil physical, chemical and biological properties are affected due to alteration in hydraulic conductivity, bulk density, osmo-deregulation, poor aeration and specific ion toxicities. A number of management and reclamation technologies are available to counter these problem but the major concern is to optimize the most economical and eco-friendly technologies. Saline soils can be cultivated growing different halophyte plants and using modern irrigation practices. Conservation and effective and efficient use of good quality water help proper leaching of soluble salts in saline soils. Saline-sodic and sodic soils can be rehabilitated with different amendments, which can provide soluble calcium to replace exchangeable sodium adsorbed on clay surfaces. Different amendments can provide calcium directly to the soil or indirectly dissolving native calcium from calcium carbonate already resent in the soil. The eroded soils can be reclaimed by providing proper soil surface cover either in the form of mulching or vegetative cover by fodder or wild shrubs. Different studies demonstrate that under adverse conditions where chemical treatments are uneconomical tree plantations provide positive net returns to investment and significant net benefit and social outcomes from these lands. These findings suggest that there is great opportunity for capital investment in afforesting abandoned degraded soils with multipurpose approaches. This chapter covers the introduction to origin, extent and sources of degraded soils, along with their management and reclamation options.

Keywords Soil pollution • Salinity stress • Soil erosion • Agroforestry • Soil management

1 Land Degradation

1.1 Introduction

Degradation means undesirable and unwanted changes brought about by human activities along with natural phenomenon. Soil degradation is among serious prevailing issues in our modern era. It is badly affecting soil's natural fertility to enhance our economic values along with ecological issues. It is being caused due to natural and anthropogenic activities. The level of degradation depends on degree of degradative processes; duration of usage of such degraded land and its management. Land degradation causes exploitation of soil resources, reduces soil productivity and alters composition of vegetations; thus influencing billions of people around the globe directly or indirectly (Ravi and D'Odorico 2005).

Degradation of soils can considerably decrease the soil's capacity to produce food and consequently about 66 % of the total world's population is malnourished (Pimentel and Burgess 2013; World Health 2000). To feed the ever increasing population of world, food production needs to be enhanced (Haub et al. 2011) It is important to reverse the land degradation to achieve this modest goal as 99.7 % of our human diet calories are fulfilled by our land resources and only 0.3 % is being contributed by aquatic ecosystems (FAO 2004). To overcome our basic food demands it is most important to maintain productivity and quality of our land. Generally, soil formation process is 10 to 40 times slower compared to soil lost (Pimentel and Burgess 2013).

It is pertinent that we should look for alternative means of intensification specially use of sustainable land management techniques (SLM). It is said that utilization of land resources is to use the water, land, plants and animals resources to fulfill our present day human demands along with enhancing their productive potential and environmental functions (Quarrie 1992). SLM focus on four land sustaining techniques, improved irrigation management, rehabilitation of degraded soil, enhance pasture and grazing processes along with maintenance of our organic soil all these steps without further degradation of our resources come to meet our present food demands (World Bank 2006). Not only to maintain but also to enhance soil natural fertility it is important to increase its carbon sequester capacity along with ability to overcome climate change (FAO 2009; FAO 2010). By using SLM technologies we meet our human food demands without further degradation of our land and water resources (IFAD 2011; Lal 1997). It is evident that allowing land degradation is expensive because it has long term effects on society as well as on land owners (Costanza et al. 2014).

1.2 Causes of Land Degradation

The cause of land degradation for a particular area can be one or combined effects of many. (Geist and Lambin 2004) classified the causes into two categories

- proximate causes (biophysical)
- overlying causes (anthropogenic)

Biophysical have direct effect on all ecosystems like drought, soil salinity, soil acidity, metal contamination, related to extreme climatic conditions while on the other hand anthropogenic causes have indirect effect on ecosystem like intensive cropping, deforestation, overgrazing or poverty, urbanization and industrialization.

Among the proximate causes, agriculture is most contributing source of land degradation but its effect on land is aggravated by inter-related relations with other causes. Severe land degradation is observed in combination like the effect of extreme climatic changes is augmented along with poor man management techniques (McIntyre and Tongway 2005; Smith et al. 2007). It is widely reported that in

Australia, repeatedly focus on annual plants can lead to soil erosion, acidity, salinization in short cause land degradation (Bouwman et al. 2005; Lavelle and Spain 2001).

1.3 Processes of Land Degradation

Following are the some processes discussed which alone or in combination effect land quality.

1.3.1 Soil Erosion

Soil erosion is one of major factor causing land degradation. Erosion not only removes upper fertile layer but also causes soil crusting or sealing, soil compaction, poor soil structure, low organic matter, poor drainage and run-off. There are two agents of soil erosion i.e. wind and water, each loss significant amount of soil and reduces its productivity and quality (Lal 1990; Troeh et al. 2004). 12 % of total land area is affected by erosion globally (Oldeman 1998).

1.3.2 Soil Salinization

Land Degradation occurs due to high concentration of soluble salts, exchangeable sodium or both in such amount that decline the plant growth and soil productivity. According to (FAO 2000), of the world's cultivated land; 3.97×10^8 ha is affected by salinity and 4.34×10^8 ha of land is affected by sodicity, thus making 6 % of total land area.

1.3.3 Water Logging

Water-logging is the rise of ground water in root zone, thus having adverse effects on plant growth. According to GLASOD assessment, 4.6 M ha area of irrigated land of Pakistan and India is affected by water logging (Bridges et al. 2002).

1.3.4 Decline in Soil Fertility

According to FAO (1994), decline in soil fertility causes land degradation by (i) lowering soil organic matter, (ii) deteriorating soil physical properties, (iii) imbalance in soil nutrient status and (iv) accumulation of toxic metals.

It is said that these processes are caused by natural (erosion, salinity etc), institutional factors (improper land policies, inadequate planning) and socio-economic activities (improper land use, exploitation of forests, contamination of resources etc) (Ezeaku and Davidson 2008). These phenomenons's have devastating impacts on human-beings and on environment.

1.4 Types of Land Degradation

Land degradation can be divided into different categories like soil erosion, soil salinity and soil acidity.

2 Soil Salinity

2.1 Salt Affected Soils

Salt-affected soils can be divided into three different categories depending upon the nature of salts.

2.1.1 Saline Soils

Saline soil means soils with excessive soluble salts that retards seed germination and plant growth (Conway 2001; Denise 2003). These soluble salts exist in soil as cations and anions. Cations are calcium (Ca^{++}), magnesium (Mg^{++}) and sodium (Na^{+}), while anions are chloride (Cl^{-}), and sulfate ($SO4^{-2}$) ions. Mostly occurring salts in saline soils are sulfates and chlorides of calcium and magnesium. Small quantities of cations potassium (K^{+}) and ($NH_4^{+)}$ and the anions bicarbonate (HCO_3^{-}), nitrate ($NO_3^{\ -}$) and carbonates (CO_3^{-2}) are also present (Appleton et al. 2009; Majerus 1996; Scianna 2002). In saline soils soluble salts are in excess while exchangeable sodium is present in small concentration thus having good physical properties, flocculated soil structure and high permeability like in normal soils (Appleton et al. 2009; Jim 2002; Majerus 1996; Scianna 2002). Such soils have electrical conductivity ≥ 4 dS m^{-1} soil reaction (pH_s) < 8.5, sodium adsorption ration (SAR) <13 (mmol $L^{-1})^{1/2}$ and exchangeable sodium percentage (ESP) < 15.

Patchy crop growth and tip burn or chlorosis of leaves of plants is observed due to salt injury in salt effected soil. These soils are also called white kallar as large quantities of soluble salts on soil surface forms efflorescence. These soils can be identified by "White alkali" which is the white crust of salts on the soil surface.

2.1.2 Saline-Sodic Soils

Soil containing both excessive soluble salts and high exchangeable sodium content to adversely affect plant growth; are known as saline-sodic soils (Majerus 1996). These soils have electrical conductivity ≥ 4 dS m^{-1} soil reaction (pH_s) > 8.5, sodium adsorption ration (SAR) >13 (mmol $L^{-1})^{1/2}$ and exchangeable sodium percentage (ESP) > 15.

Saline-sodic soils are converted into sodic soils when excess soluble salts are leached down and thus the properties of saline sodic soils changes into sodic soils having pH above 8.5, dispersed soil structure and less permeability of air and water (Denise 2003).

2.1.3 Sodic Soils

Soils having high exchangeable sodium concentration but low total soluble salts are called sodic soils (Jim 2002). Such soils are characterized by having electrical conductivity < 4 dS m^{-1} soil reaction (pH_s) > 8.5, sodium adsorption ration (SAR) > 13 (mmol $L^{-1})^{1/2}$ and exchangeable sodium percentage (ESP) > 15.

$CO3^{2-}$ and $HCO3^{-2}$ are dominant anions of sodic soils (Qadir and Schubert 2002). At high pH and in the presence of carbonate ions magnesium and calcium got precipitated, thus the concentration of sodium ion increased in soil solution compared to other cations concentration (Majerus 1996; Qadir and Schubert 2002).

Such soils often occur in semiarid and arid areas, which are frequently mentioned as "slick spots." The combination of increased sodium concentration, decreased salt concentration and increase in Ph results in the dispersion of soil particles, result in destruction of soil structure (Conway 2001; Denise 2003). Sodic soils are termed as "black alkali" due to the deposition of organic matter on soil surface by evaporation, thus darkening the soil (Denise 2003).

2.2 Origin of Salt Effected Soils

There are various interconnected sources behind the origin of salt-affected soils; nevertheless weathering of minerals and rocks is the most predominant one in accumulating soluble salts in soils. Though the salts in ocean at present occur mainly due to the weathering process of earth crust, ocean now serves as important role in distribution of salts. Soil salinization originates from one or combination of following (Chhabra 1996):

2.2.1 Soil Weathering Process

Weathering of soils and minerals leads to the accumulation of soluble salts in soil. Under humid conditions, salts leached down in the soil due to heavy rainfall. Thus in humid regions the formation of salt affected soils is rare while in arid and semi-arid regions there is not sufficient water available to leach down these salts and consequently more salts accumulate in soil surface and result in soil salinity development. This process of accumulation of salts in soil by weathering is known as primary salinization.

2.2.2 Accumulation on the Surface Due to Irrigation Under Inadequate Drainage

Improper irrigation system transports the salt on the surface of soil profile and on evaporation this salt is left behind. So, the water build up more salt on surface compared to evenly distribution of salt in the soil profile. This leads to the formation of saline soil.

2.2.3 Shallow Water Table

Inadequate water management and unsuitable drainage system are the reason behind the rise in water table of command area. In some lands it is reported that water table rises at rate of 1–2 m per year. To some extent this water is mineralized and due to increase in water table, water continues to rise upward by capillary action and on evaporation leave the salt behind. Shallow water table is the foremost reason behind developing salty soils.

2.2.4 Fossil Salts

In arid regions, salt accumulation is also derived from fossil salts, involve some entrapped solution or some former deposits in marine. Salt release is either natural or due to anthropogenic activities. Example of naturally salt release is rise in saline ground water through impermeable layer overlying saline band. Example of human induced salt release is the construction of canals or water channels in saline strata that leads to development of salinity in the area because of using this ground water for irrigation.

2.2.5 Seepage from the Upslope Containing Salts

Salinity of downslope areas is commonly observed due to water influx in the upslope areas particularly, under certain conditions when water movement in subsurface takes place through those regions which are salts rich.

2.2.6 Ocean

Soil near coastal areas usually has high salt concentration from the ocean in the course of:

- Flooding of soil surface by sea water when waves are high;
- Entrance of sea water through rivers, inlets, etc.
- Flow of groundwater
- Aerosols generated by salt-affected areas are transported many kilometers in coastal areas. It is reported that 20–100 kg/ha NaCl in land while 100–200 kg/ha NaCl in coastal areas are deposited every year. Continuous addition of this small amount of salts in soils leads to soil salinity.

2.2.7 Chemical Fertilizer and Waste Materials

Excessive use of chemical fertilizers in fields also contributes augmentation of salts in soil, yet there input in salinity development is insignificant. Nevertheless addition of some manures like sewage sludge, cow dung or slurry and industrial material like pyrites or pressmud influence the build-up of certain ions in soil that has negative effect on soil productivity (Chhabra 1996).

2.3 Causes of Salt Affected Soils

There are two main causes of sililinty

- primary salinity (Natural process)
- secondary salinity (Anthropogenically induced salinity)

Secondary salinity is primarily due to improper irrigation system and use of poor quality water.

2.3.1 Primary Salinity

Primary salinity is a naturally occurring process mostly occur in arid and semi-arid regions where rainfall is low while evapo-transpiration rate is high, thus there is not availability of sufficient water to leach salts down to avoid salinization (McDowell 2008). Due to low rainfall, high transpiration and evaporation, salinity rises as salt concentration on soil surface increases while availability of water decreases (Bridgman et al. 2008). It is estimated that 1000 million hactares of world's total land which is equal to 7% of world's area is salt affected (Rose 2004). The major contribution in salinity causes is primary salinity which is consequential of natural soil development. It mostly occurs in arid tropical areas where salinity occur naturally (Huumllsebusch 2007).

Primary salinity is also caused by natural release of some soluble salts in soil by weathering of parent material during soil development process, these soluble salts are Cl^- of Na^+, Ca^{2+} and Mg^{2+} and sometimes SO_4^{2-} and CO_3^{2-}(Ashraf and Harris 2005; Thiruchelvam and Pathmarajah 2003).

Inadequate drainage is another factor causing soil salinization; it may involve the low permeability of soil or elevated ground water. This high ground-water is often due to physiographic unevenness. The water moves from higher lands over the sloping surface towards the lower lands cause either salty lakes or temporary flooding. Under such conditions, removal of water from surface develops saline soil (Ashraf and Harris 2005). Indurate layers in soil profile and poor soil structure results in low permeability. This low permeability leads to poor drainage by restricting downward movement of water (Ashraf and Harris 2005; Thiruchelvam and Pathmarajah 2003).

2.3.2 Secondary Salinity

Secondary salinity is mainly due to disruption in hydrological cycle either through the replacement of natural vegetation with deeply rooted vegetation or through the excessive utilization or ineffective supply of water for agriculture (Beresford et al. 2004; Rose 2004). Salt affected land area is increasing day by day due to anthropogenic land-use practices (Bridgman2008). Estimated global secondary salinity rate are submitted at around 74 million hectares, with 43 million hectares irrigated land and the remaining area of non-irrigated land (Rose 2004).

Secondary salinity due to anthropogenic practices that alter the hydrologic cycle and disrupt the water balance of the soil between water irrigated and water used by crops (transpiration) (Manchanda and Garg 2008; Munns 2005). In many irrigated areas, the water table has raised due to unjustified amounts of applied water together with poor drainage. Most of the irrigation systems of the world have caused secondary salinity, sodicity or waterlogging (Manchanda and Garg 2008).

Natural salinity has been intensified from plant using more water to plant use less water cause rise in water table, when irrigation water quality is fringe or poorer (Thiruchelvam and Pathmarajah 2003). In addition, when the soil drainage may not be suitable for irrigation, the considerable rise in water table from depth of few inches to a few feet of the soil surface is occurred mainly due to irrigation. When the water table rises to 5 or 6 ft of the soil surface, ground water moves upward into the rooted area and to the soil surface. Under such circumstances, both ground water and irrigation water, contributes to the salinity. Another causes of secondary salinity are deforestation, intensive cropping, overgrazing of cattle, use of fertilizer and other amendments (Ashraf and Harris 2005; Thiruchelvam and Pathmarajah 2003).

2.3.2.1 Deforestation

Deforestation is recognized as a major cause of salinity and alkalinity of soils. Salinity is results due to migration of salts in both the upper and lower layers. That indirectly leads to the increase in temperature of surface water and reduction in average rainfall per year (Hastenrath 1991; Shukla et al. 1990). Tree covers and green vegetation's act as buffer between soil and rain. In absence of green vegetation cover top thin soil rapidly gets eroded. Rate of water run-off and sedimentation in the rivers and streams is increased due to soil erosion. That leads to flooding and soil salinization (Domroes 1991; Shukla et al. 1990).

2.3.2.2 Accumulation of Air-Borne or Water-Borne Salts in Soils

Different salts and chemicals that release from the industry and factories can enter into the soil and water and thus problem of salinity rises in the soil (Pessarakli 2010). Similarly extra water that came out from municipalities and slush are responsible for the contamination of the soil which then become the part of salinity and or alkalinity causing factors (Bond 1998).

2.3.2.3 Contamination with Chemicals

In present era use of chemicals and intense agricultural activities especially in green houses and intensive farming system playing important role in the contamination of the soil that leads to the generation of salt affected soils.

2.3.2.4 Overgrazing

Overgrazing is common where the natural soil cover is poor and hardly satisfies the fodder contents of animal husbandry mainly occur in arid and semiarid regions (Pessarakli 2010). The natural vegetation becomes scanty and salinization develops, and this process ends up in desertification due to overgrazing.

2.3.2.5 Fallowing

Soil which is uncultivated for longer period of time invites salinity because it alters the net water movement in upwards direction which results in accumulation of salts. On the other hand a soil with green top cover is useful in diverting the hydrological cycle and movements of salts downwards (Hassan et al. 2011).

Salts in the soil are electrically charged occur as ions. The main releasing sources of the ions are primary or natural sources and secondary or salinity caused by human influences (Pace and Johnson 2002).

2.4 *Impact of Salt Affected Soil on Plant*

Severe salt affected soils have influential role on plant growth both chemically like nutritional effect or toxicity and physically like osmotic effect. Thus due to these affects plant growth is delayed and quality of agricultural production is reduced (Denise 2003; Hakeem et al. 2013; Gonzalez et al. 2004).

There are three main reasons of soil salinization which can effect plant growth adversely (Conway 2001; Jim 2002).

- Osmotic effect hinders water uptake into the plants
- Specific ion effect causes nutritional imbalance in the plants
- Destruction of soil structure and reduction in permeability

2.4.1 Osmotic Deregulation

Water uptake by the plant from the root hairs is due to concentration gradient that exists among cell sap of root cells and soil solution. High salt concentration in the soil reduces water potential difference between plant cell and soil solution (BPMC 1996). High salt content in soil solution makes soil water potential more negative, this means that water is held more strongly in soils and reduces the movement of water into the cell. If salt concentration continues to increase making water potential more negative a level come when water may move out of cell to soil solution (Silvertooth and Norton 2000).

Due to this high negative potential, plants are unable to use soil water in spite of sufficient water availability in soil. Thus in this condition plant requires more energy to take water and it effects plant proper growth and development. Under drought condition especially in clayey soil, osmotic deregulation is more prominent as plant require more energy to take the water from the soil at given moisture level (Conway 2001; Gonzalez et al. 2004).

Reduction in plant growth, cell dehydration and possibly plant death are the consequences of high salinization in normal plant (less tolerant plants) but halophytes (salt tolerant plants) adopt certain physiological changes to survive under stress conditions (Scianna 2002). Salt stress symptoms appear in plants are similar to drought stress symptoms like stunted plant growth, change in color of leaves, curling of leaves and overall plant growth is suffered (Denise 2003). These indications may occur within a few days of plantation or after numerous weeks in young seedlings while in older or mature plants salt stress causes browning of leaves from tips and overall wilting of plants (BGS 2001).

Salt stress increases the production of reactive oxygen species (ROS) such as such as superoxide ($O_2^{\bullet-}$), hydrogen peroxide ($H_2O\ _2$), hydroxyl radical (OH•) and singlet oxygen ($1O\ ^2$) because in salt stress carbon fixation is limited in plants due to little carbon dioxide availability (Ahmad and Sharma 2008; Ahmad et al. 2011;

Ahmad et al. 2010; Parida and Das 2005; Hakeem et al. 2013). Oxidation of proteins, nucleic acid and lipids is done by these highly reactive species (ROS) and thus damages plants cells (Ahmad et al. 2010; Apel and Hirt 2004; Pastori and Foyer 2002).

Under saline conditions, production of ROS species in many plants is augmented (Hasegawa et al. 2000). Due to these ROS species, membrane damage was observed which leads to cellular injury and toxicity cause by salinization in various crop plants for example pea tomato, mustard, soybean and rice (Ahmad et al. 2009; Dionisio-Sese and Tobita 1998; Gueta-Dahan et al. 1997; Mittova et al. 2004; Hakeem et al. 2012).

2.4.2 Nutrition Imbalance

There are specific ions which have direct toxic effect on plants (Scianna 2002). Among these ions are boron, sodium and chloride which have negative effect on crop emergence, plant growth and crop development. Even the small quantities of these ions retard the plant growth (Gonzalez et al. 2004).

Furthermore, if sodium ions are present in high concentration it hinders the uptake of other nutrient ions which are required by the plants for proper growth by altering soil physical and chemical properties (Scianna 2002). This can cause disturbance in nutrient balance in the plants and upset plant mineral nutrition by impeding nutrient uptake (Conway 2001).

For instance calcium and potassium deficiency is because of high sodium concentration and nitrate deficiency usually occurs when sulfate and chlorides are in high concentration (BPMC 1996). At higher pH i.e. above seven, nutrient availability is less. Sodic soils having high pH are usually deficit in nutrient concentration (Denise 2003). The symptoms associated with nutrient deficiencies and toxicities of plants can be described by tip burning, necrosis, chlorosis, dieback and abscission (BPMC 1996).

Nutrient imbalances decrease the transport and availability of nutrients and effects plant growth. Nutrient deficiencies are usually due to the competitive effect among different ions like potassium, calcium and nitrate with sodium and chloride. Reduction in plant development and growth under saline conditions is due to ionic imbalance and specific ion toxicity i.e. Na^+ and Cl^-(Grattan and Grieve 1998).

It is reported that induction of Na and Cl concentrations while decrease in the concentration of other ions Ca, P, N, K and Mg due to rise in NaCl concentration (El-Wahab 2006).

As salinity directly affects the nutrient availability, uptake and its distribution or transport in plant, consequently nutrition imbalance arises. It is repeatedly reported the effect of salinity in lowering nutrient accumulation and uptake in plants (Hu and Schmidhalter 2005; Rogers et al. 2003).

2.4.3 Structure and Permeability Problem of Salts in the Soil

Soil salinity sometimes have negative effect on soil physical properties like soil structure and soil permeability and thus reducing plant growth (Scianna 2002).

Due to certain physical methods like clay swelling or dispersion, slaking and some specific conditions like hard setting and surface crusting, soil structure is disturbed in saline-sodic and sodic soils. These disturbances in soil may limits water and air movement, restricts root penetration, lowers the water holding capacity of plants, delays seed emergence and enhances the problem of erosion and run-off (Qadir et al. 2003). Sodic layer restricts roots emergence if it occurs near sodic soil surface. That's why if sodic clay layer develops on topsoil, most of roots movements are limited along with controlled movement of air and water (Fitzpatrick et al. 2003).

Seed germination is also affected by salinity problem along with but it is reported that salinity problem does not influence seed viability (Conway 2001).

2.5 Reclamation of Salt-Affected Soils

The most important category of degraded soils is salt-affected soils which had severe effects of salinity and/ or sodicity on agriculture production and increasing on a global scale with every day. Approximately, one billion hectares of land is affected with various concentration and nature of salts worldwide (Wicke et al. 2011). The contribution of anthropogenic salinization and sodication is approximately 76 million hectares (Oldeman et al. 1991). These activities are degrading the lands continuously on an estimated rate of between 0.25 and 0.5 Mha annually (FAO 2000). The continuous expansion of salt-affected area is particularly important in South Asia where there is fresh water scarcity at one hand and on the other hand arid to semi-arid climate coupled with low rainfall. The large extent of degraded soils is responsible for the low production of agriculture crops both quantitatively as well as qualitatively. This agriculture product is insufficient to feed the massively increasing population of the world. The core reason of low productivity form these soils is hampering water absorption by plant roots (osmotic deregulation), cell injury (the specific ion toxicity) along with deterioration in the physical properties of these soil (Abrol et al. 1988; Ghassemi et al. 1995; Lamond and Whitney 1992).

Saline soils are important land resources in world agriculture because salt-affected soils are usually abundant in natural resources like light and heat posing great potential to develop agriculture. Reclamation of salt-affected soils is of key importance to mitigate the pressure on every day squeezing agricultural soils. It will help in increasing the cultivated area and reducing the threats to our food security. Several methods have been experimented for the reclamation of salt-affected soils and the suitability of method depends upon physical, chemical and mineralogical characteristics of the soil including internal soil drainage, presence of hardpans in the subsoil, climatic conditions and types of salts, quality and quantity of available water, depth of ground water, replacement of excessive exchangeable Na+, lime or gypsum, cost of the amendments, topographic features of the land, and the time available for reclamation (Mashali 1991). The appropriate management of the constrained soil resources for the economic agricultural production is the main emphasis in agriculture. The prominent techniques include chemical, biological and agronomic or combination of these approaches to reduce the time of reclamation with in the economic bindings. The crop production and fertilizer use efficiency of these soils can be increased by an integrated approach, i.e. use of amendments preferably gypsum and organic/ inorganic manures which helps in maximizing and sustaining yields, improving soil health and input use efficiency (Swarp 2004).

Some of these possible techniques have been discussed in this section.

2.5.1 Physical Methods

Physical methods are those approaches which involve physical treatment of the soil without the application of any organic or inorganic chemicals. The physical methods include sub-soiling, deep ploughing, sanding, horizon mixing, profile-inversion and channeling irrigation practices like drip irrigation etc. These treatments increase the permeability of the soil, which is generally a limiting factor during the reclamation of sodic and saline-sodic soils. Deep ploughing is very useful where the subsoil has gypsum or lime (Ahmed and Qamar 2004). Salt-affected soils can be reclaimed by altering the methods of irrigation water applications for crop production may be providing adequate irrigation water or rainfall to leach down excessive salts from the root zone soil, and improving good internal soil drainage (Qadir and Schubert 2002; Zhang et al. 2008). In this regard, drip irrigation thought to be an effective approach to reclaim salt degraded soils. Research results proved that the leaching efficiency with drip irrigation remained higher compared to that with other irrigation methods (Bresler et al. 1982). It was observed that red effect drip irrigation on different soil properties on an unreclaimed salt-affected land (Tan and Kang 2009). Application of drip irrigation along with cropping significantly decreased salt concentration especially in upper 0–5 cm soil layer reducing salt concentration from 10.45 dS m^{-1} to 1.65, 3.49, and 0.94 dS m^{-1} on the 1st, 2nd, and 3rd cropping years respectively under field conditions. However, the big hindrance in this physical amelioration is availability of sufficient amount of good quality irrigation water and if available, have a high-cost in rural regions (Qadir and Schubert 2002; Zhang et al. 2006). For inland regions, ameliorating soil salinity can be achieved effectively by a plant-assisted approach than the physical approach (Li et al. 2008; Qadir and Schubert 2002; Zhang et al. 2006).

2.5.2 Chemical Process

The chemical methods include application of chemicals, such as gypsum, sulphur, sulphuric acid and hydrochloric acid. Gypsum is effective on both sodic and saline-sodic soils, while sulphur, sulphuric acid and hydrochloric acid are only effective for calcareous saline-sodic soils. These amendments remediate the soil by lowering the soil pH and react with soluble carbonates and replace the exchangeable sodium with calcium (Ahmed and Qamar 2004).

The reclamation of sodic soils is usually the most expensive compared to saline and saline-sodic soils but can be reclaimed by addition of chemical amendments, organic matter, deep tillage (Seelig et al. 1991). Gypsum has been recommended as an economical amendment for the amelioration of sodic and saline sodic soils (Elshout and Kamphorst 1990; Qadir and Schubert 2002; Shainberg et al. 1982). Gypsum has very low relative solubility being 0.2% (0.2 g in 100 mL water) that may cause hinder and prolong the reclamation process for sodic soil (Carter and Pearen 1989). The solubility and efficiency of gypsum can be enhanced with application of fine ground material and with application methods. Application of gypsum in standing water can improve the efficiency of gypsum than application on dry soil surface (Choudhary et al. 2008) due to rapid dissolution in case of standing water. Similarly, powdered form of gypsum is more efficient in reclaiming sodic soils (Ali et al. 1999; Choudhary et al. 2008; Ghafoor et al. 2001). Dut et al. (1971) claimed that 52 to 72 cm water is required to dissolve 16.5 to 23.9 Mg ha^{-1} gypsum applied on soil surface. The solubility of gypsum increases by 10 folds under sodic soil condition.

Moreover, mixing of gypsum and fast removal of Na from the soil solution will speed up the exchange process (Frenkel et al. 1989). However, if the soil is dense and has poor drainage, little or none of the exchange will be removed and gypsum application will largely be ineffective rather it can increase the soil salinity. (Ilyas et al. 1997) observed higher Na, Ca, Mg, and EC values with gypsum application that were mainly attributed to poor soil permeability where the replaced Na remained in the soil solution. However, alter one year the EC and Na started to decline. Under soil conditions deep ploughing will facilitate the process of reclamation to allow leaching of Na salts.

Application of gypsum improves physical as well as chemical properties of salt degraded soils (Ayers and Westcot 1985), soil porosity (Oster et al. 1996; Shainberg and Letey 1984) and soil hydraulic conductivity (Scotter 1978). A significant decrease in soil bulk density was recorded when surface soil was treated with phosphor gypsum (Southard and Buol 1988). Ghafoor et al. (1985) observed a significant increase in grain yield of wheat with gypsum application.

2.5.3 Organic Matter

Addition of organic amendments improves soil structure increasing soil permeability (Tejada et al. 2006). Different studies revealed that there is a positive correlation between organic matter and microbial activity (Schnürer and Rosswall 1985).

Microbial population improved soil physical properties which accelerate the ameliorative process of salt-affected soils. (McCormick and Wolf 1980) observed that alfalfa residues used as an organic amendment can reduce the deleterious effects of soil salinity. Biochar is widely used as an organic amendment now a days, has beneficial effect in ameliorating salt-affected soils. Biochar improves soil structure having positive on bulk density, pore-size distribution and particle size distribution (Roberts et al. 2009; Sohi et al. 2009). Biochar benefits biophysical properties of soils increasing availability of air and water in rhizoshere which in turn improves germination and plant survival (Lehmann et al. 2006; Zhang et al. 2014).

2.5.4 Biological Methods

By planting salt tolerant plants on salt degraded soils, water evaporation considerably decreased from surface soil (Li et al. 2010; Qadir and Schubert 2002). Many field experiments revealed that planting forages in salt degraded soil, physical properties were improved due to penetration and exclusion of an extensive and thick root system followed by leaching of excessive salts to deeper layers (Liang et al. 2007). In addition, the forage cover minimized water evaporation and salt accumulation in the surface layer soil (Ghaly 2002). Phytoremediation of salt-affected soils, the soil productivity was significantly increased compared to that with simple leaching with irrigation water (Zhang et al. 2005). Biosaline (agro) forestry's most vital prospect is the controlling soil salinity and sodicity along with the reclamation of the degraded land for high yield and other agricultural production. It is reported that agroforestry have the potential to control the salinity and sodicity (Barrett-Lennard 2002; Oster et al. 1996; Qadir and Schubert 2002; Singh 1993). Thus forestry and agroforestry systems on salt-affected soils which is referred as the biosaline agroforestry can act as the supportive land use against the salinity problem. The reason behind this is the tolerance of the some salts against salinity/sodicity and their plantation can help the soil in elimination of the salinity of the soil (Singh 1993; Turner and Lambert 2000).

2.5.5 Hydro-Technical Method

In this method a saline water of high electrolyte concentration (EC) is used by keeping in view the principle of valence-dilution effect to affect soil permeability and subsequently by successive dilutions. The valence dilution effect was first validated by (Eaton and Sokoloff 1935) for reclaiming sodic soils. After the establishment of equilibrium between monovalent and divalent cations in the soil solution and the ones which are found adsorbed, application of water to the system alter the equilibrium in such a way that it will be favorable for the adsorption of divalent cations such as Ca^{2+} after the adsorption of the monovalent cations such as Na^{+}. Contrary to this situation when the soil solution is concentrated due to evapotranspiration adsorption of monovalent cations such as Na^{+} occur first and then adsorption of divalent cations such as Ca^{2+}. The ratio of divalent to total cations when concentrations are stated in $mmol_c\ L^{-1}$ of water should be at least 0.3 and with the increase in this value water requirement for the reclamation decreases. A few natural water sources have this value of this ratio but mostly some additional Ca^{2+} is required that can be added by (1) soil application of gypsum followed by irrigation with high-salt water or (2) by placing gypsum stones in the water channels to add Ca^{2+} in the salty water through gypsum stone dissolution. The major problems with this method are limited facilities of collection, conveyance, and treatment of saline water.

2.5.6 Electro-Reclamation Method

Electro-reclamation refers to the amelioration of salt-affected soils through electrodialysis. Laboratory and field investigations have shown that treatment with electric current may simulate reclamation of saline-sodic/sodic soils, although it cannot replace the conventional procedures of soil reclamation. By this method different anions such as nitrate, sulfate, fluoride, and chloride can be removed from the soil by the method of electro-reclamation. During electro kinetic reclamation, the pH increases adjacent to the anode and decreases around the cathode. The removed cations (Ca^{2+}, Mg^{2+}, K^{+}, and Na^{+} were 19.5 %, 34.4 %, 58.9 %, and 89.6 % respectively) and anions (Cl^{-}, NO_3^{-} and SO_4^{2-} were 47.9 %, 91.5 %, and 67.6 %) from saline soils having EC = 13.7 dS m^{-1} (Kim et al. 2013). Kim et al. (2011) found a significant decrease in EC of a saline soil (EC = 7.1 dS m^{-1}) using a hexagonal two-dimensional electrode. Generally, the removed nitrate was relatively higher than either chloride or sulfate. Sulfate tends to form insoluble CaSO4, which may decrease its respond to electro reclamation. Another study showed that chloride was concentrated on the saline soil surface (EC = 7.8 dS m^{-1}). Magnesium was not removed but potassium was removed, and sulfate showed a uniform distribution (Kim et al. 2011). The removal of Ca^{2+} was increased during pulse electro remediation of saline soils with EC ranging from 6 to 21 dS m^{-1}, as the process enhances the interactions of soil water solutions (Le et al. 2003).

2.5.7 Combination of Organic and Chemical Amendments

Use of organic amendments manifolds the process of improvement of soil properties both physical and chemical as compared to the use of chemical amendments alone. Harms of salt affected soils can be lower down by the use of organic matter (organic amendment) along with gypsum (inorganic amendment). Wong et al. (2009) reported that use of organic matter improves the physico-chemical properties of soil of the salt affected areas. Addition of farm yard manure along with gypsum reduces the EC and ESP up to the great extends (Abou El-Defan et al. 2005). Solubility of the gypsum will become two times rapid with the addition of the citrate (Jones and Kochian 1996). Citrate enhances the reclamation process by causing the complexation of the Al from solution as well as from the minerals. More decrease in dispersion and EC was observed with the combined application of organic matter and gypsum (Vance et al 1998).

2.6 *Management of Salt Effected Soils*

2.6.1 Management of Saline Soils

2.6.1.1 Leaching

Salt affected soils can be reclaimed by removing the salts from the root zone area of plants either with heavy irrigation or with the drainage (Feng et al. 2005; Qadir et al. 2001; Qureshi et al. 2008). Salt affected soils can be reclaimed as well as managed by irrigating the soil with plenty of good quality water. We can determine the reclamation rate by knowing the amount of water that reaches out of root zone after passing through soil referring as leaching fraction while leaching fraction is directly related to the drainage capacity of soil. Reclamation process initiates by drainage of salts and reducing the water table. There are some cases when reducing water table will no longer be beneficial but this problem can be solved by the utilization of land for crop cultivation. Brackish water used for irrigation purposes due to shortage of good quality water is the major cause of salinity problem. Salt affected soils can be reclaimed by leaching down the salts along with irrigation sources of good quality water rather using the poor quality water. 1.5 times of the EC of the irrigation water salts can be removed from the soil while adopting the good management activities. Thus if EC of the leaching water is high we need huge quantity of water to eliminate the salts from the salt affected soils. It is general recommendation that EC can be reduced up to half with every 6 in. good quality water that can pass through the soil along with salts. That's why if we have to remove the salts 30 in. downward having EC 1.5 dS/m we need 6 in. water that will move up to the 30 in. within the soil that will EC lower the EC to 0.75 dS/m. It is proved that organic matter improves the soil properties thus with the application of the organic matter drainage capacity of the soil will be enhanced that will reduce the problem of salinity. To enhance the organic matter into the soil vegetation is very important. Growing of maximum trees can act as the buffer of the soil against the generation of the salt affected soil. Addition of salts will lower down the free energy of the water by rising the osmotic potential or solute potential. Resultantly plants feel difficulty in the uptake of the water and growth and development of the plants become less. Now it is the need of the hour to reclaim the salt affected soil to get the maximum yield as food security and sustainability are becoming major problem of the world.

2.6.1.2 Irrigation Method

It is very important that how are we irrigating the soil to check down the high concentration of the salt in the root zone. It is reported that application of the large amount of water for the irrigation purposes plays supportive role for the adequate uptake of water by plants. Sprinkler irrigation is one of the best methods for irrigation especially when water shortage and salinity are the major problem. Soluble salts leach down from the root zone when irrigation is applied to the soil for the maximum time and quantity. Thus sprinkler irrigation ranks high in efficiency as compared to the flooding. It is reported by Nielsen et al. (1966) that requirement of water becomes 3 times more in flood irrigation when compared with the sprinkler irrigation for lowering the same amount of the salts. It is also beneficial that land leveling is not required for the uniform application of the water which is the basic necessity in the flooding irrigation. Similarly drip irrigation which is sometimes also called trickle irrigation is the best method of irrigation for the perennial crops and seasonal row crops. As it supply the water the water at one point only problem of salinity become minimized. Salt concentration will become less by this method by keeping the water table low. When water table will be low risk of salinity development reduces up to great extent.

2.6.1.3 Mulching

Salts come at the surface of the soil when process of the evaporation becomes faster that application of water. Even the leached down salts can come at the surface along with water with capillary rise process when irrigation will not be applied for long time especially during the fallowing of the land. Soil salinity is the major problem when water table is shallow along with the high EC of the irrigation water. But the problem salinity can be reduced by lowering the evaporation process. Evaporation become limited when soil remain covered with vegetation. It is recommended that the salinity problem become less when process of evaporation will be lowered by mulching or covering the soil (Sandoval and Benz 1966). Thus after the fallowing of land mulching will be helpful in controlling the salinity problem.

2.6.2 Management of Sodic Soils

Excess Na^+ on the cation exchange sites causes clay particles to disperse or swell, and as a consequence these soils have poor structure, low aggregate stability, and reduced water infiltration (Rengasamy and Olsson 1991). Overall, sodic soils are a poor rooting medium for plant growth and provide lowered or insufficient nutrients. Sodic soils also have reduced biological activity and function due to the limited availability of C substrates that are likely the result of lowered net primary productivity in these soils (Rao and Pathak 1996). Remediating the effects of excess Na^+ in sodic soils can be accomplished with soil amendments and land management. Calcium amendments have been shown to reduce the effects of sodicity. Calcium flocculates clay particles leading to improvements in soil structure (Frenkel et al. 1989). Calcium also replaces Na^+ on soil exchange sites and is frequently correlated with increases in soluble Na^+ (Ilyas et al. 1997). Rates of gypsum application can be calculated by taking into account soil cation exchange capacity, target SAR, and current SAR values (Ashworth et al. 1999). After chemical treatment subsurface tile drainage may be used to remove excess sodium from the rooting zone (Pessarakli and Szabolcs 1999). Subsurface drainage can also prevent salt accumulation due to fluctuations in water table depth, capillary rise, and evaporation (Abrol et al. 1988). In order to provide advice to growers with respect to whether their management strategies have begun to bring about the changes they anticipated, a tool capable of detecting short term improvements is needed. Successful remediation of sodicity may take years and can be costly (Qadir and Oster 2002). Soil health is referred as ability of soil to perform within ecosystems and use of land to sustain high yield, good environmental quality and improve plant, animal, and human health (Doran and Parkin 1994). Soil health can be determined by the use of different indicators such as a proxy for shifts in nutrient cycling resulting from land use change, amendment application and tile drainage installation will aid in the early detection of effective remediation strategies, potentially reducing the cost and environmental impact of remediation (Ella et al. 2011; Fortuna et al. 2012). Additionally identifying soil health indicators and monitoring changes in these soil properties will aid landowners in ensuring the long–term productivity of the land. Currently, biological soil health indicators are not widely used to assess remediation progress. Reclamation of the sodic soil is very difficult and mostly expenses become high than income. By following the above procedures reclamation of the sodic soil is possible but it took many years to completely reclaim this problem while following the good crop management practices.

2.6.2.1 Drainage

Soil sodicity problem can be controlled by removing the high concentration of sodium from the root zone by good drainage practice. Low water table is helpful in reducing this problem. By the development of the tile drains and by changing the topography sodic soils can be reclaimed up to the great extent. Plantation of trees especially deep rooted is also beneficial when we want to low down the problem of sodicity. Sealing of canals or lining of canals become supportive for controlling the seepage which resultantly control the problem of the sodicity. Thus good drainage property of the soil is very important in controlling the problem of the sodicity.

2.6.2.2 Tillage and Amendments

Tillage practice is considered as the physical practice in reclaiming the problem of sodicity. Tillage cause the fragmentation of the big soil colloids having the high concentration of the sodium and amendments will become the part of the soil and reclaiming process become faster. Large organic matter which has the property of slow decomposition like straw, cornstalks, sawdust, or wood shavings used for animal bedding is reported beneficial for improving soil structure and infiltration properties of soil along with the other reclamation activities.

2.6.2.3 Supplying Calcium to Improve Water Infiltration

Refining water infiltration property of soil requires lowering of the exchangeable sodium percentage (ESP) along with raising the electrical conductivity (EC) up to more than 4 dS/m (4 mmhos/cm). It can be determined by the soil texture and irrigation method that how much exchangeable sodium percentage (ESP) is required to make the better infiltration. Sandy textured soils have the capacity to bear the exchangeable sodium percentage (ESP) upto the 12 while still having good infiltration and percolation. Surface irrigation similarly can retain good infiltration and percolation with high exchangeable sodium percentage (ESP) as compared to the sprinkler irrigation. Calcium is basic need in the reclamation process of the sodic soils as it can replace the sodium and that lowering the ESP as well as SAR.

2.6.2.4 Irrigation Water Management

Irrigation water that comes from the deep wells has great concentration of bicarbonate and thus high sodium concentration as compared to the calcium and magnesium. Irrigation with such type of water for long time creates the problem of sodicity. EC and SAR are used to evaluate the infiltration problems by the application of the irrigation water.

2.6.3 Management of Saline-Sodic Soils

To reclaim the saline-sodic soils it is the important to first reclaim the sodic soil with the use of calcium to resolve the problem of high concentration of the sodium. After reclaiming the problem of the high concentration of sodium (sodicity) problem of the high concentration of salts (salinity) can be resolved simply by the application of the high amount of irrigation water. It is the basic requirement of saline-sodic soil reclamation that to solubilize the sodium first before the leaching of all other salts. The reason behind it is that if we'll not make the sodium soluble before removing all salts from root zone problem of sodicity will left over after treating the soil for salinity problem. Thus soil structure will be deteriorated that will make infiltration process either completely stop or lower down. After this destruction remediation becomes very difficult. Therefore it is necessary to determine that how much sodium problem still remaining before applying good quality irrigation water to leach salts. High EC of irrigation water and soil supports for improving soil structure, increasing water infiltration, and resist sodium from accumulation into the soil. Except this positive effect of high EC (salt) irrigation water about soil structure it is not good for crop production.

2.6.4 Adaptations of Salt Tolerant Plants

To choose the plants that have the tolerance against the salinity is the major step in reclamation of the salt affected soils. It is because different plants have different potential to uptake and accumulation of the salts to minimize the salinity problem (Conway 2001). Different species of plants show salt tolerance against salinity by developing the mechanisms like salt exclusion, uptake and compartmentalization of salts and extrusion of salts (Holly 2004). These salt tolerance plants are also referred as the halophytes. Physiological property of halophytes is usually expressed as morphological features like salt glands, salt hairs, and succulence. Plants depend on more than one tolerance mechanism for salt tolerance (Holly 2004; Naidoo and Naidoo 1999). Halophytes can adjust osmotic effects internally by accumulating high salt concentrations or they may become able to absorb more water from saline soils (AzevedoNeto et al. 2004). Salt exclusion permits plants to maintain and reduces the quantity of salts that go to growing leaves and young fruits. A few species have adopted excluder mechanisms to tolerate the salinity stress. Through this mechanism plants filter the salts in their roots and resist against the salt uptake towards the upper parts. Salt stress is tolerated by the plants by reducing germination, growth, and reproduction to specific seasons during the year and by growing roots into non-saline soil layers, or by less uptakes of the salts from the soil (BPMC 1996). Halophytes take salts from the soil and accumulate them in to their different cells and thus maintain their water potential (Andre et al. 2004). Salt tolerant plants accumulate ions in the vacuole and produce organic solutes into their cytoplasm (Marcum 2001; Taizand Eduardo 1998). This practice of accumulation of ions in the vacuole and production of the organic solutes helps the plant to take more water with an osmotic gradient without causing harm to the salt sensitive enzymes. Plants also accumulate the salts into their vascular tissues and try to avoid the exposure of chloroplast to the salts (Misra et al. 2001). Production of organic solutes also helps the plants to retain water balance between the cytoplasm and vacuoles (Holly 2004; Marcum 2001). Plants can uptake more water from the soil when water potential of the soil will be higher than the water potential of the cells of plant (Holly 2004; Taiz and Eduardo 1998).

3 Soil Erosion

Soil erosion is the detachment of soil particles by the action of wind or water. Though soil erosion is a natural process but is accelerated by anthropogenic activities like deforestation, overgrazing, improper agricultural practices and cultivation techniques. This is a widespread problem due to which our fertile ecosystems are losing their fertility and result in degradation of all ecosystems (Lal and Stewart 1990; Troeh et al. 2004).

Loss of agricultural land is observed due to human induced erosion and other related damages. This leads to desertion and lowered productivity of agricultural lands which are also somewhat made up by addition of fertilizers over many years (Pimentel et al. 1995; Young 1998). Every year almost 10 million hectares of cropland are deserted worldwide due to soil erosion (Pimentel 2006).

3.1 *Types of Soil Erosion*

Erosion is mainly caused by two agents, water and wind and thus soils erosion is divided into main categories *viz.* water erosion and wind erosion.

3.1.1 Water Erosion

Water erosion is the detachment of soil particles by the action of water. Particles are detached by raindrop impact or transported by overland water flow. Raindrop impact breaks down the aggregate of soil particles into smaller parts, these suspended particles deposited in the pores between the aggregates and clog them resulting in 'soil crust' formation. Soil crusting seals the surface by blocking the spaces between aggregates, lowering infiltration rate and drainage while increasing the run-off. Uncovered land is more susceptible to rain drop impact compared to vegetative covered land. Thus more and more water is lost due to run-off, the upper fertile layer of the soil is lost and soil becomes less productive. As soil forming process is a time-taken process, it is measured that a soil loss of 1 t/hac/year is considered irreparable within 50–100 years of time span (Van Camp and Vauterin 2005).

Raindrop impact is the first stage of water erosion; it leads to sheet erosion. Sheet erosion removes upper fertile layer of soil. Sheet erosion develops into rill erosion, forming small channels. These channels become widen and lead to gully formation (Hillel 1998).

3.1.2 Wind Erosion

Erosion caused by wind is known as wind erosion. It is the severe problem of arid and semiarid regions but also observed in humid regions. Wind erosion transports very fine particles and lowers soil fertility and thus degradation occurs. It is reported by Global Assessment of Soil degradation that 42 million hectare of European land is wind eroded (EEA 2003).

There are three modes of soil movement, depending on particle size:

- Creep (>500µmalong the surface)
- Saltation (70-500 µm)
- Suspension (short-term suspension 20–70 µm; long term suspension<20 µm) (Shao 2000).

The extent of wind erosion is determined by soil's erodibility, climate's erosivity and also influence by various other components increasing the actual erodibility of site (Sterk and Warren 2003). Climate's erosivity depends on intensity, duration, direction and frequency of wind velocity; amount, intensity and distribution of rainfall, radiation, humidity and evaporation (Funk and Reuter 2006).

3.2 Causes of Erosion

Main cause of soil erosion when land is remaining exposed to rain or wind storms. About 60,000 kcal of raindrop energy is recorded on hitting a hectare of land in New York with 1000 mm of rainfall (Troeh 1999). This highly energized raindrop result in erosion of land even on slopes to downwards. Among all types of erosion sheet erosion is most prominent of all (Troeh 2004; Oldeman 1998). The soil erosion becomes more degradable on slopes when water flowing above them carried fertile soil particles along with it into streams and valleys. Strong wind intensifies this situation displace the upper fertile layer soil particles to different distant areas.

3.2.1 Soil Structure

Structure of the soil is an important factor in the soil erosion process. The soil with weak developed structure, medium to fine in texture and having low content of organic matter are most dominantly eroded in easy ways (Bajracharya and Lal 1992). Due to less water holding capacity and low water infiltration rates these soils are more favorable to be eroded by wind and water energy.

Texture and organic matter content of soil provides ability to the soil to resist erosive conditions. They influence the water holding capacity of soil as well as make aggregates of soil particles (Chepil 1955).

3.2.2 The Role of Vegetative Cover

Coverage of plant biomass serves as a protective blanket over the topsoil to prevent it from erosion by means of wind and water energy. Coverage of plant biomass, dead or living dissipated the rain and wind energy and as a result protects the underlying soil layer from erosion. Soil particles remains in contact and are held firmly due to upper coverage (Pimentel et al. 2005). It is reported in Utah and Montana, the rate of erosion increased 200 times as the result of ground cover decreased from 100 % to 1 % (Trimble and Mendel 1995). About 60 % of forest coverage is essential to prevent forested areas from land sliding and soil erosion. Extensive loss of forests now days for irrigation lands and pasture is followed by soil erosion (Act 2002; Singh 1993).

Degradation of topsoil is most widespread in developing countries where poor irrigation practices are being used to overcome their food demands. Crop residues are being used for heating and cooking purposes and poor agriculture practices are unable to protect land from erosion. In 1990s according to an estimate 60 % and 90 % crop residues in China and Bangladesh respectively were used for fuel purposes (Wen and Pimentel 1993). According to recent estimates still in Bangladesh 50 % of all residues from rice fields and 80 % non-rice fields are used for biomass energy (Hassan et al. 2011). In China due to availability of fossil fuels the usage of crop resides are decreased (Li et al. 2012). Even they planned to burn half of the 600 billion tons of straw residues to produce electricity annually (Lal 2007). In some areas where the fuel sources are limited the roots of shrubs and herbs are burned to use as a fuel source (Juo and Thurow 1997). All such practices exposed the land surfaces and result in degradation of land by wind and water energy and result in formation of barren lands.

3.2.3 Land Topography

The vulnerability of land to water and wind forces depends on its surface exposure as well as on its topography. It is estimated in Philippines and Jamaica where 58 % and 52 % of land respectively have slopes, face 400 times per hectare more erosion rates. Steep and marginal lands which changed from forest to crop lands are more susceptible to soil erosion (Lal and Stewart 1990). Arid lands are susceptible to soil erosion as observed in India the erosion rates are 5600 times per hectares per year (Gupta and Raina 1996). Even in developed countries like United States having massive farmlands faced 13 tons/ha/year average losses (USDA/NRCS 2007). In Europe 3 to 40 tons/ha/year loss happened after every 2 to 3 years are estimated regularly (Grimm et al. 2001; Verheijen et al. 2009).

3.2.4 Disturbances

According to estimation about three-quarters of the world's soil erosion is due to depletion of soil biomass coverage (Lal and stewart 1990). A number of factors contributing this phenomenon among which is construction sites of buildings and roads. Surface coverage of land with plant dead or living reduces level of soil erosion significantly (Pimentel 2006; Soil and Water Conservation 1993).

Along with anthropogenic activities natural processes also result in soil erosion especially along stream banks due to powerful action of flowing water. 30 % or more soil erosion is recorded on steep slopes while on flat surfaces with only 2 % of slope also faces the problem of soil erosion under powerful force of wind and water.

3.3 *Assessing Soil Erosion*

It is estimated that about 75 billion tons of fertile soil is lost every year throughout the world due to soil erosion process (Norman 1984; Padmanabhan and Hari 2008). Soil erosion due to water force is recorded to about 76 billion tons per year (Reich et al. 1999). In 2010, Central Soil Water conservation research and training institute in Dehradun, India estimated that the fertile land loss due to erosion is 16.4 billion tons per hectare annually with total annual loss estimated to be 5.334 billion tons (The Hindu 2010). Chinese Academy of Sciences, Chinese Academy of Engineering and Chinese Ministry of Water resources in 2009, after 3 years of study reported that China's 646 communities are facing about 3.75 million km^2 soil erosion issues (Stedman 2009). A further 2 year study revealed that if this rate of soil erosion continues with this rate the food production of China is reported to decline 40 % from current rate (Jie 2010). Both China and India which occupies about 13 % of total land area with sustained agricultural practices facing about 75 billion tons of soil erosion annually. The rate of soil erosion in United States agricultural areas is estimated to decline from year 1982 to 2007, 3.06 billion tons to 1.725 billion tons (USDA/NRCS 2010).

3.3.1 Worldwide Cropland

Currently it is estimated that 80 % of world land facing the problem of moderate to severe soil erosion problems only 10 % is facing the slight erosion issues (Lal 1994; Speth 1994). Erosion in cropland areas is about 30 t/ha/year on average that ranges from 0.5 to 400 t/ha/year (Pimentel et al. 1997). It is reported that about 30 % of worlds land became unproductive and barren due to the soil erosion rates from last 40 years. Such lands are unable to utilize for agricultural purposes anymore (Kendall and Pimentel 1994; World Resources 1994).

3.4 *Effects of Soil Erosion on Terrestrial Ecosystems*

Productivity of terrestrial ecosystems is significantly reduced due to soil erosion (Jones et al. 1997; Pimentel et al. 2006). Water infiltration and water holding capacity of soil decreased due to increase water runoff with erosion from soil surfaces (Torah et al. 2004). Organic matter and essential soil nutrients also degraded with erosion process along with natural biota and biodiversity also significantly affected (Torah et al. 2004; Pimentel et al. 2006). Enhanced rate of erosion where increased water runoff also results in low water holding capacity along with overall damaging effect on valuable biota and soil's biodiversity (Brevik 2009; Jones et al. 1997).

As the top soil is removed by erosion, reduction in surface soil depth is observed. Due to erosion there is not sufficient soil available to support plant and thus, crop roots are exposed and consequently more yield loss. It is reported that soil erosion can alleviate the crop yield up to 30 % (Pimentel 2006).

3.4.1 Water Availability

All sorts of vegetable and fruit production require enormous amount of water for its productivity water serves as a main limiting factor on their production (Pimentel et al. 1997). According to an estimate one hectare of corn field utilize about 7 million liters of water during its 3 months of growing season (Pimentel et al. 2004) and additional 2 million of water loss due to evaporation from soil surface as result of soil erosion due to water and strong wind forces. All of these result in less water availability for growing vegetation. Enhance water runoff significantly affect water holding capacity which result less water availability for the growing crops (Pimentel et al. 1997).

3.4.2 Nutrient Losses

Essential plant nutrients such as nitrogen, phosphorous, potassium and calcium dominantly decline due to erosion. The eroded soil has three times more nutrients as compared to remaining soil surface (Young 1989). Fertile soil estimated to have 1 to 6 kg of nitrogen contents, 1 to 3 kg of phosphorous contents and 2 to 30 kg of potassium contents whereas the left over soil having only 0.1 to 0.5 kg per ton of average nitrogen contents (Schertz et al. 1989; Langdale 1992). To overcome these problems large quantities of fertilizers are being used. The cost of loss of basic soil nutrients is estimated to be about several billion dollars each year (Torah et al. 2004). If the surface of soil is about 300 nm deep the usage of commercial fertilizers and livestock manure replace the lost of nutrients but this practice is usually expensive for poor farmers and nations. Not only expensive these synthetic fertilizers also affect the human health and also result in soil, air and water pollution (Brevik 2013; Unnevehr et al. 2003; Hakeem et al. 2011; Hakeem et al. 2014).

3.4.3 Soil Organic Matter

100 tons of organic matter per hectare is usually present in fertile soil layers (Pimentel et al. 2005; Sundquist 2010). Fertile soil's organic matter typically consist about 95 % of nitrogen and 25 % to 50 % of the phosphorous (Allison 1973). Erosion typically affects the top fertile layer of soil because most of the organic matter of dead plants and animals lies in this upper layer. Fine soil particles are mostly degraded by wind and water erosion and larger particles and stones are left behind and making it unsuitable for growing vegetation. A large number of

researches report that the eroded soil contain large amount of organic matter about 1.3 to 5 times more than left over soil (Lal 1994). The reduction of organic matter in left over soil from 0.9 % to 1.4 % results in about 50 % loss in crop production and ultimately in yield (Libert 1995; Sundquist 2010).

Depletion of the nutrients, organic matter and structure of soil results in the degradation of ecosystem and also significantly affect the crop yield. Productivity of any area is dominantly affected by the results of strong wind and water erosion. Soil erosion not only affect crop yield but also significantly affect total biomass of biota and biodiversity of any ecosystem substantially (Lazaroff 2001; Walsh and Rowe 2001).

3.4.4 Soil Depth

Suitable soil depth is required not only by plants for its roots extension but also by several soil biota‘s like earthworm (Pimentel et al. 1995; Wardle et al. 2004). Soil erosion substantially reduces the soil depth from 30 cm to less than 1 cm. The space for plant roots become declined due to which plants show stunted growth.

3.5 Conservation Technologies

Mulching, vegetation, riprap, matting, terracing, retaining walls and reforestation are common treatments to overcome soil erosion (Rivas 2006). About 30 % of the world's food production is suppressed by the agricultural land degradation for last 50 years by soil erosion (Kendall and Pimentel 1994). Biomass mulches, crop rotations, no-till, ridge-till, added grass strips, shelterbelts, contour row-crop plantation are suitable techniques for soil conservation.

All these strategies mainly focus on preventing the land from erosion by wind and water force by covering the surface layer with some sort of coverage and its residues. Not leaving the land open is the best remedy to prevent soil erosion even after harvesting the crop the coverage is provided by its remains. Thus, soil cover is the best remedy against soil erosion (Pimentel et al. 1995; Troeh et al. 2004; Pimentel 2006). The risk of wind erosion alleviates if soil cover is less than 10 %, while soil covered for more than 10 % lowers wind erosion and when soil cover is 40 % complete prevention occurs against wind erosion (Morgan and Finney 1987; Sterk 2000).

Throughout the world soil erosion is putting its disastrous effects. It is continuous, slow insidious problem. 1 mm of soil is lost every year and this is so minute amount that unnoticed by framers. About 15 t/ha of soil loss over one hectare of cropland area, rehabilitation of such soil takes about approximately 20 years even if this process continues the land is not available to support vegetation. Soil erosion severely affects if it goes unchecked and result in loss of overall biota, soil fertility, organic matter and soil water holding capacity (Pimentel and Burgess 2013).

4 Soil Acidity

Acidic soils mostly have pH values less than 7 on the pH scale (Soil Science Society 2008). Acidity of the soil mainly depends on the availability of exchangeable forms of hydrogen and aluminum ions (Brady 2001; Fageria and Baligar 2003). Higher the concentration of these exchangeable ions higher is the amount of acidity in the respective soil. Acidic soil is observed to have low fertility rates, poor in physical, biological and chemical properties. Poor management of such areas results in depressed crop yield to a significant level (He et al. 2003).

4.1 Causes of Soil Acidity

Both the natural and anthropogenic activities are responsible for soil erosion process. Natural processes happen gradually and affect the soil fertility in a gradual way but the anthropogenic effects are rapid.

4.1.1 Weathering and Leaching

The present soil is formed from the parent rocks which contain both the essential and non-essential nutrients of plants. The soil form is more acidic in nature if the parent rock and material is acidic and more alkaline in nature if the parent material is alkaline. Both the acidic and basic cations are released in soil during weathering. The influx of these nutrients is mostly overcome by leaching basic cationsthat counter act with acidic cations and the preponderance of the acidic ions enhances soil acidity. The process is more active where precipitation rate is higher than evaporation, plant's transpiration rate and high temperature boost the process of weathering and leaching (Nyarko 2012).

4.1.2 Organic Matter Decomposition

Both the plants and animals take nutrients in various forms during the course of their lives. Even after their death when the process of decomposition starts these organic matters along with many sundry chemicals are again handed over to soil. In the course of this eternal process acids are continuously formed and consumed. Usually organic matter has reactive substances like phenolics and carboxylic groups. These reactive substances on dissociation release H+ ions which result in enhanced soil acidity (Seatz and Peterson 1964). Carbonic acid also formed as reaction of CO2 which is released during process of decomposition with water. Brady (2001) reported that very little soil acidity is contributed by decaying organic matter.

4.1.3 Acid Rain

Wherever there are large cities with dense concentration of vehicles and industries, acid rain forms. Rainfall is basically acid due to deposition of oxides of sulphur and nitrogen found in atmosphere due to combustion, burning of coal/petroleum products and agricultural activities. Due to these factors pH of rainwater becomes acidic and is found between 4 and 4.5 (Brady and Weil 1984). With the excessive accumulation of these acids in the atmosphere, which if not controlled significantly affect the soil and plants growth (Brady and Weil 1984). Precipitation is also an enhancing factor in soil acidity (Donahue et al. 1983).

4.1.4 Crop Production and Removal

The main goal of any agricultural system is to produce saleable products. Soil acidification suffers as a limiting factor in this way. Respiration is necessary for both plants and microbes for their survival but it result in large amount of acid production in form of carbonic acid. Black (1968) reported that this is a very minute factor because most of carbonic acid produced during this process lost in atmosphere as CO_2 (Tang and Rengel 2003). Basic cations that are usually up-taken by plants are $Ca^{2+,}$ Mg^{2+}, K^+ and also NH^{4+}, as result more H^+ dissociation by plants for their electrical balance specially when nutrients are absorbed in form of NH^{4+}(Tisdale and Nelson 1975). More the basic cations uptake more the H^+ ions release which leads to acidity in the soil.

There are basic cations available in plant especially in leaves and stem than the grains, these basic cations neutralizes the acidic effect which is develop by different processes but when these crops are removed from field either burnt, or harvested or washed away by run-off this counter effect of basic cations is gone and ultimately soil acidity increases (Chen and Barber 1990).

Type and part of crop harvested and stage of crop at harvest basically deals with the amount of these nutrient removed. Like grain has comparatively small amount of basic cations than leaves and stem portion of the plant so forages like Hay, bermuda grass and alfalfa show more positive effects on soil acidity comparative to high-yielding grain crops.

4.1.5 Application of Acid Forming Fertilizers

The soils' inherent capacity is severely deteriorated by the result of high temperature, precipitation and incessant leaching of nutrients. This deteriorated land is unable to support any vegetative crop. Usage of agricultural land without proper management practices results in enhanced soil infertility problems. To overcome these problems most of farmers use fertilizers extensively. Mostly used chemical fertilizers are ammonium sulphate (AS), urea, muriate of potash and trisuperphosphate, etc (FAO 2004). Usage of these chemical fertilizers results in enhanced crop yield. As these fertilizers are essential for high production along with this, these chemical fertilizers significantly increase the soil acidification.

4.2 Effects of Soil Acidity on Crop Production

Soil acidity significantly affects plants yield and productivity by decaling available nutrient contents. Two major factors associated with soil infertility are presence of phytotoxic substances like Al and Mn, and P, Ca, and Mg nutrient deficiency. Mostly plants uptake the nutrient in soluble form. Soil acidification cause profusion availability of elements such as Al and Mn and result in shortage of plant's essential nutrients such as P, Ca and Mg. it is noted previously that soil acidity is associated with H^+ and $Al^{3+.}$ Surprisingly, there is no deleterious effect found on plants growth by H^+(Black 1968; Rao et al 1993). Acidic soil's most of the problems are associated with Al^{3+}. Higher Al^{3+} content in acidic soil results in reduced function and root proliferation. Roots mostly observed are stunted and club shaped. This reduces the plants availability to extract nutrients and water from soil. When aluminum is abundant it mostly fixed with phosphate in form of aluminum phosphate and making P unavailable for plant (Black 1968; Rao et al. 1993). Except molybdenum the availability of micro-nutrient boosts the soil acidity.

4.3 Management of Soil Acidity

Soil acidification is a natural ongoing phenomenon which is aggravated by human activities. With the usage of proper irrigation techniques and practices soil acidification and its harmful effects should be controlled. (Obiri-Nyarko 2012) reported techniques that how such soil acidified land should be used for sustained agricultural purposes. To overcome soil acidity issues use of organic material and lime, acid tolerant crop varieties are used. Among which use of lime and organic material combination is best in combating soil acidification problems and making soil vulnerable for irrigation practices. There is also an immense need to limit the extensive use of chemical fertilizers for combating soil acidification problems because such practices extensively enhanced soil acidity. In such areas where extensive use of lime along with organic material is a problem best remedy there is to use acid resistant crop verities.

4.3.1 Liming

Different liming materials such as dolomite lime ($CaMgCO_3$), limestone ($CaCO_3$), quick lime (CaO), slaked lime ($Ca(OH)_2$) usage are best remedies for overcoming soil acidity problems. They are used both separately and in combined forms. These liming materials along with lowering soil acidity also counteract the effect of H^+ and Al^{3+} ions (Fageria and Baligar 2005). Several other advantages of liming materials include increasing the plants essential nutrient such as Ca, P and Mg availability and reducing the toxic effect of various micro elements (Naidu et al. 1994). Liming material addition also reduce the leaching and solubility of heavy metals (Lindsay 1979; Sauve et al. 2000). Excessive nutrient availability significantly improve crop yield to substantial amounts by addition of liming materials. Soil texture, soil fertility, crop rotation, crop species and usage of organic manure are the several factors which affect the application of liming materials (Fageria and Baligar 2008).

(Sadiq and Babagana 2012) reported that application of lime material on paddy fields significantly lowers the soil acidity. In Southeast Asia acid sulfate is mostly recommended for this purpose. Application of lime in rice fields results in high Al and Fe precipitation which is responsible for their enhanced yield. Some authors also reported that high amount of Al ions contents result due to use of lime and put deleterious effects on underlying soil.

At pH 5 aluminum ions starts precipitation from soil solution. This happened due to reaction of ground magnesium Limestone GML was combined with acid sulfate soil; both of these disintegrated immediately and start releasing hydroxyl ions. Shazana et al. (2013) reported that the actual reason behind increase in soil acidity is the release of hydroxyl ions on application of ground magnesium Limestone.

(Shazana et al. 2013) reported that ground basalt is advantageous for plants as it contain plant's essential nutrients like K and P, than ground magnesium Limestone GML. The one disadvantage of ground basalt applications is it takes time to completely dissolve in soil. It is reported that in Malaysia soil content is poor in organic matter. Application of ground basalt by acid sulphate ameliorate infertile land, is highly recommended for sustained rice yield along with different organic fertilizers few months before growing season.

4.3.2 Application of Organic Materials

The organic material usage defines simply all the forms of organic materials originated from both the plants and animals. Application of organic material where improves soil's properties and fertility along with it also reduces the effect of soil acidity and aluminum ions concentration. Plants usually contain excessive amount of cations, synthesis of organic acid anions simply used for balancing cations and anions (de Wit et al. 1963). Decarboxylation of these organic acid anions results due to microbial decomposition (Tang et al. 1999; Yan et al. 1996).

It was reported that anion organic acid decarboxylation requires proton to complete its reaction during microbial decomposition (Noble et al. 1996). By up taking such proton, hydroxyl ions concentration increases which results in increase soil alkalinity. Higher the amount of cations in soil greater is the effect found on soil acidity. Plant species of legume plants such as soybean, red clover and acacia found to have higher concentration of Ca, Mg and total cations contents than non-legume crops such as maize and sorghum, also have higher content of ash alkalinity (Bessho and Bell 1992; Pocknee and Sumner 1997; Wong et al. 2000). Wong et al. (2000) also indicated that organic material associated functional groups results in increase alkalinity of soil by consuming higher content of protons.

4.3.3 Use of Acid Tolerant Crops

Several plants grow well in acidic soil due to their variant degree of acidity tolerance. Thus, we can lower the acidification rate by using acid-tolerant crops. Acid-tolerant crops are very helpful because:

- They reduce the rate of acidification by efficiently using the nitrate and soil moisture
- Limestone if not added where and when is required, the acid-tolerant crops continues the cash flow
- In liming cycle of 10–15 years, acid-tolerant crops are trying to match up with the declined pH
- On the soils having acidic sub-surface layers, it is more suitable or economical to use acid-tolerant crops rather than liming them (Upjohn 2005).

Chillies, sweet and Irish potatoes show higher degree of acidity tolerant values. They show significant growth under pH values below 5.5. Cassava and rice are best grown crops for such acidified lands (Rao et al. 1993). It is reported that most of the lowland areas in Ghana show enhanced rice yield over 1 million hectares. This land is giving outstanding yields of rice (Wakatsuki et al. 2005).

Upon the growth of most of the crops, soil acidity and aluminum toxicity put a limiting factor over their yield. Krstic et al. (2012) studied various factors of Western Serbia region such as soil pH, exchangeable acidity and aluminum ion toxicity of soil. Aluminum puts very hazardous effects on root growth and extension, significantly effecting roots water and ions uptake capacity. Usage of genetically adapted plants for aluminum toxicity is best remedy to use on such affected land under low environment impact. Availability of Aluminum tolerant germpalsm of maize had made it a suitable crop for this practice. Crops that reflect tolerance to aluminum toxicity are the best options to be chosen as acid-tolerant crops.

No doubt, acid-tolerant crops reduce the rate of acidification but they cannot stop the process. The process continues and eventually, if not properly treated, soil will become more acidic.

4.3.4 Agroforestry

Agroforestry refers to use the land for both woody perennial plants along with agricultural crops from very simple to dense systems. It comprises of very simple to complex systems. It involves a wide range of practices such as intercropping, multiple cropping, trees plantation on contours, establishing shelter belts etc. all these practices result in enhanced land yield and production by providing suitable micro-climate, permanent cover, increased infiltration, improved soil structure, organic carbon content and promoted soil fertility (WOCAT 2011).

All such practices lowers the demand of land for mineral fertilizers (Garrity et al. 2004; Schroth and Sinclair 2003).

Agroforestry's main focus is on proper land management by conserving the soil along with enhancing soil fertility and productivity especially in tropics. It focuses on the deliberate use of land for trees and crops at the same time. A large number of effects have been reported through this interaction. Essential nutrient such as Ca, Mg, and nitrate's leaching to deeper soil zones are compensated by deep rooted species and return by litter fall. (Ridley et al. 1990) reported that significant reduction of nitrates is documented by growing crops with deep rooted species such as perennial grass, multipurpose trees or shrubs.

Agroforestry plans also results in reduced rate of erosion through rain drop and by leachability of nutrients. By using proper tree species this agroforestry system also help in reducing soil acidification. It is reported that plantation of nitrogen fixing tress such as Gliricidiasepium and Albiziazygia ameliorate the acid infertile rice field in Ghana (Baggie et al. 2000). Both of these trees Albiziazygiaand Gliricidiasepium due to their high cation content increased the soil pH from 4.4 to 5.1 and 5.3 respectively.

It was reported several strategies for understanding and management of Kenyan soil for enhanced crop yield (Kisinyo et al. 2014). They improved the soil by applying organic and inorganic materials and planating crop germplasm as Al tolerant varieties. Application of lime, Phosphorous fertilizers and OMs promote soil pH and lower the Al toxicity on Kenyan acidic soil. Application of lime, P fertilizers and OMs has resulted in increased maize production from 5–57, 18–93 and 70-100 % respectively on Kenyan soils. Development of crop cultivars for Al toxicity tolerant and low Phosphorous availability also results in significant increase in crop production on Kenyan soils.

References

Abrol IP, Yadav JSP, Massoud FI (1988) Salt-affected soils and their management. Food & Agriculture Organisation of the United Nations, Rome

Act FC (2002) Riparian forest buffer panel (Bay Area Regulatory Programs)

Ahmad P, Sharma S (2008) Salt stress and phyto-biochemical responses of plants. Plant Soil Environ 54:89–99

Ahmad P, Nabi G, Ashraf M (2011) Cadmium-induced oxidative damage in mustard (Brassica juncea (L.) Czern. &Coss.) plants can be alleviated by salicylic acid. S Afr J Bot 77:36–44

Ahmad P, Jeleel CA, Azooz MM, Nabi G (2009) Generation of ROS and non-enzymatic antioxidants during abiotic stress in plants. Bot Res Intern 2:11–20

Ahmad P, Jaleel CA, Salem MA, Nabi G, Sharma S (2010) Roles of enzymatic and nonenzymatic antioxidants in plants during abiotic stress. Crit Rev Biotechnol 30:161–175

Ahmed M, Qamar I (2004) Productive rehabilitation and use of salt-affected land through afforestation (a review). Sci Vis 9:1–14

Allison FE (1973) Soil organic matter and its role in crop production. Elsevier, New York

Apel K, Hirt H (2004) Reactive oxygen species: metabolism, oxidative stress, and signal transduction. Annu Rev Plant Biol 55:373–399

Appleton BL, Greene V, Smith A, French S, Kane B, Fox L, Downing A, Gilland T (2009) Trees and shrubs that tolerate saline soils and salt spray drift.Virginia Cooperative Extension Publiction No. 430-031

Ashraf MA, Harris PJC (2005) Abiotic stresses: plant resistance through breeding and molecular approaches. Food Products Press, an imprint of Haworth Press, Binghamton

Ayers RS, Westcot DW (1985) Water quality for agriculture. Food and Agriculture Organization of the United Nations, Rome

AzevedoNeto ADd, Prisco JT, Enas-Filho J, Lacerda CFd, Silva JV, Costa PHAd, Gomes-Filho Ea (2004) Effects of salt stress on plant growth, stomatal response and solute accumulation of different maize genotypes. Braz J Plant Physiol 16:31–38

Baggie I, Zapata F, Sanginga N, Danso SKA (2000) Ameliorating acid infertile rice soil with organic residue from nitrogen fixing trees. Nutr Cycl Agroecosyst 57:183–190

Bajracharya RM, Lal R (1992) Seasonal soil loss and erodibility variation on a Miamian silt loam soil. Soil Sci Soc Am J 56:1560–1565

Bank W (2006) Sustainable land management: challenges, opportunities, and trade-offs. World Bank, Washington, DC

Barrett-Lennard EG (2002) Restoration of saline land through revegetation. Agric Water Manag 53:213–226

Beresford Q, Bekle H, Phillips H, Mulcock J (2004) The salinity crisis: landscapes, communities and politics. ERA Trial 2009:42

Bessho T, Bell LC (1992) Soil solid and solution phase changes and mung bean response during amelioration of aluminium toxicity with organic matter. Plant and Soil 140:183–196

Black CA (1968) Soil-plant relationships. Wiley, New York

Bond WJ (1998) Effuent irrigation-an environmental challenge for soil science. Aust J Soil Res 36:543–556

Bouwman AF, Van Drecht G, Van der Hoek KW (2005) Global and regional surface nitrogen balances in intensive agricultural production systems for the period 1970–2030. Pedosphere 15:137–155

Brady NC, Weil RR (1984) The nature and properties of soils. Macmillan, New York

Bresler E, Dagan G, Hanks RJ (1982) Statistical analysis of crop yield under controlled line-source irrigation. Soil Sci Soc Am J 46:841–847

Brevik EC (2009) Soil health and productivity in soils, plant growth and crop production. Encyclopedia of life support systems (EOLSS), developed under the auspices of the UNESCO. EOLSS Publishers, Oxford, UK. http://www.eolss.net. Accessed 10 Dec 2012

Brevik EC (2013) Soils and human health—an overview. In: Brevik EC, Burgess LC (eds) Soils and human health. CRC Press, Boca Raton, pp 29–56

Bridges EM, Hannam ID, Oldeman LR, Pening de Vries FWT, Scherr SJ, Sompatpanit S (eds) (2002) Responses to land degradation. Proceeding 2nd International Conference on Land Degradation and Desertification, Khon Kaen, Thailand. Oxford Press, New Delhi, India

Carter MR, Pearen JR (1989) Amelioration of a saline sodic soil with low applications of calcium and nitrogen amendments. Arid Land Res Manag 3:1–9

Chen JH, Barber SA (1990) Soil pH and phosphorus and potassium uptake by maize evaluated with an uptake model. Soil Sci Soc Am J 54:1032–1036

Chepil WS (1955) Factors that influence clod structure and erodibility of soil by wind: V. organic matter at various stages of decomposition. Soil Sci 80:413

Chhabra R (1996) Soil salinity and water quality. Balkema, Rotterdam

Choudhary MR, MunirA MS (2008) Field soil salinity distribution under furrow-bed and furrow-ridge during wheat production in irrigated environment. Pak J Water Res 12:33–40

Conway T (2001) Plant materials and techniques for brine site reclamation. Plant materials technical note

Costanza R, de Groot R, Sutton P, van der Ploeg S, Anderson SJ, Kubiszewski I, Farber S, Turner RK (2014) Changes in the global value of ecosystem services. Glob Environ Chang 26:152–158
De Wit CT, Dijkshoorn W, Noggle JC (1963) Ionic balance and growth of plants. Versl Landbouwkd Onderz 69:1–68
Denise MW (2003) Soil salinity and sodicity limits efficient plant growth and water use. Rio grande regional soil and water series guide A-140, New Mexico State University, New Mexico
Dionisio-Sese ML, Tobita S (1998) Antioxidant responses of rice seedlings to salinity stress. Plant Sci 135:1–9
Domroes M (1991) The tropical forest ecosystem: reviewing the effects of deforestation on climate and environment. In: Takeuchi K, Yoshino M (eds) The global environment. Springer-Verlag, Berlin, pp 70–80
Donahue RL, Follett RH, Tulloch RW (1983) Our soils and their management. Interstate Printers & Publishers, Danville
Doran JW, Parkin TB (1994) Defining and assessing soil quality. In: Doran JW, Coleman DC, Bezdicek DF, Stewart BA (eds) Defining soil quality for a sustainable environment. Soil science society of America special publication 35, ASA-SSSA, Madison, pp 3–21
Eaton FM, Sokoloff VP (1935) Absorbed sodium in soils as affected by the soil-water ratio. Soil Sci 40:237–248
El-Wahab A (2006) The efficiency of using saline and fresh water irrigation as alternating methods of irrigation on the productivity of Foeniculumvulgare Mill subsp. vulgare var. vulgare under North Sinai conditions. Res J Agr Biol Sci 2:571–577
Ella WM, Stenberg M, Bru D, Hellman M, Welsh A, Thomsen F, Klemedtson L, Philippot L, Hallin S (2011) Spatial distribution of ammonia-oxidizing bacteria and archaea across a 44-hectare farm related to ecosystem functioning. ISME J 5:1213–1225
Elshout S, Kamphorst A (1990) Suitability of coarse-grade gypsum for sodic soils reclamation: a laboratory experiment. Soil Sci 149:228–34
Ezeaku PI, Davidson A (2008) Analytical situations of land degradation and sustainable management strategies in Africa. J Agric Soc Sci (Pakistan) 4: 42–52
Fageria NK, Baligar VC (2008) Ameliorating soil acidity of tropical Oxisols by liming for sustainable crop production. Adv Agron 99:345–399
FAO (2004) Food and Agriculture Organization of the United Nations; 2011. FAOSTAT database
FAO (2009) Food security and agricultural mitigation in developing countries: options for capturing synergies. FAO, Rome
Food and Agriculture Organization of the United Nations (FAO) (2010) Climate-smart agriculture: policies, practice and financing for food security, adaptation and migration. FAO, Rome
Feng ZZ, Wang XK, Feng ZW (2005) Soil N and salinity leaching after the autumn irrigation and its impact on groundwater in Hetao Irrigation District, China. Agric Water Manag 71:131–143
Fitzpatrick RW, Merry RH, Cox JW, Rengasamy P, Davies PJ (2003) Assessment of physicochemical changes in dryland saline soils when drained or disturbed for developing management options. CSIRO Land and Water Technical Report 2/03. South Australia, Australia
Fortuna L, Nunnari G, Gallo A (2012) Model order reduction techniques with applications in electrical engineering. Springer Science & Business Media.
Funk R, Reuter HI (2006) Wind erosion. In: Boardman J, Poeson J (eds) Soil erosion Eur, Wiley, pp, 563–582
Garrity GM, Bell JA, Lilburn TG (2004) Taxonomic outline of the prokaryotes. Bergey's manual of systematic bacteriology. Springer, New York/Berlin/Heidelberg
Geist HJ, Lambin EF (2004) Dynamic causal patterns of desertification. Bioscience 54:817–829. (ed) 2001. National workshop on agricultural use of gypsum. LRRI, NARC, PARC, Islamabad, Pakistan
Ghaly FM (2002) Role of natural vegetation in improving salt affected soil in northern Egypt. Soil Tillage Res 64:173–178

Ghassemi F, Jakeman AJ, Nix HA (1995) Salinisation of land and water resources: human causes, extent, management and case studies. CAB international, Wallingford
Gonzalez NLM, Toth T, Garcia D (2004) Integrated management for the sustainable use of salt-affected soils in Cuba. Ecosistemas y Recursos Agropecuarios 20:85–102
Grattan SR, Grieve CM (1998) Salinity; "mineral nutrient relations in horticultural crops. Scientia horticulturae 78:127–157
Grimm M, Jones R, Montanarella L (2001) Soil erosion risk in Europe. EUR 19939 EN. Joint Research Centre, European Commission, Luxembourg.
Gueta-Dahan Y, Yaniv Z, Zilinskas BA, Ben-Hayyim G (1997) Salt and oxidative stress: similar and specific responses and their relation to salt tolerance in citrus. Planta 203:460–469
Gupta JP, Raina P (1996) Wind erosion and its control in hot arid areas of Rajasthan. In: Wind erosion in West Africa: the problem and its control. Margraf Verlag, Berlin, pp 209–218
Hakeem KR, Chandna R, Rehman R, Tahir I, Sabir M, Iqbal M (2013) Unraveling salt stress in plants through proteomics. In: Parvaiz A, Azooz MM, Prasad MNV (eds) Salt stress in plants: signelling, omics and adaptations. Springer science+Business Media LLC, New York, pp 47–61
Hakeem KR, Khan F, Chandna R, Siddiqui TO, Iqbal M (2012) Genotypic variability among soybean genotypes under NaCl stress and proteome analysis of salt tolerant genotype. Appl Biochem Biotech 168:2309–2329
Hakeem KR, Ahmad A, Iqbal M, Gucel S, Ozturk M (2011) Nitrogen-efficient cultivars can reduce nitrate pollution. Environ Sci Pollut Res 18:1184–1193
Hakeem KR, Sabir M, Ozturk M, Mermut A (2014) Soil remediation and plants: prospects and challenges. Academic Press/Elsevier, New York, p 724
Hasegawa PM, Bressan RA, Zhu JK, Bohnert HJ (2000) Plant cellular and molecular responses to high salinity. Annu Rev Plant Biol 51:463–499
Hassan MK, Pelkonen P, Pappinen A (2011) Assessment of bioenergy potential from major crop residues and wood fuels in Bangladesh. J Basic Appl Sci Res 1:103
Hastenrath S (1991) Climate dynamics of the tropics. Kluwer, Dordrecht
Haub C, Gribble J, Jacobsen L (2011) World population data sheet 2011. Population Reference Bureau, Washington, DC
He ZL, Yang XE, Baligar VC, Calvert DV (2003) Microbiological and biochemical indexing systems for assessing quality of acid soils. Adv Agron 78:89–138
Hillel D (1998) Environmental soil physics: fundamentals, applications, and environmental considerations. Academic, San Diego
Hu Y, Schmidhalter U (2005) Drought and salinity: a comparison of their effects on mineral nutrition of plants. J Plant Nutri Soil Sci 168:541–549
Huumllsebusch C (2007) Organic agriculture in the tropics and subtropics: current status and perspectives. Kassel University Press GmbH, Kassel
IFAD (2011) Rural poverty report 2011. New realities, new challenges: new opportunities for tomorrow's generation. IFAD, Roma (Italia)
Jie D (2010) Chinese soil experts warn of massive threat to food security. SciDevNet
Jim M (2002) Managing salt affected soils. NRCS, South Dakota
Jones AJ, Lal R, Huggins DR (1997) Soil erosion and productivity research: a regional approach. Am J Altern Agric 12:185
Juo ASR, Thurow TL (1997) Sustainable technologies for use and conservation of steeplands. Food & Fertilizer Technology Center for the Asian and Pacific Region, Taipei City
Kendall HW, Pimentel D (1994) Constraints on the expansion of the global food supply. Ambio 23:198–205
Kim S, Rayburn AL, Voigt T, Parrish A, Lee DK (2011) Salinity effects on germination and plant growth of prairie cordgrass and switchgrass. Bioenergy Res 5:225–235
Kisinyo PO, Othieno CO, Gudu SO, Okalebo JR, Opala PA, Ngetich WK, Nyambati RO, Ouma EO, Agalo JJ, Kebeney SJ (2014) Immediate and residual effects of lime and phosphorus fertilizer on soil acidity and maize production in western Kenya. Exp Agric 50:128–143

Lal R (1990) Soil erosion and land degradation: the global risks. Advances in soil science, Springer, New York
Lal R (1994) Water management in various crop production systems related to soil tillage. Soil Tillage Res 30:169–185
Lal R (1997) Degradation and resilience of soils. Philos Trans R Soc Lond B Biol Sci 352:997–1010
Lamond RE, Whitney DA (1992) Management of saline and sodic soils. Kansas State University/ Cooperative Extension Service, Manhattan
Lavelle P, Spain AV (2001) Soil ecology. Kluwer Academic Pub, Dordrecht
Lazaroff C (2001) Biodiversity gives carbon sinks a boost Environment News Service 13
Lehmann J, Gaunt J, Rondon M (2006) Bio-char sequestration in terrestrial ecosystems; a review. Mitig Adapt Strat Glob Chang 11:395–419
Li L, Ishikawa Y, Mihara M (2012) Effects of burning crop residues on soil quality inWenshui, Shanxi of China. Int J Environ Rural Dev 3:30–36
Li P, Song A, Li Z, Fan F, Liang Y (2008) Silicon ameliorates manganese toxicity by regulating manganese transport and antioxidant reactions in rice (Oryza sativa L.). Plant and Soil 354:407–419
Li Z, Sun WY, Ma WP, Yin HJ, Li HJ, Qu SG, Cao WD (2010) Review and prospects of improvement technology for saline-alkali soil. Shandong Agric Sci 2:273–277
Liang Y, Sun W, Zhu YG, Christie P (2007) Mechanisms of silicon-mediated alleviation of abiotic stresses in higher plants: a review. Environ Pollut 147:422–428
Libert B (1995) The environmental heritage of Soviet agriculture. CAB International, Wallingford
Lindsay WL (1979) Chemical equilibria in soils. Wiley, New York
Majerus M (1996) Plant materials for saline-alkaline soils. US Department of Agriculture, Natural Resources Conservation Service. Plant materials center, Bridger, Mont. Montana technical note:5
Manchanda G, Garg N (2008) Salinity and its effects on the functional biology of legumes. ActaPhysiologiaePlantarum 30:595–618.(ed) 1991. In: Choukar-alla R (ed) Proceedings of the international conference on agriculture management of salt-affected area in Agadir Morocco. Publ. IAV Hassan, Morocco
McCormick RW, Wolf DC (1980) Effect of sodium chloride on CO 2 evolution, ammonification, and nitrification in a Sassafras sandy loam. Soil Biol Biochem 12:153–157
McDowell RW (2008) Environmental impacts of pasture-based farming. CABI, Cambridge
McIntyre BS, Tongway D (2005) Grassland structure in native pastures: links to soil surface condition. Ecol Manage Restor 6:43–50
Misra AN, Srivastava A, Strasser RJ (2001) Utilization of fast chlorophyll a fluorescence technique in assessing the salt/ion sensitivity of mung bean and Brassica seedlings. J Plant Physiol 158:1173–1181
Mittova V, Guy M, Tal M, Volokita M (2004) Salinity up-regulates the antioxidative system in root mitochondria and peroxisomes of the wild salt-tolerant tomato species Lycopersiconpennellii. Journal of experimental botany 55:1105–1113.(ed) 1987. International geomorphology, 1986: proceedings of the first international conference on geomorphology/ed on behalf of the British Geomorphological Res Group by V. Gardiner and sectional ed, MG Anderson…[et al.]. Wiley, Chichester, c1987
Munns R (2005) Genes and salt tolerance: bringing them together. New Phytol 167:645–663
Naidu R, Bolan NS, Kookana RS, Tiller KG (1994) Ionic strength and pH effects on the sorption of cadmium and the surface charge of soils. Eur J Soil Sci 45:419–429
Noble AD, Zenneck I, Randall PJ (1996) Leaf litter ash alkalinity and neutralisation of soil acidity. Plant and Soil 179:293–302
Norman M (1984) GAIA: an atlas of planet management. Anchor press-doubleday, Garden city (ed) 2012. In: Proceedings of the international conference on agriculture science and engineering (ICASE2012)
Oldeman LR (1998) Soil degradation: a threat to food security. Report.

Oldeman LR, Hakkeling RTA, Sombroek WG (1991) World map of the status of human-induced soil degradation: an explanatory note, 2nd. rev 9066720468. ISRIC [etc.].
Oster JD, Shainberg I, Abrol IP (1996) Reclamation of salt-affected soil. In: Soil erosion, conservation and rehabilitation., pp 315–352
Pace M, Johnson P (2002) Growing turf on salt-affected sites. HG-519, Cooperative Extension Service, Utah State University
Padmanabhan E, Hari E (2008) Impact of microvariability on classification and management of peatlands, Curtin University, Bentley
Parida AK, Das AB (2005) Salt tolerance and salinity effects on plants: a review. Ecotoxicol Environ Saf 60:324–349
Pastori GM, Foyer CH (2002) Common components, networks, and pathways of cross-tolerance to stress. The central role of "redox" and abscisic acid-mediated controls. Plant Physiol 129:460–468
Pessarakli M (2010) Handbook of plant and crop stress. CRC Press, Boca Raton
Pessarakli M, Szabolcs I (1999) Soil salinity and sodicity as particular plant-crop stress factors. Handbook of plant and crop stress. Rev and expanded. Marcel Dekker, Inc., New York
Pimentel D (2006) Soil erosion: a food and environmental threat. Environ Dev Sustain 8:119–137
Pimentel D, Burgess M (2013) Soil erosion threatens food production. Agriculture 3:443–463
Pimentel D, Harvey C, Resosudarmo P, Sinclair K (1995) Environmental and economic costs of soil erosion and conservation benefits. Science 267:1117
Pimentel D, Houser J, Preiss E, White O, Fang H, Mesnick L, Barsky T, Tariche S, Schreck J, Alpert S (1997) Water resources: agriculture, the environment, and society. Bioscience 47:97–106
Pocknee S, Sumner ME (1997) Cation and nitrogen contents of organic matter determine its soil liming potential. Soil Sci Soc Am J 61:86–92
Qadir M, Schubert S (2002) Degradation processes and nutrient constraints insodic soils. Land Degrad Dev 13:275–294
Qadir M, Schubert S, Ghafoor A, Murtaza G (2001) Amelioration strategies for sodic soils: a review. Land Degrad Dev 12:357–386
Qadir M, Steffens D, Yan F, Schubert S (2003) Sodium removal from a calcareous saline-sodic soil through leaching and plant uptake during phytoremediation. Land Degrad Dev 14:301–307
Quarrie J (1992) Earth Summit'92. The United Nations conference on environment and development, Rio de Janeiro 1992.
Qureshi AS, McCornick PG, Qadir M, Aslam Z (2008) Managing salinity and waterlogging in the Indus Basin of Pakistan. Agric Water Manag 95:1–10
Rao DLN, Pathak H (1996) Ameliorative influence of organic matter on biological activity of salt-affected soils. Arid Land Res Manage 10:311–319
Rao IM, Zeigler RS, Vera R, Sarkarung S (1993) Selection and breeding for acid-soil tolerance in crops. Bioscience 43:454–465
Ravi S, D'Odorico P (2005) A field-scale analysis of the dependence of wind erosion threshold velocity on air humidity. Geophys Res Lett 32 (ed) 1999. Sustaining the global farm. Selected papers from the 10th international soil conservation organization meeting
Rengasamy P, Olsson KA (1991) Sodicity and soil structure. Soil Res 29:935–952
Ridley AM, Slattery WJ, Helyar KR, Cowling A (1990) The importance of the carbon cycle to acidification of a grazed annual pasture. Anim Prod Sci 30:529–537
Rivas T (2006) Erosion control treatment selection guide. Gen. Tech Rep. 0677 1203 – SDTDC, USDA forest service. San Dimas Technology and Development Center, San Dimas, p 64
Roberts KG, Gloy BA, Joseph S, Scott NR, Lehmann J (2009) Life cycle assessment of biochar systems: estimating the energetic, economic and climate change potential. Environ Sci Technol 44:827–833
Rogers ME, Grieve CM, Shannon MC (2003) Plant growth and ion relations in lucerne (Medicago sativa L.) in response to the combined effects of NaCl and P. Plant and Soil 253:187–194

Rose CW (2004) An introduction to the environmental physics of soil, water and watersheds. Cambridge University Press, Cambridge

Sadiq AA, Babagana U (2012) Influence of lime materials to ameliorate acidity on irrigated paddy fields: a review. Acad Res Int 3:413

Sandoval FM, Benz LC (1966) Effect of bare fallow, barley, and grass on salinity of a soil over a saline water table. Soil Sci Soc Am J 30:392–396

Sauve S, Hendershot W, Allen HE (2000) Solid-solution partitioning of metals in contaminated soils: dependence on pH, total metal burden, and organic matter. Environ Sci Technol 34:1125–1131

Schnürer MC, Rosswall T (1985) Microbial biomass and activity in an agricultural soil with different organic matter contents. Soil Biol Biochem 17:611–618

Schroth G, Sinclair FL (2003) Trees, crops, and soil fertility: concepts and research methods. CABI, Wallingford, UK

Scianna J (2002) Salt-affected soils: their causes, measure, and classification. Res Method Hort. note 5

Scotter DR (1978) Preferential solute movement through larger soil voids. I. Some computations using simple theory. Soil Res 16:257–267

Seatz LF, Peterson HB (1964) Acid, alkaline, saline and sodic soils. In: Bear FE (ed) Chemistry of the soil. Reinhold Publishing Corporation, New York, pp 292–319

Seelig BD, Richardson JL, Knighton RE (1991) Comparison of statistical and standard techniques to classify and delineate sodic soils. Soil Sci Soc Am J 55:1042–1048

Shainberg I, Letey J (1984) Response of soils to sodic and saline conditions. Hilgardia 52:1–57

Shainberg I, Keren R, Frenkel H (1982) Response of sodic soils to gypsum and calcium chloride application. Soil Sci Soc Am J 46:113–117

Shao Y (2000) Physics and modelling of wind erosion, vol 23, Atmospheric and oceanographic sciences library. Kluwer Academic Publishers, Dordrecht

Shazana M, Shamshuddin J, Fauziahand CI, Syed Omar SR (2013) Alleviating the infertility of an acid sulphate soil by using ground basalt with or without lime and organic fertilizer under submerged conditions. Land Degrad Dev 24:129–140

Shukla J, Nobre C, Sellers P. (1990) Amazon deforestation and climate change. Science (Washington) 247:1322–1325

Silvertooth JC, Norton ER (2000) Evaluation of a calcium-based soil conditioner in irrigated cotton. Cotton, A College of Agriculture Report

Singh TV (1993) Development of tourism in the Himalayan environment: the problem of sustainability. In: Rawat MSS (ed) Himalaya, a regional perspective: resources, environment and development, Dehli, pp 63–77

Smith DMS, McKeon GM, Watson IW, Henry BK, Stone GS, Hall WB, Howden SM (2007) Learning from episodes of degradation and recovery in variable Australian rangelands. Proc Natl Acad Sci 104:20690–20695

Sohi S, Lopez-Capel E, Krull E, Bol R (2009) Biochar, climate change and soil: a review to guide future research. CSIRO Glen Osmond, Australia

Soil and Water Conservation (1993) Alabama handbook for erosion control, sediment control, and stromwater management on construction sites and urban areas Alabama soil and water conservation committee

Soil Science Society of Agriculture (2008) Glossary of soil science terms 2008 ASA-CSSA-SSSA

Southard RJ, Buol SW (1988) Subsoil saturated hydraulic conductivity in relation to soil properties in the North Carolina Coastal Plain. Soil Sci Soc Am J 52:1091–1094. (ed.) 1994. (Unpublished) 1994. Presented at the Third INC-D session of the intergovernmental negotiating committee on the international convention on desertification United Nations New York 17 January 1994

Sterk G (2000) Flattened residue effects on wind speed and sediment transport. Soil Sci Soc Am J 64:852–858

Sterk G, Warren A (2003) European conference on wind erosion on agricultural land, 30 April 2 May 2001, Thetford, UK. Catena 52:171–172

Sundquist B (2010) Chapter 9, Food supply from soil. Topsoil loss and degradation: causes, effects and implications (ed) 2004. Advances in sodic land reclamation. INT. conf. on sustainable management of sodic lands, Lucknow

Tan J, Kang Y (2009) Changes in soil properties under the influences of cropping and drip irrigation during the reclamation of severe salt-affected soils. Agric Sci China 8:1228–1237

Tang C, Rengel Z (2003) Role of plant cation/anion uptake ratio in soil acidification. Handbook of soil acidity. Marcel Dekker, New York, pp 57–81

Tang C, Sparling GP, McLay CDA, Raphael C (1999) Effect of short-term legume residue decomposition on soil acidity. Aus J Soil Res 37:561–561

Tejada M, Garcia C, Gonzalez JL, Hernandez MT (2006) Use of organic amendment as a strategy for saline soil remediation: influence on the physical, chemical and biological properties of soil. Soil Biol Biochem 38:1413–1421

Thiruchelvam S, Pathmarajah S (2003) An economic analysis of salinity problems in the Mahaweli River System H Irrigation Scheme in Sri Lanka Economy and environment program for Southeast Asia (EEPSEA)

Tisdale SL, Nelson NL (1975) Soil fertility and fertilizers. Mac Millan Co. Inc, New York

Trimble SW, Mendel AC (1995) The cow as a geomorphic agent-A critical review. Geomorphology 13:233–253

Troeh FR, Hobbs JA, Donahue RL (2004) Soil and water conservation for productivity and environmental protection. Prentice-Hall Inc, Upper Saddle River

Turner J, Lambert M (2000) Change in organic carbon in forest plantation soils in eastern Australia. For Ecol Manage 133:231–247

Unnevehr LJ, Lowe FM, Pimentel D, Brooks CB, Baldwin RL, Beachy RN, Chornesky EA, Hiler EA, Huffman WE, King LJ (2003) Frontiers in agricultural research: food, health, environment, and communities. National Academies of Science, Washington, DC

Upjohn B, Fenton G, Conyers M (2005) Soil acidity and liming. Agfact AC.19, 3rd edn. State of New South Wales, Department of Primary Industries

Van Camp M, Vauterin P (2005) Tsoft: graphical and interactive software for the analysis of time series and Earth tides. Comput Geosci 31:631–640

Verheijen FGA, Jones RJA, Rickson RJ, Smith CJ (2009) Tolerable versus actual soil erosion rates in Europe. Earth-Sci Rev 94:23–38

Wakatsuki T, Buri MM, Fashola OO (2005) Ecological engineering for sustainable rice production and the restoration of degraded watersheds in West Africa. Rice is life: scientific perspectives for the 21st century:363–366.

Walsh KN, Rowe MS (2001) Biodiversity increases ecosystems ™ ability to absorb CO2 and Nitrogen. Brookhaven National Laboratory

Wardle DA, Bardgett RD, Klironomos JN, Setala H, Van Der Putten WH, Wall DH (2004) Ecological linkages between aboveground and belowground biota. Science 304:1629–1633

Wen DH, Pimentel D (1993) Soil erosion and conservation in China. In: Pimentel D (ed) World soil erosion and conservation. Cambridge University, pp 63–86

Wong MTF, Gibbs P, Nortcliff S, Swift RS (2000) Measurement of the acid neutralizing capacity of agroforestry tree prunings added to tropical soils. J Agric Sci 134:269–276

World Health Organization (2000) The world health report 2000: health systems: improving performance. World Health Organization, Geneva/Switzerland

Yan F, Schubert S, Mengel K (1996) Soil pH increase due to biological decarboxylation of organic anions. Soil Biol Biochem 28:617–624

Young A (1989) Agroforestry for soil conservation. CAB International, Wallingford

Zhang C, Wang L, Nie Q, Zhang W, Zhang F (2008) Long-term effects of exogenous silicon on cadmium translocation and toxicity in rice (Oryza sativa L.). Environ Exp Bot 62:300–307

Zhang JH, Liu YP, Pan QH, Zhan JC, Wang XQ, Huang WD (2006) Changes in membrane-associated H+-ATPase activities and amounts in young grape plants during the cross adaptation to temperature stresses. Plant Sci 170:768–777

Zhang L, Sun X, Tian Y, Gong X (2014) Biochar and humic acid amendments improve the quality of composted green waste as a growth medium for the ornamental plant Calatheainsignis. Scientia horticulturae 176:70–78

Nitrogen Management in Rice-Wheat Cropping System in Salt-Affected Soils

Behzad Murtaza, Ghulam Murtaza, Muhammad Sabir, Muhammad Amjad, and Muhammad Imran

Contents

Abstract Soil salinity is a complex abiotic stress factor that limits plant growth and productivity by affecting soil physical and chemical properties, and causing disturbances in plants including osmotic stress, ionic toxicity and nutrient imbalance. Nitrogen is among the macro-nutrients whose uptake and assimilation is being severely affected by salinity/sodicity in arid and semi-arid areas of the world. Efficient nutrient management especially of N under salinity is indispensable for survival of crops species and economical yield. Rice and wheat are the two most important crops and rice-wheat cropping system is common in many parts of the world including Pakistan. Production of these crops is severely affected by salinity

B. Murtaza (✉) • M. Amjad • M. Imran
Department of Environmental Sciences, COMSATS Institute of Information and Technology, Vehari, Pakistan
e-mail: behzadmurtaza@ciitvehari.edu.pk

G. Murtaza • M. Sabir
Institute of Soil and Environmental Sciences, University of Agriculture Faisalabad, Faisalabad 38040, Pakistan

K.R. Hakeem et al. (eds.), *Soil Science: Agricultural and Environmental Prospectives*, DOI 10.1007/978-3-319-34451-5_3

by decreasing the N-use efficiency (NUE). Salinity affects the NUE in these crops by competition between Cl^- and NO_3^- and also by decreasing the plant internal NUE via affecting the enzymes of N-assimilation [Glutamate synthase (GS) or glutamine-2-oxoglutarate amino transferase (GOGAT), and glutamate dehydrogenase (GDH)]. NUE under salinity is also affected by source of N i.e. NO_3^- and NH_4^+, and literature suggests that combination of these both sources increases NUE in wheat and rice. Gypsum as an ameliorant (source of Ca^{2+}) is commonly used in salt-affected soils and plays important role in reclaiming these problematic soils and increasing nutrient use efficiency. Calcium improves NUE by increasing N uptake and ensures the structural and functional integrity of plant membranes, regulates ion transport and control ion exchange behavior as well as cell enzyme activity. This chapter emphasizes the impact of gypsum on NUE in rice-wheat cropping system in salt-affected soils for economical crop production.

Keywords Nitrogen use efficiency • Soil stress • Calcium • Rice-wheat cropping system • Plant nutrients

1 Introduction

Salinity has severely affected crop growth and development in irrigated regions of the world mainly in arid and semi-arid areas due to higher evapo-transpiration rate than the precipitation. Saline and/or sodic soils are characterized by excessive soluble salts and exchangeable sodium (Na^+) percentage which adversely affect the physical and chemical properties of soil. Fluctuations in chemical composition of soil solution and soil pH deteriorate the soil leading to osmotic stress and specific ion effects coupled with plant nutrient disorder (Marchuk and Rengasamy 2010). Physical processes developed structural problems in saline/sodic soils which may influence air and water mobility, water-holding capacity, mechanical operations, root penetration, seedlings growth and runoff processes (Qadir and Schubert 2002). Such, physical and chemical processes have an impact on soil microbial activity and plant roots, and ultimately on crop productivity.

In salt-affected soils, availability of plant nutrient is severely limited to sustain high crop production, especially of nitrogen (N) due to leaching, denitrification and volatilization losses (Shi et al. 2010). Modern agriculture depends upon the commercial N fertilization for getting higher crop yields. The worldwide utilization of N has now crossed 100×10^6 tons, but more than 50 % of it being used by Asian countries (Faostat 2009; Heffer and Prud'homme 2010). It is projected that 53×10^6 tons of N was used in cereals, signifying 55.3 % of total N consumption (Heffer and Prud'homme 2013). Almost 80 % of N fertilizer is being used as urea from which the risk of volatilization losses is much higher specifically in saline soil conditions due to which fertilizer use efficiency (FUE) by crops is greatly affected. Wheat (*Triticum aestivum* L.) and maize (*Zea mays* L.) both contribute to 17.3 % of the global use, followed by rice i.e. 15.8 % of the N fertilizer (Heffer and Prud'homme 2013). Many researchers have examined that the lowest nutrient use efficiency is largely due to NH_3 volatilization and high pH of soil. As the nitrogenous fertilizer used contain N as ammonical nitrogen or becomes ammonical upon hydrolysis, therefore, on saline/sodic soils the NH_3 volatilization losses might be much higher than normal soils.

Successful crop production under saline/sodic soils demands the optimum use of plant nutrients, particularly N fertilizers, in addition to other agronomic practices. Therefore, N application and management to increase yield and NUE from these soils seems to be important (Hakeem et al. 2011). The following sections review the possible options and efforts made for the reclamation of salt-affected soils and effective use of gypsum to enhance the NUE.

2 Extent and Nature of Salt-Affected Soils

Primary salinity is more widespread as compared to secondary salinity (Rengasamy 2010). On global basis, about 1×10^9 ha of land is saline and ≈ 50 % of the irrigated lands are severely affected due to secondary salinity. According to an estimate, over 8×10^8 ha of land in the world are salt-affected, either by salinity (3.97×10^8 ha) or sodicity (4.34×10^8 ha) (Faostat 2009). Of the current 2.30×10^8 ha of irrigated land, 4.5×10^7 ha are salt-affected (20 %). Moreover, the salt-affected areas are increasing @ of 10 % each year due to low rainfall, high surface evapotranspiration, irrigation with brackish water, and poor tillage practices. It is assessed that 1.6×10^6 ha of arable land is lost every year (Ghassemi et al. 1995).

A large portion of irrigated land is in Asia, where salinity has already affected an alarming and ever growing fraction of the area (Rains and Goyal 2003). In Pakistan, the estimated salt-affected area is about 6.68×10^6 ha (Khan 1998). The whole salt-affected area is categorized as 0.598×10^6 ha slightly saline-sodic, 1.33×10^6 ha moderately saline-sodic, 2.3×10^6 ha severely saline-sodic, and 1.96×10^6 ha very severely saline-sodic (Book 2003). Increasing salinity reduced the average production of main crops by more than 50 % (Bray et al. 2000). These losses are of great concern for the countries, like Pakistan and India, the financial prudence of which depend on mainly on agriculture industry. A current loss of US $12 billion annually due to salinity in the United States is expected to rise in future as soils are expected to be further affected (Munns et al. 2006; Pitman and Läuchli 2002).

Salinity is also widespread in various parts of the world associated with dry land agriculture (Rengasamy 2010). Salt-affected soils contain elevated concentrations of numerous salts (Zhang et al. 2010). It is followed that the detrimental effects of saline/saline-sodic environments on most species are mainly associated with combination of osmotic stress and ionic stress exerted by Na^+ and Cl^- components of NaCl (Blumwald 2000; Hasegawa et al. 2000; Munns and Tester 2008). High osmotic pressure of soluble salts affects plant growth by retardation of water and essential nutrients (Jouyban 2012; Tester and Davenport 2003). The problem is categorically inexorable in irrigated areas (Flowers and Colmer 2008; Zhu 2001) with 1/3rd of the global food production (Munns and Tester 2008; Zhang et al. 2010).

Salt-affected soils are formed due to primary and secondary salinity. Primary salinization involves the accumulation of soluble salts in soil due to high salt content in the parent material. Secondary salinization is caused by anthropogenic activities e.g. salty irrigation water and inadequate drainage (Ahmad and Chang 2002; Hasegawa et al. 2000; Qadir et al. 2007; Sharma and Minhas 2005; Yuan et al. 2007). Salt affected soils with considerable amounts of soluble salts in the rhizosphere influenced the crop yield and water and nutrients uptake (Tanji 2002). The presence of salts results in physiological and metabolic changes through adverse soil properties and detrimental to crop growth (Chinnusamy et al. 2005). Soil salinity not only limits the crop growth by various means such as salt content, water/ nutrient availability but also damage the soil physical conditions (Qadir et al. 2007).

Salt stress disturbed the soil physical structure and affected the nutrients availability, transport and imbalance/ partitioning and caused nutrient deficiencies. The competition of essential nutrients with dominant ions such as Na^+ and Cl^- reduced nutrients uptake and plant growth. Salt stress caused ionic imbalance and specific ion toxicity which affected the metabolic changes/ biophysical and physiology of plants. Accumulation of excessive salts induced deficiencies of N, P, K, Ca, Mg and thus hampers crop growth (Baghalian et al. 2008; Tabatabaie and Nazari 2007).

In salt-affected soils during reclamation with chemical amendments, considerable leaching losses of N may occur along with soluble salts including desorbed Na^+ as well as other macro and micro-nutrients (Murtaza et al. 2006). In Pakistan, the efficiency of different fertilizers having N in ammonium form or which produce NH_3 in the soil upon application is discouragingly low. High soil pH and hot climatic conditions in our country are conducive to N losses through ammonia volatilization. Ammonia volatilization losses ranging from 22 to 53 % have been reported from calcareous soils of Pakistan (Hussain and Naqvi 1998). So, it is the need of the hour to optimize the NUE for achieving the challenging yield on salt-affected and normal soils.

3 Relationship Between Soil Properties and Salinity/Sodicity

Soil texture plays a convincing role in crop irrigation, as well as the role of soil texture with regards to the impact of salinity/sodicity is not an exemption. Texture is directly correlated with a soils, ability to infiltrate water i.e. permeability and infiltration, available water holding capacity (WHC), and ion exchange soil capacity (Warrence et al. 2002). Clayey soils have comparatively high WHC and are slow to drain due to their small sized pore diameters. Conversely, sandy soils hold less water and are faster to drain. In normal practices of irrigation, leaching fraction (loss of water from the root zone) of the sandy soils have found to be greater than clayey soils when both soils are flooded with the same volumes of water. Similarly, high saline water can be used for longer periods of time on sandy soils.

Another important aspect of soil texture is the highest cation exchange sites in soils due to their small sized particle and high surface area. Consequently, clayey soils have the greatest risk for Na^+ binding and soil dispersion. Conversely, sand having greater particle size, low surface area and constitute smaller exchange sites. Silt falls somewhere in between clayey and sandy soils (Andrews and Martin 2000).

Sodium adsorption onto the soil exchange sites is controlled by the solution chemistry. When irrigation water having higher levels regarding Na^+ relative to Ca^{2+} and Mg^{2+} enters the soil solution, Na^+ dominates on the exchange sites of the soil particles (Van de Graaff and Patterson 2001). Ultimately, complete balanced chemistry between the exchange sites and soil solution will probably be achieved. As clay content increases, so does the time necessary to accomplish this new equilibrium. Correspondingly, the time to reverse the Na^+ accumulation increases. Reduced hydraulic conductivity (HC) and permeability due to sodicity further aggravated this problem. Therefore, it is indicated that soils with higher fraction of clay will not only be more at risk of dispersion, but will also be more resistant to attempts to stand this augmented soil ESP (Ghafoor et al. 2004; Mostafazadeh-Fard et al. 2008; Shainberg et al. 2001).

Overall, clays are characteristically more prone to dispersion than are sands and silts. More salts, including Na^+, will accumulate in clayey soils as compared to sandy soils because of clay's fundamentally lower leaching fraction (LF) and greater surface area. Secondly, the risk for Na^+ permeability problems is inherently higher in clayey soils due to their structure. Third, the higher the clay content, the more time is required to desorb Na^+. Finally, it should be observed that clayey soils are inherently more problematic to work with than sandy soils like mechanical alteration, tillage, mechanical alteration, drainage and leaching, particularly when they become dispersed (Association 2010). In general, more dispersion takes place in clay soils than in silt and sand. Sandy soils have more LF as compared to clay and silt dominated soils. Clay soil has a large exposed surface and lower LF. Clay soils are difficult to manipulate, till, leach and drain when they are dispersed. Ayers and Westcot (1985) concluded that soils dominated by montmorillonite have less permeability than illite-vermiculite and kaolinite-sesquioxide clays. It is because of structure and crystal lattices of clay.

4 Ionic and Osmotic Stresses

There is a pre-dominance of nonessential nutrients in salt-affected soil environments. Essential nutrients must be absorbed by the plants from a diluted source even in the presence of saturated solution with superfluous nutrients. It demands more energy and plants are incapable to fulfill their energy requirement. Plant growth is severely affected by salinity due to ionic and osmotic stress. Osmotic stress is enforced by the increased osmotic potential in the root zone.

Primary and secondary effects are developed due to the highest salt concentration in apoplast that drastically overcome survival, growth and its yield. Primary effects are ion toxicity, instability and hyperosmolarity. In saline/sodic soils, various salt contaminants are there but particularly NaCl freely diffuses into the water to give Na^+ and Cl^- ions (Castillo et al. 2007; Pagter et al. 2009). The Na^+ is the smallest molecule absorbed by higher plants and transported to other parts of the plant. It inhibits different cellular and sub-cellular processes which are responsible for ion damage, osmotic stress and nutritional imbalance (Cha-um et al. 2007; Siringam et al. 2009; Ahmad et al. 2012). Such changes cause depression in plant growth and abnormal development leading to plant death (Davenport et al. 2005; Quintero et al. 2007). Halophytes have defense mechanisms for homeostasis of ions, regulation of hormones and antioxidant (Sairam and Tyagi 2004).

The Na^+ toxicity in plants might be due to its storage in the plant shoot which transported from the soil solution through the transpiration pull (Tester and Davenport 2003). Its deposition in different organs and tissues occurs due to distinctive mechanism of Na^+ transport. Moreover, 0.1 M Na^+ level is cytotoxic representing that Na^+ directly affects specific physiological and biochemical processes (Serrano 1996). Hyperosmotic shock is imposed due to high salt concentrations by lowering the water potential and restricts cell expansion (Hasegawa et al. 2000; Zhu 2001). During severe drought, the lowest water potential of apoplast can advance to cellular water loss posing dryness.

Salinity affects metabolic processes in plants and induces enzymatic changes (Khan and Panda 2008; Kumar et al. 2000; Pessarakli 2010; Ahmad et al. 2012; Hakeem et al. 2012). As a result, prime (hyperosmotic stresses and ionic imbalance) and trivial effects (oxidative destruction) of salinity stress may develop. Carbon reduction is inhibited during stress environment by the Calvin cycle and decrease in oxidized $NADP^+$ to function as an electron acceptor in photosystems. During photosynthetic electron transfer, ferrodoxin is over reduced; electrons may be transferred by Mehler reaction from photosystem I (PSI) to O_2 to form superoxide radicals (O_2^-), which activates chain reactions that produce more energetic reactive oxygen species (ROS). Any instability during cellular redox homeostasis can be titled as an oxidative stress that lead to the production of ROS owing to univalent reduction of O_2. Salinity stress enhances the production of ROS like superoxide radical (O_2), alkoxyl radical (RO), hydrogen peroxide (H_2O_2), and singlet oxygen ($_1O^2$) formation through increased transfer of electron to O_2. It is well stated that cytotoxic ROS, which are formed in the mitochondria and peroxisomes during metabolic processes, can damage typical metabolism through oxidative destruction of proteins, nucleic acids and lipids (Grant and Loake 2000). Lipid peroxidation, brought by ROS, is similarly important in membrane deterioration (Demiral and Türkan 2005; Mandhania et al. 2006) (Fig. 1).

Photosystem II (PSII) plays a vital role in photosynthetic response to environmental perturbations. The salinity effects on PSII have been studied comprehensively and photochemistry of PSII is contradictory. Salt stress could inhibit PSII activity (Ashraf and Harris 2004; Munns 2002), while others have showed that

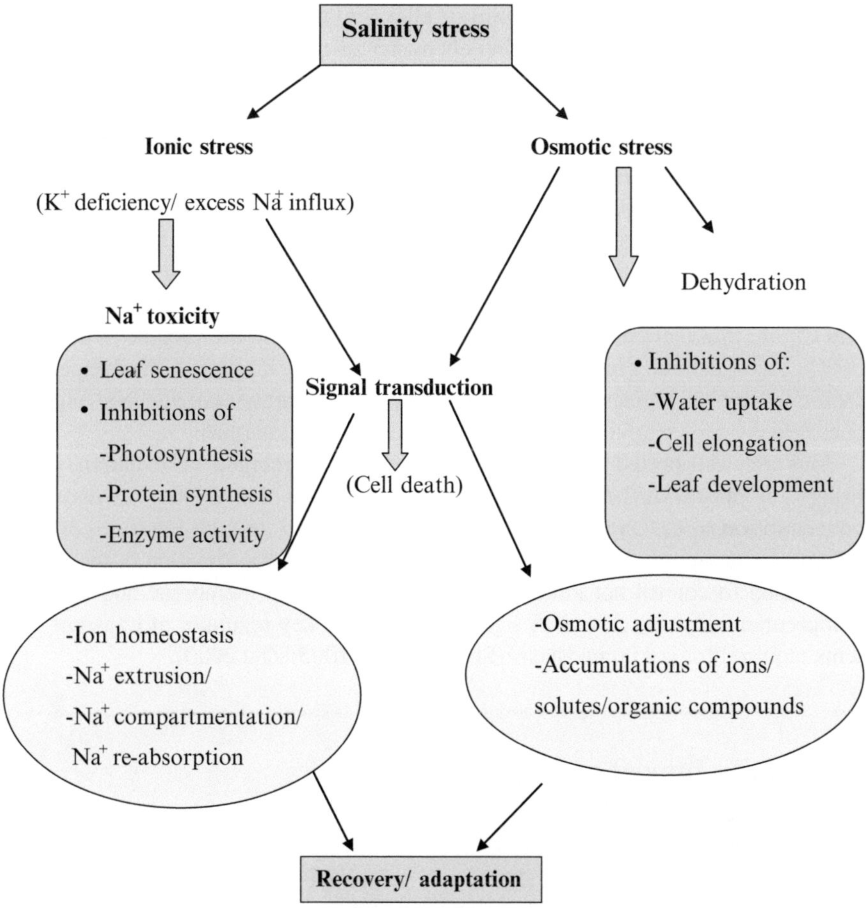

Fig. 1 Schematic representation of ionic and osmotic stress, and plant recovery mechanism

salinity has no effect on this parameter (Abadía et al. 1999). It has been examined that there were no alterations in photochemistry of PSII in salt stressed wheat, barley and sorghum barley when they were cultivated in relatively normal conditions. Lu and Zhang (1998), indicating that salinity stress has no impact on PSII photochemistry but it is the interaction between normal and other environmental stresses which damaged to PSII.

5 Salinity Stress at Cellular Level

Secondary effects are noticed by ion disequilibrium and hyperosmotic stress under salinity stress (Hasegawa et al. 2000; Zhu 2001). In saline condition, plants are either undeveloped during the salt stress or doing cellular adjustment to tolerate this environment. Tolerance mechanisms can be characterized as those that minimize the osmotic stress/ion disequilibrium or alleviate/overcome the secondary effects due to salinity stresses. The solution chemistry of the saline environment initially creates a water potential difference between the apoplast and symplast which decrease turgor pressure, and ultimately decrease plant growth (Bohnert et al. 1995; Taiz et al. 2015).

The osmotic adjustment is response at cellular level due to reduction in turgor pressure. In halophytes and glycophytes, the cytosolic and organellar machinery is equally sensitive to Na^+ and Cl^-; so osmotic adjustment is done by accumulating osmoprotectants and compatible osmolytes in these compartments (Bohnert and Sheveleva 1998). However, for osmotic adjustment and compartmentalization, Na^+ and Cl^- are proficient osmolytes to reduce cytotoxicity into the vacuole (Blumwald 2000). Since, cellular growth of plants occurs principally through vertical expansion facilitated by an increase in vacuolar volume, compartmentalizing Na^+ and Cl^- accelerates osmotic adjustment that is necessary for cellular growth.

Ions are transported into the vacuole through membrane vesiculation which brings tonoplast near the plasma membrane (Hasegawa et al. 2000). The excessive concentration of Cl^- and Na^+ transport mostly occurred through ion transport systems from the apoplast to the vacuole. Probably, strong coordination of the ions is established to control net ionic influx across the plasma membrane and vacuolar compartmentalization. The SOS signal pathway is a key regulator of transport systems required for ion homeostasis (Shabala et al. 2005; Zhu 2000).

6 Salinity Impact on N-Metabolism

Worldwide soil salinity and water scarcity are severe threats for plant productivity and growth (Golldack et al. 2011; Hu and Schmidhalter 2005). Key physiological processes responsible for plant growth and development are carbon (C) and N metabolisms which are adversely affected by salinity (Iqbal et al. 2015; Touchette and Burkholder 2007). Fertilizer consumption may be increased by around twofolds in the next two decades to get the increased demand in food production (Byrnes and Bumb 1998). In near future, plant-nutrition-related research will be a high-priority to achieve more crop production and sustain the soil fertility. Moreover, it has been observed that total shoot nitrogen absorption (SNA) is mostly decreased under salinity stress ambient conditions because of the antagonistic interaction of NH_4^+, Na^+, or Cl^- and NO_3^- (Hu and Schmidhalter 2005).

Salt-affected agricultural soils in the world are mostly nitrogen-deficient. In such soils, N fertilization and regular irrigation with good quality water are required to improve crop productivity and soil fertility. Salts diffused in the irrigation water could interfere with the N available for plants uptake, as salinity is found to be a major threat for N availability to plants. Salinity stress impairs basic metabolic processes especially photosynthesis (Fatma et al. 2013; 2014; Nazar et al. 2014) modifying the hormonal status leading to growth and yield inhibition (Khan and Khan 2014; Paranychianakis et al. 2004). Salinity induced biochemical alterations in plants have mostly been emphasized for fruit productivity and its quality, osmotic adjustments and ion compartmentation. However, rare data are found with reference to salinity effects on N metabolic pathways (Berrios et al. 2000). Dissipation of essential nutrients and NO_3, particularly, can induce fluctuations in gene expression and enzyme activity under stress conditions (Forde 2002). Nitrate uptake and transport appear very susceptible to salinity yielding severe ramification for NO_3^- assimilation in the plants (Berrios et al. 2000).

Salt stress exhibited metabolic changes in the plants by the production of low molecular weight organic solutes such as linear/cyclic polyols (glycerol, mannitol or inositol or derivatives of methylated inositol, amino acids (proline or glutamate) and betaines (glycine or alanine) (Jouyban 2012; Munns and Tester 2008).

Plants with N constraint produce sulfonium compounds such as dimethylsulfonium propionate proportionate to N enrich betaines. Organic solutes are considered as compatible osmolytes because at high concentration these are not detrimental to enzymes/cellular structures and function in osmotic adjustments. The cytosol with compatible osmolytes balances the high concentration of salts in the vacuole and outside the cell. Salt stress causes metabolic changes in plants and the most susceptible plant species is ice plant. Salt stress educes the mode of photosynthesis to crassulacean acid metabolism (CAM) pathway from C3 mechanism. The expression of CAM pathway in salt stress environment induces phosphoenolpyruvate (PEP) carboxylase and decreases water loss due to opening of stomata at night (Jouyban 2012).

Plants growth in salt-affected soils undergoes various physiological, morphological, biochemical and developmental modifications (Ashraf and Harris 2004; Wang et al. 2008). Plants opt different physiological processes such as selective ion transport, compartmentation and compatible osmolytes responsible for osmotic adjustments (Munns and Termaat 1986). On the basis of N which is a key element of the compatible osmolytes viz. amides, amino acids and betaines etc., N metabolism has prime significance and must be focused under salt-affected conditions (Munns and Tester 2008; Parida and Das 2005). High salt content damaged the ionic balance at cellular, tissue and plant level (Niu et al. 1995; Surabhi et al. 2008).

Phytoavailability of N is generally in the form of NO_3^- and NH_4^+ (Ruiz et al. 2007) with their assimilation to glutamine and glutamate (Wang et al. 2012). The principal enzymes responsible for translocation are glutamine-2-oxoglutarate amino transferase (GOGAT) or glutamate synthase and glutamate dehydrogenase (GDH) (Esposito et al. 2005; Wickert et al. 2007). It was suggested that nitrate is reduced to nitrite by nitrate reductase (NR) while nitrite to ammonium by nitrite reductase (NiR)). Shi et al. (2010) investigated that glutamine synthetase (GS), glutamate dehydrogenase (GDH) and glutamate synthase (Fd/ NADH-GOGAT) accumulate NH_4^+ into organic molecules. The cytosol with GS1 and chloroplasts with GS2 are subsisted by glutamine synthetase (GS) (Kusano et al. 2011; Witte 2011).

It was investigated that some gene expressions were involved in N metabolism under salt stress. Na^+ and Cl^- were accumulated in rice seedlings under salinity stress while concentration of K^+ was reduced under such conditions. The leaves and roots of rice showed accelerated gene expression of *OsGS1, OsGDH1-3, OsNADH-*

GOGAT, when subjected to 4 h salinity stress. The reduction in NO_3^- in rice leaves and roots suppressed the *OSNR1* and resultantly NH_4^+ was reduced (Wang et al. 2012). Nitrate reductase the foremost enzyme in the N metabolism is very sensitive to NaCl (Gouia et al. 1994). Phytochromes and production of compatible solutes like proline help in the N assimilation and salt tolerance thereby facilitating water uptake, maintaining osmotic balance, and protection of the cell against ROS (Iqbal et al. 2014). The importance of proline cannot be denied and is one of the twenty DNA-encoded amino acids, and salt stressed plants accumulates significant amount of proline (Mane et al. 2010; Wang et al. 2011). Proline is principally a regulatory amino acid and activates the metabolic rate and responsible for plant adaptation to stress. Salt stress results in imbalance of nutrients by the nutrient competition/ transport, salinity effect on nutrients by physiological inactivation of certain nutrient and affecting the growth and yield of crops (Grattan and Grieve 1999; Jacoby and Moran 2001).

7 Interaction of Salinity and N Fertilization

On global scale, soil salinity/sodicity is one of the major problems for agriculture because salinity decreases the rate of photosynthesis and plant growth to various levels. Salinity and N interaction is very complicated because it is affected by different levels of salinity/sodicity, form of N, plant age, plant species, genotypes, the composition and concentration of nutrients and prevailing climatic conditions.

Nitrogen, in one form or another, accounts for about 80 % of the total mineral nutrients absorbed by plants (Marschner 1995). It is one of the most yield limiting nutrients in crop production under most agro-ecological conditions and its efficient use is important for the economic sustainability of cropping systems (Fageria et al. 2010). Uptake of N by rice was inhibited under high sodium chloride (NaCl) and sodium sulfate (Na_2SO_4) concentration in the roots, and the excess amount of absorbed Na^+ depressed NH_4^+ absorption (Britto et al. 2004). Impaired N absorption by plants reduces root permeability and the subsequent decrease in water and nutrient uptake under high salt concentrations (Martínez-Ballesta et al. 2006).

Grattan and Grieve (1999) found that salinity and N interaction can reduce N accumulation in crop plants since an increase in Cl^- uptake is often accompanied by a decrease in shoot NO_3^- level which may be associated with competition between NO_3^- and Cl^- ions. Britto et al. (2004) reported that the interaction between NO_3^- and Cl^- might be related to the salt tolerance of the cultivar under investigation and found that salt tolerant cultivars had higher NO_3^- influx rates than the sensitive cultivars. The NH_4^+ fed plants were more sensitive to salinity than NO_3^- fed plants in nutrient solutions (Grattan and Grieve 1999). Some studies indicate that increased NO_3^- in solution decreased Cl^- uptake and its accumulation (Martinez and Cerda 1989). Munns and Tester (2008) reported that NaCl decreased total N in wheat plants because NO_3^- has an antagonistic relation with Cl^- produced through ionization of NaCl in the growth medium. Chloride salts of both Na^+ and K^+ inhibited

NO_3^- uptake indicating that the process was more sensitive to anionic salinity than to cationic salinity (Russell 2000).

Application of N fertilizers in N deficient soils improved growth and yield of cotton (Grattan and Grieve 1999), wheat (Murtaza 2011), corn (Absalan et al. 2011), rice (Murtaza 2011), millet and sorghum (Esmaili et al. 2008). Nutrients uptake and their use efficiency in salt-affected soils is low due to antagonistic cationic and anionic interactions (Fageria et al. 2011). Hence, a higher dose of nutrients is required in salt-affected soils as compared with normal soils.

In solution culture, wheat and maize were highly sensitive to NH_4^+ during NaCl applications instead of NO_3^- (Tshivhandekano and Lewis 1993). Ion activity of Na^+ and Cl^-, is highly affected by higher salt concentration in soil solution, creating a nutritional disorder in plants (de Lacerda et al. 2003). Crop yield or quality is affected by nutrient-salinity interactions depending on the crop species, the nutrient status in plant tissues, salinity level, composition of the salts and environmental factors (Fageria et al. 2011). In salt-stress plants accumulation of Na^+ and Cl^- in leaves through the transpiration flow is a general and long-term process (Møller and Tester 2007). Plants prefer to absorb K^+ over Na^+ resulting in Na^+-induced K^+ deficiency under saline-sodic conditions (Maas et al. 1999). Grattan and Grieve (1999) conducted field scale and solution culture experiments to understand the nutritional status of plants under salinity stress.

8 Interactive Effect of Salinity and Ca^{2+}

Cell division, membrane integrity and ion homeostasis are significantly affected by calcium (Hu and Schmidhalter 2005). The Ca^{2+} uptake by plants depends on its concentration, soil pH and crop species. Passive nature of calcium in the plants develops its weakness symptoms first on the younger parts of the plants. In case of increased salt concentration in the rhizosphere, Ca^{2+} requirement of plants is increased (Tuna et al. 2007). Moreover, Ca^{2+} uptake is decreased by ionic interactions and precipitation in salt-affected soils (Grattan and Grieve 1999).

Significant changes in morphological, anatomical and hydraulic properties occur during corn growth when Na^+/Ca^{2+} ratio is high (Xiao-Hua et al. 2010). Addition of Ca^{2+} may partially restore the inhibition of corn root cell production and its prolongation in NaCl saline conditions. Ca^{2+} application tended to reverse these changes and improved root hydraulic conductivity (Qadar 2009). Adequate supply of Ca^{2+} in saline soils is a critical step in controlling the severity of specific ion toxicity, particularly in crops susceptible to sodium and chloride injury (Tuna et al. 2007).

Calcium supplementation improved germination and growth of bean in salinity stress by increased root membrane integrity (Awada et al. 1995). Ameliorative role of supplemental calcium in alleviating the adverse effects of Na^+- salinity depends on the calcium concentration, the crop and Na^+ source (Hu and Schmidhalter 2005).

9 NUE in Wheat and Rice Under Saline Conditions

The world's top three cereal grains in 2008 graded on the monetary basis were rice, superseded by wheat and corn (FAO, Stat data).

9.1 *Wheat (***Triticum aestivum** *L.)*

Wheat is the major staple food crop in Pakistan. It has a pivotal role in agricultural guidelinrs and country's economy relies heavily it. It contributes to the value-added agriculture (14.4 %) and GDP (3.1 %) (Anonymous 2009–2010). Wheat plants were grown in a growth chamber under contained environment to evaluate the impact of N sources and salts (ammonium sulfate or calcium nitrate) on the growth, survival and N uptake (Botella et al. 1997). It was perceived that efficacy of salinity on N uptake was influenced by the employed nitrogenous source, and NO_3^- uptake was noticeably inhibited as compared to NH_4^+ uptake. Salinity increased the harmony for NH_4^+ and reduced the affection for NO_3^-. Chloride strongly inhibited the net NO_3^- uptake (Kronzucker et al. 1999), affected its efflux (Deane-Drummond and Glass 1982) but had no influence on its influx (Glass et al. 1985). Nitrate reductases or molybdoenzymes reduced NO_3^- to NO_2. Ali et al. (2001) observed that N metabolism in wheat seedlings was inhibited by NaCl stress inducind the activity of NR enzyme. Ionic interactions rely on the total salt-tolerance of the plants (Leidi et al. 1991). In addition, it was found in their study that NO_3^- uptake was hampered to a larger extent than NH_4^+ uptake. Shoot growth was raised by NH_4^+ as compared with NO_3^-. A blend of NO_3^- and NH_4^+ appears the best N source for wheat growth. The nitrogenous sources should be supplemented with NH_4^+ to improve plant growth in saline conditions since NO_3^- uptake is more affected than NH_4^+ uptake.

9.2 *Rice (***Oryza sativa** *L.)*

Rice is an important cereal crop in tropical and temperate localities of the world. In Pakistan, it contributes in value added agriculture (6.4 %) and % in GDP (1.4) (Anonymous 2009–2010). Soil salinity is an utmost component that declines rice yield and its productiveness (Ghafoor et al. 2004). Rice is vulnerable to salts at seedling and reproductive stages (Moradi and Ismail 2007; Zeng et al. 2001). Rice plants exposed to 6.65 dS m^{-1} EC_e showed more than 50 % reduction in yield (Zeng and Shannon 2000). The germination, growth, solute accumulation and gene expression of rice plants are influenced considerably in NaCl stress conditions (Hoai et al. 2003; Nemati et al. 2011). Nitrogen metabolism and ion balance in the stressed rice seedlings was studied by Wang et al. (2012). Roots and leaves varied in sensitivity and response time under salinity stress. The total amino acids, NO_3^- and inorganic ions were measured after 4 h of exposure. Na^+ and Cl^- build-up was stimulated in roots with reduced K^+ content. However, same behavior was observed in leaves in more time (5 days).

10 Nitrate Leaching

Nitrate is common ion naturally produced during N cycle. Worldwide groundwater reservoirs (aquifers) are contaminated by nitrate leaching from vadose zone to the saturated zone depending upon the hydraulic properties of geological material. It may be a slow process but its effects are very drastic (Baghvand et al. 2010). Nitrate contamination of groundwater also depends on precipitation, inundation, recharge rate, geological heterogeneities and depth of phreatic surface from the earth surface. Thus, groundwater quality is adversely affected by chemical pollutants (Attoui et al. 2012), organic pesticides (Worrall and Kolpin 2004) viruses or bacteria (Schijven et al. 2010). Immoderate application of nitrogenous sources has adversely perturbed the groundwater quality in agricultural areas (Huang et al. 2013). Contamination can render groundwater and makes it unfavorable for human consumption. NO_3^- leaching from agricultural zones leads to environmental problems as high NO_3^- in water has potential for ecological damage and health hazards. NO_3^- levels exceeding the permissible limits (10 mg L^{-1}) can cause methemoglobinemia, stomach cancer, miscarriage and Lymphoma (Ward et al. 1994). The enrichment of waters with N and P triggers the growth of aquatic vegetation which subsequently results in eutrophication and O_2 depletion in water bodies (Vitousek et al. 1997).

However, anthropogenic activities have greatly increased the NO_3^- concentration, particularly in groundwater (Chand et al. 2011). Nitrate leaching due to non-point agricultural sources (fertilizers) is recognized as one of the serious threats for groundwater pollution (Salemi et al. 2012). Jalali (2005) of groundwater in Europe and found that NO_3^- concentrations in groundwater exceeded limits (WHO) for drinking water (Lægreid et al. 1999). Liu et al. (2005) studied that NO_3^- pollution of groundwater in northern China and described that over 50 % of the studied positions contained NO_3^- higher than threshold level.

During last few decades, elevated use of N has been associated with enhanced N losses. In arid and semi-arid regions, water is applied in excess of evapo-transpiration and NO_3^- leaching may occur (Jalali and Rowell 2003) resulting into contaminated groundwater. Nitrogen leaching in agricultural parts should receive substantial attention because a large number of people in urban and rural areas rely on groundwater for drinking.

11 Salinity/Sodicity and N Management

Several methods are employed for reclamation of salt-affected soils. Suitability of each method depends on physical, chemical and mineralogical characteristics of the soil such as internal soil drainage, presence of pans in the subsoil, climatic conditions and types of salts, quality and quantity of water, depth of groundwater, exchangeable Na^+, lime or gypsum, amendment cost, topography of the land, and time for reclamation (Huertas et al. 2008). Chemical, biological and agronomic practices are prominent employed reclamation methods. The combination of these approaches increases the efficiency and reduces the reclamation period. The crop productivity and efficient use of fertilizer in soils can be enhanced using a mixed approach, i.e. usefulness of organic/inorganic amendments preferably gypsum for sustaining yields, improving soil health and input use efficiency (Wallender and Tanji 2011).

Chemical amendments employed for reclamation of saline-sodic soils are either soluble Ca^{2+} salts like $CaCl_2.2H_2O$, $CaSO_4.2H_2O$ and phosphogypsum or relatively less soluble ground $CaCO_3$. Common Ca^{2+} mobilizers in calcareous soils by enhancing the conversion of $CaCO_3$ to more soluble $CaSO_4$, $Ca(HCO_3)_2$, $Ca(NO_3)_2$ or $CaCl_2$ include H_2SO_4, HCl, HNO_3, S, FeS_2, CaS_5, $FeSO_4$ and $Al_2(SO_4)_3$ (Ghafoor et al. 2004). Application of chemical amendments and subsequently their drainage with concentrated electrolyte may also bring adverse effects on environment Qadir et al. (2007). Some polymers and by-products of certain industries are useful material as soil amendments. Addition of organic matter (farm manure, slaughter house wastes, poultry excreta, green manure, etc.) alone can reclaim the saline-sodic soils but at a slow rate (Qadir et al. 2001).

Some chemical fertilizers may also supply soluble Ca^{2+} directly (calcium nitrate and single superphosphate) or indirectly by producing physiological acidity within the zones of their application (ammonium sulfate and urea). The application of fertilizers in the usual economical doses cannot reduce the soil sodicity to a large extent. Moreover, amendments also improve soil physical conditions (porosity, infiltration rate, permeability) by eliminating the Na^+ dominance that causes undesirable changes in sodic soils. In packed soil columns, improvement in hydraulic conductivity (HC) was observed with gypsum. From scanning electron microscope observation, it was concluded that increase in HC was linked with an increase in pore space reducing soil dispersion. Similarly, a three-fold increase in HC was observed in gypsum-applied saline-sodic soil as compared to distilled water application. A remarkable decline was observed in soil bulk density when treated with phosphogypsum. Adequate infiltration rate in saline-sodic soils was achieved after gypsum application (Armstrong and Tanton 1992). Progression in physical properties of saline-sodic soils during reclamation is attributed to the raised levels of Ca^{2+} both in soil solution and on exchange complex which help in flocculation of the dispersed soil grains thereby improving hydraulic properties of the soil.

Several studies have indicated that calcareous sodic soils can be improved by growing crops or trees with or without the application of chemical amendments (Mishra et al. 2004). The review on choice of the amendments, their conjunctive use, methods of application, their particle-size and sources showed some inconsistent results on the potency of various amendments for amelioration of salt-affected soils. Generally gypsum and H_2SO_4 were found the most effective reclaimants. But, low price and freight, general availability and easy application of gypsum make it the most commonly used and highly cost-effective Ca^{2+} source for reclaiming both the calcareous and non-calcareous sodic/saline-sodic soils (Ghafoor et al. 2001).

Gypsum ($CaSO_4.2H_2O$) is calcium-rich, dissolves at high pH, and does not contain element or compounds that might interfere during reclamation. Calcium from applied gypsum in salt-affected soils can help in decreasing volatilization losses of NH_4^+-N from ammonium nitrate, UAN, urea, ammonium sulfate, or other ammonium phosphates. Calcium can decrease the pH by precipitating carbonates and forming complex of Ca^{2+} salt with NH_4OH which hampers NH_3 loss to the atmosphere. According to Ghafoor et al. (2008), physiological disorders occur when the ratio of Ca^{2+} to total cations fall below a certain threshold level. However, halophytes have strong tendency for absorption of Ca^{2+} from the growth medium with low calcium.

Gypsum application to soils play important role in improving the N fertilizer use efficiency. Due to calcareous salt-affected soils and severe climatic conditions in Pakistan, NH_3^-N volatilization losses are predominant and there is a close relationship between the extent of volatilized NH_3^-N and the soil pH (Zhenghu and Honglang 2000). Ammonia volatilization losses ranging from 22–53 % have been reported from Pakistani soils (Iqbal et al. 1998). NH_3 volatilization was a significant pathway of N losses from Vietnamese paddy fields throughout the cropping period (Watanabe et al. 2009).

Urea is the major N fertilizer used for crop production and it is susceptible to hydrolysis followed by NH_3 volatilization due to conversion to ammonium bicarbonate, which may temporarily raise soil pH to above 9.0 in both acidic and calcareous soils (Harrison and Webb 2001) leading to release of NH_3^-N Similarly, other N fertilizers when added to calcareous saline-sodic or sodic soils may liberate NH_3 depending upon the degree of calcareousness and alkalinity of the soil. Addition of soluble calcium salts to N fertilizers can reduce NH_3^-N volatilization as much as 90 % (Fenn and Hossner 1985).

Rengasamy (2010) investigated the partial replacement of mineral N-fertilization in salt-affected soils along with improving the poor physicochemical and biological soil characteristics by the application of some soil amendments (sulphur and gypsum). The application of amendments promotes the use efficiency of N fertilization. Ca^{2+} supplementation to rice plants under salt stress increased chlorophyll contents, proline and soluble sugar of the leaves (Zhu et al. 2004). The net photosynthetic rate and stomatal conductance increased more significantly with Ca^{2+} which was significantly reflected on grain and straw yields due to better nutrients provision than the control plants receiving the recommended dose of N without calcium.

Efficient use of N can be achieved by decreasing salinity or N rates. A significant gain in salt tolerance was noted when plants were fertilized with organic N sources as compared with inorganic-N sources. Its deficiency is very familiar in crops and causes hasty retardation of plant growth in both normal and salt stress conditions. Nitrogen application into N deficient soils improved growth and production of corn at moderate salt stress (Absalan et al. 2011), wheat and rice (Murtaza 2011). In most of the scenarios, adequate N conditions are favorable for decreasing overall N uptake but its dry weight concentration was raised or remains ineffective (Hu and Schmidhalter 2005).

On the contrary, decline in N storage by plants has been observed in a number of studies (Pessarakli 2010). Declined N storage is understandable in the soil as Cl^- and NO_3^- have antagonistic interactions for their absorption. Salinity induced reduction in NO_3^- concentration in wheat leaves was found without affecting the total N content (Hu and Schmidhalter 1997). Increased concentrations of NO_3^- linearly decreased Cl^- concentrations in several crops (Garg and Chandel 2011; Maggio et al. 2007). In plants, NO_3^- concentration effective in reducing Cl^- concentration should be high enough in soil solution.

A few studies in Pakistan (Murtaza 2011) and some in India (Singh et al. 2009) have shown that crops grown in saline and/or sodic soils yielded better with higher N fertilizer rates than recommended for non-saline/sodic soils mainly through growth dilution (Sairam and Tyagi 2004; Woyema et al. 2012). Singh et al. (2009) reviewed a series of experiments and recommended that rice and wheat crops grown in sodic soils should receive 25 % N above the recommended rates for non-saline non-sodic soils. Excess amount of N fertilizer application could cause environmental problems (Choudhury and Kennedy 2005).

12 Conclusion

Worldwide salinity is one of the most crucial abiotic factors in plants that induce ion toxicity and osmotic stress along with changes in hydraulic properties (soil porosity, permeability and hydraulic conductivity). Impact of salinity and N deficiency on mineral constituents was tremendously crop specific. However, salinity disrupts N application in two ways; firstly, nutrient uptake and translocation are influenced by the ionic strength of the substrate regardless of its composition. Secondly, minerals interaction in plants is decreased by salinity through encounter with major ions in the substrate (Na^+ and Cl^-). Therefore, numerous experiments have been conducted to calculate appropriate doses of N fertilizer for soils varying in texture to protect the environment. Crop requirements for N and other mineral nutrients vary on salt-affected and normal soils due to changes in soil physical and chemical characteristics. Therefore, farmers can economically benefit from judicious applications of N fertilizer for adequate crop production from their marginal land resources without harming the biosphere equilibrium.

References

Abadía A, Belkhodja R, Morales F, Abadía J (1999) Effects of salinity on the photosynthetic pigment composition of barley (*Hordeum vulgare* L.) grown under a triple-line-source sprinkler system in the field. J Plant Physiol 154(3):392–400

Absalan AA, Armin M, Asghripour MR, Karimi-Yazdi S (2011) Effects of different forms of nitrogen application on yield response of corn under saline conditions. Adv Environ Biol 5(4):719–724

Ahmad R, Chang M (2002) Salinity control and environmental protection through halophytes. J Drain Water Manag 6(2):17–25

Ahmad P, Hakeem KR, Kumar A, Ashraf M, Akram NA (2012) Salt induced changes in photosynthetic activity and oxidative defense system of three cultivars of mustard (Brassica juncea L.). Afr J Biotechnol 11:2694–2703

Ali A, Tucker T, Thompson T, Salim M (2001) Effects of salinity and mixed ammonium and nitrate nutrition on the growth and nitrogen utilization of barley. J Agron Crop Sci 186(4):223–228

Andrews DC, Martin GR (2000) Criteria for liquefaction of silty soils. In: Proc 12th World conf. on earthquake engineering. NZ Soc. for EQ Engrg. Upper Hutt, New Zealand

Anonymous (2009–10) Economic survey of Pakistan. Economic Wing, Ministry of Food, Agri. and Livestock, Govt. Pakistan, Islamabad, Pakistan

Armstrong A, Tanton T (1992) Gypsum applications to aggregated saline—sodic clay topsoils. J Soil Sci 43(2):249–260

Ashraf M, Harris P (2004) Potential biochemical indicators of salinity tolerance in plants. Plant Sci 166(1):3–16

Association WPH (2010) Soil improvement committee of the California Fertilizer Association. Interstate Publishers, Inc, Sacramento

Attoui B, Kherci N, Bousnoubra H (2012) State of vulnerability to pollution of the big reservoirs of groundwater in the region of Annaba-Bouteldja (NE Algeria). Geographia Tech 2:1–13

Awada S, Campbell W, Dudley L, Jurinak J, Khan M (1995) Interactive effects of sodium chloride, sodium sulfate, calcium sulfate, and calcium chloride on snapbean growth, photosynthesis, and ion uptake. J Plant Nutr 18(5):889–900

Ayers RS, Westcot DW (1985) Water quality for agriculture. Food and Agriculture Organization FAO Irrigation and drainage paper 29 Rev 1 FAO, UNO, New York, USA

Baghalian K, Haghiry A, Naghavi MR, Mohammadi A (2008) Effect of saline irrigation water on agronomical and phytochemical characters of chamomile (*Matricaria recutita* L.). Sci Hortic 116(4):437–441

Baghvand A, Nasrabadi T, Bidhendi GN, Vosoogh A, Karbassi A, Mehrdadi N (2010) Groundwater quality degradation of an aquifer in Iran central desert. Desalination 260(1):264–275

Berrios EF, Gentzbittel L, Kayyal H, Alibert G, Sarrafi A (2000) AFLP mapping of QTLs for in vitro organogenesis traits using recombinant inbred lines in sunflower (*Helianthus annuus* L.). Theoret Appl Genet 101(8):1299–1306

Blumwald E (2000) Sodium transport and salt tolerance in plants. Curr Opin Cell Biol 12(4):431–434

Bohnert HJ, Nelson DE, Jensen RG (1995) Adaptations to environmental stresses. Plant Cell 7:1099–1111

Bohnert HJ, Sheveleva E (1998) Plant stress adaptations—making metabolism move. Curr Opin Plant Biol 1(3):267–274

Book PSY (2003) Federal bureau of statistics. Federal Bureau of Statistics, Statistics Division, Government of Pakistan, Islamabad, Pakistan

Botella MA, Martínez V, Nieves M, Cerdá A (1997) Effect of salinity on the growth and nitrogen uptake by wheat seedlings. J Plant Nutr 20(6):793–804

Bray EA, Bailey-Serres J, Weretilnyk E (2000) Responses to abiotic stresses. In: Buchanan B, Gruissem W, Jones R (eds) Biochemistry and molecular biology of plants. American Society of Plant Physiology, Rockville, pp 1158–1203

Britto DT, Ruth TJ, Lapi S, Kronzucker HJ (2004) Cellular and whole-plant chloride dynamics in barley: insights into chloride–nitrogen interactions and salinity responses. Planta 218(4):615–622

Byrnes BH, Bumb BL (1998) Population growth, food production and nutrient requirements. J Crop Prod 1(2):1–27
Castillo E, Tuong TP, Ismail A, Inubushi K (2007) Response to salinity in rice: comparative effects of osmotic and ionic stresses. Plant Prod Sci 10(2):159–170
Cha-um S, Vejchasarn P, Kirdmanee C (2007) An effective defensive response in Thai aromatic rice varieties (Oryza sativa L. spp. indica) to salinity. J Crop Sci Biotechnol 10:257–264
Chand S, Ashif M, Zargar M, Ayub BM (2011) Nitrate pollution: a menace to human, soil, water and plant. Univ J Environ Res Technol 1:22–32
Chinnusamy V, Jagendorf A, Zhu J-K (2005) Understanding and improving salt tolerance in plants. Crop Sci 45(2):437–448
Choudhury A, Kennedy I (2005) Nitrogen fertilizer losses from rice soils and control of environmental pollution problems. Commun Soil Sci Plant Anal 36(11–12):1625–1639
Davenport R, James RA, Zakrisson-Plogander A, Tester M, Munns R (2005) Control of sodium transport in durum wheat. Plant Physiol 137(3):807–818
de Lacerda CF, Cambraia J, Oliva MA, Ruiz HA, Prisco JT (2003) Solute accumulation and distribution during shoot and leaf development in two sorghum genotypes under salt stress. Environ Exper Bot 49(2):107–120
Deane-Drummond CE, Glass AD (1982) Nitrate uptake into barley (Hordeum vulgare) plants a new approach using $^{36}ClO_3^-$ as an analog for NO_3^-. Plant Physiol 70(1):50–54
Demiral T, Türkan I (2005) Comparative lipid peroxidation, antioxidant defense systems and proline content in roots of two rice cultivars differing in salt tolerance. Environ Exper Bot 53(3):247–257
Esmaili E, Kapourchal SA, Malakouti MJ, Homaee M (2008) Interactive effect of salinity and two nitrogen fertilizers on growth and composition of sorghum. Plant Soil Environ 54(12):537–546
Esposito S, Guerriero G, Vona V, Rigano VDM, Carfagna S, Rigano C (2005) Glutamate synthase activities and protein changes in relation to nitrogen nutrition in barley: the dependence on different plastidic glucose-6P dehydrogenase isoforms. J Exper Bot 56(409):55–64
Fageria N, Moreira A, Coelho A (2011) Yield and yield components of upland rice as influenced by nitrogen sources. J Plant Nutr 34(3):361–370
Fageria NK, Baligar VC, Jones CA (2010) Growth and mineral nutrition of field crops. CRC Press, Boca Raton/Florida
Faostat F (2009) Statistical databases. Food and Agriculture Organization of the United Nations. Statistical Database, Rome
Fatma M, Asgher M, Masood A, Khan NA (2014) Excess sulfur supplementation improves photosynthesis and growth in mustard under salt stress through increased production of glutathione. Environ Exper Bot 107:55–63
Fatma M, Khan MIR, Masood A, Khan NA (2013) Coordinate changes in assimilatory sulfate reduction are correlated to salt tolerance: involvement of phytohormones. Annu Rev Res Biol 3(3):267–295
Fenn L, Hossner L (1985) Ammonia volatilization from ammonium or ammonium-forming nitrogen fertilizers. In: Advances in Soil Science. Springer-Verlag, New York, pp 123–169
Flowers TJ, Colmer TD (2008) Salinity tolerance in halophytes. New Phytol 179(4):945–963
Forde BG (2002) The role of long-distance signalling in plant responses to nitrate and other nutrients. J Exper Bot 53(366):39–43
Garg N, Chandel S (2011) Effect of mycorrhizal inoculation on growth, nitrogen fixation, and nutrient uptake in *Cicer arietinum* (L.) under salt stress. Turk J Agric For 35(2):205–214
Ghafoor A, Gill M, Hassan A, Murtaza G, Qadir M (2001) Gypsum: an economical amendment for amelioration of saline-sodic waters and soils and for improving crop yields. In: National workshop on agricultural use of gypsum. LRRI, NARC, PARC, Islamabad, Pakistan
Ghafoor A, Murtaza G, Ahmad B, Boers TM (2008) Evaluation of amelioration treatments and economic aspects of using saline–sodic water for rice and wheat production on salt-affected soils under arid land conditions. Irrig Drain 57(4):424–434

Ghafoor A, Qadir M, Murtaza G (2004) Salt-affected soils: principles of management. Allied Book Centre, Lahore

Ghassemi F, Jakeman AJ, Nix HA (1995) Salinisation of land and water resources: human causes, extent, management and case studies. CAB international, Wallingford

Glass AD, Thompson RG, Bordeleau L (1985) Regulation of NO_3^- influx in barley studies using $^{13}NO_3^-$. Plant Physiol 77(2):379–381

Golldack D, Lüking I, Yang O (2011) Plant tolerance to drought and salinity: stress regulating transcription factors and their functional significance in the cellular transcriptional network. Plant Cell Rep 30(8):1383–1391

Gouia H, Ghorbal MH, Touraine B (1994) Effects of NaCl on flows of N and mineral ions and on NO_3^-reduction rate within whole plants of salt-sensitive bean and salt-tolerant cotton. Plant Physiol 105(4):1409–1418

Grant JJ, Loake GJ (2000) Role of reactive oxygen intermediates and cognate redox signaling in disease resistance. Plant Physiol 124(1):21–30

Grattan S, Grieve C (1999) Mineral nutrient acquisition and response by plants grown in saline environments. Handb Plant Crop Stress 2:203–229

Hakeem KR, Ahmad A, Iqbal M, Gucel S, Ozturk M (2011) Nitrogen-efficient cultivars can reduce nitrate pollution. Environ Sci Pollut Res 18:1184–1193

Hakeem KR, Khan F, Chandna R, Siddiqui TO, Iqbal M (2012) Genotypic variability among soybean genotypes under NaCl stress and proteome analysis of salt tolerant genotype. Appl Biochem Biotechnol 168:2309–2329

Harrison R, Webb J (2001) A review of the effect of N fertilizer type on gaseous emissions. Advan Agron 73:65–108

Hasegawa PM, Bressan RA, Zhu J-K, Bohnert HJ (2000) Plant cellular and molecular responses to high salinity. Ann Rev Plant Biol 51(1):463–499

Heffer P, Prud'homme M (2013) Nutrients as limited resources. Fertilizer outlook 2013–2017, 81st IFA annual conference, Chicago, USA

Heffer P, Prud'homme M (2010) Fertilizer outlook 2010–2014. In: 78th IFA annual conference, Paris

Hoai NTT, Shim IS, Kobayashi K, Kenji U (2003) Accumulation of some nitrogen compounds in response to salt stress and their relationships with salt tolerance in rice (*Oryza sativa* L.) seedlings. Plant Growth Regul 41(2):159–164

Hu Y, Schmidhalter U (1997) Interactive effects of salinity and macronutrient level on wheat. II. Composition J Plant Nutr 20(9):1169–1182

Hu Y, Schmidhalter U (2005) Drought and salinity: a comparison of their effects on mineral nutrition of plants. J Plant Nutr Soil Sci 168(4):541–549

Huang T, Pang Z, Edmunds WM (2013) Soil profile evolution following land-use change: implications for groundwater quantity and quality. Hydrol Proc 27(8):1238–1252

Huertas E, Salgot M, Hollender J, Weber S, Dott W, Khan S, Schäfer A, Messalem R, Bis B, Aharoni A (2008) Key objectives for water reuse concepts. Desalination 218(1):120–131

Hussain F, Naqvi M (1998) Ammonia losses from salt affected soils. Pak J Soil Sci 14:1–2

Iqbal N, Umar S, Khan NA (2015) Nitrogen availability regulates proline and ethylene production and alleviates salinity stress in mustard (Brassica juncea). J Plant Physiol 178:84–91

Iqbal N, Umar S, Khan NA, Khan MIR (2014) A new perspective of phytohormones in salinity tolerance: regulation of proline metabolism. Environ Exp Bot 100:34–42

Iqbal Z, Hussain F, Naqvi M (1998) Ammonia volatilization from urea applied to salt-affected soils under flooded and non-flooded conditions. Pak J Biol Sci 1(4):372–375

Jacoby B, Moran N (2001) Mineral nutrient transport in plants. Handbook of plant and crop physiol. The Hebrew University of Jerusalem, Rehovot, Israel

Jalali M (2005) Nitrates leaching from agricultural land in Hamadan, western Iran. Agric Ecosyst Environ 110(3):210–218

Jalali M, Rowell D (2003) The role of calcite and gypsum in the leaching of potassium in a sandy soil. Exp Agric 39(04):379–394

Jouyban Z (2012) The effects of salt stress on plant growth. Tech J Engin App Sci 2(1):7–10
Khan G (1998) Soil salinity/sodicity status in Pakistan. Soil survey of Pakistan, Lahore 59
Khan M, Panda S (2008) Alterations in root lipid peroxidation and antioxidative responses in two rice cultivars under NaCl-salinity stress. Acta Physiol Plant 30(1):81–89
Khan N, Khan M (2014) The ethylene: from sensescence hormone to key player in plant metabolism. J Plant Biochem Physiol 2:124
Kronzucker HJ, Glass AD, Siddiqi MY (1999) Inhibition of nitrate uptake by ammonium in barley. Analysis of component fluxes. Plant Physiol 120(1):283–292
Kumar RG, Shah K, Dubey R (2000) Salinity induced behavioural changes in malate dehydrogenase and glutamate dehydrogenase activities in rice seedlings of differing salt tolerance. Plant Sci 156(1):23–34
Kusano M, Tohge T, Fukushima A, Kobayashi M, Hayashi N, Otsuki H, Kondou Y, Goto H, Kawashima M, Matsuda F (2011) Metabolomics reveals comprehensive reprogramming involving two independent metabolic responses of Arabidopsis to UV-B light. The Plant J 67(2):354–369
Lægreid M, Bockman OC, Kaarstad O (1999) Agriculture, fertilizers and the environment. CABI publishing/Norsk Hydro ASA, Wallingford/Oslo
Leidi E, Silberbush M, Lips S (1991) Wheat growth as affected by nitrogen type, pH and salinity. I. Biomass production and mineral composition. J Plant Nutr 14(3):235–246
Liu G, Wu W, Zhang J (2005) Regional differentiation of non-point source pollution of agriculture-derived nitrate nitrogen in groundwater in northern China. Agric Ecosys Environ 107(2):211–220
Lu C, Zhang J (1998) Thermostability of photosystem II is increased in salt-stressed sorghum. Funct Plant Biol 25(3):317–324
Maas EV, Grattan S, Skaggs R, Schilfgaarde JV (1999) Crop yields as affected by salinity. Agric Drain 38:55–108
Maggio A, Raimondi G, Martino A, De Pascale S (2007) Salt stress response in tomato beyond the salinity tolerance threshold. Environ Exper Bot 59(3):276–282
Mandhania S, Madan S, Sawhney V (2006) Antioxidant defense mechanism under salt stress in wheat seedlings. Biologia Plantarum 50(2):227–231
Mane A, Karadge B, Samant J (2010) Salinity induced changes in photosynthetic pigments and polyphenols of Cymbopogon Nardus (L.) Rendle. J Chem Pharmaceut Res 2(3):338–347
Marchuk AG, Rengasamy P (2010) Cation ratio of soil structural stability (CROSS). In: Proceedings 19th world congress of soil science 2010. CSIRO Publishing, Brisbane, pp 5981–5983
Marschner H (1995) Functions of mineral nutrients: macronutrients. In: Mineral nutrition of higher plants, 2nd edn. Academic, New York, pp 299–312
Martínez-Ballesta MC, Silva C, Lopez-Berenguer C, Cabañero F, Carvajal M (2006) Plant aquaporins: new perspectives on water and nutrient uptake in saline environment. Plant Biol 8(5):535–546
Martinez V, Cerda A (1989) Influence of N source on rate of Cl, N, Na and K uptake by cucumber seedlings grown in saline condition. J Plant Nutr 12(8):971–983
Mishra A, Sharma S, Pandey R, Mishra L (2004) Amelioration of a highly alkaline soil by trees in northern India. Soil Use Manage 20(3):325–332
Møller IS, Tester M (2007) Salinity tolerance of Arabidopsis: a good model for cereals? Trends Plant Sci 12(12):534–540
Moradi F, Ismail AM (2007) Responses of photosynthesis, chlorophyll fluorescence and ROS-scavenging systems to salt stress during seedling and reproductive stages in rice. Ann Bot 99(6):1161–1173
Mostafazadeh-Fard B, Heidarpour M, Aghakhani A, Feizi M (2008) Effects of leaching on soil desalinization for wheat crop in an arid region. Plant Soil Environ 54(1):20
Munns R (2002) Comparative physiology of salt and water stress. Plant Cell Environ 25(2):239–250

Munns R, James RA, Läuchli A (2006) Approaches to increasing the salt tolerance of wheat and other cereals. J Exper Bot 57(5):1025–1043
Munns R, Termaat A (1986) Whole-plant responses to salinity. Funct Plant Biol 13(1):143–160
Munns R, Tester M (2008) Mechanisms of salinity tolerance. Annu Rev Plant Biol 59:651–681
Murtaza G (2011) Optimization of nitrogen use-efficiency in rice-wheat cropping system on tile drained salt-affected soils with different textures. First annual technical progress report for higher education commission (HEC) funded project (July 2010 to June 2013). Institute of Soil and Environmental Sciences University of Agriculture, Faisalabad, Pakistan
Murtaza G, Ghafoor A, Qadir M (2006) Irrigation and soil management strategies for using saline-sodic water in a cotton–wheat rotation. Agric Water Manage 81(1):98–114
Nazar R, Khan MIR, Iqbal N, Masood A, Khan NA (2014) Involvement of ethylene in reversal of salt-inhibited photosynthesis by sulfur in mustard. Physiol Plant 152(2):331–344
Nemati I, Moradi F, Gholizadeh S, Esmaeili M, Bihamta M (2011) The effect of salinity stress on ions and soluble sugars distribution in leaves, leaf sheaths and roots of rice (*Oryza sativa* L.) seedlings. Plant Soil Environ 57(1):26–33
Niu X, Bressan RA, Hasegawa PM, Pardo JM (1995) Ion homeostasis in NaCl stress environments. Plant Physiol 109(3):735
Pagter M, Bragato C, Malagoli M, Brix H (2009) Osmotic and ionic effects of NaCl and Na 2 SO 4 salinity on Phragmites australis. Aquat Bot 90(1):43–51
Paranychianakis NV, Aggelides S, Angelakis AN (2004) Influence of rootstock, irrigation level and recycled water on growth and yield of Soultanina grapevines. Agric Water Manage 69(1):13–27
Parida AK, Das AB (2005) Salt tolerance and salinity effects on plants: a review. Ecotoxicol Environ Saf 60(3):324–349
Pessarakli M (2010) Handbook of plant and crop stress. CRC Press, Boca Raton/Florida
Pitman MG, Läuchli A (2002) Global impact of salinity and agricultural ecosystems. In: Salinity: environment-plants-molecules. Springer, New York, pp 3–20
Qadar A (2009) Alleviation of salt stress through nutrient management. Enhancing nutrient use efficiency in problem soils. Central Soil Salinity Res Inst, Karnal
Qadir M, Oster J, Schubert S, Noble A, Sahrawat K (2007) Phytoremediation of sodic and saline-sodic soils. Adv Agron 96:197–247
Qadir M, Schubert S (2002) Degradation processes and nutrient constraints in sodic soils. Land Degrad Dev 13(4):275–294
Qadir M, Schubert S, Ghafoor A, Murtaza G (2001) Amelioration strategies for sodic soils: a review. Land Degrad Dev 12(4):357–386
Quintero JM, Fournier JM, Benlloch M (2007) Na^+ accumulation in shoot is related to water transport in K^+-starved sunflower plants but not in plants with a normal K^+ status. J Plant Physiol 164(1):60–67
Rains D, Goyal S (2003) Strategies for managing crop production in saline environments: an overview. J Crop Prod 7(1–2):1–10
Rengasamy P (2010) Soil processes affecting crop production in salt-affected soils. Funct Plant Biol 37(7):613–620
Ruiz JM, Rivero RM, Romero L (2007) Comparative effect of Al, Se, and Mo toxicity on NO_3^- assimilation in sunflower (*Helianthus annuus* L.) plants. J Environ Manage 83(2):207–212
Russell JM (2000) Sodium-potassium-chloride cotransport. Physiol Rev 80(1):211–276
Sairam R, Tyagi A (2004) Physiology and molecular biology of salinity stress tolerance in plants. Curr Sci-Bangalore 86(3):407–421
Salemi E, Mastrocicco M, Colombani N, Aschonitis V (2012) Comparison of three different methods for groundwater intrinsic vulnerability mapping in the Ferrara province, Italy
Schijven J, Hassanizadeh SM, de Roda Husman AM (2010) Vulnerability of unconfined aquifers to virus contamination. Water Res 44(4):1170–1181
Serrano R (1996) Salt tolerance in plants and microorganisms: toxicity targets and defense responses. Int Rev Cytol 165:1–52

Shabala L, Cuin TA, Newman IA, Shabala S (2005) Salinity-induced ion flux patterns from the excised roots of Arabidopsis sos mutants. Planta 222(6):1041–1050

Shainberg I, Levy G, Goldstein D, Mamedov A, Letey J (2001) Prewetting rate and sodicity effects on the hydraulic conductivity of soils. Soil Res 39(6):1279–1291

Sharma BR, Minhas P (2005) Strategies for managing saline/alkali waters for sustainable agricultural production in South Asia. Agric Water Manag 78(1):136–151

Shi WM, Xu WF, Li SM, Zhao XQ, Dong GQ (2010) Responses of two rice cultivars differing in seedling-stage nitrogen use efficiency to growth under low-nitrogen conditions. Plant Soil 326(1–2):291–302

Singh G, Qadar A, Yaduvanshi N, Dey P (2009) Enhancing nutrient use efficiency in problem soils. Central Soil Salinity Research Institute, Karnal

Siringam K, Juntawong N, Cha-um S, Kirdmanee C (2009) Relationships between sodium ion accumulation and physiological characteristics in rice (Oryza sativa L. spp. indica) seedlings grown under iso-osmotic salinity stress. Pak J Bot 41(4):1837–1850

Surabhi G-K, Reddy AM, Kumari GJ, Sudhakar C (2008) Modulations in key enzymes of nitrogen metabolism in two high yielding genotypes of mulberry (Morus alba L.) with differential sensitivity to salt stress. Environ Exp Bot 64(2):171–179

Tabatabaie SJ, Nazari J (2007) Influence of nutrient concentrations and NaCl salinity on the growth, photosynthesis, and essential oil content of peppermint and lemon verbena. Turk J Agric For 31(4):245–253

Taiz L, Zeiger E, Møller IM, Murphy AS (2015) Plant physiology and development. Sinauer Associates, Sunderland

Tanji KK (2002) Salinity in the soil environment. In: Salinity: environment-plants-molecules. Springer, Dordrecht, pp 21–51

Tester M, Davenport R (2003) Na+tolerance and Na+transport in higher plants. Ann Bot 91(5):503–527

Touchette BW, Burkholder JM (2007) Carbon and nitrogen metabolism in the seagrass, Zostera marina L.: environmental control of enzymes involved in carbon allocation and nitrogen assimilation. J Exp Marine Biol Ecol 350(1):216–233

Tshivhandekano T, Lewis O (1993) Differences in response between nitrate-and ammonium-fed maize to salinity stress and its amelioration by potassium. S Afr J Bot 59:597–597

Tuna AL, Kaya C, Ashraf M, Altunlu H, Yokas I, Yagmur B (2007) The effects of calcium sulphate on growth, membrane stability and nutrient uptake of tomato plants grown under salt stress. Environ Exp Bot 59(2):173–178

Van de Graaff R, Patterson RA (2001) Explaining the mysteries of salinity, sodicity, SAR and ESP in on-site practice. In: On-site '01 conference: advancing on-site wastewater systems, University of Armidale, New England, September. pp 25–27

Vitousek PM, Aber JD, Howarth RW, Likens GE, Matson PA, Schindler DW, Schlesinger WH, Tilman DG (1997) Human alteration of the global nitrogen cycle: sources and consequences. Ecol Appl 7(3):737–750

Wallender WW, Tanji KK (2011) Agricultural salinity assessment and management. American Society of Civil Engineers (ASCE), Reston

Wang H, Wu Z, Chen Y, Yang C, Shi D (2011) Effects of salt and alkali stresses on growth and ion balance in rice (*Oryza sativa* L.). Plant Soil Environ 57(6):286–294

Wang H, Wu Z, Zhou Y, Han J, Shi D (2012) Effects of salt stress on ion balance and nitrogen metabolism in rice. Plant Soil Environ 58(2):62–67

Wang W-H, Köhler B, Cao F-Q, Liu L-H (2008) Molecular and physiological aspects of urea transport in higher plants. Plant Sci 175(4):467–477

Ward MH, Zahm SH, Weisenburger MDD, Gridley MG, Cantor KP, Saal MRC, Blair A (1994) Dietary factors and non-Hodgkin's lymphoma in Nebraska (United States). Cancer Causes Control 5(5):422–432

Warrence NJ, Bauder JW, Pearson KE (2002) Basics of salinity and sodicity effects on soil physical properties. Departement of Land Resources and Environmental Sciences, Montana State University-Bozeman, Bozeman

Watanabe T, Son TT, Hung NN, Van Truong N, Giau TQ, Hayashi K, Ito O (2009) Measurement of ammonia volatilization from flooded paddy fields in Vietnam. Soil Sci Plant Nutr 55(6):793–799

Wickert E, Marcondes J, Lemos MV, Lemos EG (2007) Nitrogen assimilation in Citrus based on CitEST data mining. Genet Mol Biol 30(3):810–818

Witte C-P (2011) Urea metabolism in plants. Plant Sci 180(3):431–438

Worrall F, Kolpin DW (2004) Aquifer vulnerability to pesticide pollution—combining soil, land-use and aquifer properties with molecular descriptors. J Hydrol 293(1):191–204

Woyema A, Bultosa G, Taa A (2012) Effect of different nitrogen fertilizer rates on yield and yield related traits for seven durum wheat (*Triticum turgidum* L. var Durum) cultivars grown at Sinana, South Eastern Ethiopia. Afr J Food Agric Nutr Dev 12(3):6079–6094

Xiao-Hua L, Huang Z-R, Huang Y-L, Jian K, Zhang Z-H, Zhao-Pu L (2010) Response of two Jerusalem artichoke (*Helianthus tuberosus*) cultivars differing in tolerance to salt treatment. Pedosphere 20(4):515–524

Yuan B-C, Li Z-Z, Liu H, Gao M, Zhang Y-Y (2007) Microbial biomass and activity in salt affected soils under arid conditions. Appl Soil Ecol 35(2):319–328

Zeng L, Shannon MC (2000) Salinity effects on seedling growth and yield components of rice. Crop Sci 40:996–1003

Zeng L, Shannon MC, Lesch SM (2001) Timing of salinity stress affects rice growth and yield components. Agric Water Manag 48(3):191–206

Zhang J-L, Flowers TJ, Wang S-M (2010) Mechanisms of sodium uptake by roots of higher plants. Plant Soil 326(1–2):45–60

Zhenghu D, Honglang X (2000) Effects of soil properties on ammonia volatilization. Soil Sci Plant Nutr 46(4):845–852

Zhu J-K (2000) Genetic analysis of plant salt tolerance using Arabidopsis. Plant Physiol 124(3):941–948

Zhu J-K (2001) Plant salt tolerance. Trends Plant Sci 6(2):66–71

Zhu X, Yang J, Liang Y, Lou Y, Yang X (2004) Effects of exogenous calcium on photosynthesis and its related physiological characteristics of rice seedlings under salt stress. Sci Agri Sin 37(10):1497–1503

Management of Acid Sulfate Soils for Sustainable Rice Cultivation in Malaysia

Qurban Ali Panhwar, Umme Aminun Naher, Jusop Shamshuddin, Othman Radziah, and Khalid Rehman Hakeem

Contents

Abstract Acid sulfate soils are an alternative aspect for the sustainable rice cultivation. Mostly these soils are not suitable for the crop production unless are effectively improved. Acid sulfate soils have toxicity due to the presence of high Aluminum (Al), and iron (Fe) with low pH (<4.0). Usually these types of soils have nutrient deficiency especially in phosphorus, which causes poor plant growth and development. The soils need to be improved with some soil amendments like application of basalt, ground magnesium limestone and organic materials (biofertilizer) that can increase the soil pH, improve soil nutrients and reduce the Al and Fe toxicity. The application of biofertilizer can enhance the rice plant growth, and yield by producing plant growth hormones (IAA) and organic acids that can chelate Al toxicity and solublize insoluble form of phosphorus in the soil.

Q.A. Panhwar (✉)
Soil Science Division, Nuclear Institute of Agriculture, Tandojam, 70060 Sindh, Pakistan

Department of Land Management, Faculty of Agriculture, Universiti Putra Malaysia, 43400 Serdang, Selangor, Malaysia
e-mail: pawhar107@yahoo.com

U.A. Naher
Soil Science Division, Bangladesh Rice Research Institute, Gazipur 1701, Bangladesh

J. Shamshuddin • O. Radziah
Department of Land Management, Faculty of Agriculture, Universiti Putra Malaysia, 43400 Serdang, Selangor, Malaysia

K.R. Hakeem
Faculty of Forestry, Universiti Putra Malaysia, 43400 Serdang, Selangor, Malaysia

K.R. Hakeem et al. (eds.), *Soil Science: Agricultural and Environmental Prospectives*, DOI 10.1007/978-3-319-34451-5_4

Hence, using ground magnesium limestone, basalt and biofertilizer, rice cultivated on acid sulfate soils can produce yield equivalent to that of the granary areas of Malaysia.

Keywords Aluminum • Basalt • Biofertilizer • Ground magnesium limestone • Iron toxicity

1 Introduction

Rice is a staple food of Malaysia and currently its self-sufficiency level is 72 %. It is required that the production of rice can be increased for the reason of food security (Shamshuddin et al. 2014; Hakeem et al. 2011). Rice production can be enhanced by increasing the area planted to rice or by increasing the yield per unit area. With no expansion in area to be expected and the slowing down in yield increase, the growth of rice production has fallen below market demand. Farmers have to depend partly on the productivity of marginal soils for their rice supply. One of the soils targeted for rice cultivation in Peninsular Malaysia is acid sulfate soil (Fig. 1). The soils which are acidic have soil infertility is a main cause to crop production in both tropical and temperate regions of the world (Sanchez 1976; Von Uexküll and Mutert 1995). In Malaysia, acid sulfate soils mainly occur in the coastal plains of the west coast states of Peninsular Malaysia and Sarawak (Borneo Island). Acid sulfate soils are normally not suitable for the crop production unless they are properly ameliorated and their fertility improved (Shamshuddin et al. 2014). Due to the presence of excess amounts of Al and/or Fe in the soils, available phosphorus is mostly fixed

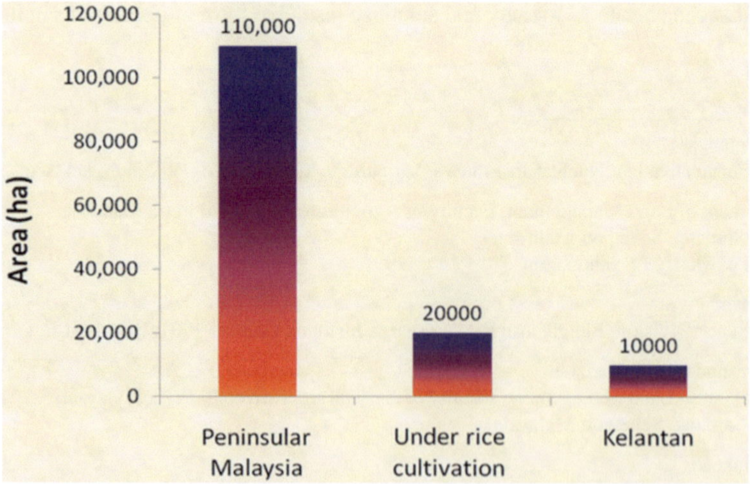

Fig. 1 Acid sulfate soils in Malaysia

(Dent 1986). The soil is characterized by the presence of pyrite (FeS_2), which produces high acidity when the areas having the soils are open up for development. Many of these soils have nutrient deficiency, especially phosphorus which causes poor plant growth. Their pH is low (<3.5) and they contain toxic amounts of Al and/or Fe, which are toxic to crop, such as rice (Bian et al. 2013).

2 Constrains in Acid Sulfate Soils

Acid sulfate soils have mainly the Al and Fe in high quantities and due to its access amount they become toxic for many of the plants including rice. Therefore, plays can not grow well and soils will be not suitable for the crop cultivation unless could be developed.

2.1 Aluminum Toxicity

Aluminum toxicity is one of the main factors that influence the plant root growth. In acid sulfate soils, Al^{3+} restricts the growth of roots either by inhibition of cell division, cell elongation or due to both (Marschner 1991). The Al ions concentration

Fig. 2 Aluminum toxicity symptoms on rice plants

mostly high in the soil solution unless the pH is <5.0 (Zhang et al. 2007); these soils contains more than 40% of the potentially arable soils in the world (Delhaize et al. 2004). The soil acidity problem is intensified by the use of NH_4 fertilizers and acidic rain water (von Uexkull and Murtert 1995). The major symptom of the Al toxicity can be shown by rapid inhibition of root growth that may result to a reduction in vigor and crop yields (Kochian et al. 2005).

The high exchangeable Al can be a cause for the shallow rooting, poor usage of soil nutrients (Fig. 2) and toxicity of Al (Wright1989). Furthermore, it was also perceived that exchangeable Al in a managed Ultisol is lower compared to a virgin soil (Hardy et al. 1990). Neutralization of this kind of Al has been entirely investigated worldwide (Guadalix and Pardo 1995). The plants have different mechanisms to resist or tolerate the toxic effect of Al under Al toxicity condition. These resistance type of mechanisms in plants have been external or exclusion through exudation of organic acids from the radical apexes. The subsequent chelation of the roots in the rhizosphere and internal or Al-tolerant since Al chelation is produced inside the cell and then later stored and classified in organelles like the vacuole (Kochian 1995; Ramgareeb et al. 2004).

2.1.1 Aluminum Tolerance Mechanisms in Plants

Aluminum tolerance mechanisms have been identified into two types: (i) those that exclude Al from the root cells; and (ii) those that allow Al to be tolerated once it has entered the plant cells (Barceló and Poschenrieder 2002). Plant species in tropical places are very resistant to Al stress and few of these species can be able to accumulate high concentrations of Al in the leaves, greater than 1% of their dry weight (Jones and Ryan 2003). On the other hand, cereals like Secale cereale, *Zea mays*, *Hordeum vulgare*, *Triticum aestivum*, *X Triticosecale*, *Sorghum bicolor* and *Avena sativa* do not accumulate internally high Al concentrations, but rather use the Al exclusion mechanism through organic acid exudation (Ma et al. 2004; Caniato et al. 2007). This would be one of the greatest widely used mechanisms by many of the plant species studied. However, how Al causes rapid inhibition of cell elongation and many different mechanisms of Al toxicity have been predicted (Kochian 1995). The Al may interact with the root cell wall, disrupt the plasma membrane and inhibit transport processes on the plasma membrane (Fig. 3). Whereas the mechanisms responsible for Al-induced inhibition of root elongation are difficult, all these inhibitory effects result from the binding of Al with extracellular and intracellular substances.

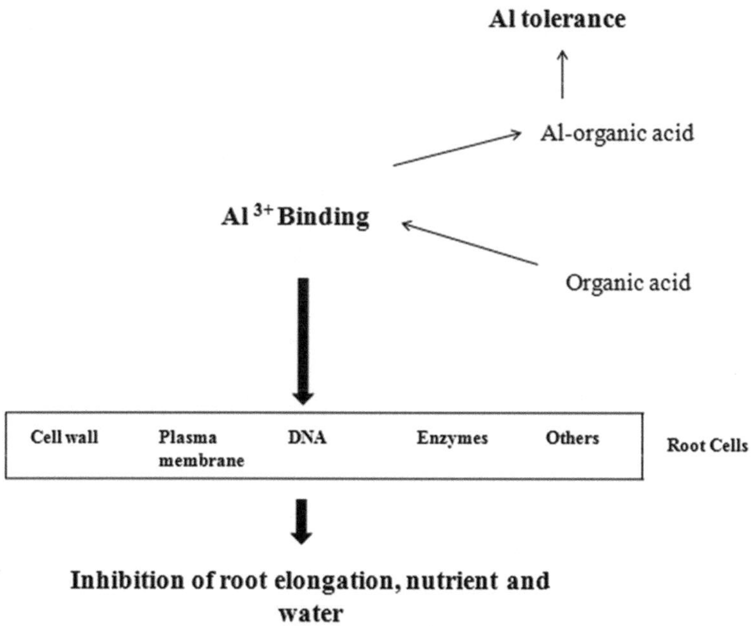

Fig. 3 Schematic representation of Al toxicity and the role of Al-chelating substance such as organic acids in detoxifying Al

2.2 *Iron Toxicity*

Among the toxicities, iron toxicity is well-recognized as the most widely distributed nutritional disorder in lowland-rice production. Critical Fe concentration varies from 0.05 to 5.37 $cmol_c\ kg^{-1}$ soil (Dobermann and Fairhurst 2000), implying that Fe may not be the only source soil toxicity that causes reduction in yield. Furthermore, in acid soils, it is one of the important restrictions particularly for rice production with Zn deficiency. Whereas, the most commonly observed micronutrient disorder in wetland rice (Neue et al. 1998).

There are several soil types that can be iron-toxic, including acid sulfate soils (Tinh 1999), acid clay soils (Alaily et al. 1998), peat soils (Deturck 1994), and valley-bottom soils receiving interflow water from contiguous slopes (Sahrawat and Diatta 1995). The Fe^{2+} concentrations in the soil solution that reportedly affect lowland-rice yields from 10 to >2000 mg L^{-1}. Iron-induced yield reduction is frequently related with a poor soil nutrient status (Benckiser et al. 1983) or with accumulation of respiration inhibitors. Hence, iron toxicity may be described as a multiple nutritional disorder hastened by, but also increasing conditions of P, K, and Zn deficiency and H_2S toxicity (Ottow et al. 1993).

The entrance of iron-toxicity symptoms in rice needs excessive uptake of Fe^{2+} by roots and its acropetal translocation via xylem flow into the leaves. Inside the leaf,

Fig. 4 Iron toxicity on rice crop cultivated under acid sulfate soils

excess amounts of Fe^{2+} cause an prominent production of radicals that can irreversibly damage cell structural components (Thompson and Ledge 1987) and lead to an accumulation of oxidized polyphenols (Yamauchi and Peng 1993). The typical visual iron toxic symptom related with those processes like "bronzing" of the rice leaves (Howeler 1973). Rice-yield losses connected with the presence of bronzing symptoms commonly range from 15 to 30 %. Nevertheless, in the case of severe Fe^{2+} toxicity, may affect the complete crop failure can occur (Audebert and Sahrawat 2000). Overcoming rice-yield reductions caused by iron toxicity requires both adapted and tolerant varieties as well as appropriate management interventions.

The iron toxicity is a form that can be expressed in rice plants and joined with an excessive amount of Fe^{2+} in the soil (Fig. 4). While it may happen in a wide range of soil types, general characteristics shared by most Fe-toxic soils are great amounts of reducible Fe, low pH, and low CEC and exchangeable K content (Ottow et al. 1982). These may be associated with P and Zn deficiency and H_2S toxicity. Furthermost, Fe toxicity is linked to water logging and only occurs under anoxic soil conditions (Ponnamperuma et al. 1967). Paddy rice soils are exposed to periodic changes between oxic and anoxic conditions. Since oxygen diffuses in air about 10^3 to 10^4 times faster than in water or in water-saturated soils (Armstrong 1979). The oxygen depleted rapidly by the respiration of soil microorganisms and plant roots in waterlogged soils (Prade et al. 1990). With the depletion in oxygen, NO, Mn^{4+}, Fe and SO_4^{2-} can play as an electron acceptors for microbial respiration and sequentially become reduced in a flooded rice field. Oxygen and nitrate are used within the first hours or days of flood. The subsequent reduction of Mn^{4+} also occurs quickly, since the Mn content of soils is frequently low (Ponnamperuma 1977). Within a few days after flood or at a redox potential of <180–150 mV, the reduction of Fe^{3+} begins (Patrick and Reddy 1978).

In the presence of organic material, Fe^{2+} toxicity can occur due to the consequent reduction of Fe^{3+} under flooded soil conditions (Tran and Vo 2004). The plants mostly rice have the capability of taking up more iron than most of the other plants and Fe^{2+} is the iron species prevailing in paddy fields. Subsequently, Fe^{2+} is easily taken up, the uptake mechanism of Fe^{2+} is probably of less important in the flooded environments (Mengel 1979). After uptake into the root cortex, the reduced Fe^{2+} can get enter in the xylem after a symplastic passage via Casparian strip. However, a huge share of Fe^{2+} may enter in the xylem directly via an apoplastic bypass (Asch 1997) or after the cause of root injury from the pulling and transplanting of the seedling. In the xylem, Fe^{2+} follows the transpiration stream in the acropetal long-distance transport.

3 Management of Acid Sulfate Soils

Acid sulfate soils are difficult to manage for crop production. A part from being acidic, they contain toxic amount of Al and Fe. Utilization of acid sulfate soils requires approval of the regulatory authorities so that there is no unacceptable environmental harm occurs (Sullivan 2012). In recent years, food security has become a big concern in the less developed part of the world. The supply of rice, the staple food in Asia, is sometimes interrupted by the environmental conditions. Rice production can be increased by expanding the area cropped to rice or by improving its productivity. With no expansion in area under production in sight and the slowing down in yield increase, the growth of rice production has fallen below demand, and this is exacerbated by the population increase. As such, farmers have resorted to growing rice on the marginal or degraded lands. One of the areas that can be used for rice production is acid sulfate soils which are abundant in Southeast Asia, occurring almost exclusively in its coastal plains (Shamshuddin 2006; Anda et al. 2009). For sustainable cultivation of rice on acid sulfate soils, proper water management is very crucial (Hanhart and Duong Van Ni 1993; Elisa Azura 2012).

Irrigation with fresh water and leaching can help eliminate acid water from the area under rice cultivation. Mostly farmers are cultivating MR 219 rice variety and its yield is about 3.80 t ha^{-1} in acid sulfate soil, using typical farmers' practice, this is far behind from the national demand due to some of the paddy fields in the area are deserted by the farmers (Enio et al. 2011).

A study was conducted in acid sulfate soils at IRRI Philippines in continuous flooding and flooding with soil drying to determine the effect of iron toxicity in rice (IR-32). There was no significant difference found in plant growth up to the second crop under both water treatments due to iron toxicity. There was better growth of rice found under the continuous water regime from third cropping. This might be due to the continuous flooding reduced Fe toxicity level. On other hand, in drying and re-flooding soils contained higher amounts of Fe in soil solution that becomes toxic to the rice plant (Sahrawat 1979).

Rice can be grown very well at soil pH above 6. Acid sulfate soils are normally not suitable for crop production unless they are properly ameliorated. This soil fertility restriction can be improved effectively by applying amendments (Suswanto et al. 2007; Shazana et al. 2013), such as dolomitic limestone (GML), basalt and organic materials (bio-fertilizer).

The normal agronomic practice worldwide to succeed acid sulfate soils for crop production is the liming. In Malaysia, some parts of acid sulfate soils have been reclaimed for rice. The addition of this lime [$(Ca,Mg)(CO_3)_2$] would enhance soil pH accordingly, with concomitant addition of Ca and Mg into the soil. For 2nd season of the trial, soil pH increased linearly with increasing exchangeable Ca (Shazana et al. 2013).

GML ameliorated in the soil according to the following reactions:

$$(Ca,Mg)(CO_3)_2 \rightarrow Ca^{2+} + Mg^{2+} + CO_3^{2-} \quad (1)$$

$$CO_3^{2} + H_2O \rightarrow HCO_3^- + OH^- \quad (2)$$

$$Al^{3+} + 3OH^- \rightarrow Al(OH)_3 \quad (3)$$

The GML dissolved quickly on applying it into the acidic soil, releasing Ca and Mg (Eq. 1) and these macronutrients can be taken up by the growing rice plants. Consequently, hydrolysis of CO_3^- (Eq. 2) would produce hydroxyls that neutralized Al by forming inert Al-hydroxides (Eq. 3). Soil pH is highly correlated with exchangeable Ca (Fig. 5). Soil pH increased significantly following reduction in exchangeable Al (Fig. 6).

As for basalt, it dissolves according to the following equation (De Coninck 1978):

$$4Mg^{2+} + 4SiO_4^{4-} + 4H_2O \rightarrow 4Mg^{2+} + Si(OH)_4 + 4OH^-$$

The above equation shows that the hydrolysis of silicate produces high amount of hydroxyls that increases soil pH. However, it takes time for ground basalt to dissolve completely even under acid sulfate soil condition (Shazana et al. 2013). That means basalt should be applied way ahead of seeding/transplanting. One of the eventual products of basalt dissolution is silicic acid, the form of Si that rice plant can take up (Bokhtiar et al. 2010). The basalt (containing silicate) has potential to increase rice yield significantly (Nagabovanalli et al. 2009) by the presence of calcium silicate. Additionally, for rice cultivation silicate is better in terms of ameliorative effects compared to that of calcium carbonate.

Moreover, acid sulfate soils can also be ameliorated, to some extent, by the addition of organic materials (Muhrizal et al. 2003), like, biofertilizer. Application of

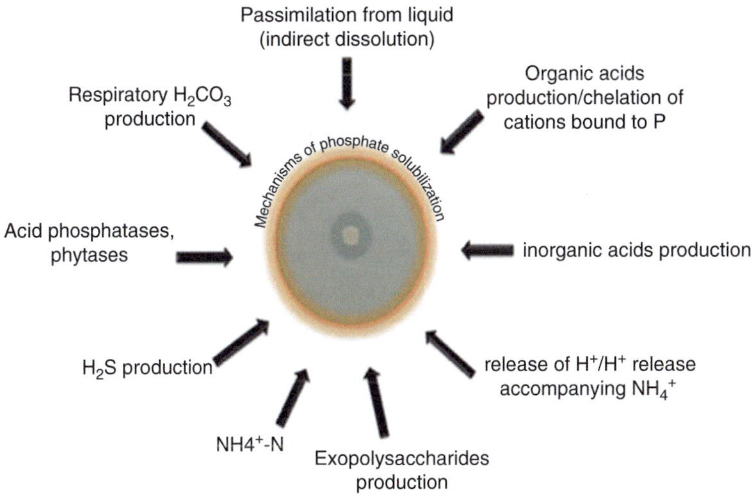

Fig. 5 Mechanisms of P-solubilization by PGPR (Source: Adapted from Zaidi et al. (2009))

organic matter in Al-toxic soils upsurges yield by detoxification of Al through the increase of soil pH and complexation of Al by organic matter (Hue and Amien 1989). The minor increase in soil pH can be due to release of NH_3 through decomposition of organic matter as being described (Hoyt and Turner 1975). The use of organic-based biofertilizer can supply essential nutrient elements to the crop plant and improve soil organic carbon reserve (Panhwar et al. 2014a). Farmers are using huge amount of chemical fertilizers to booster crop production, which is a risk for environment and one of the important factors for causing climate change. Among the major nutrient elements, nitrogen fertilizer is applied in large quantities. However, plants are able to use about 50 % of the applied N fertilizer, while 25 % is lost from the soil-plant system causing environment pollution such as depletion of stratospheric ozone due to NO_2 emission and ground water contamination through NO_3 leaching (UNESCO-SCOPE 2013).

Acid sulfate soils contain low amount of total microorganisms, with the population varying according to vegetation and the amount of amelioration. Hence, addition of potential plant growths promoting rhizobacteria (PGPR) might be effective for the enhancing crop yield. It is proven that high number of PGPR is associated with plant rhizosphere (Naher et al. 2009; Panhwar et al. 2011).

Application of PGPR in combination with PR application increased P in soybean tissue, resulting in higher plant height, nutrient uptake compared with PR alone (Jintaridth et al. 2006). In addition, the application of these beneficial microorganisms

Table 1 Effect of different amendments on rice plants

Treatments	Plant height (cm)	Chlorophyll (SPAD value)	Tillers plant^{-1}	Plant biomass (g pot^{-1})	Weight of grains pot^{-1} (g)
Control	78.17c	34.53c	4b	14.30d	8.10d
Biofertilizer	90.83a	42.30a	6a	26.53a	17.92a
GML	85.83b	41.73b	5b	18.98b	14.80b
Basalt	88.50a	41.13b	5b	16.07c	9.09c

Means within the same column followed by the same letters are not significantly different at $p<0.05$
Source: Modified from Panhwar et al. (2013)

Fig. 6 Rice cultivated with the biofertilizer and ground magnesium limestone (GML) application under acid sulfate soils at Semerak, Kelantan, Malaysia

Fig. 7 Rice cultivated without amendments under acid sulfate soils at Semerak, Kelantan, (MalaysiaSource: Adapted from Zaidi et al. (2009))

enhanced the economic efficiency in terms of reduced production cost of P fertilizers (Ngoc Son et al. 2006). The mechanism for P solubilization is shown in Fig. 5.

Application of bio-fertilizer and other amendments has been applied on rice soils (acid sulfate) at Semerak, Kelantan, Malaysia. Application of GML, basalt and bio-fertilizer significantly increased plant height, leaf chlorophyll content and yield of rice (Table 1). This might be the resultant effect of increasing soil pH which reduced Al and/Fe toxicity. The applied bio-fertilizer contained a consortium of nitrogen fixing and phosphate solubilizing bacteria which were able to produce significant amount of organic acids and growth promoting phytohormone (Panhwar et al. 2014b). The organic acids were able to chelate the Al and reduced Al-toxicity. Phosphate solubilizing bacteria made the insoluble P to become soluble and available to plant, while the phytohormone (indoleacetic acid) improved root growth and increased nutrient uptake which increased the yield of rice (Panhwar et al. 2013). It is expected that after application of bio-fertilizer, microbes produce high amounts of organic acids at plant rhizosphere that will increase P uptake and simultaneously will reduce Al toxicity surrounding the root tissue (Figs. 6, 7).

4 Summary and Future Prospect

The acid sulfate soils are difficult to manage because they have low pH and contain the higher concentrations of Al and/or Fe. Rice yield on acid sulfate soils in Malaysia is very low, presumably due to Al and/or Fe toxicity. Multiple mechanisms of Al tolerance in rice have been documented and the secretion of organic acids with Al-chelating capacity from the root tips has been considered as an important one. Using ground magnesium limestone, basalt and organic materials like biofertilizer, rice cultivated on acid sulfate soils can produce yield equivalent to that of the granary areas of Malaysia.

Acknowledgements We wish to acknowledge Universiti Putra Malaysia and the Ministry of Education, Malaysia (the research is funded by LRGS – Food Security) for financial and technical supports.

References

Alaily F, Sib J, Madrid K, Soubeiga Z, Roth C (1998) Paddy soils derived from basaltic colluvium in the highland of Madagascar. J Plant Nutr Soil Sci 161:633–637

Anda M, Shamshuddin J, Fauziah CI, Syed Omar SR (2009) Dissolution of ground basalt and its effect on an Oxisol chemical properties and cocoa growth. Soil Sci 174:264–271

Armstrong W (1979) Aeration in higher plants. Adv Bot Res 7:226–332

Asch F (1997) Physiologische Faktoren der Salzresistenz bei Reis (*Oryza sativa* L.). Ph.D thesis, University of Hamburg, Germany

Audebert A, Sahrawat KL (2000) Mechanisms for iron toxicity tolerance in lowland rice. J Plant Nutr 23:1877–1885

Barcelo J, Poschenrieder C (2002) Fast root growth responses, root exudates, and internal detoxification as clues to the mechanisms of aluminium toxicity and resistance: a review. Environ Exp Bot 48:75–92

Benckiser G, Ottow JCG, Santiago S, Watanabe I (1983) Iron toxicity – Effect of phosphorus, potassium, calcium and magnesium fertilization on rhizoflora, redox potential and iron uptake by different rice varieties (*Oryza sativa* L.). Landw. Forschung 36:285–299

Bian M, Zhou M, Sun D, Li C (2013) Molecular approaches unravel the mechanism of acid soil tolerance in plants. Crop J 1:91–104

Bokhtiar SM, Hai-Rong H, Yang-Rui Li, Delvi VA (2010) Effect of silicon on yield contributing parameters and its accumulation in abaxial epidermis of sugarcane leaf blades using energy dispersive X-ray analysis. In: Proceedings of the international conference on balanced nutrient management for tropical agriculture. April 12–16, 2010; Pahang: Malaysia

Caniato FF, Guimaraes CT, Shaffert RE, Alves VMC, Kochian LV, Borem A, Klein PE, Magalhaes JV (2007) Genetic diversity for aluminum tolerance in sorghum. Theor Appl Genet 114:863–876

De Coninck F (1978) Physico-chemical aspects of pedogenesis. International Training Center for Post-Graduate Soil Scientists, Ghent University, Ghent

Delhaize E, Ryan PR, Hebb DM, Yamamoto Y, Sasaki T, Matsumoto H (2004) Engineering high-level aluminum tolerance in barley with the ALMT1 gene. Proc Natl Acad Sci U S A 101:15249–15254

Dent DL (1986) Acid sulfate soil: A baseline for research and development. International Institute for Land Reclamation and Improvement. Volume 39, ILRI Publication. Wageningen, The Netherlands

Deturck P (1994) Iron toxicity to rainfed lowland rice in Sri Lanka. KULFLTB, Leuven

Dobennann A, Fairhurst T (2000) Rice: nutrient disorders and nutrient management. Phosphate Institute of Canada and International Rice Research Institute, Los Banos

Elisa Azura A (2012) Improving the fertility of an acid sulfate soil in Merbok, Kedah, Malaysia for rice cultivation. MS thesis, Universiti Putra Malaysia

Enio Kang MS, Shamshuddin J, Fauziah CI, Husni MHA (2011) Pyritization of the coastal sediments in the Kelantan plains in the Malay peninsula during the Holocene. Am J Agri Biol Sci 134:1–20

Guadalix ME, Pardo MT (1995) Effect of liming on soil properties and P solubility in some spanish Andosol. Geoderma 63:53–61

Hanhart K, Duong Van N (1993) Water management on rice fields at Hoa An in the Mekong Delta. In: Dent DL, Mensvoort MEF (eds) Selected papers of the Ho Chi Minh city symposium on acid sulfate soils. ILRI Publication 53, Wageningen, The Netherlands, pp 161–176

Hardy DH, David Raper Jr C, Miner GS (1990) Chemical restrictions of root in Ultisol subsoils lessened by long-term management. Soil Sci Soc Am J 54:1657–1660

Hakeem KR, Ahmad A, Iqbal M, Gucel S, Ozturk M (2011) Nitrogen-efficient cultivars can reduce nitrate pollution. Environ Sci Pollut Res 18:1184–1193

Howeler RH (1973) Iron-induced oranging disease of rice in relation to physiochemical changes in a flooded Oxisol. Soil Sci Soc Am Proc 37:898–903

Hoyt PB, Turner RC (1975) Effects of organic materials added to very acid soil on pH, aluminum, exchangeable bases, ammonium, and crop yield. Soil Sci 119:227–237

Hue NV, Amien I (1989) Aluminum detoxification with green manures. Commun Soil Sci Plant Anal 20:1499–1511

Jintaridth B, Limthong P, Inthasothi T, Charoenchamratcheep C (2006) Role and the effectiveness of phospho-microorganisms with rock phosphate. In: The 18th world congresss of soil science abstract. July 9–15. Philadelphia, Pennsylvania USA, p 681

Jones DL, Ryan PR (2003) Nutrition. Aluminum toxicity. Encyclopedia of applied plant science, Elsevier Academic Press, London, pp 656–664

Kochian LV (1995) Cellular mechanisms of aluminum toxicity and resistance in plants. Annu Rev Plant Physiol Plant Mol Biol 46:237–260
Kochian LV, Pineros MA, Hoekenga OA (2005) The physiology, genetics and molecular biology of plant aluminum resistance and toxicity. Plant Soil 274:175–195
Ma JF, Nagao S, Sato K, Ito H, Furukawa J, Takaeda K (2004) Molecular mapping of a gene responsible for Al-activated secretion of citrate in barley. J Exp Bot 55:1335–1341
Marschner H (1991) Mechanisms of adaptation of plants to acid soils. Plant Soil 134:1–20
Mengel K (1979) Einführung in den Stoffwechsel der Pflanze. G. Fischer Verlag, Stuttgart/New York
Muhrizal S, Shamshuddin J, Fauziah I, Husni MHA (2003) Alleviation of aluminum toxicity in acid sulfate soils in Malaysia using organic materials. Commun Soil Sci Plant Anal 34:2993–3012
Nagabovanalli PB, Chowdappa N, Nagapa C, Abbinaholalu MC, Thubinakere HH, Siddanagouda PU (2009) Effects of calcium silicate as a source of silicon and as amendment on growth and yield of rice in coastal zone soils of Karnataka, South India. In: Liao H, Xian X, Kochian L (eds) Proceedings of 7th international symposium on plant-soil at Low pH. South China University of Technology Press, pp 170–171
Naher UA, Radziah O, Shamsuddin ZH, Halimi MS, Mohd Razi I (2009) Isolation of diazotrophs from different soils of Tanjong Karang rice growing area in Malaysia. Inter J Agri Biol 11(5):547–552
Neue HU, Quijano C, Senadhira D, Setter T (1998) Strategies for dealing with micronutrient disorders and salinity in lowland rice systems. Field Crop Res 56:139–155
Ngoc Son T, Diep CN, Giang TTM (2006) Effect of bradyrhizobia and phosphate solubilizing bacteria application on soybean in rotational system in the mekong delta. Omonrice 14:48–57
Ottow JCG, Benckiser G, Watanabe I (1982) Iron toxicity of rice as a multiple nutritional soil stress. In: Proceedings of international symposium on distribution, characteristics and utilization of problem soils. Tropical Agriculture Research Center, Ibaraki, pp 167–179
Ottow JCG, Prade K, Bertenbreiter W, Jacq VA (1993) Iron toxicity mechanisms of flooded rice (*Oryza sativa* L.) in Senegal and Indonesia. In: Raunet M (ed) Bas-Fonds et Riziculture. CIRAD-CA, Montpellier, pp 231–241
Panhwar QA, Radziah O, Zaharah AR, Sariah M, Mohd Razi I, Naher UA (2011) Phosphate solubilization and growth enhancement of aerobic rice by phosphate-solubilizing bacteria. Span J Agric Res 9(3):810–820
Panhwar QA, Naher UA, Shamshuddin J (2013) Improvement of acid sulfate soils for rice cultivation in the reason of food security. Frontiers in conservation agriculture in South Asia and beyond (F-CASA), March 26, 2013, Kathmandu, Nepal. pp 5–6
Panhwar QA, Shamshuddin J, Naher UA, Radziah O, Mohd Razi I (2014a) Changes in the chemical properties of an acid sulfate soil and the growth of rice as affected by bio-fertilizer, ground magnesium limestone and basalt application. Pedosphere 24(6):827–835
Panhwar QA, Shamshuddin J, Naher UA, Radziah O, Mohd Razi I (2014b) Biochemical and molecular characterization of potential phosphate-solubilizing bacteria in acid sulfate soils and their beneficial effects on rice production. PLoS One 14:1–10. doi:10.1371/journal.pone.0097241
Patrick WH Jr, Reddy CN (1978) Chemical changes in rice soils, in IRRI: soils and rice. The International Rice Research Institute, Manila, pp 361–379
Ponnamperuma FN (1977) Physico-chemical properties of submerged soil in relation to fertility. IRRI research paper series no. 5. The International Rice Research Institute, Manila, Philippines
Ponnamperuma FN, Tianco EM, Loy T (1967) Redox equilibria in flooded soils: the iron hydroxide systems. Soil Sci 103:374–382
Prade K, Ottow JCG, Jacq VA, Malouf G, Loyer JY (1990) Relationships between the properties of flooded rice soils and iron toxicity in Lower Casamance (Senegal). Studies, review and summary of previous work. Cahiers ORSTOM, Serie Pedologie, IRD, Montpelliers, France, pp 453–474

Ramgareeb S, Cooke JA, Watt MP (2004) Responses of meristematic callus cells of two Cynodon dactylon genotypes to aluminium. J Plant Physiol 161:1245–1258
Sahrawat KL (1979) Iron toxicity to rice in an acid sulfate soil as influenced by water regimes. Plant Soil 51(1):143–144
Sahrawat KL, Diatta S (1995) Nutrient management and season affect soil iron toxicity. Annual report 1994. West Africa Rice Development Association, Bouaké, Côte d'Ivoire, pp 34–35
Sanchez PA (1976) Properties and management of acid soils in the tropics. Wiley, New York
Shamshuddin J (2006) Acid sulfate soil in Malaysia. UPM Press, Serdang
Shamshuddin J, Elisa AA, Shazana MARS, Fauziah CI, Panhwar QA, Naher UA (2014) Properties and management of acid sulphate soils in Southeast Asia for sustainable cultivation of rice, oil palm and cocoa. Adv Agron 124:91–142
Shazana MARS, Shamshuddin J, Fauziah CI, Syed Omar SR (2013) Alleviating the infertility of an acid sulphate soil by using ground basalt with or without lime and organic fertilizer under submerged condition. Land Degrad Dev 24(2):129–140
Sullivan L (2012) Acid sulfate soils and their management: a global perspective. In: Proceedings of 7th acid sulfate soil conference. Vassa, Finland, pp 127–129
Suswanto T, Shamshuddin J, Syed Omar SR, Mat P, Christofer CBS (2007) The alleviating an acid sulfate soil cultivated to rice (*Oryza sativa*) using ground magnesium limestone and organic fertilizer. Jurnal Tanah dan Lingkungan 9(1):1–9
Thompson JE, Legge RL (1987) The role of free radicals in senescence and wounding. New Phytol 105:317–344
Tinh TK (1999) Reduction chemistry of acid sulphate soils: reduction rates and influence of rice cropping. Acta Universitatis Agriculturae Sueciae, Agraria 206, Uppsala, Sweden
Tran KT, Vo TG (2004) Effects of mixed organic and inorganic fertilizers on rice yield and soil chemistry of the 8 th crop on heavy acid sulfate soil (*Hydraquentic Sulfaquepts*) in the Mekong Delta of Vietnam. A paper presented at the 6th international symposium on plantsoil at low pH. August 1–5, 2004; Sendai, Japan
UNESCO-SCOPE (2013) Policy briefs "Human alteration of the nitrogen cycle, threats, benefits and opportunities".Accessed January, 2013, http://www.green.wikia.com/wiki/Fertilizer. 2013-01-18 T04:43:22Z 1.0
Von Uexküll HR, Mutert E (1995) Global extent, development and economic impact of acid soils. In: Date RA, Grundon NJ, Rayment GE, Probert ME (eds) Plant–soil interactions at low pH: principles and management. Kluwer Academic Publishers, Dordrecht, pp 5–19
Wright RJ (1989) Soil aluminum toxicity and plant growth. Commun Soil Sci Plant Anal 20:1479–1497
Yamauchi M, Peng XX (1993) Ethylene production in rice bronzing leaves induced by ferrous iron. Plant Soil 149:227–234
Zaidi A, Khan MS, Ahemad M, Oves M, Wani PA (2009) Recent advances in plant growth promotion by Phosphate-solubilizing microbes. In: Han MS et al (eds) Microbial strategies for crop improvement. Springer, Berlin/Heidelberg
Zhang J, He Z, Tian H, Zhu G, Peng X (2007) Identification of aluminium-responsive genes in rice cultivars with different aluminum sensitiveness. J Exp Bot 58:2269–2278

Petroleum Hydrocarbons-Contaminated Soils: Remediation Approaches

Hafiz Naeem Asghar, Hafiz Muhammad Rafique, Zahir Ahmad Zahir, Muhammad Yahya Khan, Muhammad Javed Akhtar, Muhammad Naveed, and Muhammad Saleem

Contents

H.N. Asghar (✉) • H.M. Rafique • Z.A. Zahir • M.J. Akhtar • M. Naveed • M. Saleem
Institute of Soil and Environmental Sciences, University of Agriculture Faisalabad, Faisalabad, Pakistan
e-mail: naeemasghar@yahoo.com

M.Y. Khan
Institute of Soil and Environmental Sciences, University of Agriculture Faisalabad, Faisalabad, Pakistan

University of Agriculture Faisalabad, Sub-Campus Burewala, Vehari, Pakistan

K.R. Hakeem et al. (eds.), *Soil Science: Agricultural and Environmental Prospectives*, DOI 10.1007/978-3-319-34451-5_5

Abstract Petroleum, the backbone of today's mechanized society, now has become a threat to environment due to extraction and transportation. Accidental oil spills occur regularly at many locations throughout the world. Contamination of soil and water resources with petroleum oil and its products has become a serious problem due to carcinogenic and mutagenic compounds. Efforts are now focused on seeking potential remediation techniques for cleanup of petroleum hydrocarbons-contaminated soils in a cost effective and eco-friendly way. Various physical, chemical and biological remediation strategies have been used to restore contaminated soils. However, plant assisted bioremediation of petroleum hydrocarbons-contaminated soil is getting more attention as compared to sole use of either microorganisms or plants. The challenging task for such efforts to be successful is not only the survival of microorganisms upon their inoculation into hostile contaminated environment but also positive plant-microbe interactions. Bacteria having ACC-deaminase enzymes are considered helpful for plants in stressed environment. We have discussed that use of bacteria equipped with dual traits of bioremediation potential and ACC-deaminase activity in association with plants can be a good approach for remediation of petroleum hydrocarbons-contaminated soil.

Keywords Petroleum hydrocarbons • Bioremediation • Soil contamination • Phytoremediation • Soil pollution

1 Introduction

All over the world, petroleum oil is considered the most precious resource that is extracted from the ground (IHRDC 2010). Huge annual consumption of oil by the major countries of the world like United States of America (USA), Japan and China has indicated the heavy reliance of the world on petroleum oil (European Energy Portal 2010). However, extraction and transportation of petroleum products are posing an inevitable risk to environment. Both human as well as mechanical errors could be the main cause of terrestrial environment pollution with petroleum products. Contamination of soil and water resources with petroleum hydrocarbons has been increasing over the time in many parts of the world due to accidental oil spills. Oil spills and consequent pollution in sea and beaches has become alarming since the seaborne trade of oil begins (Hayes et al. 2010; Defeo et al. 2009). Terrestrial environment is also contaminated by accidental spills such as the Gulf war oil spill being the worst oil spill in the history caused detrimental addition of 780,000 to 1,500,000 tons of oil to the environment (Mughal 2013). In 2010, beaches and marshes of Gulf faced pollution spillage of millions of gallons of oil because of Deepwater Horizon (DP oil spill) pipe line leakage due to explosion in a well called Macondo (Jernelov 2010). Niger delta is commonly being hit by oil pollution due to poor infrastructure, spills during refining and theft of oil (Amnesty International 2009). An independent experts group estimated a total release of 9 to 13 million barrels in the last 50 years in Niger delta (Jernelov 2010). Another famous incident of oil spill that alarmed the world about petroleum pollution was Exxon-Valdez crash in March 1989 that caused release of 11 million gallons of crude oil (Downs et al. 1993). The oil pollution is not only limited with developed or industrialized countries but it also happened in developing countries. For example, crash of a Greek ship Tasman Spirit near coastal line of Karachi, Pakistan resulted in spill of 28,000 tons of crude oil in July, 2003 (The Daily Times Pakistan, August, 16, 2003). Recently, in 2014, an oil spill of 92,000 US gallons occurred in Bangladesh (UNEP/OCHA 2015).

Due to ecological and environmental hazards linked with petroleum hydrocarbons-contamination, various physical, chemical and biological approaches have been proposed and used for remediation of petroleum hydrocarbons-contaminated soil and water resources. No doubt, physical and chemical approaches are quick and more effective but these are usually environmentally destructive, costly, may produce secondary pollutants, and cleanup is limited to small area. Alternative to these physical and chemical remediation techniques, biological approaches are environment friendly, cost effective, efficient even at low concentration of pollutants, relatively easier to adopt and do not generate perilous secondary products. The biological ways to remediate contaminated sites could be microorganisms based remediation (microbial bioremediation), plant based remediation (phytoremediation) and/or combined use of microorganisms and plants (microbe-assisted phytoremediation or rhizoremediation).

2 Health Hazards Linked with Petroleum Contamination

Petroleum hydrocarbons contamination of air, water and soil is posing serious threat to all living organisms. Some serious effects of petroleum contamination on human health are carcinogenicity, genotoxicity, (Aguilera et al. 2010) rashes on skin, childhood leukemia (Gudzenko et al. 2015), miscarriage in women (Hurtig and Sebastián 2002), irritation in skin and respiratory system disorders. For example, amongst the mono-aromatic compounds present in gasoline like benzene, toluene, ethylbenzene and xylene commonly known as BTEX; benzene is carcinogenic, toluene damages central nervous system, ethylbenzene causes skin irritation and long exposure to xylene may lead to aplastic anemia in human (Singla et al. 2012; Santiago et al. 2015). Studies conducted in Ecuador by Hurtig and Sebastián (2004) revealed that the victims of the oil pollution have been struck by cancer of skin, liver, stomach, kidney, soft tissues, lymph node and hematopoietic tissues. Inhalation of gasoline may cause nausea, numbness and drowsiness. Potential toxicity of benzene may impair nerves and cause anemia while toluene may cause impairment in central nervous system of human beings. Janjua et al. (2006) surveyed the hazardous effect of oil spill on eyes, respiratory system, skin and central nervous system of the victims of Tasman Spirit shipwreck in 2003 occurred in Arabian Sea, Karachi, Pakistan. The authors found strong linear correlation between oil pollution exposure and symptoms of eye and skin irritation and lungs problems in people while such correlation was not recorded in non-exposed group and thereby concluded that the symptoms were due to oil exposure. Similarly, Khurshid et al. (2008) recorded hematological and biochemical changes in the people residing and/or working in the vicinity of Karachi coastline and the authors observed increased level of lymphocytes and eosinophiles. Diesel is one of the most persistent constituent of petroleum oil, which is mutagenic and carcinogenic in nature (CCME 2001). Lemiere et al. (2005) investigated the mutagenicity of polyaromatic hydrocarbons (PAHs) on rats by feeding them for 2 to 4 weeks with food contaminated with polyaromatic hydrocarbons. They observed mutagenic changes in DNA and bone marrow and concluded that petroleum oil-contaminated food sources were potential genotoxic for consumer. There are 16 PAHs compounds which are listed as carcinogenic to human beings by Environmental Protection Agency of United States of America (Lampi et al. 2006). Kuwait faced one of the biggest oil polluted environment in the history as all compartments of the environment were heavily polluted with petroleum. The Kuwait's crude oil is considered sour due to higher quantity of sulfur and thereby poses more toxic effects on health of people (Husain 1994). Many people suffered from infections, nutritional disorders, damaged nervous system, respiratory system, asthma and bronchial diseases (Gastañaga et al. 2002). Of the 159 respondents of Bangladesh oil spill in 2014, 72 % reported no health effect of oil spill while respondents who reported health problems, among them, 55 % were facing difficulty in breathing (UNEP/OCHA 2015). The problem of petroleum pollution is worldwide and it is increasing significantly day by day. Therefore, worldwide struggles have been increasing to find out economical, publically acceptable and self-sustained technologies to remediate petroleum contamination to mitigate health and environmental problems (Vidali 2001).

3 Approaches to Remediate Petroleum Contamination

Various physical, chemical and biological techniques have been used for remediation of petroleum hydrocarbons-contaminated soil and water resources. Physical remediation techniques may include use of different kinds of booms, inorganic absorbent, skimmers and solidifier to avoid spread of spilled oil especially in water. Thermal remediation methods include *in-situ* thermal desorption, burning and incineration. While, oxidation-reductions, encapsulation, solvent extraction and use of dispersants are listed in chemical remediation options for petroleum oil pollution. Among biological approaches that have been adopted so far are bioremediation, phytoremediation, rhizoremediation, bio-augmentation, plant assisted bioremediation, chemical oxidation coupled with bioremediation, biopiling, land farming, and infiltration galleries etc. Each strategy has its own pros and cons and is discussed in the following section.

3.1 Physical Approaches

Physical treatments to remediate petroleum contaminated water include use of booms, skimmers and adsorbents. While, physical methods adopted for treatment of petroleum oil contamination of terrestrial lands include *in-situ* and *ex-situ* techniques. *In-situ* technique includes soil aeration while *ex-situ* technique involves shifting of contaminated soil to a chemical treatment unit such as solvent/water extraction and/or bringing to a thermal treatment unit such as low temperature thermal unit or high temperature thermal unit (incineration).

3.1.1 Booms

Booms are physical barrier to the movement of spilled oil especially when spills occur in water. Booms are used as first response to oil spill to avoid spreading of oil with turbulence of water. Generally, three types of booms are used which are fence boom, curtain booms and fire-resistant booms (Potter and Morrison 2008).

3.1.2 Skimmers

Skimmers are used in connection with booms for recovering spilled oil from water surface without bringing any change in properties of oil (Hammoud 2001). Types of skimmers generally used are weir, oleophillic and suction skimmers. Choice of skimmer depends on thickness and type of oil. All types of skimmers are made from material that contains oleophillic properties.

3.1.3 Adsorbents

As a final step in clean up of oil spills after skimming, adsorbents whether natural or synthetic are used to convert liquid oil into semi-solid phase for complete removal of oil (Adebajo et al. 2003). The adsorbent may be organic, natural inorganic and synthetic.

3.1.4 Soil Washing

Soil washing is an *ex-situ* technique in which separation of fine soil particles (clay and silt) from coarse particles using water and/or solvent is done as contaminants tend to sorb on fine particles. This technique can be applied both for organic and inorganic pollutants. Selection of solvents depends on their ability to dissolve target contaminant (Madadian et al. 2014). Separated fine particles are concentrated and further treated by following other suitable approaches.

3.1.5 Thermal Treatment

Burning of oil spill on site is quick and simple way to get rid of petroleum contamination without requirement of any specialized material. However, *in-situ* burning of oil is restricted to calm conditions of wind, fresh oil spill, sufficient supply of oxygen (Buist et al. 1999) and light products of petroleum as these are burnt quickly without harming aquatic life. For remediation of oil pollution in soil environment, thermal treatments include electrical resistant heating, steam injection and extraction. In electrical resistance heating system, heat produced by moisture in soil pores in response to applied electric current is used to heat soil to vaporize the contaminant and hence steam is produced (Beyke and Fleming 2005). Steam injection and extraction also called as steam enhanced extraction is the process of injecting steam into dug well by which non-aqueous phase liquids are displaced. Upon injection, steam offers its latent heat of vaporization to the soil and when steam heat is lost it changes into hot water which moves through the soil pores. Continued process of injecting steam brings temperature of soil near to temperature required for vaporization of target compounds and produces vapors. The vapors are transported and condensed in contaminant condensate storage. The application of this technique depends on permeability of soil, depth of contaminant location and type of contamination. This technique is alternative where excavation followed by incineration is not possible such as under surface of ground and underground tanks (USEPA 1996). The mechanism involved behind extraction technique is that upon heating organic contaminant like petroleum hydrocarbons lose its density and adsorption on solid and vapor pressure and diffusion into aqueous and gaseous phase is increased (Isherwood et al. 1992). Eventually, all changes that take place due to steam temperature increase the recovery of oil from underground. Among thermal treatments, thermal desorption and incineration are *ex-situ* treatments in which contaminated soil is treated under controlled conditions. In thermal desorption technique, petroleum sludge is thermally destructed by using high temperature oxidation under controlled conditions and the waste sludge is converted into gasses such carbon dioxide, water, sulfur dioxide and oxides of nitrogen. This technique has been used as a yardstick for comparison of other remediation technologies as it serves complete destruction of contaminants (Jadidi et al. 2014). Thermal desorption is distinguished from incineration due to its ability to desorb volatile components from polluted soil without incineration of soil. As thermal desorption is characterized for the removal of volatile compounds, so the process is completely dependent upon the volatility of the contaminants. Therefore, molecules with higher molecular weight such as polycyclic aromatic hydrocarbons are difficult to remediate with this technique (Bansal and Sugiarto 1999). Incineration on the other hand is a use of high temperature (730°C–1200°C) for complete combustion of oily sludge in the presence of air and auxiliary fuels (Hu et al. 2013). Incineration process of treatment is mostly adopted in refineries for sludge disposal. The incineration process is dependent on time, temperature of combustion and rates of sludge feed. Incineration is not only remedial technology but also provides source of energy to run turbines (Naranbhai and Sanjay 1999).

3.1.6 Advantages of Physical Treatment

(a) These can be applied for all kinds of oil
(b) These are generally non destructives strategies and recovery of oil is possible
(c) These techniques are simple and easy to handle where no extensive expertise is required.
(d) Thermal treatment has advantage of time effectiveness over other technologies
(e) Efficiency of incineration and thermal desorption is approximately 99 % which is much higher than other remediation technologies

3.1.7 Disadvantages of Physical Treatment

(a) Extensive labor is required
(b) Not self-sustained
(c) Weak stability in strong wind and currents of water.
(d) Can only be used in conjunction with other techniques
(e) Prior treatment of oily sludge with high moisture content is required for incineration
(f) Thermal treatment is source of secondary pollution as thermal desorption and incineration are the sources of emission of low molecular polycyclic aromatic hydrocarbons
(g) Incomplete combustion may cause pollution in atmosphere
(h) Initial installment and running cost is very high

3.2 Chemical Approaches

Chemical approaches to remediate petroleum hydrocarbons-contaminated soil include the use of dispersants, encapsulation, and chemical oxidation.

3.2.1 Dispersants

When oil spillage occurs on water surface, it disperses into water column naturally. However, dispersion depends on the viscosity of oil. Less viscous oil disperses more due to natural energy of water currents but oil with high viscosity is less amendable with energy of sea. Dispersants are used for dispersion and dilution of highly viscous oil. These may be defined as the mixture of surface active agents dissolved in solvents and stabilizers (Dave and Ghaly 2011). Dispersants contains two parts; one is oleophilic, the other is hydrophilic; these are used for reducing surface tension and consequently increase dispersion of oil. The solvents play role of carrier for dispersant to targeted oil and water interface where it re-arranges molecules by connecting oleophilic part to oil and hydrophilic part to water molecule. Dispersant reduces the size of droplet and thereby increase dispersion of oil on water surface (Lessard and Demarco 2000).

3.2.2 Solvent Extraction

It is an *ex-situ* technique where oil contamination is separated from the media through separation followed by concentration process by using non-aqueous liquid. This method is not widely used for remediation of large area due to its expensiveness as solvents used in this technique are too costly. Soils must be dried before applying this technique and solvents themselves may be source of secondary pollution. The extraction also depends on soil type and composition, for instance, agglomeration occurs when the contaminated soil is clayey in nature and requires longer time of contact of solvent with clay particles (Berset et al. 1999). Solvent extraction, however, has advantage of being used to extract vast range of pollutants. Moreover, pollutants with higher concentration can be treated.

3.2.3 Encapsulation

In this technology, the contaminated material is physically isolated and contained by covering it with low permeable material such as textile and clay caps to avoid infiltration, leaching and resultantly migration (Robertson et al. 2003). This technology is highly dependent on the permeability of the site and saturating capacity of the covering material. This technique cannot be long lasting as sooner or later the covering material gets saturated. Also this technology is not cost effective especially in case when the contaminant depth is higher (Khan et al. 2004).

3.2.4 Chemical Oxidation

Comparing to other remedial strategies, chemical oxidation for decomposition of petroleum hydrocarbons is a short term technology which takes months or even weeks to oxidize contaminants. Chemical oxidants irreversibly reduce petroleum hydrocarbons into CO_2 and H_2O depending upon the contact time with contaminants. Mostly chemical oxidation is carried out to decontaminate sites where area is smaller and concentration of contamination is high. There are different kinds of chemical oxidants in use, however, choice of chemical oxidants depends upon the understanding of hydrogeological condition of the targeted area. The most commonly used chemical oxidants are Fenton's reagent, hydrogen peroxide, permanganate of sodium and potassium and ozone. Effectiveness of chemical oxidation can be enhanced when it is used in conjunction with ultraviolet light (Liang et al. 2003). The success of chemical oxidation technology depends on prior information about the site (soil permeability, texture of soil, soil reactivity), choice of appropriate chemical oxidant and solubility characteristics of solvents etc.

3.2.5 Advantages of Chemical Treatment

(a) It can be used as an in-situ treatment
(b) Destruction of contaminant is rapid (weeks or months)
(c) Oxidation of contaminant is complete (except Fenton's Reagent)
(d) Possesses compatibility with post treatment such as enhancement in aerobic degradation
(e) Being an *in-situ* strategy, it causes minimum disruption to other site operations

3.2.6 Disadvantages of Chemical Treatment

(a) Dispersants are of inflammable nature and can adversely affect human health during spray and also damage the sea life
(b) Higher initial and overall cost compared to other technologies
(c) In case of low permeability soils, oxidant may not get contact with contaminants
(d) Significant loss of chemical oxidant may result due to reaction with soil instead of target contaminants
(e) Clogging of capillary fringe may occur due to precipitation of minerals
(f) May be a source of secondary pollution

3.3 Biological Approaches

Remediation technologies other than biological are expensive, ecologically disruptive and demand substantial input of energy and heavy machinery. In this context, biological options for remediation are more appealing and environment friendly.

3.3.1 Bioremediation

Remediation of petroleum-contaminated soil and water resources by the use of microorganisms is known as bioremediation and is a promising option for complete conversion of petroleum hydrocarbons into carbon dioxide, water and inorganic compound (Kuiper et al. 2004; Barea and Pozo 2005). It is environment friendly and cost efficient compared to traditional physico-chemical remediation strategies (Gallego et al. 2001; Bundy et al. 2002; Mulligan and Yong 2004; Bento et al. 2005; Joo et al. 2008; Gargouri et al. 2014; Fuentes et al. 2014; Pizarro et al. 2014). Numerous bacterial genera reported for biodegradation abilities include *Sphingomonos*, *Cycloclasticus* (Ho et al. 2002), *Burkholderia* (Caballero-Mellado et al. 2007), *Bacillus*, *Bravibacterium*, (Xiao et al. 2012) and *Pseudomonos* (Liu et al. 2013). Microbes have been manipulated for bioremediation of petroleum hydrocarbons in different ways which include biostimulation, biopiling, land farming, bioventing and bioaugmentation (Zhou and Hua 2004).

Stimulation of indigenous microflora by provision of inorganic nutrients like nitrogen and phosphorus to expedite the degradation of pollutants is referred to as biostimulation (Perfumo et al. 2007). Degradation of organic pollutants has been reported by many scientists because of increased microbial growth in response to addition of inorganic nutrients (Sarkar et al. 2005; Ron and Rosenberg 2014). However, blending of inorganic nutrients may also inactivate microbial population and could result in decreased bioremediation process (Mani and Kumar 2014).

Biopile is an *ex-situ* process of remediating organic contaminants in which contaminated soil is piled up where air, fertilizer and other amendments are injected in the contaminated pile to accentuate microbial oxidation of hydrocarbons (Hazen 1997). Contaminated soils are blended and mounded while aeration and moisturizing system is installed prior to piling. The amendments added are bulking agent, chemicals to adjust pH of biopile and periodical addition of nutrients and moisture. To avoid leaching of contaminants, biopile is constructed on impermeable layer of soil (USEPA 2012). Biopiles, also called biocells are constructed where growth conditions for aerobic bacteria are optimized for biodegradation of contaminants. The volatile fractions of petroleum hydrocarbons are evaporated during aeration process while heavier fraction of petroleum hydrocarbon is broken down by biodegradation process.

Land farming is *in-situ* remediation method adopted particularly for remote areas where minimum equipment are required and passive aeration is carried out by tilling periodically (Paudyn et al. 2008). Like biopiling, land farming also requires addition of amendments like bulking agent, chemicals for adjusting pH and inorganic nutrients to speed up bioremediation process (McCarthy et al. 2004). Land farming is adversely affected by environmental conditions such as rainfall and low ambient temperature that consequently affect the rate of biodegradation (Gan et al. 2009).

In bioventing oxygen is provided in contaminated soil and/or water to increase the redox potential for enhancing the bioremediation process. Bioventing is not cost effective technique and to make it cost effective biosparging have been innovated (Mani and Kumar 2014).

Bioaugmentation is addition of microorganisms which have ability to degrade petroleum oil in polluted soil or water resources to enhance the process of bioremediation. Microorganisms used for bioaugmentation could be pre-acclimated and/or genetically engineered. In this process, genes relevant to biodegradation could also be conjugated in the indigenous microbial population (El-Fantroussi and Agathos 2005). It is relatively simple and easy process of bioremediation. Alisi et al. (2009) reported that due to bioaugmentation, about 75 % of diesel contents were decreased in 42 days. However, success of bioaugmentation is limited due to inoculum failure which could be due to competition with indigenous microflora (Vidali 2001). Other possible reasons for inoculum failure could be low redox potential, unavailability of inorganic nutrients, bioavailability of target compounds and improper soil conditions like moisture, temperature and pH etc. (Suja et al. 2014).

3.3.1.1 Degradation of Petroleum Hydrocarbons

Crude oil is complex mixture of aliphatic and cyclic hydrocarbons. On refining, crude oil yields different compounds such as gasoline, kerosene, diesel and lubricating oil. Straight chain alkane and alkene are called aliphatic hydrocarbons while cyclic hydrocarbons are of two kinds; cycloalkanes (saturated hydrocarbons) and aromatic hydrocarbons (unsaturated hydrocarbons). Alkane compounds are of three types; linear alkane (n-alkanes), branched alkanes and cycloalkane. One or more rings of carbon atoms are present in cycloalkanes; however, these rings are not benzene rings because the hydrocarbon molecules characterized by the presence of one or more benzene or aromatic rings comprise separate class which is called aromatic hydrocarbons. These compounds are further categorized into mono, di and polyaromatic hydrocarbons. In crude oil the major portion is linear alkane or n-alkanes, if biodegradation has not been happened earlier (Ollivier and Magot 2005). Generally, it is considered that alkanes are degraded more rapidly and wide range of microorganisms is capable of biodegrading alkanes both short and long chain. The alkanes are degraded mono-terminally by addition of oxygen and converted into alcohols, aldehyde and fatty acids. The resistance to degradation increases with increasing length of chain (Atlas and Bartha 1981). Three mechanisms of alkane degradation are proposed by Huguenot et al. (2015) given as Fig. 1.

Cycloalkanes are recalcitrant constituent of petroleum hydrocarbons and are found abundantly in petroleum products. These are mostly degraded through co-metabolism by alkane degraders. This process of co-metabolism is initiated by conversion of cycloalkanes into alcohol or ketone by monooxygenase (Sayyed and Patel 2011).

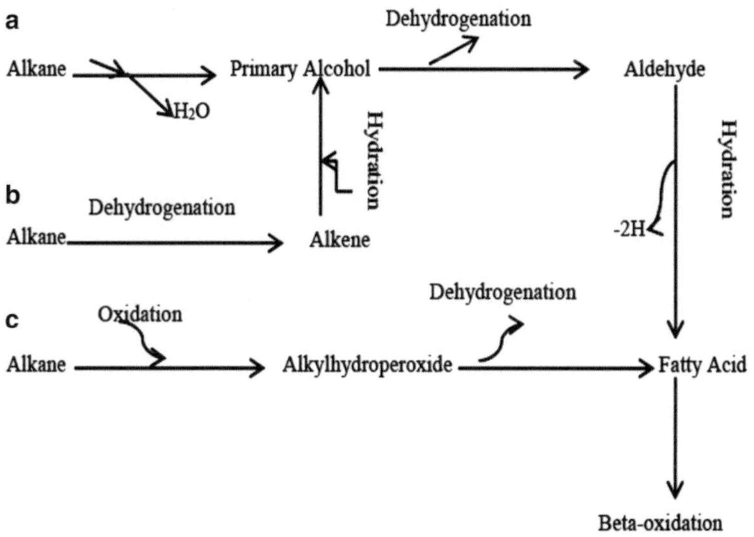

Fig. 1 Terminal methyl oxidation pathways adapted by alkane degrading microorganisms (Extracted from Atlas and Bartha 1981; Huguenot et al. 2015)

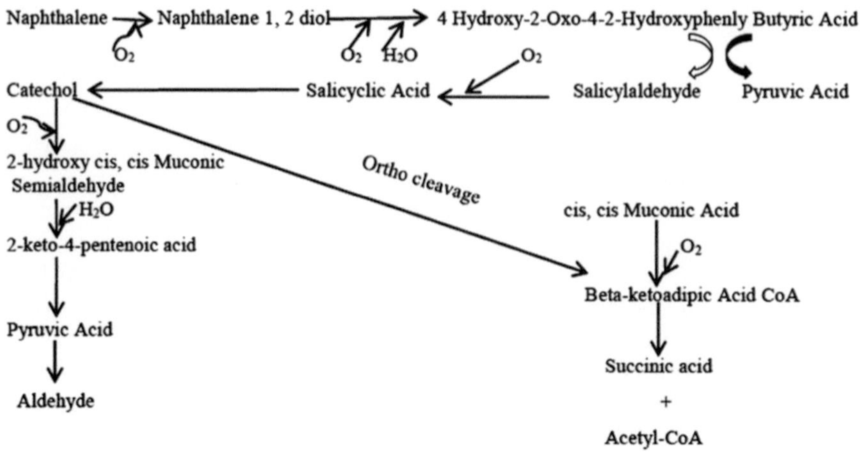

Fig. 2 Degradation of aromatic ring by microorganism showing ortho- and meta- cleavage of catechol (Extracted from Atlas and Bartha 1981; Huguenot et al. 2015)

Degradation of aromatic hydrocarbons especially of polycyclic aromatic hydrocarbons is slow as compared to alkane because one or more oxidation steps are required to convert into catechol which is further opened by oxidation on ortho or meta points of the ring (Atlas and Bartha 1981). The initial oxidative step is mediated by monooxygenase or hydroxylating dioxygenase and this dioxygenase mediated opening of catechol results into the production of cis,cis muconate and unsaturated dicarboxylic acid (Fig. 2). Finally acetyl-CoAs are produced from this product through beta-oxidation.

3.3.1.2 Constraints in Bioremediation

Bioremediation process faces a number of serious constraints including petroleum contamination in more than one medium like soil, water and gaseous phase and complex matrix of inorganic and organic contaminants with variety of toxicological behavior. Moreover, heterogenic subsurface conditions, uncontrollable sub-optimal environmental conditions, costly analytical procedure and difficulty in monitoring process could cause limitation in bioremediation (Pollard et al. 1994). Some of these important constraints have been discussed below.

3.3.1.2.1 Composition of Petroleum Waste

Petroleum hydrocarbons are a complex mixture of thousands of molecules with unique chemical structure and behavior that make the bioremediation process uncertain as molecules with different chemistry make them either easily biodegradable or very difficult to break down (American Academy of Microbiology 2011). Microbial degradation order of petroleum hydrocarbons is alkane > monoaromatic hydrocarbons > cycloalkane > polycyclic aromatic hydrocarbons > asphalthene (van Hamme et al. 2003).

3.3.1.2.2 Weathering of Petroleum Waste

Weathering process limits the susceptibility of petroleum hydrocarbons to biodegradation as easily biodegradable fractions of petroleum hydrocarbons are weathered due to evaporation, reaction of sunlight leaving behind the recalcitrant portion of contamination (Bossert and Bartha 1984). An aqueous phase is required for microorganism to grow on petroleum hydrocarbons which is limited due to weathering as hydrophobicity/octanol water portioning co-efficient increases during weathering. In other words, the bioavailability of compounds is restricted and consequently the process of bioremediation is constrained (Leahy and Colwell 1990; Cerniglia 1993).

3.3.1.2.3 Climatic Condition

Climatic conditions controlling biodegradation are temperature, soil moisture, redox potential, oxygen, pH and nutrient status.

Temperature is probably the most crucial factor that affects bioremediation process as the solubility and bioavailability of hydrophobic compounds directly depends upon temperature. The temperature controls the viscosity of petroleum hydrocarbons and thereby increases or decreases the distribution and diffusion of petroleum hydrocarbons (Gibb et al. 2001). Also, the microbial metabolism is directly affected by the temperature and consequently the activity of microorganisms in the environment is affected.

Bioremediation of various component of oil may be influenced due to inorganic nutrients. Soil, contaminated with hydrocarbons, suffers from deficiency of inorganic nutrients due to the higher concentration of carbon in contaminants. Generally, nitrogen (N) and phosphorus (P) are limiting in hydrocarbon-contaminated soils and affect the rate of bioremediation by suppressing the proliferation of microorganism because N and P are major cellular components of bacteria (Alamri 2009; King et al. 1998). Most of the petroleum hydrocarbons are degraded aerobically with the exception of heavier molecules like asphalthene and polycyclic aromatic hydrocarbons. For aerobic degradation, proper provision of oxygen is of immense importance. Parallel to oxygen supply for bioremediation process, soil moisture is of same importance for growth of microbial population and bioavailability of low molecular weight compounds. Hence the delicate balance between soil oxygen and soil moisture is necessary for successful bioremediation process.

3.3.1.2.4 Bioavailability

In bioremediation context, bioavailability is called the portion of petroleum hydrocarbons available for microbial degradation (Pollard et al. 1994). Bioavailability of different fractions of petroleum hydrocarbons is very important regulating factor for acceptable rate of bioremediation. Main reason of decreased bioavailability of contaminants is the ageing process which is attributed to sequestration of contaminants into solid phase either on clay minerals or on organic matter (Stewart et al. 2003). Break down of petroleum hydrocarbon compounds is either by direct attachment of microorganism to non-aqueous phase, liquid-water interface or by mass transfer of non-aqueous phase liquid to aqueous phase (Mohanty and Mukerji 2008). Problem of bioavailability may be overcome by the addition of surface active agent or the bacteria with ability to secrete bio-surfactants.

Although, bioremediation is very appealing strategy for remediation of petroleum hydrocarbons-contaminated soils. However, its efficiency is limited because of limited growth of microbes in oil contaminated soil environment due to large molecules of hydrocarbons, improper moisture and anaerobic conditions etc. If, microbes are used in association with plants, the remediation processes could be expedited many times than their separate application.

3.3.2 Phytoremediation

Phytoremediation is use of plants for remediation and/or restoration of contaminated soil or ground water resources. There are various mechanisms involved in phytoremediation like phytoextraction, rhizofiltration, phytotransformation, phytostabilization, phytodegradation, rhizoremediation, and phytovolatilization for the remediation of contaminated soil and ground water (Glick 2010; Hakeem et al. 2014). Degradation of contaminants by plants could be either intracellular metabolic process after uptake of contaminants in their cells or by releasing extracellular enzymes that degrade contaminants (Mougin 2002; Liu et al. 2013; Sabir et al. 2014). However, there are some limitations for metabolic degradation of organic pollutants like polycyclic aromatic hydrocarbons and other components of petroleum hydrocarbons due to limited uptake by plants of such type of contaminants (Mougin 2002; Newman and Reynolds 2005; Martin et al. 2014). The hydrophobic nature of organic pollutants is the main hindrance for plant uptake and it has been reported that plants can only uptake, translocate and metabolize the organic compounds with octanol-water partition coefficient (Kow) log Kow $\leq$ 1 (Limmer and Burken 2014). The petroleum hydrocarbons with octanol water partitioning coefficient (Kow) greater than 3 are not soluble in water thus cannot be taken up by plants (Rojo 2009). Different types/mechanisms of phytoremediation are elaborated precisely in following section:

3.3.2.1 Phytodegradation/Phytotransformation

Breakdown of organic contaminants by plants either internally by their metabolic process or externally by releasing enzymes is called phytodegradation or phytotransformation (Vaziri et al. 2013). Plants possess several enzymes such as cytochrome P450s and glutathione-s-transferase by which plants are able to transform toxic chemicals (Mougin 2002). There are very few reports of direct degradation of petroleum hydrocarbons by plants (Mougin 2002; Newman and Reynolds 2005). Soybeans (*Glycine max*) has been reported for uptake and degradation of 14C-anthracene (Edwards et al. 1982).

3.3.2.2 Phytostabilization

Containment or immobilization of contaminant in soil or ground water by plant roots is called phytostabilization. It involves the use of plants to reduce the bioavailability and migration of contaminants in soil (Germida et al. 2002). Stabilization of contaminants by plants occurs due to adsorption on the root surface, accumulation by the roots, or isolation within the root zone using plants as organic pumps (Adam and Duncan 1999; Pilon-Smits 2005). Schnoor (2002) suggests that organic chemicals with log Kow values greater than 3.0 are strongly sorbed to plant roots. Schwab et al. (1998) reported up to 30 % of naphthalene adsorption to the roots of alfalfa, while 15 % on roots of tall fescue. Conclusion drawn by the authors was that adsorption of lipophilic compounds onto the surface of roots may be an important sink for PAHs in soils and an initial step in phytoremediation. Binet et al. (2000) reported 0.006 and 0.11 % of extractable PAHs by ryegrass (Lolium perenne) as determined by GC-MS. In case of petroleum hydrocarbons, phytostabilization may simply involve the establishment of a vegetative cover to minimize potential migration of the contaminant through soil erosion or leaching (Germida et al. 2002).

3.3.2.3 Phytovolatilization

Phytovolatilization is a process in which contaminants are taken up by plants and subsequently moved into atmosphere by the process of evapotranspiration through stomata into the atmosphere (Pilon-Smits 2005). Watkins et al. (1994) assessed phytovolatilization of radiolabelled [7-14C] naphthalene with the Rhodes grass (*Chloris gayana*). The authors concluded that the pollutant was taken up by the grass roots, translocated within the plant, and volatilized through the above ground biomass. Overall, plant uptake and accumulation of hydrocarbons from contaminated soil is quite small and limited to low molecular weight compounds. Many petroleum hydrocarbons are large, high molecular weight molecules which are also lipophilic, thus excluding them from the plant root (Qui et al. 1997). Phytodegradation and phytovolatilization are therefore considered minor pathways of hydrocarbon removal from soil systems.

3.3.2.4 Advantages of Phytoremediation

(a) Minimum disturbance to the environment
(b) On-site remediation technology
(c) Cost effective

(d) No excavation and transportation is required
(e) Can be applied to remediate different kinds of hazardous materials
(f) Large area can be covered
(g) It is aesthetic to environment and therefore has favorable public perception.

3.3.2.5 Disadvantages of Phytoremediation

(a) The contaminated area must be large enough to grow the plants.
(b) Phytoremediation is limited to low concentration of contamination
(c) Phytoremediation is limited to root zone depth
(d) Problem to handle contaminated plants

No doubt, phytoremediation is a promising strategy to remediate polluted soil and ground water with xenobiotic compounds, but it has limitations due to toxicity of pollutants and it has been observed in several studies that under petroleum hydrocarbon stress plants fail to attain sufficient biomass for meaningful remediation and reduced plant growth with decreased root and shoot lengths has been observed (Merkl et al. 2005; Germaine et al. 2009).

3.3.3 Plant Assisted Bioremediation/Microbial Assisted Phytoremediation

Surely, both plants and microbes have potential to biodegrade petroleum hydrocarbons but their individual success is limited. Plants cannot absorb and degrade the hydrophobic compounds with high molecular weight and microbes may face problem in biodegradation due to deficiency of nutrients and oxygen (Pilon-Smits 2005). However, partnership between plants and microbes expedites the biodegradation of hydrophobic compounds (Kuiper et al. 2004). Remediation of petroleum hydrocarbons in rhizosphere of plants by the assistance of microbes is known as rhizoremediation. However, endophytic microbial degradation of petroleum hydrocarbons has also been observed which refers to the symbiotic relationship between plants and endophytic microorganism (Newman and Reynolds 2005). Work by van Aken et al. (2004) revealed that symbiotic relationship between *Methylobacterium* sp. strain BJ001 and hybrid poplar approximately degraded 60 % of explosives such as TNT (2,4,6-trinitrotoluene) and HMX (octahydro-1,3,5,7-tetranitro-1,3,5,7-tetrazocine) into carbon dioxide in 2 months. However, in case of PAHs and petroleum compounds, rhizosphere degradation (rhizoremediation) is more prevalent than direct uptake and metabolism by the plant or its associated symbionts (Hutchinson et al. 2003). Plant assisted bioremediation is more pleasing and publically accepted because of its low cost and meaningful remediation of organic compounds (Gurska et al. 2009). Both plants and microbes must have ability to tolerate stress imposed by petroleum hydrocarbons for the success of this strategy (Germida et al. 2002 and Bona et al. 2011). Bioremediation assisted by plants is useful to overcome constraints that limit alone application of plants or microorganisms and thereby enhances degradation of recalcitrant soil contaminant like polyaromatic hydrocarbons (Huang et al. 2004). Roots of higher plants play important role in assisting microbial population to degrade pollutants in many ways such as provision of molecular oxygen, inorganic nutrients through sloughed off cells, soluble exudates and lysates (Phillips 2008; Martin et al. 2014). Plants improve soil aeration by directly giving off oxygen to the root zone as well as allowing improved entry of oxygen into the soil by diffusion along old root channels (Marschner 2012). If microbes do not have degradative pathways for petroleum hydrocarbons and cannot use them as carbon source, even then petroleum hydrocarbons can be mineralized through the process of co-metabolism (Gojgic-Cvijovic et al. 2012). Co-metabolism is mechanism by which plants assist microbes to co-metabolize a contaminant in the soil using the root exudates as an energy source. Plant enzymes degrade the compounds, and then further degradation is carried out by microbes ultimately into carbon dioxide and water. Some structural analogues of polyaromatic hydrocarbons such as phenols, terpenes and flavonoids are released by some plant roots as exudates and thus promote the growth of petroleum hydrocarbons degrading bacteria (Rentz et al. 2005; Martin et al. 2014) and can act as trigger of PAHs degradation-pathway (Singer et al. 2003). Although cooperation of plant with microorganisms is true yet organic pollutants present in the system cause impairment in plant growth and limit the benefits of adding vegetation to a contaminated soil (Gartler et al. 2014).

Plant growth impairment is a serious limitation in petroleum hydrocarbons-contaminated soils because plant species are sensitive to contaminants and fail to produce sufficient biomass and root activities (Huang et al. 2001). Successful plant-assisted microbial remediation of petroleum hydrocarbons-contaminated soil is highly dependent on elimination of root inhibition. This inhibition of root growth in contaminated soils is mainly due to stress-induced ethylene; an established stress phytohormones (Arshad et al. 2007). Synthesis of ethylene in plant at a rate where it exerts inhibitory effects on plant roots is in response to contaminant-induced stress. Thus, a preferential target is to regulate/limit the biosynthesis of ethylene so that normal or more root biomass can be achieved in stressed environment. Some plant growth promoting rhizobacteria (PGPR) are reported to be equipped with ACC-deaminase enzyme that hydrolize ACC (an immediate precursor of ethylene in plants) into α-ketobutyric acid and ammonia and thus, regulate the biosynthesis of ethylene when plants are inoculated with such bacteria (Glick 2005). This reveals the fact that just as plant can affect microbial growth, microorganism can affect and protect plant growth in contaminated soils. Thus rhizobacteria containing ACC deaminase activity regulate plant growth in both biotic and abiotic stresses and dramatically increase the biomass of plant especially roots which is a desirable parameter for plants to be used to assist the process of biodegradation of petroleum hydrocarbons-contaminated soils. But, unfortunately, the isolation and subsequent inoculation with bacteria having ACC-deaminase activity does not fulfill the job of remediating the petroleum hydrocarbons contamination unless they are acclimated to such contaminants. According to suggestions made by Glick (2010) for the enhancement of degradation of organic contaminants in soil that the bacteria inoculated for remediation of organic contaminants must possess twin nature of plant growth promotion as well as degrader of soil contaminant. In plant assisted bioremediation process, biomass production in considerable quantity is of key importance as root biomass provides multi-fold benefits to bacteria in facilitating the degradation of petroleum hydrocarbon compounds and shoot biomass provides cover to soil which not only accentuates degradation process but is also aesthetically more pleasant. To aid plants in production of higher biomass both under normal and stress conditions, bacteria which are usually called plant growth promoting bacteria (PGPB) play their role directly or indirectly (Ahemad and Kibret 2014). Plant growth promoting bacteria benefit directly by solubilizing mineral nutrients such as phosphorus (Hussain et al. 2013), fixing atmospheric nitrogen symbiotically or in rhizosphere of plants in associative manner, siderophore production, phytohormones production (Vessey 2003) or regulating stress induced ethylene (Shaharoona et al. 2006; Arshad et al. 2007) and production of volatile compounds (Ryu et al. 2003; Blom et al. 2011). Indirect mechanisms of plant growth promotion by PGPB include inducing defense mechanism of plant (Ryu et al. 2004), biocontrol by producing antibiotic against pathogenic microorganism and induced systemic resistance (Ryu et al. 2004).

Plant-assisted bioremediation occurs naturally, however it can be accentuated by exploiting suitable plant and microbe synergism especially PGPB with bioremediation potential may be beneficial as these bacteria not only degrade pollutants of the interest but also get rid of the plants from toxic effect of pollutants (Kuiper et al. 2004). In case of plant-assisted bioremediation of mixture of petroleum hydrocarbons, plants are unable to take up petroleum hydrocarbons, however, PGPB equipped with bioremediation potential break these large molecular weight compounds by their catalyzing action (Huang et al. 2004). Bacteria with ability to utilize 1-aminocyclopropane-1-carboxylate (ACC) as sole nitrogen source are of crucial importance when used in combination with plants under stress conditions (Arshad et al. 2007). Huang et al. (2004) conducted study to evaluate different processes such as bioremediation, phytoremediation, land farming and combination of plant growth promoting bacteria having ACC-deaminase activity for remediation of PAHs-polluted contaminated site. Their results revealed that most efficient process was multi process system rather than bioremediation or phytoremediation alone and the authors concluded that success of the multi process system was due to tolerance of plants to contaminants and PGPR that enhanced the tolerance of plants by reducing stress induced ethylene. ACC-deaminase containing bacteria hydrolyze the ACC exuded from germinating seed and roots. To maintain the equilibrium of ACC inside and outside of the roots or seed, ACC move outside the roots or seed as exuded ACC is being cleaved by the bacteria into ammonia and α-ketobutyrate and thus lowers ethylene production inside plant which results into better germination and root growth (Glick 2005; 2007). However, the survival and proliferation of inoculated ACC-deaminase bacteria in the polluted environment is of special consideration. Glick (2010) suggested bacteria with twin nature of plant growth promotion and biodegradation potential as better option instead of only plant growth promoting bacteria to be used in plant-assisted bioremediation process. Some examples of successful remediation of petroleum hydrocarbons by plant assisted bioremediation process are given in Table 1.

Table 1 Plant growth-promoting bacteria in plant-assisted bioremediation

Plant growth-promoting bacteria	Plant	Type of compound	Reference
Methylobacterium populi VP2	*Lycopersicon esculentum*	Phenanthrene	Ventorino et al. 2014
Indigenous Bacterial Consortia	*Trigonella foenumgraecum*	Mixture of Petrol, Diesel and Engine oil	Shanker et al. 2014
PGPR	*Avena sativa*	Petroleum oil	Xun et al. 2015
Serratia marcescens BC-3	*Avena sativa*	Petroleum hydrocarbon	Dong et al. 2014
Pseudomonas putida	Pasture Plants	Polycyclic aromatic hydrocarbons	Pizarro et al. 2014
Burkholderia sp.	*Axonopus affinis*	Diesel	Tara et al. 2014
Pseudomonas sp.	*Testuca arundinacea* L.	Petroleum hydrocarbon	Liu et al. 2013
PGPR	Indian mustard	Petroleum hydrocarbon	Graj et al. 2013

4 Conclusion

Plant assisted bioremediation could be an efficient mean for remediation of petroleum hydrocarbons-contaminated soil and water resources as compared to other physical, chemical and biological approaches. However, for successful establishment of plant-microbe partnership, it is necessary that microbes with biodegradation potential must have the ability to effectively colonize and survive in the rhizosphere of plants. Moreover, it is also of prime importance that plants must produce sufficient biomass with prolific root system. Bacteria having ability to produce ACC deaminase help the plants to tolerate contaminants stress by improving root growth in contaminated soils which ultimately results in improved uptake of water and nutrients. Better establishment of plants in contamination results in enhanced microbial survival and activities due to more availability of root-exuded nutrients. Therefore, co-existing plants and microbes could be more effective for remediation of petroleum hydrocarbons-contaminated soil as compared to their individual application.

References

Adebajo MO, Frost RL, Kloprogge JT, Carmody O, Kokot S (2003) Porous materials for oil spill cleanup: a review of synthesis and absorbing properties. J Porous Mater 10(3):159–170

Adam G, Duncan HJ (1999) Effect of diesel fuel on growth of selected plant species. Environ Geochem Heal 21:353–357

Aguilera F, Laffona B, Méndez J, Pásaroa E (2010) Review on the effects of exposure to spilled oils on human health. Appl Toxicol 30:291–301

Ahemad M, Kibret M (2014) Mechanisms and applications of plant growth promoting rhizobacteria: current perspective. J King Saud Univ Sci 26:1–20

Alamri SA (2009) Use of microbiological and chemical methods for assessment of enhanced hydrocarbon bioremediation. J Biol Sci 9:37–43

Alis C, Musella R, Tasso F, Ubaldi C, Manzo S, Cremisini C, Sprocati A (2009) Bioremediation of diesel oil in a co-contaminated soil by bioaugmentation with a microbial formula tailored with native strains selected for heavy metals resistance. Sci Total Environ 407:3024–3032

American Academy of Microbiology (2011) Microbes and oil spills. FAQ series. http://www.dfompo.gc.ca/science/publications/microbes/pdf/microbes-eng.pdf

Amnesty International (2009) Nigeria: petroleum, pollution, and poverty in the Niger delta. Amnesty International Publications, London

Arshad M, Saleem M, Hussain S (2007) Perspective of bacterial ACC deaminase in phytoremediation. Trends Biotechnol 25:356–362

Atlas RM, Bartha R (1981) Microorganisms and some novel pollution problems. In: Microbial ecology: fundamentals and applications. Addison-Wesley Publishing Company, Inc., Philippines, pp 422–427

Bansal KM, Sugiarto (1999) Exploration and production operations – waste management a comparative overview: US and Indonesia cases. In: SPE Asia Pacific oil and gas conference and exhibition. Society of Petroleum Engineers, Jakarta

Barea IM, Pozo MI (2005) Microbial co-operation in the rhizosphere. J Exp Bot 56:1761–1778

Bento FM, Camargo FAO, Okeke BC, Frankenberger WT (2005) Comparative bioremediation of soils contaminated with diesel oil by natural attenuation, biostimulation and bioaugmentation. Bioresour Technol 96:1049–1055

Berset JD, Ejem M, Holzer R, Lischer P (1999) Comparison of different drying, extraction and detection techniques for the determination of priority polycyclic aromatic hydrocarbons in background contaminated soil samples. Anal Chim Acta 383:263–275

Beyke G, Fleming D (2005) In situ thermal remediation of DNAPL and LNAPL using electrical resistance heating. Remediation 15:5–22

Binet P, Portal J, Leyval C (2000) Dissipation of 3-6-ring polycyclic aromatic hydrocarbons in the rhizosphere of ryegrass. Soil Biol Biochem 32:2011–2017

Blom D, Fabbri C, Connor EC, Schiestl FP, Klauser DR, Boller T, Eberl L, Weisskopf L (2011) Production of plant growth modulating volatiles is widespread among rhizosphere bacteria and strongly depends on culture conditions. Environ Microbiol 13:3047–3058

Bona C, Rezende I, Santos G, Souza L (2011) Effect of soil contaminated by diesel oil on the germination of seeds and the growth of *Schinus terebinthifoilus*. Braz Arch Biol Technol 54:1379–1387

Bossert I, Bartha R (1984) The fate of petroleum in soil ecosystems. In: Atlas RM (ed) Petroleum microbiology. Macmillan Publ, New York, pp 435–473

Buist I, McCourt J, Potter S, Ross S, Trudel K (1999) *In situ* burning. Pure Appl Chem 71:43–65

Bundy JG, Paton GI, Campbell CD (2002) Microbial communities in different soils types do not converge after diesel contamination. J Appl Microbiol 92:276–288

Caballero-Mellado J, Onofre-Lemus J, Estrada-de P, Santos L, Martinez-Aguilar L (2007) The tomato rhizosphere, an environment rich in nitrogen-fixing *Burkholderia* species with capabilities of interest for agriculture and bioremediation. Appl Environ Microbiol 73:5308–5319

CCME (2001) Canada-wide standards for petroleum hydrocarbons (PHC) in soil user guidance. Canadian Council of Ministers of the Environment (CMME), Winnipeg, Manitoba

Cerniglia CE (1993) Biodegradation of polycyclic aromatic hydrocarbons. Biodegrad 3:351–368

Dave D, Ghaly AE (2011) Remediation technologies for marine oil spills: a critical review and comparative analysis. Am J Environ Sci 7:423–440

Defeo O, MacLachlan A, Schoeman DS, Schlancher TA, Dugan J, Jones A, Lastra M, Scapini F (2009) Threats to sandy beach ecosystem: a review. Estuar Coast Shelf Sci 81:1–12

Dong R, Gu L, Guo C, Xun F, Liu J (2014) Effect of PGPR *Serratia marcescens* BC-3 and AMF *Glomus intraradices* on phytoremediation of petroleum contaminated soil. Ecotoxicology 23:674–680

Downs MA, Palinkas LA, Petterson JS, Russell J (1993) Social, cultural, and psychological impacts of the Exxon-Valdez oil spill. Hum Organ 52:1–13

Edwards N, Todd BR, Garver E (1982) Uptake and metabolism of ^{14}C anthracene by soybean (*Glycine max*). Environ Exp Bot 22:349–357
El-Fantroussi SI, Agathos SN (2005) Is bioaugmentation a feasible strategy for pollutant removal and site remediation? Curr Opin Microbiol 8:268–275
EPA US Environmental Protection Agency (2012) Biopiles. Cleaning up UST system releases. Online at: http://www.epa.gov/OUST/cat/biopiles.htm
EPA US Environmental Protection Agency (1996) EPA environmental engineering source book. Ann Arber Press, Inc., Chelsea
European Energy Portal (2010) Non-renewables. Available at: http://www.energy.eu/#non-renewable
Fuentes S, Mendez V, Aguila P, Seeger M (2014) Bioremediation of petroleum hydrocarbons: catabolic genes, microbial communities, and applications. Appl Microbiol Biotechnol 98:4781–4794
Gallego JR, Loredo J, Llamas JF, Vazquez F, Sanchez J (2001) Bioremediation of diesel-contaminated soils: evaluation of potential *in situ* techniques by study of bacterial degradation. Biodegradation 12:325–335
Gan S, Lau EV, Ng HK (2009) Remediation of soils contaminated with polycyclic aromatic hydrocarbons (PAHs). J Hazard Mat 172:532–549
Gargouri B, Karray F, Mhiri N, Aloui FM, Sayadi S (2014) Bioremediation of petroleum hydrocarbons-contaminated soil by bacterial consortium isolated from an industrial wastewater treatment plant. J Chem Technol Biotechnol 89:978–987
Gartler J, Wimmer B, Soja G, Reichenauer TG (2014) Effects of rapeseed oil on the rhizodegradation of polyaromatic hydrocarbons in contaminated soil. Int J Phytorem 16:671–683
Gastañaga VM, Gray G, Kaiser KS, Reed RJ, Smith TC (2002) Self-reported symptoms and medical conditions among 11,868 Gulf War-era Veterans. Am J Epidemiol 155:1033–1044
Germaine KJ, Keogh E, Ryan D, Dowling DN (2009) Bacterial endophyte-mediated naphthalene phytoprotection and phytoremediation. FEMS Microbiol Lett 296:226–234
Germida J, Frick C, Farrell R (2002) Phytoremediation of oil contaminated soils. Dev Soil Sci 28:169–186
Gibb A, Chu A, Wong RCK, Goodman RH (2001) Bioremediation kinetics of crude oil at 5°C. J Environ Eng 127:818–824
Glick BR (2005) Modulation of plant ethylene levels by the bacterial enzyme ACC deaminase. FEMS Microbiol Lett 251:1–7
Glick BR (2010) Using soil bacteria to facilitate phytoremediation. Biotechnol Adv 28:367–374
Glick BR, Todorovic B, Czarny J, Cheng ZY, Duan J, McConkey B (2007) Promotion of plant growth by bacterial ACC deaminase. Critic Rev Plant Sci 26:227–242
Gojgic-Cvijovic GD, Milic JS, Solevic TM, Beskoski VP, Ilic MV, Djokic LS, Vrvic MM (2012) Biodegradation of petroleum sludge and petroleum polluted soil by a bacterial consortium: a laboratory study. Biodegradation 23:1–14
Graj W, Lisiecki P, Szulc A, Chrzanowski Wojtera-Kwiczor J (2013) Bioaugmentation with petroleum-degrading consortia has a selective growth-promoting impact on crop plants germinated in diesel oil-contaminated soil. Water Air Soil Pollut 224:1–15
Gudzenko N, Hatch M, Bazyka D, Dyagila I, Reissc RF, Brenner A, Chumak V, Babkina N, Zablotska LB, Mabuchi K (2015) Non-radiation risk factors for leukemia: a case-control study among chornobyl cleanup workers in Ukraine. Environ Res 142:72–76
Gurska J, Wang WX, Gerhardt KE, Khalid AM, Isherwood DM, Huang XD, Glick BR, Greenberg BM (2009) Three years field test of a plant growth promoting rhizobacteria enhanced phytoremediation system at a land farm for treatment of hydrocarbon waste. Environ Sci Technol 43:4472–4479
Hakeem KR, Sabir M, Ozturk M, Mermut A (2014) Soil remediation and plants: prospects and challenges. Academic, Elsevier/New York, p 724
Hammoud AH (2001) Enhanced oil spill recovery rate using the weir skimmer. Retrieved from http://www.epa.gov/oem/docs/oil/fss/fss06/hammoud.pdf

Hayes MO, Michael J, Betenbaugh DV (2010) The intermittently exposed, coarse-grained beaches of Prince William Sound, Alaska: comparison with open-ocean beaches. J Coastal Res 26:4–30
Hazen TC (1997) Bioremediation. In: Amy P, Haldeman D (eds) Microbiology of the terrestrial subsurface. CRC Press, Boca Raton, pp 247–266
Ho KF, Lee SC, Gloria MYC (2002) Characterization of selected volatile organic compounds, polycyclic aromatic hydrocarbons and carbonyl compounds at a roadside monitoring station. Atm Environ 36:57–65
Hu G, Li J, Zeng G (2013) Recent development in the treatment of oily sludge from petroleum industry: a review. J Hazard Mat 261:470–490
Huang XD, Glick BR, Greenberg MB (2001). Combining remediation technologies increases kinetics for removal of persistent organic contaminants from soil. In: Greenberg BM, Hull RN, Roberts Jr. MH, Gensemer RW (eds) Environmental toxicology and risk assessment, vol 10, ASTM, Philadelphia, p 271–278
Huang XD, El-Alawi Y, Penrose DM, Glick BR, Greenberg BM (2004) A multi-process phytoremediation system for removal of polycyclic aromatic hydrocarbons from contaminated soils. Environ Pollut 130:465–476
Huguenot D, Mousset E, van Hullebusch ED, Oturan MA (2015) Combination of surfactant enhanced soil washing and electro-fenton process for the treatment of soils contaminated by petroleum hydrocarbons. J Environ Manag 153:40–47
Hurtig AK, Sebastián S (2002) Geographical differences in cancer incidence in the Amazon Basin of Ecuador in relation to residence near oil fields. Int J Epidemiol 31:1021–1027
Hurtig AK, Sebastián MS (2004) Incidence of childhood leukemia and oil exploitation in the Amazon Basin of Ecuador. Int J Occup Environ Health 10:245–250
Husain T (1994) Kuwaiti oil fires: source estimates and the plume characterization. Atm Environ 28:2149–2158
Hussain MI, Asghar HN, Arshad M, Shahbaz M (2013) Screening of multi-traits rhizobacteria to improve maize growth under axenic conditions. J Anim Plant Sci 23:514–520
Hutchinson S, Schwab A, Banks M (2003) Biodegradation of petroleum hydrocarbons in the rhizosphere. Phytoremediation. In: McCutcheon S, Schnoor J (eds) Transformation and control of contaminants. Wiley, Hoboken, pp 355–386
IHRDC (2010) Petroleum online. Retrieved on October 8, 2010, from the International Human Resource Development Corporation website: http://www.ihrdc.com.
Isherwood WF, Ziagos J, Nichols E, Rice D, Krauter P (1992) Enhancing aquifer cleanup with reinjection. In: Proceedings of the 1992 petroleum hydrocarbons and organic chemicals in ground water: prevention, detection, and remediation conference, Houston, p 103–109
Jadidi N, Roozbehani B, Saadat A (2014) The most recent research in oily sludge remediation process. Am J Oil Chem Technol 2:340–348
Janjua NZ, Kasi PM, Nawaz H, Farooqui SZ, Khuwaja UB, Hassan N, Jafri SN, Lutfi SA, Kadir MM, Sathiakumar N (2006) Acute health effects of the Tasman Spirit oil spill on residents of Karachi, Pakistan. BMC Public Health 6:84
Jernelöv A (2010) The threats from oil spills: now, then, and in the future. Ambio 39:353–366
Joo HS, Ndegwa PM, Shodaand M, Phae CG (2008) Bioremediation of oil-contaminated soil using *Candida catenulata* and food waste. Environ Pollut 156:891–896
Khan FI, Husain T, Hejazi R (2004) An overview and analysis of site remediation technologies. J Environ Manage 71:95–122
Khurshid R, Sheikh MA, Iqbal S (2008) Health of people working/living in the vicinity of an oil polluted beach near Karachi Pakistan. East Mediterr Health J 14:179–182
King RB, Long GM, Sheldon JK (1998) Practical environmental bioremediation: the field guide. CRC Press/Lewis Publishers, Boca Raton
Kuiper I, Lagendijk E, Bloemberg G, Lugtenberg B (2004) Rhizoremediation: a beneficial plant-microbe interaction. Mol Plant Microb Interact 17:6–15

Lampi MA, Gurska J, McDonald KIC, Xie FL, Huang XD, Dixon DG, Greenberg BM (2006) Photoinduced toxicity of polycyclic aromatic hydrocarbons to Daphnia magna: ultraviolet-mediated effects and the toxicity of polycyclic aromatic hydrocarbon photoproducts. Environ Toxicol Chem 25:1079–1087

Leahy JG, Colwell RR (1990) Microbial degradation of hydrocarbons in the environment. Microbiol Rev 54:305–315

Lemiere S, Cossu-Leguille A, Bispo MJ, Jourdain MC, Lanhers D, Burnel Vasseur P (2005) DNA damage measured by the single-cell gel electrophoresis (comet) assay in mammals fed with mussels contaminated by the 'Erika' oil-spill. Mutat Res 581:11–21

Lessard RR, Demarco G (2000) The significance of oil spill dispersants. Spill Sci Technol Bull 6:59–68

Liang S, Min JH, Davis MK, Green JF, Remer DS (2003) Use of pulsed-UV processes to destroy NDMA. J Am Water Works Assoc 95:121–131

Limmer MA, Burken JG (2014) Plant translocation of organic compounds: molecular and physicochemical predictors. Environ Sci Technol Lett 1:156–161

Liu W, Sun J, Ding L, Luo Y, Chen M, Tang C (2013) Rhizobacteria (*Pseudomonas* sp. SB) assists phytoremediation of oily-sludge-contaminated soil by tall fescue (*Testuca arundinacea* L.). Plant Soil 371:533–542

Madadian E, Gitipour S, Amiri L, Alimohammadi M, Santloo J (2014) The application of soil washing for treatment of polycyclic aromatic hydrocarbons contaminated soil. A case study in petrochemical complex. Environ Prog Sus Energy 33:107–114

Mani D, Kumar C (2014) Biotechnological advances in bioremediation of heavy metals contaminated ecosystems: an overview with special reference to phytoremediation. Int J Environ Sci Technol 11:843–872

Marschner H (2012) Marschner's mineral nutrition of higher plants. Academic, London

Martin BC, Suman JG, Charles AP, Megan HR, Mark T (2014) The role of root exuded low molecular weight organic anions in facilitating petroleum hydrocarbon degradation: current knowledge and future directions. Sci Total Environ 472:642–653

McCarthy K, Walker L, Vigoren L, Bartel J (2004) Remediation of spilled petroleum hydrocarbons by land farming at an Arctic site. Cold Reg Sci Technol 32:93–105

Merkl N, Schultze-Kraft R, Arias M (2005) Influence of fertilizer levels on phytoremediation of crude oil-contaminated soils with the tropical pasture grass *Brachiaria brizantha* (Hochst. Ex A. Rich.) stapf. Int J Phytorem 7:217–230

Mohanty G, Mukherji S (2008) Enhancement of NAPL bioavailability by induction of cell-surface hydrophobicity in *Exiguobacterium aurantiacum* and *Burkholderia cepacia*. Ind J Biotechnol 3:295–306

Mougin C (2002) Bioremediation and phytoremediation of industrial PAH polluted soils. Polycycl Aromat Comp 22:1011–1043Mughal MAZ (2013) Persian Gulf desert and semi-desert. In: Robert Warren Howarth (ed.) Biomass and Ecosystems, vol. 3. Ipswich, MA: Salem Press, pp. 1000–1002

Mughal MAZ (2013) Persian Gulf desert and semi-desert. In: Robert Warren H (ed) Biomass and Ecosystems, vol 3. Salem Press, Ipswich, pp 1000–1002

Mulligan CN, Yong RN (2004) Natural attenuation of contaminated soils. Environ Int 30:587–601

Naranbhai KP, Sanjay SS (1999) Hazardous, toxic and mixed chemical waste's disposal by incineration. In: 15th international conference on solid waste technology management, Philadelphia, USA, 1999.

Newman L, Reynolds C (2005) Phytodegradation of organic compounds. Curr Opin Biotechnol 15:225–230

Ollivier B, Magot M (2005) Petroleum microbiology. ASM Press, Washington, DC

Paudyn K, Rutter A, Rowe RK, Poland JS (2008) Remediation of hydrocarbon contaminated Canadian Arctic by land farming. Cold Reg Sci Technol 53:102–114

Perfumo A, Banat I, Merchant R, Vezzulli L (2007) Thermally enhanced approaches for bioremediation of hydrocarbon contaminated soils. Chemosphere 66:179–184

Phillips AL (2008) The relationship between plants and their root-associated microbial communities in hydrocarbon phytoremediation systems. Ph.D. thesis. Dept. Soil Sci. Univ. Saskatchewan, Saskatoon, Canada

Pilon-Smits E (2005) Phytoremediation. Ann Rev Plant Biol 56:15–39

Pizarro TP, Niqui JL, Solano ARJ, Bastida MFF, Garcia C, Ramos JL (2014) Field trial on removal of petroleum hydrocarbon pollutants using a microbial consortium for bioremediation and rhizoremediation. Environ Microbiol Rep. doi:10.1111/1758-2229.12174

Pollard SJT, Hrudey SE, Fedorak PM (1994) Bioremediation of petroleum and creosote-contaminated soils: a review of constraints. Waste Manag Res 12:173–194

Potter S, Morrison J (2008) World catalogue of oil spill response products, 9th Edn, S.L. Ross Environental Research Ltd. Ottawa, Canada, p 1–42

Qui X, Leland T, Shah S, Sorensen D, Kendall E (1997) Field study: grass remediation for clay soil contaminated with polycyclic aromatic hydrocarbons. In: Kruger E, Anderson T, Coats J (eds) Phytoremediation of soil and water contaminants. Am. Chem. Soc, Washington DC

Rentz JA, Alvarez PJJ, Schnoor JL (2005) Benzo[a]pyrene co-metabolism in the presence of plant root extracts and exudates: implications for phytoremediation. Environ Pollut 136:477–484

Robertson D, Burnely S, Barratt R (2003) The immobilization of flue gas treatment residues through the use of single staged wash and crystalline encapsulation (CME) treatment process. In: Proceedings of 11th annual North American waste to energy conferences, Tampa, p 135–143.

Rojo F (2009) Degradation of alkanes by bacteria. Environ Microbiol 11:2477–2490

Ron EZ, Rosenberg E (2014) Enhanced bioremediation of oil spills in the sea. Curr Opin Biotechnol 27:191–194

Ryu CM, Farag MA, Hu CH, Reddy MS, Wei HX, Pare PW, Kloepper JW (2003) Bacterial volatiles promote growth in Arabidopsis. Proc Nat Acad Sci 100:4927–4932

Ryu CM, Farag MA, Hu CH, Reddy MS, Kloepper JW, Pare PW (2004) Bacterial volatiles induce systemic resistance in Arabidopsis. Plant Physiol 134:1017–1026

Sabir M, Zia-ur-Rehman M, Hakeem K R, Saif Ullah (2014) Phytoremediation: Mechanisms and adaptations. In: Hakeem KR, Sabir M, Ozturk M, Murmut A (eds) Soil remediation and plants. Academic press, Elsevier, USA

Santiago F, Alves G, Otero UB, Tabalipa MM, Scherrer LR, Kosyakova N (2015) Monitoring of gas station attendants exposure to benzene, toluene, xylene (BTX) using three-color chromosome painting. Mol Cytogenet 7:1–15

Sarkar D, Ferguson M, Datta R, Birnbaum S (2005) Bioremediation of petroleum hydrocarbons in contaminated soils: comparison of biosolids addition, carbon supplementation, and monitored natural attenuation. Environ Pollut 136:187–195

Sayyed RZ, Patel PR (2011) Soil microorganisms and environmental health. Int J Biotech Biosci 1:41–66

Schnoor JL (2002) Technology evaluation report: phytoremediation of soil and groundwater. GWRTAC series TE-02-01

Schwab A, Al-Assi ABanks M (1998) Adsorption of naphthalene onto plant roots. J Environ Qual 27:220–224

Shaharoona B, Arshad M, Zahir ZA (2006) Effect of plant growth promoting rhizobacteria containing ACC-deaminase on maize (*Zea mays* L.) growth under axenic conditions and on nodulation in mung bean (*Vigna radiata* L.). Lett Appl Microbiol 42:155–159

Shankar S, Kansrajh C, Dinesh MG, Satyan RS, Kiruthika S, Tharanipriya A (2014) Application of indigenous microbial consortia in bioremediation of oil-contaminated soils. Int J Environ Sci Technol 11:367–376

Singer AC, Smith D, Jury WA, Hathuc K, Crowley DE (2003) Impact of the plant rhizosphere and augmentation on remediation of polychlorinated biphenyl contaminated soil. Environ Toxicol Chem 22:1998–2004

Singla V, Pachauri T, Satsangi A, Kumari KM, Lakhani A (2012) Comparison of BTX profiles and their mutagenicity assessment at two sites of Agra. Ind Sci World J 2012:1–11. doi:10.1100/2012/272853

Stewart MA, Jardine PM, Brandt CC, Barnetto MO, Fendori SE, Mackay LD, Mehlhorn TL, Paul K (2003) Effects of concentration, aging and soil properties on bio accessibility of Cr (III) and Cr (VI) on soil. Soil Sedi Contam 12:1–21

Suja F, Rahim F, Taha MR, Hambali N, Razali MR, Khalid A, Hamzah A (2014) Effects of local microbial bioaugmentation and biostimulation on the bioremediation of total petroleum hydrocarbons (TPH) in crude oil contaminated soil based on laboratory and field observations. Int Biodeter Biodegrad 90:115–122

Tara N, Afzal M, Ansari TM, Tahseen R, Iqbal S, Khan QM (2014) Combined use of alkane-degrading and plant growth-promoting bacteria enhanced phytoremediation of diesel contaminated soil. Int J Phytorem 16:1268–1277

The Daily Times Pakistan (2003) Editorial: a tragic but avoidable disaster. The Daily Times Pakistan, August, 16, 2003. Available at http://archives.dailytimes.com.pk/editorial/16-Aug-2003/editorial-a-tragic-but-avoidable-disaster

UNEP/OCHA (2015) Sundarbans oil spill assessment. Joint United Nations/Government of Bangladesh Mission. Published by Joint UNEP/OCHA Environment Unit Emergency Preparedness and Environment Section Emergency Services Branch, Office for the Coordination of Humanitarian Affairs (OCHA), Switzerland

van Aken B, Yoon J, Schnoor JL (2004) Biodegradation of nitro-substituted explosives 2,4,6-trinitrotoluene, hexahydro-1,3,5-trinitro-1,3,5-triazine, and octahydro-1,3,5,7-tetranitro-1,3,5-tetrazocine by a photosymbiotic *Methylobacterium sp.* associated with poplar tissues (*Populus deltoides* x nigra DN34). App Environ Microbiol 70:508–517

van Hamme JD, Singh A, Ward OP (2003) Recent advances in petroleum microbiology. Microbiol Mol Biol Rev 67:503–549

Vaziri A, Panahpour E, Beni MHM (2013) Phytoremediation, a method for treatment of petroleum hydrocarbon contaminated soils. Int J Farm Allied Sci 21:909–913

Ventorino, Sannino F, Piccolo A, Cafaro V, Carotenuto R, Pepe O (2014) *Methylobacterium populi* VP2: plant growth-promoting bacterium isolated from a highly polluted environment for polycyclic aromatic hydrocarbon (PAH) biodegradation. Sci World J:1–11. doi:10.1155/2014/931793

Vessey JK (2003) Plant growth promoting rhizobacteria as biofertilizers. Plant Soil 255:571–586

Vidali M (2001) Bioremediation. An overview. Pure Appl Chem 73:1163–1172

Watkins J, Sorensen D, Sims R (1994) Volatilization and mineralization of naphthalene in soil grass microcosms. In: Anderson T, Coats J (eds) Bioremediation through rhizosphere technology. American Chemical Society, Washington D.C. 563, p 123–131

Xiao X, Chen H, Si C, Wu L (2012) Influence of biosurfactant-producing strain *Bacillus subtilis* BS1 on the mycoremediation of soils contaminated with phenanthrene. Intl Biodeter Biodegr 75:36–42

Xun F, Xie B, Liu S, Guo C (2015) Effect of plant growth-promoting bacteria (PGPR) and arbuscular mycorrhizal fungi (AMF) inoculation on oats in saline-alkali soil contaminated by petroleum to enhance phytoremediation. Environ Sci Pollut Res 22:598–608

Zhou QX, Hua T (2004) Bioremediation: a review of applications and problems to be resolved. Prog Nat Sci 14:937–944

Environmental Impacts of Nitrogen Use in Agriculture, Nitrate Leaching and Mitigation Strategies

Sadia Bibi, Saifullah, Asif Naeem, and Saad Dahlawi

Contents

S. Bibi (✉)
Institute of Soil and Environmental Sciences, University of Agriculture Faisalabad, Faisalabad 38040, Pakistan
e-mail: sadia_1565@yahoo.com

Saifullah • S. Dahlawi
Department of Environmental Health, College of Applied Medical Sciences, University of Dammam, Dammam, Saudi Arabia

A. Naeem
Soil Science Division, Nuclear Institute for Agriculture and Biology (NIAB), Faisalabad, Pakistan

K.R. Hakeem et al. (eds.), *Soil Science: Agricultural and Environmental Prospectives*, DOI 10.1007/978-3-319-34451-5_6

Abstract Nitrogen (N) fertilization has been found a powerful tool for increasing crop production since the last six decades. Except for legumes, which fix their N biologically by rhizobium, majority of the crops require N for the production of seed and forage. Ammonium (NH_4^+) and nitrate (NO_3^+) are major plant available forms of N in soil, the later having six times higher movement in and is therefore prone to leaching loss. Nitrate leaching down the soil profile results in low N use efficiency and contamination of underground water stream which is a major route of NO_3 entry into food chain. Nitrate related regulations, its health and ecological issues, contribution of food and water to nitrate ingestion and its main mechanisms of movement in soil hve been described for better understanding of the factors affecting NO_3 leaching. Nitrate leaching is governed by a number of factors that affect accumulation and movement of residual NO_3 in soil. These factors including plant characteristics, seasonal fluctuations, climate changes and soil properties are discussed in detail. Management of NO_3 leaching, which has been the main focus of this chapter is categorized into fertilizer, soil and irrigation based management options. Fertilizer management options take into account the balance use of fertilizers, right dose and time of application and controlled release of N through using nitrification inhibitors and slow release fertilizers. Organic agriculture, conservation tillage and growing of crops in high leaching risk associated season are proposed as soil management options. Irrigation management mainly comes around evapotranspiration based irrigation scheduling and wise use of deficit irrigation. In short, the chapter is an effort to make a comprehensive understanding of the reader about NO_3 leaching problem, its possible effect on human health and ecology and measures to manage NO_3 leaching without compromise on crop yields.

Keywords Nitrogen • Nitrogen use efficiency • Nitrate leaching • Evapotranspiration • Soil science

1 Introduction

Nitrogen (N) fertilization has been found a powerful tool for increasing crop production since the last six decades. Since N is a constituent of chlorophyll and many enzymes, it performs a significant part in different growth process of plants. Nitrogen induced increase in yield may be related to increased production of panicles in cereals and pods in legumes (Fageria et al. 2006; Fageria and Baligar 2007). Nitrogen also reduces grain sterility and improves grain or seed weights (Fageria et al. 2006; Fageria and Baligar 2007). Moreover, grain to straw ratio and harvest index of N (N uptake in the grain/N uptake in grain plus straw) which are positively associated with yield in field crops are improved by N application (Fageria et al. 2006; Hakeem et al. 2011).

Depending upon texture, surface soils (upper 15 cm layer) can naturally contain 0.1–0.6 % N (Cameron et al. 2013) representing 2000–12,000 kg N ha^{-1}. Except for legumes which fix their N biologically by rhizobium, majority of the crops require N for the production of seed and forage. Further, the N uptake and assimilation varies among different plant species and their parts. Recovery efficiency of the applied N lies around 30–40 %; the remaining part is lost by leaching as nitrate ($NO_{3)}$, denitrification to gaseous forms, volatilization from surface of high pH soils, surface runoff and immobilization by soil microbes.

Ammonium (NH_4^+) and NO_3 are major plant available forms of N in soil. Owing to its positive charge, NH_4^+ has very poor mobility in negative charged soils of the subtropical climate (Richter and Roelcke 2000). On the other end, NO_3^- movement in soil profile is six times higher than NH^{4+} with flowing water and is therefore prone to leaching loss (Dinnes et al. 2002). In addition to low nutrient use efficiency (NUE), NO_3^- leaching below soil profile results in contamination of underground water stream. Since cereal grains, the most consumed food, contain negligible nitrate, drinking of nitrate polluted water or its use for growing crops is the main route of its entry into food chain. Upon ingestion of NO_3^-, it is acted upon by some bacteria or enzymes in the digestive system that reduces it to nitrite (NO_2^-). Nitrite then absorbs in blood where it oxidizes Fe^{+2} to Fe^{+3} to convert hemoglobin into methemoglobin as a result of which its level increased in blood than normal. Methemoglobin has more affinity for oxygen due to which capacity of red blood cells to release oxygen to tissue decreases. The resultant hypoxic condition is known as methemoglobinemia or blue baby syndrome. At very high concentration, NO_2^- may react with amines and amides and form cancer causing compounds (nitrosamide and nitrosamines). The death rate of gastric cancer patients was found in strong correlation with daily nitrate intake rate in 12 countries of the world (Fine et al. 1982).

There are a number of factors that can affect accumulation and movement of residual NO_3^- in soil. Among these, dose and time of fertilizer application, irrigation schedule and tillage practices are the most important to be considered in order of their significance. Heavy application of N to soil could result in high NO_3^- leaching, low NUE and high risk of water contamination. Dose of applied nitrogen has positive correlation with leaching of NO_3^- away from active root zone (Paramasivam et al. 2002; Jalali 2005). Similarly, high leaching losses of NO_3^- were reported by Fan et al. (2010) at 225–300 kg ha^{-1} N application compared to that below 150 kg ha^{-1} N application. Therefore, it seems that application at optimal level can minimize leaching losses of NO_3^- (Sexton et al. 1996). However, under applying N results in malnourished plants which at later growth stage would not be able to metabolize NO_3^- within their body and efficiently utilize N from well nourished

(NO_3^- sufficient) soil. Application of farmyard manure together with chemical fertilizer can also increase NO_3^- buildup for leaching loss.

In addition to heavy fertilization, problem of NO_3^- abundance in soil profile may also occur when time of application does not synchronize with plant demanding stage. In most of the cereal crops surface distribution of solid fertilizer is difficult at mid-stage because of their tall stature and traffic difficulty. So, presence of high N in soil when there is no crop or when the crop's demand is very low (e.g. before emergence and/or at harvesting) would result in high leaching losses (Shi et al. 2012). Moreover, NO_3^- leaching seems to be higher whenever the abundance of NO_3^- in soil profile coincides with or followed by a period of high rainfall/heavy irrigation. Split application and avoiding fertilizer application during heavy rainfall period (monsoon season) could enhance NUE and reduce NO_3^- leaching losses (Jia et al. 2014). The common practice for N fertilizer application in cereals is to apply half at sowing and remaining half in two or three equal splits at critical growth stages.

Around the globe, surface and sub-surface waters were found to contain NO_3^- at levels exceeding the maximum permissible limit (MCL) recommended by WHO (2004) (Tahir and Rasheed 2008; Iqbal et al. 2013). Some reports say that no tillage (NT) or reduced tillage practices favour the formation of continuous soil macropores which may enhances preferential flow of NO_3. Conversely, lower NO_3^- leaching under NT system than common tillage practice has been reported and found to be associated with decreased mineralization or denitrification of N under the former system (Randall and Iragavarapu 1995; Patni et al. 1998). These reports has begun the debate if intensive agricultural activities like high rates of N fertilizers, repeated application of organic manures, tillage practices and/or high levels of irrigation are responsible for high levels of NO_3^- in water? Due to entirely different climatic condition and management practices at each and every sphere of the world, this review discuss the main causes and management options specific to particular climate and soil. It is also a need of time to identify water saving irrigation practices along with proper rate and time of N application to improve the yield as well as decrease leaching losses of N. Further, it seems necessary to determine the movement and buildup of residual NO_3^- in response to different tillage and fertilization practices under arid to semi-arid climatic conditions.

Optimal N management in agroecosystem is yet a debatable issue. This chapter discuss the fate of N in response to different management strategies like nitrogen source, rate and timing of application, irrigation and tillage systems. Further, this chapter mainly provides a discussion of practical aspects of N management to reduce its surface and subsurface water pollution.

2 Nitrogen in the Environment

Gain and loss of N in the agroecosystem system is associated with many complex and interlinked processes. In agricultural systems, the main routes for N loss are: (a) Gaseous emissions as ammonia volatilization and denitrification (b) leaching (i.e., removal below root zone with percolating water) (c) Plant uptake (d) surface runoff. The N cycle can be easily understood with the help of simple mathematical equation as follows:

$$N_{net} = N[e + bf + c + om + min.] - N[pl + g + i + l + r]$$

The positive sign indicates the addition of N and negative sign indicates the depletion of N from soil. Where, N_{net} is net N added into the soil, e is electrical discharge, bf is biological fixation, c is chemical fertilizer, om is organic manure, min is mineralization, pl is uptake by plants, g is emissions as volatilization or denitrification, i is immobilization, l is leaching and r is surface runoff.

Plant uptake and surface runoff losses are minimal. The losses through volatilization is significant at pH usually above 8.0, high temperature and low CEC soils. Anoxic conditions are favorable for denitrification. Leaching of NO_3^- in ground water make it more detrimental for human health compared to other chemical elements (Garcia et al. 2012).

3 Nitrate Leaching from Soils

Nitrogen fertilizer is a worrisome source of NO_3^- leaching to groundwater. The amount of N leached through the soil profile depends upon the quantity of N present in soil solution and the one that drained over a prescribed period of time (Cameron et al. 2013). Four major forms of N are present in soil, (a) contained in organic matter, (b) part of microbial bodies, (c) NH_4^+ ions bind on the clay surface and organic matter (d) mineral forms of N (NH_4^+, NO_3 and traces of NO_2^-) in soil solution. Because there is no significant adsorption of NO_3 onto the soil surface and the NO_3 is water soluble, it moves downward rapidly with water passing through soil profile which is economically and environmentally undesirable. When it enters in ground water there is a little chance for denitrification. Although, in groundwater there are anaerobic conditions but absence of organic carbon (C) reduces denitrification rate. The problem of NO_3 leaching is more severe in developed countries where N fertilizer and organic wastes are applied at higher rates. In recent years, NO_3 was also detected in ground water of some developing countries including Pakistan due to increased use of fertilizer and raw manure (Tahir and Rasheed 2008).

4 Nitrate Related Regulations

The NO_3 and NO_2 are considered hazardous and legal limits are set for their safe concentration in drinking water and food. The maximum concertation of NO_3 in drinking water set by U.S. Environmental Protection Agency (U.S. EPA 1991) and World Health Organization (WHO 2004) is 50 mg L^{-1} (equivalent to 10 mg N L^{-1}). Moreover, 3.7 mg NO_3 per kg^{-1} of body has been set as the maximum acceptable daily intake (ADI) level of NO_3 by the European Commission's Scientific Committee on Food and The Joint Expert Committee on Food Additives (JECFA) of the Food and Agriculture Organization of the United Nations/World Health Organization. Intake below the above prescribed level is considered as safe for healthy children and adults. Concentration of NO_3 is almost fifty times higher in vegetables than drinking water (US EPA 1991); vegetables often contain >2000–3000 mg NO_3 per kg fresh weight. Intake of dietary NO_3, however, is less likely to increase nitrosation, because of the presence of nitrate reductase enzymes in vegetables.

4.1 Primary Health Issue: Methemoglobinemia

The primary risk of nitrate loaded drinking water is the development of methemoglobinemia from NO_3^- derived NO_2^-. In red blood cells (RBCs), iron (Fe) is usually present in reduced state, i.e. ferrous (Fe^{2+}). The NO_2^- oxidized it to ferric (Fe^{3+}) state which has reduced or negligible capacity to carry oxygen to vital organs of body. The hemoglobin (Hb) containing Fe in Fe^{3+} state is known as methemoglobin (MHb). Although, the production of MHb is a normal process human metabolism but its concentration remains within safe limits. However, when level of MHb is too high not to transfer oxygen to cells, is a condition known as methemoglobinemia. When MHb is 1 and 2 % of the total Hb in adults and infants, respectively, it is considered in safer zone (Denshaw-Burke et al. 2014). The RBCs do contain mechanisms to stop this oxidation process and reverse the reaction to form Hb. However, the RBCs have finite life span and later on are unable to resist against oxidation process. Oxidative stress results in ageing of cells due to production of MHb that is not removed from blood circulation (Denshaw-Burke et al. 2014).

The early manifestation of methemoglobinemia is Cyanosis. It is evident only at 5–10 % conversion of Hb to MHb and is indicated by bluish lips and nails (Denshaw-Burke et al. 2014). The blood color of methemoglobinemia patient is chocolate brown and other indicators include sleepiness, vomiting and diarrhea, and in severe conditions even lead to death due to deprivation of oxygen to body cells.

4.2 Secondary Health Issues

The secondary health problems associated with ingestion of excess NO_3 include acute respiratory infection, thyroid problems, birth defects and colon cancer etc. In addition, the scientific research suggests that the ingestion of NO_3 may cause transmissible changes in the structure of the genetic material of cells that contribute to bladder and ovarian cancers and also to risk of non-Hodgkin's lymphoma. It could also be a reason for the development of thyroid hypertrophy, insulin-dependent diabetes mellitus and respiratory tract infections or causes spontaneous abortions. The NO_3 polluted drinking water causes severe problems in already sick people (malaria and cholera) and develop symptoms of vomiting, pneumonia, nausea, diarrhea, hepatoenteritis, gastroenteritis, muscular cramps, and several poisoning syndromes.

5 Contribution of Water and Food to NO_3 Ingestion

The occurrence of methemoglobinemia is related with the ingestion of drinking water, with most common cases associated with well water. Most victims of NO_3 were reported with water source having NO_3 level up to 50 mg L^{-1}. However, some reports indicate high level of NO_3 in sterilized/boiled water due to having been concentrated by evaporation (Bruning-Fann and Kaneene 1993). Bacterial contamination of high NO_3 water also causes more conversion of NO_3 to NO_2 before its entry to stomach and poses serious risks.

In humans, the intake of NO_3 through food often do not cause toxicity. This might be due to the reason that ascorbic acid like compounds present in foods chelate the NO_3 and minimizes its reduction to NO_2 in the gut. The most common food related NO_3 toxicity was reported in infants consuming formula milk prepared in well water (Bruning-Fann and Kaneene 1993). Moreover, grains can accumulate negligible amount of NO_3 which is even useful to fulfill human protein needs. Many vegetables like spinach, lettuce and root vegetables contain high levels of nitrates. Further, the processing and handling of these vegetables may also increase the risk to consumer. For example, spinach and carrots stored at room temperature contained more NO_3 as compared to fresh spinach (Fomon 1993).

6 Nitrate Related Ecological Issues in Aquatic Ecosystems

Leaching of NO_3 into surface and ground waters is one of the pathways by which it enters into aquatic ecosystems. This inorganic N can disturb the aquatic ecosystems by three ways: First, it can decreases the pH of fresh water by increasing the concentration of H^+ ions which ultimately reduces the acid-neutralizing capacity of lakes. Second, it can result in eutrophication of lakes by enhancing growth and proliferation of primary producers. Third, high concentration may be too toxic to impair the ability of aquatic life to survive and reproduce.

The decrease in pH of water could results in the production of mobile aluminum (Al^{3+}) and other heavy metals like cadmium (Cd), copper (Cu), lead (Pb) and zinc (Zn) (Nelson and Campbell 1991). This dissolved Al^{3+} then reduced the availability of orthophosphate and disturbs the P cycling in water system. A pH range of 5.5–6.0 is considered as a threshold limit below which organisms cannot survive. Some nutrients like silicon (Si) and iron (Fe) can significantly increase the mass of algae; however, according to recent literature NO_3^- is considered the primary cause of cultural eutrophication of aquatic ecosystems (Anderson et al. 2002; Smith 2003). Human activities which according to estimates have increased N fluxes into the coastal waters of the Gulf of Mexico by 4- to 5-fold, into the coastal waters of the northeastern USA by 6- to 8-fold, and into the European rivers draining to the North Sea region by 6- to 20-fold are the main cause of NO_3 linked eutrophication (Smith 2003).

The regular monitoring of aquatic ecosystems to prevent the eutrophication problem has been previously based on P and chlorophyll-a concentrations; in recent criteria, N has also been involved. The upper limits of N suggested for eutrophic temperate lakes and streams are 1260 and 1500 μg/L, respectively (Smith 2003). The well-known cases of hypoxic (or anoxic) water bodies due to cultural eutrophication are the Baltic and Black Seas, Chesapeake Bay and the northern Gulf of Mexico (Anderson et al. 2002; Smith 2003). In these water bodies, suitable habitat for food, growth and reproduction of both invertebrates and fishes (sensitive benthic species, particularly) significantly reduced and their excessive deaths were recorded (Anderson et al. 2002). This adverse effect of hypoxic condition on aquatic animals could further be aggravated by the formation of reduced ionic species, such as hydrogen sulphide (H_2S) (Breitburg 2002). Hydrogen sulphide affects the nervous system even at very low concentrations, leading to mortalities in aquatic animals.

7 Physical Transport Mechanisms of NO_3

The movement of any dissolved ion like NO_3 in the field is governed by the following physical mechanisms:

7.1 Convective/Mass Flow

This occurs because of the movement of NO_3 along with the actual movement of water through the soil during drainage events. Under saturated zone, the convective flux of NO_3 is in steady state and hence can be described by Darcy's law as follows:

$$Jc = J_w C = -K_s \ dH / dz \tag{1}$$

Where, J_c is the mass of NO_3 per unit area per unit time transported by convection, C is the nitrate concentration in mass per solution volume, J_w is the water flux, K_s is the saturated hydraulic conductivity and dH/dz is the hydraulic gradient. The distance transported per unit time by convection depends on the average pore water velocity, v, where:

$v = q/\theta_v$ and θ_v is the volumetric water content.

Under unsaturated conditions, the transient flow of NO_3 along with water can be explained by Richards's equation:

$$\partial\theta / \partial t = \partial / \partial z [K_h (\partial h / \partial z + 1)] \tag{2}$$

Here, Kh is unsaturated hydraulic conductivity, $\partial\theta/\partial t$ is change in water contents with time and $\partial h/\partial z$ is change in matric potential with space. As such this equation is not solved because of two variables h and θ.

Convective transport implies uniform displacement of the pulse of NO_3 just like a piston which is true only for structureless soils. But, in reality, processes of diffusion and hydrodynamic dispersion tends to spread the NO_3 pulse throughout soil profile. In field, the convective flow paths of the solution are never estimated exactly, another volume-averaged expression is used to describe convective flow. A separate solute transport mechanism, called hydrodynamic dispersion is included as an average of three dimensional convection.

7.2 Diffusion

Dissolved/water soluble NO_3 spread out as a result of random thermal motion of ions/molecules, a process known as molecular diffusion. This net movement of NO_3 is generally proportional to concentration gradient, the cross-sectional area for diffusion and the time available for diffusion. In one dimension, Fick's first law for steady state transport is explained as:

$$J_{NO3} = -D\partial C / \partial z \tag{3}$$

Where, J is NO_3 flux density (kg/m^2.s), D is molecular diffusion coefficient of solute in solution (m^2/s) and dC/dz is the concentration gradient. When value of concentration gradient becomes one, the value of D and solute moved will be more. Diffusion coefficient depends upon soil texture, physical and chemical properties of ion/molecule, temperature, salt and water content. In actual field conditions, the expression must be modified due to decreased cross-sectional area and increased actual path length by solid and air spaces. The modified form of Fick's law is:

$$J = -\epsilon(\theta) D\partial C / \partial z \quad (4)$$

Here, € is the tortuosity factor whose value is < 1 and according to studies tend to decrease in non-linear fashion with decreasing θ

7.3 Hydro-dynamic Dispersion

The NO_3 not only moves with the flowing water as shown in Eq. 1, but also mixes with the soil solution of different chemical composition. Being non-adsorbing solute NO_3 has no interaction with the soil surfaces (no sorption) and produces solute concentration an "S shape curve" that varies with time. The process of dispersion can easily be visualized with the help of simple cylindrical column which is already filled with water. By the law of water conservation, each segment of the column must have same one dimensional flux density in steady state conditions. The NO_3 which is introduced from the inlet end not only diffuses and be convicted with Jw, but also spread out around the solid barriers. This velocity distribution is three dimensional due to three boundary effects: one boundary effect is due to more velocity of solute at the center of pores than along the edges, second is due to the pore size distribution and the third boundary effect occurs when actual flow path of solute fluctuates with respect to mean direction of flow. The mathematical equation for hydrodynamic dispersion is as follows:

$$J_h = -D_h \partial C_l / \partial z$$

7.4 Sorption

Sorption of anions (negatively-charged ions) like halides and NO_3 is less likely to occur in groundwater but most commonly noted in soils that contain allophone, imogolite and other poorly-crystallized oxide or hydroxide materials.

8 Factors Affecting NO_3 Leaching in NO_3 Leaching Environments

8.1 Plant Characteristics

Root systems play a very important role in NO_3 leaching. Root length and surface area are the two most important parameters indicative of nutrient acquisition (Thorup-Kristensen 2001). If the rooting zone is small and shallow, the highly mobile NO_3 ion can easily search the way to groundwater. However, on the basis of diffusion theory NO_3 uptake is more even in low rooting density crops. But, it does not account for the ease with which NO_3 can escape low rooting density crops. The difference is only due to root density distributions in the top soil and subsoil. Generally, the rooting densities are more in top soil as compared to sub soil. The increase of rooting density in this region may enhance water and NO_3 uptake, reduces the chances of leaching. It was found true for field-grown maize having direct relationship with subsoil root growth and NO_3 uptake (Wiesler and Horst 1994). Catch crop species showed a strong correlation with NO_3 depletion from that zone and subsoil root proliferation (Thorup-Kristensen 2001). Habib and La Folie (1991) findings contradicted to that of above scientists and concluded that the top soil rooting densities have pivot role in reducing NO_3 leaching because N source is in the surface layer. The plants having more rooting densities along with deep tap roots exploit more the mineralization zone, reduce the downward displacement of water. This type of root architecture helps the plants to store water and NO_3 and reduce the risk of drought stress and nitrate pollution. The root length of a specific plant depends upon its genetic makeup and temporal and spatial distribution of nutrients. Further, in non-homogenous field situations less branched rooting pattern is optimal for nutrient acquisition in one environment; but more branched pattern for second environment. Model root architecture is required for ions of different mobility over a wide range of environments and soil textures.

The thinner roots are more efficient to capture NO_3 compared to coarse roots because of infinitely small diameter and more reactive surface area. But, the finer roots have some disadvantages too; they are more susceptible to herbivore attack, less capable of exploring compact soils and limited growth potential and transport capacity.

8.2 Seasonal Fluctuations

Seasonal fluctuations in climatic conditions along with the drainage events are one of the promising factors affecting NO_3 leaching. The greatest NO_3 leaching losses were recorded during autumn, early winter and late summer months because of slower plant N uptake due to cooler weather conditions with high amounts of drainage (Wild and Cameron 1980). In autumn and winter, drainage is high because of slower evapotranspiration. The autumn applied N fertilizer had nitrate leaching losses between 15 and 19 % than spring applied fertilizer (8–11 %). Autumn rainfall after crop harvest can also cause mineralization of organic N and leaching of residual soil NO_3 (Cameron et al. 2013). So, the efficiency of autumn applied N (NUE) is low than spring applied.

But in tropical regions like Pakistan, where summers are not dry and 60–70 % rains are concentrated in monsoon season, the leaching of NO_3 is often higher in the summer than the winter. Cameron et al. (2013) concluded that there were greater leaching of NO_3 in summer monsoon (400–500 mm rainfall) than winter (100 mm rainfall).

8.3 Climate Change

Climate change is also a contributing factor towards the leaching of NO_3 down into ground water by modifying key soil processes that control crop growth (MAFF 2000). Increased CO_2 enhances rate of photosynthesis and it can demand application of additional fertilizers, however, rainfall and temperature can be both detrimental and beneficial. The factors governing the rate of mineralization of organic N are nature and abundance of the organic matter, humidity, temperature, pH and faunal activity. According to Leiros et al. (1999), increased ambient temperature is expected to decrease soil organic matter content which in turn will affect its hydraulic properties. Nitrate build up in soil are linearly affected by temperature and soil organic matter content (Leiros et al. 1999) leading to an increased risk of leaching (Olesen et al. 2002).

Although the mineralization and nitrification are directly related to temperature and indirectly to rainfall (Emmett et al. 2004). However, the changes in soil moisture during the summer season (Leiros et al. 1999) and uptake of NO_3 uptake by vegetation (Ineson et al. 1998) determine the extent of overall effects caused by these agents. Generally, microbial and enzyme activities are low when the soil is either too dry or saturated (Sardans et al. 2008). While reviewing the effects of wetting and drying cycles on mineralization, Borken and Matzner (2009) concluded that increasing summer precipitation could enhance N and C fluxes whereas increasing summer droughts will reduce them. The commonly observed pulse in net mineralization of N and C following wetting of dry soil is short-lived because it is derived from release of solutes and exposure of hidden organic matter, accumulated plant necromass and microbial cell lysates. To simulate the effect of increasing temperature on mineralization of organic N in soil, Rustad et al. (2001) simulated the effect of increased temperature on net N mineralization rate and plant productivity and reported increased of 46 and 19 %, respectively as a result of artificial warming in the range 0.3–6 °C over a period of 2–9 years. Overall, climate change scenarios is appearing to enhance mineralization of N in soil.

Proportioning between run-off and infiltration is an important control on N leaching to groundwater. Changes in rainfall intensity and hydraulic properties will lead to change in partitioning of NO_3 between run off and recharge. In a test farm of Netherlands, weather-induced fluctuations in NO_3 are found to be in the range of 55–153 % of average field concentration (Rozemeijer et al. 2009). Callesen et al. (2007) showed that periods of frost results in large losses of NO_3 and attributed this change to increased mineralization and ammonification. Contrarily, Matzner and Borken (2008) suggested that post-frost changes in NO_3 pulse are more likely to be associated with reduced uptake rather than increased mineralization. Under alpine, arctic and forest vegetation elevated nitrate losses from soils occurred only in the year following exceptional soil frost. Decrease in losses with short-term repeated events evidenced that pool of N susceptible to freeze-thaw events is rather limited. Different attempts undertaken to model the impact of climate change on N leaching and crop has generated variable results. For example, Eckersten et al. (2001) simulated the possible consequences of both elevated atmospheric CO_2 and temperature using two linked process-oriented models (SOIL/SOILN). The model predicted an increase of 10–20 % in present value of winter wheat production by the year 2050. Precipitation and drainage was expected to increase with the consequent increased in N leaching flux by 10 kg ha^{-1} $year^{-1}$. However, Ulen and Johansson (2009), based upon the simulation using tile drain and piezometer, reported this value to be only 0.06 kg ha^{-1} $year^{-1}$ due to an increased temperature of 2 °C during the growing season from 1993 to 2005 (April to September). They also predicted an increase in precipitation by 16 mm, mainly in June. The projected increased frequency of droughts and decrease in summer recharge will lead to an increased requirement for agricultural irrigation.

8.4 *Soil Properties*

Soil parameters such as soil texture, hydraulic conductivity, residual water content, porosity and cation exchange capacity (CEC), predict leaching potential of soil (Vachaud and Chen 2002). Most of the alkaline and calcareous soils having pH around 8 are negatively charged, NO_3 cannot be retained in these soil. The mean content of N differs among the soil texture; heavy soils have good water holding capacity due to high porosity which resulted in low leaching. So, because of slower drainage and the greater potential for denitrification, fine texture soils exhibit less NO_3 leaching than coarse textured ones (Di and Cameron 2002; Fan et al. 2010). On the same grounds, Liu et al. (1998) reported higher NO_3 leaching in sandy soil (200 cm depth) compared to clayey soil (100 cm depth) in Loess Plateau of northern China. The extremes of NO_3 loss in one growing season on coarse textured soil reported by Tong et al. (2005) might be due to excess N supply beyond the crop needs for optimum growth/yield.

There is a great variability in pore-size and continuity in spatial distribution of pores that will contribute to irregular movement of water down to the soil profile. For example, macropores created by wetting and drying cycle and activity of roots/earth worms can allow NO_3 to leach down into deeper soil layers (Silva et al. 2000). When there is a heavy rainfall after a long dry spell, residual N present in soil can be washed through large soil cracks or channels by irrigation water or rainfall, by-passing through the fine pores. While macropores may only constitute 5 % of the total porosity of a soil, they allow ready movement of water, NO_3 and other solutes (Bouma et al. 1981). This transport phenomenon is very difficult to explain because of its high spatial and temporal variability.

pH, organic C, potentially available N, NO_3-N, sand content and hydraulic conductivity has significant positive correlations with NO_3 concentration in soil water, while bulk density and clay content had significant negative correlation. The soils having negative charges on their exchange sites have high potential to loose NO_3 than positively charged acid tropical soils. High N and organic C load and low clay content in soil profiles with high hydraulic conductivity would ensure NO_3 in ground water. The type of soil govern the rate of water infiltration and hence the processes of nitrification and denitrification.

9 Management Options to Minimize NO_3 Leaching

As seen in the above discussions, the NO_3 leaching is multifaceted problem and there is no single magical cure that can solve it. An integrated approach is required to minimize leaching losses and increase. In broader sense, options to minimize NO_3 leaching are fertilizer, soil and irrigation based management strategies which are discussed in detail as follows;

9.1 Fertilizer Based Management Options

9.1.1 Balanced Fertilization

The pressing need to feed the growing population propelled the farmers to apply N fertilizer at higher rates, which is a root cause of NO_3 accumulation in soil. To increase crop yield while keeping NO_3 at minimum, balanced fertilization is an effective method. For example, the application of P fertilizer along with N may decrease nitrate leaching (Fan et al. 2003). Zhang et al. (2004) reported NO_3 accumulation in soil in a wheat-maize cropping system over 9-years period in the following order: N>NK>NPK>NP>CK>PK. Yuan et al. (2000a) reported much lower accumulation of NO_3 with NP where N was applied at 220 kg ha^{-1} compared to NK or N alone with N application at 1171 and 1075 kg ha^{-1}, respectively. The most probable reason of low NO_3 accumulation in soil was much higher N uptake (1360 kg N ha^{-1}) by plants in NP treatment compared to NK and N alone resulting in uptake of 720 and 800 kg of N ha^{-1}. Alone applied N resulted in not only high concentration of soil NO_3, but also in its movement to deeper layers in the soil profile (100–180 cm) compared with NP treatment (80–120 cm layer). Manure along with NPK further decreases NO_3 leaching. But, manure applied at higher rates can enhance nitrate leaching e.g., by applying poultry manure which contain higher proportion of N, 40–75 % of accumulated N leached to 200 cm depth and NO_3-N concentration in water of 50 % wells exceeded 10 mg N L^{-1}. Therefore, manure should be applied at lower rates to minimize its negative impacts on the environment. In short, balanced fertilization at proper rates may decrease NO_3 accumulation in the soil profile and further its water contamination.

9.1.2 Right Dose of N Fertilizer Application

Due to rising food demand, grain yield goal is currently considered as the sole independent variable for determining plant N recommendations all over the world. The realistic yield goal should not be more than 10 % of the current or recent average yield of the farm. It has enforced the agriculturists to adopt intensive agriculture, including increased application of water and fertilizer (Rong and Xuefeng 2011). It is believed that if the Haber-Bosch process for industrial fixation of N had not been invented, 40 % of the current human population would not be alive. However, the increase N fertilizer inputs and crop yields are not concomitant; low nitrogen use efficiency resulting from leaching potential of NO_3 might be the reason. The fertilizer N applied in excess of crop demands may happen when residual soil inorganic N content is not properly considered or when estimated yield goals are larger than the expected yields under particular soil types and climate (Keeney 1997).

Worldwide, dominant and main source of N input in the crop production systems is the application of chemical fertilizers. About 60 % of global N fertilizer is used for producing the world's three major cereals viz. rice, wheat and maize contributing to reliance of 50 % of the human population on N fertilizer for food production. The extent of quantity N loss and its depth varies with soil, crop and experimental conditions. Abbasi et al. (2011) concluding from a field experiment on maize that there was non-significant difference between 60 and 80 % of fertilizer level regarding NO_3 losses but, 80 % treatment was better when considering crop yield along with NO_3 losses. At 100 % application of N (400 kg ha^{-1}), NO_3 losses were maximum. In a sandy farmland of North-West China, application of 225, 300 and 375 kg N ha^{-1} caused higher NO_3-N accumulation in soil compared to 0 and 150 kg N ha^{-1} (Rong and Xuefeng 2011). No NO_3 leaching was observed when N fertilizer was applied below 150 kg N ha^{-1} whereas, the N rate above 400 kg ha^{-1} caused NO_3 leaching and decreases fertilizer use efficiency (Barraclough et al. 1992. Although rate of N varied with crop, application at higher than the recommended for wheat and maize resulted in increased residual NO_3 in soil and its leaching below soil profile at the later growth stages with the movement of water (Wang et al. 2010; Jia et al. 2014).

9.1.3 Right Time of Fertilizer N Application

Usually, too much N is applied at early developmental stages where crop needs are minimal. The limited N application at the end of crop growing season and before the next crop favors the establishment of extensive rooting system and reduces losses (Al-Kaisi and Yin 2003). N application in four splits either in the urea or manure form resulted in less nitrate leaching than that in two splits. According to Isidoro et al. (2006) and Claret et al. (2011), application of N with first irrigation shows highest leaching losses (35–43 %) which could be attributed to less biomass at seedling stage, poorly developed root systems and less assimilation by plants. That's why, higher agronomic efficiency of applied N with winter wheat as obtained when the first N dose was applied 90 days after seeding. Reported that when total application of 102 N ha^{-1} was splitted in ten equal doses, only 6 % of the applied urea-N lost by leaching while corresponding value for single dose applied at transplanting was 13 %. Hence, single dose application can reduce NUE by 50 % and ultimately results in reduced yield which was also witnessed by Dunbabin et al. 2009.

The ill-timed application may disturb N balance of soil resulting in increased residual N build up directly proportional to the rate applied of fertilizer. No doubt, this is a positive sign of soil fertility gain for the upcoming crop, but farmers again apply N to the next crop irrespective of the current status of the soil. (Claret et al. 2011). When the sowing of winter-autumn crop is delayed, then topdressing of N fertilizer should be preferred than fertilization at the time of sowing to minimize NO_3 loss. It has been shown that recovery of N was more when fertilizer was applied at tillering rather than at emergence under similar agro-environmental conditions (Kirda et al. 2001; López-Bellido et al. 2005). Reported that optimum ratio of base to topdressing was 50:50 and suitable top-dress developmental stages were jointing and anthesis. Shi et al. (2007) concluded from their study that irrespective of rate and type of fertilizer applied, increased topdressing of N clearly elevates NUE by reducing NO_3-N losses and shows no difference in nitrate accumulation in plants. The sound management to reduce NO_3 leaching in corn was to side-dress N at six and twelve leaf stage but delayed application at sixteen leaf stage minimized yield benefits (Jaynes 2013).

9.1.4 Nitrification Inhibitors and Controlled-Release Fertilizers

The NH_4^+ has the ability to be sorbed on high CEC soils but NO^{3-} can be leached down. This leaching process can be slowed down by lowering the population and activity of *Nitrosomonas* bacteria in the first step of nitrification process. The N fertilizers are used along with different chemicals like nitropyrene. However, the most practiced technique is to retard entry of water into the fertilizer particle and exit of N out by coating of water soluble N fertilizer with less soluble materials. Generally, three types of materials have been used for the purpose: (1) Encapsulated urea; coating of urea with impermeable material to allow slow entry of water and exit of soluble N, (2) Encapsulated urea needing disintegration; coating of urea with impermeable material that needs to be broken physically, chemically, or biologically before the N is dissolved, and (3) Semi-permeable coated urea: By diffusion, water get inside creating sufficient internal pressure that destroy coating.

Sulfur and neem-extract coated urea (SCU) has been recommended as a promised technique with characteristics of slow-N release. Elemental sulfur (S) was preferred because of its relatively low cost and easy handling. Now a days, variety of polyolefin resin coated slow-release N fertilizer are available in the market. The strength of coatings can be adjusted to provide a range of N release rates that are suitable for a variety of cropping systems. However, it needs further research to optimize the application of these newer materials for different crops and soil types.

9.2 *Soil Based Management Options*

9.2.1 Shift to Organic Agriculture (Merits vs Demerits)

The solid wastes contain large amount of N and usually applied to the soil in their present form without going to any preparation procedure like conversion in to composting, biochar etc. As we are already aware, in these manures N is present in organic form and converted to inorganic form through mineralization process which ultimately is a serious risk to environment.

It is evident from literature that mineralization can release up to 50 % of manure based organic-N (Power and Doran 1984). After a period of 3 months, about 13 % of the N mineralized is from non-composted aged cattle manure (Hartz et al. 2000). Klausner et al. (1994) recorded that decomposition of organic N form dairy manure was 21, 9, 3 and 2 % over a 4-year period. In another study, it was concluded that over an application period of 21-year, mineralization rate of cattle manure was 56 % (Chang and Janzen 1996). This uncertainty and variability of organic N mineralization increases the risk of over and under application. In nitrate vulnerable zones, the maximum permissible limit of N from manure is 170 kg ha^{-1} $year^{-1}$ (Mantovi et al. 2006).

The soils in arid zone of world are deficient in organic matter which enforce the farmers to add high inputs of external fertilizers (either organic or inorganic) into nutrient poor drylands. Organic matter is a main pool of N which it is released from organic matter by mineralization. There is a growing interest in organic agriculture as an environmentally friendly alternative to conventional agriculture because it is a practice of choice around the dynamic world. But, along with benefits there are some demerits if the organic wastes when not applied without proper evaluation.

Manure from dairy production can serve as a valuable source of N for agricultural fields but, efficient use of animal manure is a greater challenge than mineral fertilizer. From environmental point of view, repeated and heavy application of manure in agriculture is questionable. Mostly, the solid manure is spread over the surface of soil in fields just before planting and is either left on the surface or incorporated (Tarkalson et al. 2006) without considering the soil and manure specifications regarding N. The large proportion of N in cattle manure collected from dairy farms is organic fraction. However, the proportion of organic N that can be absorbed by plant roots during the first and subsequent growing periods is called as plant available N (Tarkalson et al. 2006). Mineralization of organic material as a result of microbial activity is influenced by several factors like type of manure (age, feed and sex of animal), C/N ratios, water soluble and recalcitrant compounds, soil moisture, temperature, pH and oxygen availability (Sistani et al. 2008). These factors may differ both spatially and temporally making it difficult to determine exact availability factor from site to site and through years.

The effects of independent application of mineral fertilizer and manure at agronomically optimum rates are highly discordant (Diacono and Montemurro 2010). One group of researchers thought that manure is considered the root of all evil. It is metaphorical for environmental degradation. For example, Basso and Ritchie (2005) observed that the total amount of NO_3 leached was 681 kg ha^{-1} in the manure treatment followed by the compost and then chemical fertilizer with values of 390 and 348 kg ha^{-1}, respectively. Because, in manure whole of the N is applied at the one time, so more leaching occur as compared to chemical fertilizer (mostly urea) which is normally applied in three splits. Dividing application of fertilizer in more splits causes more adaptation between plant and fertilizer. This management helps to use fertilizer when plant needs fertilizer. When half of the total fertilizer is applied before planting and field is repeatedly irrigate, large fraction of N is leached as plant uptake is low at this stage.

A long duration study (135 years) was carried out to check the effect of farm yard manure compared to N fertilizer at Rothamsted Experimental Station (Powlson et al. 1989). The continued use of FYM equivalent to 238 kg N ha^{-1} increased the total soil N content (0–23 cm) to 7680 kg N ha^{-1} compared with 2570 kg N ha^{-1} for the N fertilizer treatment equivalent to 144 kg N ha^{-1}. The NO_3 leaching losses were five times greater in FYM treatment compared with fertilizer treatment leading to conclusion that mineralization of organic N would have contributed a significant part to the NO_3 loss. Stoddard et al. (2005) also recorded significant increase in NO_3 concentration in manured soils as compared to inorganic treatment. The leaching was greater in winter as compared to summer which was mainly due to late fall and early spring mineralization of organic N resulting in excess of crop N uptake in summer and also in the fall leading to elevated levels of NO_3 in leachate during winter. Long term studies show that up to 50 % more NO_3 leaching occurs due to annual manuring relative to control soils, because of the gradual buildup of mineralizable N in manured soils and the loss of soil organic matter in un-manured soils (Shepherd and Newell-Price 2013; Pang and Letey 2000). This occurs only when soils are manured both in winter and summer annually without considering its residual effect. Continued manure applied organic N that is not mineralized in the first year is added into cumulative organic N pool that raises future N availability. This is most commonly experienced in sandy soils than clayey soils (Hassink 1995). Shepherd and Withers (1999) suggested that manure should be applied in rotation, for example, once in every 3 years because manure application may increase NO_3 loss if there is no synchronicity between N mineralization and crop N uptake. Chadwick et al. (2000) did not recorded extra leaching of NO_3 in the first year after manure application, but significant leaching in the following years. The manure based nitrate-N leaching could also be increased if a large amount of N is supplied to crop from manure without adjusting subsequent inorganic fertilizer application resulting in post-harvest residue of soil mineral N. Taking into consideration of all the above discussion, one solution to slow the build-up of N in the soil is to apply manure at low rates and more frequently.

9.2.2 Conservation Tillage

Tillage systems significantly affect dynamics of N in soil through their effect on N pools in the soil system. Tillage increases soil aeration, porosity and hydraulic conductivity which can increase residue decomposition. This process can lead to build up of high quantity of readily plant available N in soil (Dinnes et al. 2002) which increases its potential for leaching into shallow water tables. Halvorson et al. (1999) reported more accumulation of soil NO_3 down to 150 cm depth with conventional tillage compared with no-tillage system in spring wheat (Triticum aestivum L.) - fallow cropping system. They found it to be associated with the higher mineralization of N at the soil surface induced by soil disturbance. However, no tillage is rarely practiced because of sudden attack of weeds during growth phase of crop. However, Randall and Iragavarapu (1995) reported that 11-year average of NO_3 losses for moldboard plowing and no tillage were 43 and 41 kg ha^{-1}, respectively under continuous corn. The greater length of the study, which caused greater variability in the soil and environmental conditions, was attributed as the cause of the narrow difference in NO_3–N loss.

Stoddard et al. (2005) reported a lack of difference in NO_3 leaching between no tillage and minimum tillage (chisel plough + discing) which might be due to insufficient disturbance of the soil to affect physical and biological properties. It suggests that minimum tillage could be adopted to avoid disadvantages of both conventional tillage and notillage. Contrarily, Mkhabela et al. (2008) suggested that denitrification is significantly lower under conventional tillage than no tillage which in part, could be the reason of lower NO_3 concentrations observed under no tilled corn field. Evaluated the effect of no tillage, minimum tillage and deep tillage on nitrate leaching on silt loam soil at Lincoln, Canterbury, New Zealand. The average cumulative NO_3-N leached from winter cover crops was 208, 192 and 200 kg ha^{-1} for deep tilled, the minimum and no tillage treatments, respectively. Patni et al. (1998) also reported no significant difference between no tillage and conventional tillage, but, reported lower NO_3 concentrations under former practice. Usman et al. (2013) attributed low NO_3 leaching under no tillage system to more evenly distributed mineralization throughout the crop life span, while under conventional tillage there is rapid mineralization after cultivation and more chances of leaching than its availability to crop. In case of no tillage, crop residues present on soil surface has less plant available N because of its wider C:N ratio. Bellido et al. (2013) concluded from an 18-year field study on Vertisol that during most of the years NO_3 concentration was higher under conventionally tilled plots as compared to non-tilled plots. They supported their results by suggesting that under un-ploughed soil there was less decomposition of crop residues which was responsible for low net N mineralization and more N immobilization and nitrification differences.

9.2.3 Growing of Cover Crops in High Leaching Season

The growing of cover crops is the best option to minimize NO_3 losses in post-harvest seasons. In most of the areas after one crop harvest there is a gap or fallow period for next crop to grow. If some cover crops are not grown then prairies become the part of that land. The function of the cover crops, between the main crop seasons, is to accumulate inorganic N and thus reduce the chance of leaching. The N is then slowly released for the next growing season after residue decomposition. The other beneficial aspects of cover crops are to prevent soil erosion, enhance SOM and act as herbicide. Cover crops must have the ability to grow on less fertile, cool weather without inhibiting the growth of row crops. It not only reduces the NO_3 concentration in soil but also increase the growth of row and or following crops (Mei-singer and Delgado 2002). The crops which are more effective in reducing NO_3 leaching are grasses and brassicas than legumes. The plant species that can be used as cover crops vary from region to region depending upon climatic conditions. Rye was successfully used as a cover crop but the major drawbacks to grow rye as a cover crop are; it overwinters early, consumes more water and immobilize more N. Reported that the growing of cover crops like forages in winter decreased 25 kg N ha^{-1} $year^{-1}$ NO_3 leaching. Reported that after maize harvest covering of land with rye decreased 80 % NO_3 leaching losses as compared to winter fallow.

9.3 *Irrigation Based Management Strategies*

9.3.1 Significance of Evapotranspiration (ET) Based Irrigation Scheduling

Evapotranspiration (ET), the main mode of water loss from agricultural lands, is a critical component of hydrologic cycle. Across the hydrosphere, biosphere and atmosphere, ET is a form of continuous energy flow (Wang et al. 2012), and every aspect of productivity of the ecosystem is virtually influenced by it. Thus, sustainable management of water resources and balanced water supply among industrial, domestic, ecological and agricultural sectors necessitates having adequate knowledge on ET (Wang et al. 2012).

According to ET is extremely important for accurate predictions of crop productivity in dynamic resource environments. However, heterogeneity of vegetation and difficulties in measuring hydrological processes at comparable scales make the estimation of ET usually complicated. The most investigated variables of ET are evaporation and transpiration (Wang et al. 2012), however the contributions of precipitation, irrigation and soil factors remains relatively less investigated aspects. Thus, partitioning ET into its fractions could improve our understandings not only on the management of water resources, but also on recent global climatic variations (Fig. 1).

Increased depletion of ground water with intensifying irrigation is severely limiting crop productivity, food security, social stability and economic growth. In arid/semi-arid regions, soil water gained from precipitation and irrigation could easily be depleted by ET due to high temperatures. In developing countries, water productivity is far below than that in the developed world. This is particularly true for Pakistan, a developing country with over approximately 190 million people, where flood irrigation with pumped groundwater drive crop production and the scientific method for irrigation scheduling of crops are not followed. Flood irrigation may cause

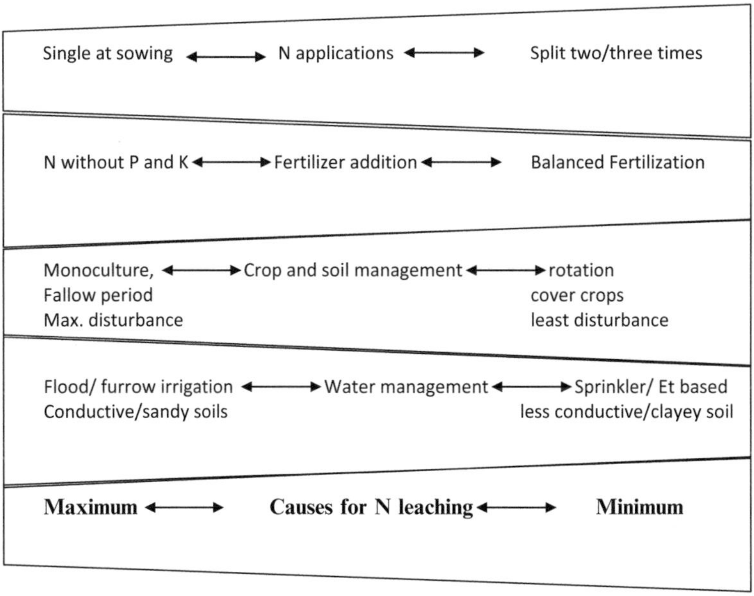

Fig. 1 Effect of N fertilizer, soil, crop and water management practices on N leaching

uneven distribution of nutrients, high rate of nutrient loss due to leaching and in some cases injurious to crops at early stages. The continuous pumping of ground water due to irrigation has resulted in the lowering of water table in several regions of Pakistan and many other countries.

Model data collection and simulation processes are exposed to errors by the current hydro-agronomic model parameters. Therefore, more efficient ways of accounting for the component fractions of the water budget are needed. In this regard, computer based softwares like CROPWAT 8.0 which use baseline meteorological data and specific crop coefficients to estimates parameters like ET are highly important (Hogue et al. 2005). In this way, water productivity could be enhanced by quantifying the contributions of precipitation, irrigation and soil water fractions to ET (Fig. 2).

9.3.2 Deficit Irrigation

In the past, NO_3 leaching has been paid a very little attention in arid and semi-arid regions. Although total annual rainfall is very low in these regions, but 60–70 % of the precipitation is generally concentrated in monsoon season (July–September). Heavy rainfall in monsoon season transports surface NO_3 deep into the soil profile. This phenomenon is more prevalent in areas where summer fallow procedure is

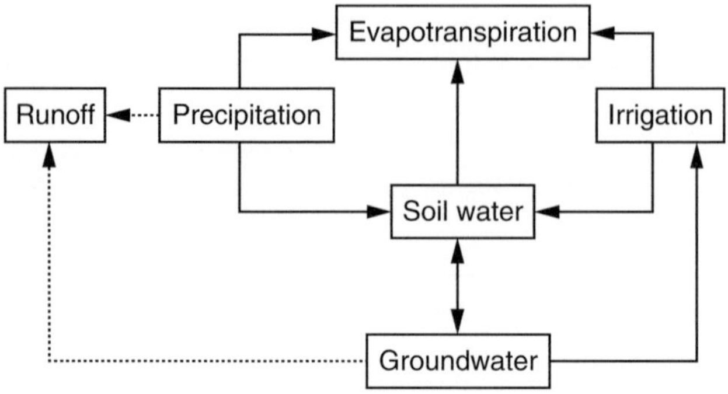

Fig. 2 Flow diagram of the soil water balance model

practiced. Additionally, flood irrigation being the common farming practice could also cause NO_3 transport to deeper soil layers (Cameron et al. 2013).

Over irrigation in case of maize crop caused 47 % increase in NO_3 leaching as compared to optimum irrigation (Gheysari et al. 2009). Heavy rainfall or irrigation immediate to N fertilizer application is likely to aggravate NO_3 leaching due to possible bypass flow through macropores as well as lower ammonia volatilization (Di and Cameron 2002). Yuan et al. (2000a, b) reported 64 % higher accumulation of NO_3 in 0–400 cm layer over a period of 8 years under irrigated than dryland conditions. The leakage of NO_3 may also occur below 400 cm depth. So, soil analysis further to 100 cm depth, the most emphasized depth previously, is needed. Wang et al. (2010) evaluated that higher rates of irrigation drastically increased drainage and NO_3 in drainage water. On the other hand, deficit irrigation seems an efficient irrigation management practice to reduce NO_3 leaching without compromising on crop yield. There are different modern ways to regulate deficit moisture conditions in the soil like furrow irrigation system, partial root zone drying irrigation. Djman et al. (2013) found less residual N with 50 % of full irrigation treatment than rainfed condition. Abbasi et al. (2011) irrigated corn fields at 60, 80, 100 and 120 % of crop water requirement (CWR) and observed 15, 29 and 35 % of NO_3 leaching, respectively. Wang et al. (2012) studied the effect of deficit irrigation i.e. 0.6, 0.8 and 1.0 of ET_c on NO_3 distribution in soil and concluded that no NO_3 accumulation was recorded in medium (0.8 ET_c) and low irrigation (0.6 ET_c) levels up to 200 cm soil profile. Skinner et al. (1999) recorded increased N uptake in furrow irrigation system which resulted in less NO_3 leaching. Similarly, Kirda et al. (2005) reported improved RE_N and lower buildup of N in the soil profile with partial root zone drying as compared to full and deficit irrigation that might be due to increased surface area of roots for water and nutrient uptake through lateral branching under the former irrigation practice (Mingo et al. 2004). Tafteh and Sepaskhah (2012) used three irrigation techniques viz., ordinary furrow irrigation, variable alternate furrow irrigation and fixed alternate furrow irrigation for maize. They observed that less water was required in variable alternate furrow irrigation causing less drainage water and NO_3 concentration in soil. However, high application of manure may dilute or completely nullify the beneficial effect of deficit irrigation in reducing leaching losses of NO_3, reported by Tarkalson et al. (2006) from an experiment on maize crop. The preferential flow of water from macropores after precipitation and irrigation are the main reasons for losses of N under limited water conditions.

References

Abbasi MS, Sharif M, Kazmi T, Sultan AM (2011) Isolation of plant growth promoting rhizobacteria from wheat rhizosphere and their effect on improving growth, yield and nutrient uptake of plants. Plant Biosyst 145:159–168

Al-Kaisi MM, Yin X (2003) Effects of nitrogen rate, irrigation rate, and plant population on corn yield and water use efficiency. Agron J 95:1475–1482

Anderson DM, Gilbert PM, Burkholder JM (2002) Harmful algal blooms and eutrophication: nutrient sources, composition, and consequences. Estuaries 25:704–726

Barraclough D, Jarvis S, Davies G, Williams J (1992) The relation between fertilizer nitrogen applications and nitrate leaching from grazed grassland. Soil Use Manage 8:51–55

Basso B, Ritchie JT (2005) Impact of compost, manure and inorganic fertilizer on nitrate leaching and yield for a 6-year maize–alfalfa rotation in Michigan. Agric Ecosyst Environ 108:329–341

Bellido L, Muñoz-Romero V, López-Bellido RJ (2013) Nitrate accumulation in the soil profile: long-term effects of tillage, rotation and N rate in a Mediterranean Vertisol. Soil Till Res 130:18–23

Borken W, Matzner E (2009) Reappraisal of drying and wetting effects on C and N mineralization and fluxes in soils. Glob Chang Biol 15:808–824

Bouma J, Dekker L, Muilwijk C (1981) A field method for measuring short-circuiting in clay soils. J Hydrol 52:347–354

Breitburg D (2002) Effects of hypoxia, and the balance between hypoxia and enrichment, on coastal fishes and fisheries. Estuaries 25:767–781

Bruning-Fann C, Kaneene JB (1993) The effects of nitrate, nitrite, and N-nitroso compounds on animal health. Vet Hum Toxicol 35:237–253

Callesen I, Borken W, Kalbitz K, Matzner E (2007) Long-term development of nitrogen fluxes in a coniferous ecosystem: does soil freezing trigger nitrate leaching? J Plant Nutr Soil Sci 170:189–196

Cameron KC, Di HJ, Moir JL (2013) Nitrogen losses from the soil/plant system: a review. Ann Appl Biol 162:145–173

Chadwick D, John F, Pain B, Chambers B, Williams J (2000) Plant uptake of nitrogen from the organic nitrogen fraction of animal manures: a laboratory experiment. J Agric Sci 134:159–168

Chang C, Janzen H (1996) Long-term fate of nitrogen from annual feedlot manure applications. J Environ Qual 25:785–790

Claret M, Urrutia R, Ortega R, Best S, Valderrama N (2011) Quantifying nitrate leaching in irrigated wheat with different nitrogen fertilization strategies in an Alfisol. Chilean J Agric Res 71:148–156

Denshaw-Burke M et al (2014) Methemoglobinema. http://emedicine.medscape.com/article/204178-overview. Retrieved on 27 November 2015

Di H, Cameron K (2002) Nitrate leaching in temperate agroecosystems: sources, factors and mitigating strategies. Nutr Cycl Agroecosyst 64:237–256

Diacono M, Montemurro F (2010) Long-term effects of organic amendments on soil fertility. A review. Agron Sustain Dev 30:401–422

Dinnes DL, Dl K, Jaynes DB, Kaspar TC, Hatfield JL, Colvin TS, Cambardella CA (2002) Nitrogen management strategies to reduce nitrate leaching in tile drained Mid-Western soils. Agron J 94:153–171

Djman K, Irmak S, Martin DL, Ferguson RB, Bernards ML (2013) Plant nutrient uptake and soil nutrient dynamics under full and limited irrigation and rainfed maize production. Agron J 105:527–538

Dunbabin V, Armstrong R, Officer S, Norton R (2009) Identifying fertilizer management strategies to maximize nitrogen and phosphorus acquisition by wheat in two contrasting soils from Victoria, Australia. Soil Res 47:74–90

Eckersten H, Blombäck K, Kätterer T, Nyman P (2001) Modelling C, N, water and heat dynamics in winter wheat under climate change in southern Sweden. Agric Ecosyst Environ 86:221–235

Emmett BA, Beier C, Estiarte M, Tietema A, Kristensen HL, Williams D, Peñuelas J, Schmidt I, Sowerby A (2004) The response of soil processes to climate change: results from manipulation studies of shrub lands across an environmental gradient. Ecosystems 7:625–637

Fageria NK, Baligar VC (2007) Agronomy and physiology of tropical cover crops. J Plant Nutr 30:1287–1339

Fageria NK, Baligar VC, Clark RB (2006) Physiology of crop production. Haworth, New York

Fan J, Hao MD, Shao MA (2003) Nitrate accumulation in soil profile of dry land farming in Northwest China. Pedosphere 13:367–374

Fan J, Hao M, Malhi SS (2010) Accumulation of nitrate N in the soil profile and its implications for the environment under dryland agriculture in northern China: a review. Can J Soil Sci 90:429–440

Fine DH, Challis BC, Hartman P, Van Ryzin J (1982) Endogenous synthesis of volatile nitrosamines: model calculations and risk assessment. IARC Sci Publ 41:379–396

Fomon SJ (1993) Nutrition of normal infants. Mosby, Baltimore

García-Garizábal I, Causapé J, Abrahao R (2012) Nitrate contamination and its relationship with flood irrigation management. J Hydrol 442–443:15–22

Gheysari M, Mirlatifi SM, Homaee M, Asadi ME, Hoogenboom G (2009) Nitrate leaching in a silage maize field under different irrigation and nitrogen fertilizer rates. Agric. Water Manage 96:946–954

Habib R, Lafolie F (1991) Water and nitrate redistribution in soil as affected by root distribution and absorption. In: Atkinson D (ed) Plant root growth an ecological perspective. Blackwell Scientific Publications, Oxford, pp 131–146

Hakeem KR, Ahmad A, Iqbal M, Gucel S, Ozturk M (2011) Nitrogen-efficient cultivars can reduce nitrate pollution. Environ Sci Pollut Res 18:1184–1193

Halvorson AD, Reule CA, Follett RF (1999) Nitrogen fertilization effects on soil carbon and nitrogen in a dryland cropping system. Soil Sci Soc Am J 63:912–917

Hartz T, Mitchell J, Giannini C (2000) Nitrogen and carbon mineralization dynamics of manures and composts. Hort Sci 35:209–212

Hassink J (1995) Density fractions of soil macroorganic matter and microbial biomass as predictors of C and N mineralization. Soil Biol Biochem 27:1099–1108

Hogue T, Bastidas L, Gupta H, Soroshian S, Mitchell K, Emmerich W (2005) Evaluation and transferability of the Noah land surface model in semiarid environments. J Hydrometeorol 6:68–84

Ineson P, Benham DG, Poskitt J, Harrison AF, Taylor K, Woods C (1998) Effects of climate change on nitrogen dynamics in upland soils. 2. A soil warming study. Glob Chang Biol 4:153–161

Iqbal F, Ali S, Tauqeer HM, Shakoor MB, Farid M, Iftikhar U, Nazir MM (2013) Assessment of ground water contamination by various pollutants from sewage water in Chakera village, Faisalabad. Int J Environ Monit Anal 1:182–187

Isidoro D, Quílez D, Aragüés R (2006) Environmental impact of irrigation in La Violada district (Spain). J Environ Qual 35:766–775

Jalali M (2005) Nitrates leaching from agricultural land in Hamadan, western Iran. Agric Ecosyst Environ 110:210–218

Jaynes D (2013) Nitrate loss in subsurface drainage and corn yield as affected by timing of side-dress nitrogen. Agric Water Manage 130:52–60
Jia X, Shao L, Liu P, Zhao B, Gu L, Dong S, Zhao B (2014) Effect of different nitrogen and irrigation treatments on yield and nitrate leaching of summer maize (Zea mays L.) under lysimeter conditions. Agric Water Manage 137:92–103
Keeney D (1997) What goes around comes around – the nitrogen issues cycle. In: Mortwedt JJ, Shaviv A (eds) Third Int. Dahlia Greidinger Sym. on Fertilization and the Environment. Technion, Haifa
Kirda C, Derici M, Schepers J (2001) Yield response and N-fertilizer recovery of rainfed wheat growing in the Mediterranean region. Field Crops Res 71:113–122
Kirda C, Topcu S, Kaman H, Ulger A, Yazici A, Cetin M, Derici M (2005) Grain yield response and N-fertilizer recovery of maize under deficit irrigation. Field Crops Res 93:132–141
Klausner SD, Kanneganti VR, Bouldin DR (1994) An approach for estimating a decay series for organic nitrogen in animal manure. Agron J 86:897–903
Leiros MC, Trasar-Cepeda C, Seoane S, Gil-Sotres F (1999) Dependence of mineralization of soil organic matter on temperature and moisture. Soil Biol Biochem 31:327–335
Liu DQ, Tong YA, Sun BH, Emteryd O (1998) Study on effect of nitrogen fertilizer use on environmental pollution. Plant Nutr Fertil Sci 4:8–15 [in Chinese]
López-Bellido L, López-Bellido RJ, Redondo R (2005) Nitrogen efficiency in wheat under rainfed Mediterranean conditions as affected by split nitrogen application. Field Crops Res 94:86–97
MAFF (2000) Fertilizer recommendations for arable and horticultural crops, RB209. Revised 2000. MAFF Publications, London
Mantovi P, Fumagalli L, Beretta GP, Guermandi M (2006) Nitrate leaching through the unsaturated zone following pig slurry applications. J Hydrol 316:195–212
Matzner E, Borken W (2008) Do freeze-thaw events enhance C and N losses from soils of different ecosystems? A review. Eur J Soil Sci 59:274–284
Mei-singer JJ, Delgado JA (2002) Principles for managing nitrogen leaching. J Soil Water Conserv 57:485–498
Mingo DM, Theobald JC, Bacon MA, Davies WJ, Dodd IC (2004) Biomass allocation in tomato (Lycopersicon esculentum) plants grown under partial rootzone drying: enhancement of root growth. Funct Plant Biol 31:971–978
Mkhabela M, Madani A, Gordon R, Burton D, Cudmore D, Elmi A, Hart W (2008) Gaseous and leaching nitrogen losses from no-tillage and conventional tillage systems following surface application of cattle manure. Soil Till Res 98:187–199
Nelson WO, Campbell PGC (1991) The effects of acidification on the geochemistry of Al, Cd, Pb and Hg in freshwater environments: a literature review. Environ Pollut 71:91–130
Olesen JE, Petersen BM, Berntsen J, Hansen S, Jamieson P, Thomsen A (2002) Comparison of methods for simulating effects of nitrogen on green area index and dry matter growth in winter wheat. Field Crops Res 74:131–149
Pang X, Letey J (2000) Organic farming challenge of timing nitrogen availability to crop nitrogen requirements. Soil Sci Soc Am J 64:247–253
Paramasivam S, Alva AK, Fares A, Sajwan KS (2002) Fate of nitrate and bromide in an unsaturated zone of a sandy soil under citrus production. J Environ Qual 31:671–681
Patni NK, Masse L, Jui PY (1998) Groundwater quality under conventional and no-tillage: I. Nitrate, electrical conductivity, and pH. J Environ Qual 27:869–877
Power JF, Doran JW (1984) Nitrogen use in organic farming. In: Hauck RD (ed) Nitrogen in crop production. ASA CSSA and SSSA, Madison, pp 585–600
Powlson DS, Poulton PR, Addiscott TM, McCann DS (1989) Leaching of nitrate from soils receiving organic or inorganic fertilizers continuously for 135 years. In: Hansen, Henriksen (eds) Nitrogen in organic wastes applied to soils. Academic, London, pp 334–345
Randall G, Iragavarapu T (1995) Impact of long-term tillage systems for continuous corn on nitrate leaching to tile drainage. J Environ Qual 24:360–366

Richter J, Roelcke M (2000) The N-cycle as determined by intensive agriculture–examples from central Europe and China. Nutr Cycl Agroecosyst 57:33–46
Rong Y, Xuefeng W (2011) Effects of nitrogen fertilizer and irrigation rate on nitrate present in the profile of a sandy farmland in Northwest China. Proc Environ Sci 11:726–732
Rozemeijer JC, Broers HP, Geer FCV, Bierkens MFP (2009) Weather-induced temporal variations in nitrate concentrations in shallow groundwater. J Hydrol 378:119–127
Rustad LE, Campbell JL, Marion GM, Norby RJ, Mitchell MJ, Hartley AE, Cornelissen JHC, Gurevitch J, GCTE-NEWS (2001) A metaanalysis of the response of soil respiration, net nitrogen mineralization, and aboveground plant growth to experimental ecosystem warming. Oecologia 126:543–562
Sardans J, Peñuelas J, Prieto P, Estiarte M (2008) Drought and warming induced changes in P and K concentration and accumulation in plant biomass and soil in a Mediterranean shrubland. Plant Soil 306:261–271
Sexton BT, Moncrief JF, Rosen CJ, Gupta SC, Cheng HH (1996) Optimizing nitrogen and irrigation inputs for corn based on nitrate leaching and yield on a coarse-textured soil. J Environ Qual 25:982–992
Shepherd M, Newell-Price P (2013) Manure management practices applied to a seven-course rotation on a sandy soil: effects on nitrate leaching. Soil Use Manage 29:210–219
Shepherd M, Withers P (1999) Applications of poultry litter and triple superphosphate fertilizer to a sandy soil: effects on soil phosphorus status and profile distribution. Nutr Cycl Agroecosyst 54:233–242
Shi Y, Yu Z, Li Y, Wang X (2007) Effects of nitrogen fertilizer rate and ratio of base and topdressing on winter wheat yield and fate of fertilizer nitrogen. Sci Agric Sin 40:54–62
Shi Z, Lia D, Jing Q, Kia J, Jianga D, Cao W, Dai T (2012) Effects of nitrogen applications on soil nitrogen balance and nitrogen utilization of winter wheat in a rice-wheat rotation. Field Crop Res 127:241–247
Silva S, Kendall C, Wilkison D, Ziegler A, Chang C, Avanzino R (2000) A new method for collection of nitrate from fresh water and the analysis of nitrogen and oxygen isotope ratios. J Hydrol 228:22–36
Sistani K, Adeli A, Mcgowen S, Tewolde H, Brink G (2008) Laboratory and field Tzevaluation of broiler litter nitrogen mineralization. Bioresour Technol 99:2603–2611
Skinner RH, Hanson JD, Benjamin JG (1999) Nitrogen uptake and partitioning under alternate- and every-furrow irrigation. Plant Soil 210:11–20
Smith VH (2003) Eutrophication of freshwater and coastal marine ecosystems: a global problem. Environ Sci Pollut Res 10:126–139
Stoddard C, Grove JH, Coyne MS, Thom WO (2005) Fertilizer, tillage, and dairy manure contributions to nitrate and herbicide leaching. J Environ Qual 34:1354–1362
Tafteh A, Sepaskhah A (2012) Yield and nitrogen leaching in maize field under different nitrogen rates and partial root drying irrigation. Int J Plant Prod 6:93–114
Tahir MA, Rasheed H (2008) Distribution of nitrate in the water resources of Pakistan. Afr J Environ Sci Technol 2:397–403
Tarkalson D, Payero J, Ensley S, Shapiro CA (2006) Nitrate accumulation and movement under deficit irrigation in soil receiving cattle manure and commercial fertilizer. Agric. Water Manage 85:201–210
Thorup-Kristensen K (2001) Are differences in root growth of nitrogen catch crops important for their ability to reduce soil nitrate-N content, and how can this be measured? Plant Soil 230:185–195
Tong YA, Shi W, Lu DQ, Emteryd O (2005) Relationship between soil texture and nitrate distribution and accumulation in three types of soil profile in Shaanxi. Plant Nutr Fertil Sci 11:435–441 [in Chinese]
Ulen B, Johansson G (2009) Long-term nutrient leaching from a Swedish arable field with intensified crop production against a background of climate change. Acta Agric Scand Sect B Soil Plant 59:157–169

US EPA (Environmental Protection Agency) (1991) National primary drinking water regulations, final rule. Fed Reg 56:3526–3597

Usman K, Khan EA, Khan N, Khan MA, Ghulam S, Khan S, Baloch J (2013) Effect of tillage and nitrogen on wheat production, economics, and soil fertility in rice-wheat cropping system. Am J Plant Sci 4:17–25

Vachaud G, Chen T (2002) Sensitivity of computed values of water balance and nitrate leaching to within soil- class variability of transport parameters. J Hydrol 264:87–100

Wang Q, Li F, Zhao L, Zhang E, Shi S, Zhao W, Song W, Vance MM (2010) Effects of irrigation and nitrogen application rates on nitrate nitrogen distribution and fertilizer nitrogen loss, wheat yield and nitrogen uptake on a recently reclaimed sandy farmland. Plant Soil 337:325–339

Wang Z, Liu F, Kang S, Jensen CR (2012) Alternate partial root zone drying irrigation improves nitrogen nutrition in maize (Zea mays L.) leaves. Environ Exp Bot 75:36–40

Wiesler F, Horst WJ (1994) Root growth and nitrate utilization of maize cultivars under field conditions. Plant Soil 163:267–277

Wild A, Cameron K (1980) Nitrate leaching through soils and environmental considerations with special reference to recent work in the United Kingdom. In: Proceedings of the meeting on soil nitrogen as fertilizer or pollutant, pp 289–306

World Health Organization (2004) Guidelines for drinking-water quality: recommendations (vol 1). World Health Organization, Geneva

Yuan S, Wei S, Chang B (2000a) Biodegradation of polycyclic aromatic hydrocarbons by a mixed culture. Chemosphere 41:1463–1468

Yuan XM, Tong YA, Yang XY, Li XL, Zhang FS (2000b) Effect of phosphate application on soil nitrate nitrogen accumulation. Plant Nutr Fertil Sci 6:397–403

Zhang WS, Ji WH, Kolbe H (2004) The current situation and controlling manures of non-point source pollutions in China. Sci Agric Sin 37:1008–1017

Potassium for Sustainable Agriculture

Abdul Wakeel, Mehreen Gul, and Christian Zörb

Contents

A. Wakeel (✉)
Institute of Soil and Environmental Science, University of Agriculture Faisalabad, Fauisalabad 38040, Pakistan

International Potash Institute, Industriesstrasse 31, CH-6300 Zug, Switzerland
e-mail: abdulwakeel77@gmail.com

M. Gul
Institute of Soil and Environmental Science, University of Agriculture Faisalabad, Fauisalabad 38040, Pakistan

C. Zörb
Institut für Kulturpflanzenwissenschaften, University of Hohenheim, Stuttgart 70593, Germany

K.R. Hakeem et al. (eds.), *Soil Science: Agricultural and Environmental Prospectives*, DOI 10.1007/978-3-319-34451-5_7

Abstract Potassium (K) is the third most important plant nutrient required in higher amounts by plants. It plays role in charge balancing, osmotic adjustments and enzyme activation in plant cells. It is highly mobile in the plant because it is not the structural part of plant tissue, but is present in ionic form in the plants. Although about 2.3 % of the earth's crust consist of K, but most of its parts bound with clay minerals and is not available to plants. Release of K from micaceous soils not only provides an essential plant element, but also converts mica into illite and vermiculite with more sites for K absorption. Therefore, K fertilization in such soils may cause K fixation and most of K applied become unavailable or slowly available to plants. Potassium is also named as a quality element because it improves the agricultural-product quality. Potassium also develops resistance against different environmental stresses and has ability to mitigate biotic and abiotic stresses developing immunity in the plants. In human and animal nutrition, K has significant role and its deficiency in humans can cause cardiovascular and nervous diseases. The ultimate source of K in human nutrition is the soil which provides K to plants and is then used by humans and animals. Potassium fertilizers need to be applied to the soils to replenish the exchangeable K and non-exchangeable K for sustainable soil fertility, otherwise severe deficiency of K may happen which will be even more difficult to cure. As K has strong interaction with clay minerals, therefore it is intensely recommended to apply K fertilizers based on soil mineralogy and K dynamics in the soil. Although K in soil is evaluated by three standard methods, including visual observations, soil analysis and plant tissue analysis, however, for sustainable agriculture site-specific K recommendations based on soil mineralogy are direly needed to have a better crop response and economical agricultural productions.

Keywords Sustainable agriculture • Plant nutrients • Potassium • Crop yield • Fertilizers

1 Introduction

Potassium (K) plays a key role as inorganic osmoticum and its impact on leaf movement, stomatal regulation, axial growth and tropisms are well recognized (Marschner 1995; Shabala 2003; Ahmad et al. 2013). Among the mineral nutrients, K is the 2nd most abundant element in plants and contributes 2–10 % of plant dry weight (Tisdale et al. 1993). Potassium is a counter ion for the charge balance and ion transport across the plasma and intra-organelle membranes (Dreyer and Uozumi 2011; Shabala 2003). Potassium also controls the process of photo assimilates loading into the phloem (Ache et al. 2001; Gajdanowicz et al. 2011). There are over 70 enzymes, the activities of which were shown to be sensitive to K (Marschner 1995).

Over 250 years, K fertilizers were derived from wood ashes and livestock manure, traditionally. However, since 2000, global potash demand has grown by 40 %, with over 50 million tons of potash produced in 2012. Globally, above ground parts of annual crops contain 60 million tons of K.

1.1 Potassium Dynamics in Soil

Most of the soils are abundant in K because 2.3 % of the earth's crust is occupied with the K-minerals; however, this part of the K is mostly bound to clay minerals and is not available for plant growth. Soil solution concentration represents the most available form of K to plants. Potassium released from exchangeable as well as non-exchangeable to solution form improves soil solution concentration. Capacity of soil represents the total K amount in soil which is readily available to plants, while soil intensity represents concentration of K in soil solution which is readily available for plants. Mica minerals contain 6–9 % K while alkali feldspars contain 3.5–12 % K and both represent major natural reserves for plant available K after weathering. Potassium uptake by plants reduces its concentration in rhizosphere in the roots vicinity which ultimately stimulates the K-release from the minerals (Kuchenbuch and Jungk 1984). That release of K not only provides nutrition to plants but also change micas to illite and then to vermiculite as explained in Fig. 1 (Havlin et al. 1999; Wakeel et al. 2013). Application of K fertilizers to such soils may lead to fixation of K to those soil minerals, making them slowly available or unavailable for plants (Scott and Smith 1987). Nevertheless, fixed form of K is again released and is available to plants when K concentration of K in soil solution is lowered (Cox et al. 1999). Smectites are also K-fixing clay minerals with layer charge of 0.2–0.6 per half unit-cell, CEC 0.8–1.2 mol (+) kg^{-1} soil and layer thickness 1.0–2.0 nm (Bohn et al. 2001).

Therefore, in K fixing soils, sometimes, higher recommended levels of K fertilizer do not show improvements in plant growth; but a much higher fertilizer dose is required to get a response (Doll and Lucas 1973; Mengel and Kirkby 2001). Such

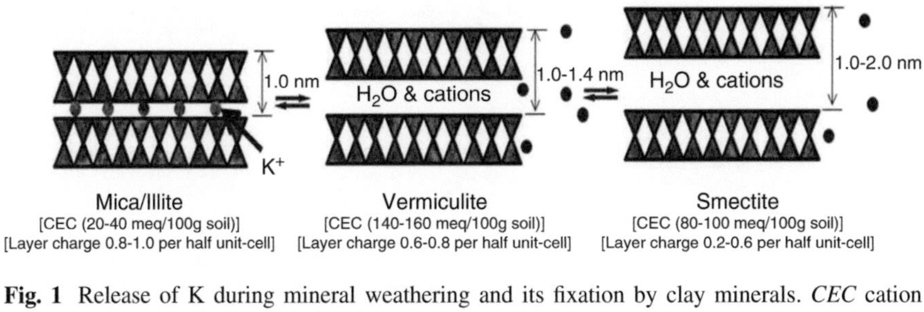

Fig. 1 Release of K during mineral weathering and its fixation by clay minerals. *CEC* cation exchange capacity (Havlin et al. 1999)

soils are rich with vermiculite and illite clay minerals with high CEC and fixed most part of applied K which is unavailable to plants.

1.2 Potassium Dynamics in Plants

Potassium is an essential plant element and higher plant required in large amounts. It is highly mobile in the plant, as it is not the structural part of plant and is present in ionic form in plant tissue. In the plant cell, cytosol maintains highest K concentration in the range of 130–150 mM (Leigh 2001), while the K range is 20–100 mM in the vacuole, which also indicate supply of K to the plants (Fernando et al. 1992). Potassium is performing three main functions *viz.* osmoregulation, enzyme activation and charge balance in the plants (Schubert 2006).

Potassium ions assumed to bind to the enzyme surfaces to activate them by changing their conformations. These enzymes basically involved in many vital functions of plant cell, including ribosomal polypeptide synthesis (Jones and Pollard 1983). It has been reported that K is centered with six oxygen atoms in the corner in an octahedron of enzyme dialkyl-glycine decarboxylase (Miller 1993; Fig. 2).

The electrochemical difference between the outer medium and cytosol is basically a main process for ion transport. Plant membranes are relatively permeable for K due to the presence of K selective channels, developing ease for K to play its role in charge balancing to maintain various activities in the cells. When an electrochemical potential is lower in the cytosol than outer medium, K ion move into the cell to maintain electrical balance. K ion import improves the electrochemical potential and attains the final equilibrium (Mengel and Kirkby 2001). The cytosolic

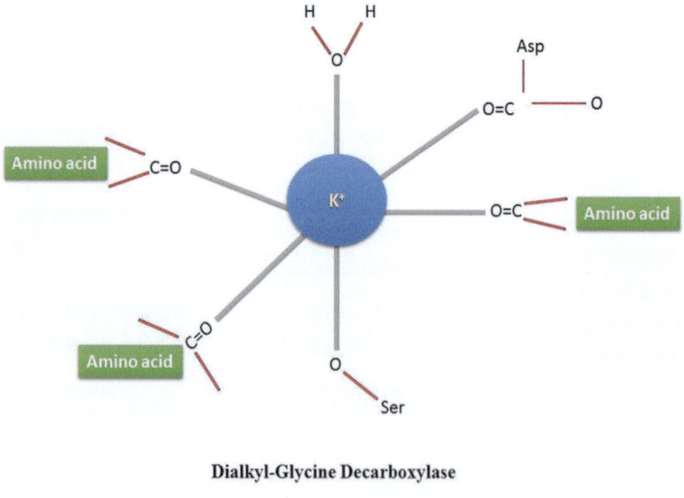

Fig. 2 Potassium complexed by organic molecules of which the oxygen atoms are oriented to positive charge of K^+ (Adapted from Miller 1993)

negative charge is sustained by the activity of proton pumps. Released H^+ ions by plasmalemma H^+ pumps from the cytosol into the apoplast sustain the negative charge in the cytosol maintaining it in the range of 120–200 mV.

Similarly, K enters into the stroma for charge balance using the charge gradient produced by the H^+ pumped out of the stroma (Berkowitz and Peters 1993).

Potassium may accumulate in the vacuoles at high concentration (Hsiao and Lauchli 1986), where K also functions as an osmoticum and maintain turgor potential in the guard cell by water uptake from the soil (Moran et al. 1988) and in the phloem (Mengel and Haeder 1977; Smith and Milburn 1980).

1.3 Potassium vs. Other Cations

The main cations essential for plant growth include ammonium, calcium, magnesium, and K. Among others, sodium (Na) and aluminum (Al) have significant impact on crop growth by influencing soil pH. Plants have cation transport systems at plasma membrane which are cation specific as well as non-specific. In soil a certain ratio between K and other cations such as Ca, Mg, and Zn etc. should be maintained for better nutrient uptake. Application of too much Ca and Mg can cause a K deficiency due to the competition between Ca, Mg and K. A realistic ratio between K and Ca or Mg is 2 or above for optimum K uptake by plants. However, excessive application of K can also decrease the uptake of Ca and Mg, as well as micronutrients such as Zn. Therefore, K/Ca or K/Mg ratio should be less than 10 for optimum Ca and Mg uptake. It is particularly important to take into account of this interaction when using hard water with a high calcium and magnesium concentration.

Potassium and ammonium (NH_4^+) both have the same size and valence properties, consequently always compete with each other for non-exchangeable and exchangeable soil-particles sites. Nitrogen fertilizer may affect the availability of K^+ in long as well as short term usage. In case of long term activities, NH_4^+ fertilization depleted non-exchangeable and exchangeable K^+ from the soil, while in short term activity, soil solution K^+ concentration may increase with increase in NH_4^+ fertilization. K transport through plant membranes is mainly affected by NH_4^+ because of direct competition between NH_4^+ and K^+ for different transporters (Zhang et al. 2010).

Numerous investigations have concluded that only K is the monovalent cation which is essential in all higher plants; however, Na may also promote the biomass production of some plant species without meeting the criteria of essentiality (Flowers et al. 1977). Elements having similar physicochemical properties may perform equally well in a metabolic process and it appears that K and Na may replace each other partially for few nonspecific osmotic or unspecific metabolically active functions. Various studies have revealed that plants may require K in lower amount particularly for cytoplasmic activities which cannot be replaced by any other cation, however, about 90 % of total K localized in vacuoles performing osmotic functions can be substituted by some other cations having similar physicochemical properties

(MacRobbie 1977; Leigh and Jones 1986). In some halophytic plant species, Na may stimulate the plant growth by cell expansion and replace K partially for osmotic functions (Nunes et al. 1983).

1.4 Ionic Uptake and Homeostasis

Water and nutrients move towards stele through root cells by two ways, (1) through apoplast and/or (2) via symplast (Fig. 3). Although, before being part of stele, casparian strip due to its hindrance may enhance ions to enter into the symplast. Then these translocated ions enter into epidermal or cortex cell through symplastic way, and loaded into the stele which made up of living cells of parenchyma and elements of dead trachea for further translocation to the shoot. Because dead elements have no cytoplasm continuity and through the crossing plasma membrane, ions must leave the symplast (Taiz and Zeiger 2006).

The Na influx distorted the plants ionic ratio through K transporters or pathways. This similarity of both monovalent K and Na ions make it tough for cells to discriminate between each other. Sodium enters into the plant cell through two pathways i.e. either through non-selective cation channels (NSCC; Amtmann and Sanders 1999) or through high-affinity K transporters HKT1 (Rus et al. 2001). NSCC also allows calcium to pass through. Transport of cations in maize to xylem may control through an outward-rectifying cation channels (ORCC) via root stellar cells (Roberts and Tester 1997). Although outward-rectifying cation channels (ORCC) are highly K selective channels, but also allows Na and Ca to move through in lesser extents (Cramer et al. 1994). Sodium also interferes with Ca uptake in

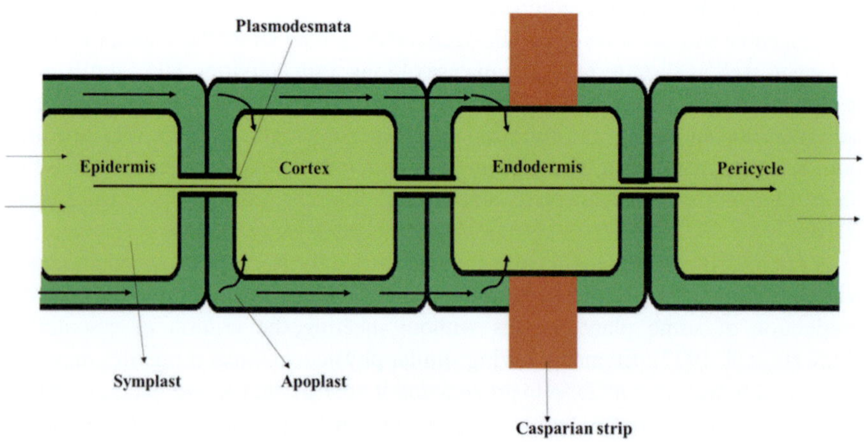

Fig. 3 Two pathways of nutrients movement within root cells to the stele

gramineous plants, and sugar beet plants have also shown Ca deficiency symptoms when Na was increased in the soil, even in a very small concentration (Abd-El-Motagally 2004; Wakeel et al. 2010).

2 Potassium in Agriculture

Potassium functions and requirements are specific for all crops. It maintains cell turgor by increased water content, movement of water and nutrients relationship through the plant and sugar transport to storage organs i.e. beet roots, grains, fruits, tubers etc. Potassium is also named as 'quality element' which ensures the quality agricultural produce. Here we review and discuss the role of K for crop yield enhancement, quality crop produce, resistance against environmental stresses and sustainability of agricultural lands.

2.1 Crop Yield Enhancement

Potassium fertilization affected plant density, which consist of leaf length, width, total number of leaves, stem diameter, plant height, dry leaf yield and yield attributes of tobacco affecting its leaf length, width, stem diameter, total number of leaves, plant height, and total biomass. Foliar and soil application of K together increased wheat grain yield (Arabi et al. 2002). Only a foliar application may not provide the amounts required for a macro nutrient such as K to show high effects. In K deficient soils, the increased grain yield may be attributed to a significant increase in the number of tillers, plant height, leaf area index, spike length, 1000 grain weight, and water use efficiency (Mesbah 2009). It was concluded that K application at different growth stages with different rates may improve growth and yield of wheat crop.

Potassium also promoted plant growth, development and biomass yield of maize (Davis et al. 1996). Balanced fertilization can improve the plant growth; however, most of the farmers use higher amounts of nitrogen and phosphorus with lower K fertilizers. Application of K increased 17 % rice yield, generally reported from different parts of the rice grown area (IPI 1986). Bhatti et al. (1981) determined that K fertilizer applied at the rate of 90 kg ha^{-1} for paddy rice proved economical dose for better crop growth. Recommended economical dose of K fertilizer at the rate of 60 kg ha^{-1} was also recommended for rice grown on different soil (Rehman et al. 1982; Krishnan 1977). Khattak and Bhatti (1986) studied the effect of K and phosphorus on wheat and maize yield and reported that phosphorus was applied during first season, which was sufficient for further 2–3 seasons but potash was sufficient for 4th and even 5th crop. Siddique et al. (1997) reported that a K fertilizer dose of 110 kg ha^{-1} was optimum for wheat yield. Potassium fertilization of 40 kg ha^{-1} can increase 21–35 % of paddy and wheat yield (Hussain and Yasin 2003). Foliar application of K fertilizer also increased different parameters of wheat crop, and increase straw- and grain yield as was observed with the application of 1 % or 5 % KCl solution (Abdi et al. 2002).

Most K uptake in wheat takes place as the shoot is undergoing its rapid phase of growth (Gregory et al. 1979). For cereals, more than 70 % of K remains in the straw. Data from a long-term fertilization experiment demonstrate that this value is not fixed, but depends on cereal species and fertilization level. The K concentration in straw was strongly increased by K fertilization in barley and in wheat (Fig. 3). In practical agriculture, straw and thus significant amounts of K may remain on the field or be removed as a source of biomass. How crop residues are handled is, therefore, very important when calculating the fertilizer requirements.

2.2 Product Quality

The crop quality is a versatile parameter and might combine complex parameters such as a certain metabolite concentration, nutritional value, taste, processing properties, appearance, shelf life and technological quality. Plant product quality is the main focus of the producer's revenue, it serves as a major point for the price of the product in market and quality of the product is more important for consumers (Pettigrew 2008). All these quality parameters are controlled by genetic makeup and physiological impact; however, K also plays a special role in quality formation because of its involvement in protein synthesis and transport of Photo-assimilates from source to sink (Allen and David 2007). It is necessary to balance other nutrients with K for better product quality. Although, product quantity is the principal function driving the producer's revenue stream, product quality is important for consumers and often serves as a criterion for the market price of the crop (Pettigrew 2008). This is most evident for highly processed products such as wine, but also for other crops (Zörb et al. 2014). Although, a huge world population is striving for only food without considering the quality, but at the same time consumers nowadays also require better quality produce, and are willing to pay even more for good quality food. Balanced nutrition with adequate supply of K fertilizers depicts beneficial effect on the harvested product quality and improves nutritional value such as protein content, vitamins and oil contents. Optimum supply of K in crops may lead to better taste, flavor and appearance, and food production without- or less disease and pests attack. Application of K ensures the quality and yield of cotton, tobacco, banana, potato and many food crops. For example, K nutrition substantially influences the concentration of reducing sugars in potato, which are critical precursors for acrylamide formation during frying. An inverse relationship of reducing sugar concentration and K supply has been observed in most cases (Marschner and Krauss 1980). Gerendas et al. (2007) clearly demonstrated that tubers grown with high N and low K fertilization accumulated most precursors for acrylamide formation during frying. As reviewed in Zörb et al. (2014), K plays an essential role for grapevine growth and yield as well as for wine quality (Mpelasoka et al. 2003). Grape berries function as a strong sink for K, especially during the onset of ripening and volume increase, thus their juice contains this ion as major cation. However, the too high K concentration decreases free organic acids, increases the overall pH, lowers the tartrate-to-malate ratio, increases the tartrate precipitation during winemaking and reduces the overall wine quality (Somers 1977). In cereals it increases grain size and improves starch quality. Potassium improves color and flavor in vegetables and increases sugar contents, vitamin C and size of fruits. In addition, it helps in the maintenance of quality of storage, transportation and prolongs the shelf life (e-ifc, 2007). For export to developed countries, quality production is very critical, because the quality parameters include nutritive as well as healthy products. Therefore, K has significance for healthy agricultural productions.

2.3 *Potassium and Environmental Stresses*

Environmental stresses are great contributors to yield reductions of different crops. Potassium develops the abilities in plants to resist against biotic and abiotic stresses like pathogens, insects, salinity, drought and other abiotic stresses. The role of K is very significant and clear to develop immunity in plants and is discussed briefly.

2.3.1 Biotic Stresses

Biotic stress is stress caused by living organisms such as bacteria, viruses, fungi and insects which damage the plant. Pathogen attack activates the superoxide-generating NADPH-oxidases (Bolwell and Wojtaszek 1997; Lamb and Dixon 1997); however, K nutrition has a significant impact on the plant to develop resistance against these pathogens and decrease the incidence of diseases (Prabhu et al. 2007). Adversely, K deficiency can cause defense signaling reduces the crop growth (Amtmann et al. 2008), therefore, K supply during early growth could strengthen the inherent defense potential of plants to pathogens. Potassium transport systems are considered to be very important for plant responses to viral infection and the virus-host recognition. In recent studies, significant changes in K fluxes have been observed in response to microbial inoculation and Ca^{2+} signaling seems to be less essential in recognition of the early stages of viral infection (Shabala et al. 2010). Such K fluxes are assumed to be performed by outward rectifying channels abundantly formed in the plasma membrane in response to viral infection. The physiological rationale behind may be the ejection of viral DNA into the host cells by reduced cell turgor pressure. Potassium fertilization decreases the disease and insect attack in plants. Perrenoud (1990) reviewed that use of K fertilizers reduces the insects and mite infestation by 63 % and increase the yield up to 36 %. Increased synthesis of high-molecular-weight compounds in K-sufficient plants and decreased production of low-molecular-weight compounds inhibit the insect attack (Marschner 2012). Higher K^+ concentrations also decrease the chance of insect infestation by developing stronger cell walls (Mengel 2001).

2.3.2 Abiotic Stresses

Potassium concentration may vary in different parts of plant-cell such as cytoplasm, vacuole etc. (Leigh and Jones 1984). K concentration in the cytoplasm is maintained and remain constant (Leigh 2001) but in vacuole its concentration substantially varies depending on the provision of K^+ to plants. Plant performance was better under cold stress, when K concentration was a luxury in consumption in plant (Bergmann and Bergmann 1985). Before stress initiation, K concentration was not a luxury in plants, but plant allows surviving with different strategies under sudden hidden abiotic stresses (Kafkafi 1990). When K supply was low in early vegetative growth of plants, its structure suffers completely and express frost damage and lodging to unhealthy plants as compared to healthy.

(a) *Drought stress*

Soil moisture availability in arid and semi-arid regions is a major limiting factor for crop production. The root elongation rate is crucial for uptake of nutrients and water which have usually low concentration in soil-solution and are adsorbed strongly to the soil (Kafkafi 1991). Drought and heat stress effects can be mitigated by increasing K concentration in soil (Kafkafi 1990; Cakmak 1994). Under low moisture condition, K fertilization increased growth and yield (Sharma et al. 1996; Yadav et al. 1999; Egilla et al. 2001). Drought stress affects the protein synthesis, enzyme activation, water relation and stomatal conductance in plants (Marschner 1995). Optimum K fertilization increased water potential and minimize leaf osmotic potential of *Vigna radiata* (Table 1) as compared to untreated plants (Nandwal et al. 1998), wheat (Pier and Berkowitz 1987; Sen-Gupta et al. 1989) and maize (Premachandra et al. 1991) grown under drought. Nitrogenase activity, dry matter and nodulation increased with a supply of incremental K at ¼ of field capacity moisture level in broad beans (Abd-Alla and Wahab 1995). Potassium plays a major role in osmotic adjustment and accumulated higher amount of solutes in soybean, tropical grasses and maize under drought condition (Ford and Wilson 1981; Itoh and Kumara 1987; Premachandra et al. 1991). Stomata control water plant water loss through transpiration stream. Under K-deficient condition, stomatal function disturbed and water losses cause injurious effects on the plant at damaging levels (Gething 1990).

Under water stress, stomata close and efficiency of carboxylase activity reduced in chloroplasts. This long time closure of stomata leads to photo-reduction of O_2

Table 1 Effect of K^+ on water and osmotic potentials and relative water content in *Vigna radiata* leaves under water stress condition

	Water potential (⁻MPa)		Osmotic potent (⁻MPa)		Relative water content	
K^+ application (mmol dm^{-3})	Control	Stress	Control	Stress	Control	Stress
0.65	0.55	0.84	1.34	1.55	80.9	72.3
3.20	0.47	0.78	1.30	1.76	84.2	74.7
4.50	0.50	0.79	1.22	1.73	86.1	76.9

into toxic species. Though, drought stress causes a more severe effect during the inadequate K supply (Humble and Raschke 1971). Water stressed plants require more K because K play protective role against stress induced photo-oxidative damage. The K in the protective role under drought stress is well documented (Pier and Berkowitz 1987; Sen-Gupta et al. 1989). Plants photosynthetic efficiency is drastically (Table 2) reduced due to dehydration of chloroplast under drought stress (Berkowitz and Kroll 1988).

(b) Salt stress

Salt-affected soils have a higher concentration of Na^+ as compared to Ca^{2+} and K^+ ions, which results in Na^+ passive accumulation in shoot and root (Bohra and Doerffling 1993). Higher Na levels, disturbed cell membrane integrity and K ions selectivity uptake displace Ca through root membranes (Cramer et al. 1985, 1987). K ion loading into xylem is generally regulated by uptake of K from external K solution (Engels and Marschner 1992). This specifically reduced K uptake under salinity stress interferes K movement towards the shoot resulting in lower K uptake in the shoot and increased K concentration in the root (Table 3). This inhibitory effect on K translocation within plant was reduced by K supply at higher concentration; yet in maize grown in nutrient solution was stronger with application of low K fertilization (Botella et al. 1997).

Similarly, application of K fertilizer increased K concentration in spinach tissues, reduced growth differences among plants grown in high and low salt concentration (Chow et al. 1990). The salt concentration induced shoot-growth inhibition

Table 2 Effect of water stress and K^+ supply on net photosynthesis rate in wheat leaves

	Photosynthesis ($\mu mol\ CO_2\ m^{-2}\ s^{-1}$)		
K^+ application (mM)	*Low water stress*	*Mild water stress*	*Severe water stress*
0.2	38	11	2
2.0	47	42	21
6.0	46	45	28

Source: Sen-Gupta et al. 1989

Table 3 Concentration of Na^+ and K^+ (mmol kg^{-1} DW) in shoot and root of 19 days maize seedlings in response to salinity and potassium applications

		NaCl (1 mM)		NaCl (100 mM)	
		0.1 mM KNO_3	1 mM KNO_3	0.1 mM KNO_3	1 mM KNO_3
Shoot	K^+	630	1310	426	731
	Na^+	33.8	34.4	2128	1625
	K^+/Na^+ ratio	20.8	42.5	0.21	0.45
Root	K^+	396	689	630	813
	Na^+	89	72	1447	1203
	K^+/Na^+ ratio	4.48	9.82	0.44	0.68

Source: Botella et al. 1997

and cause Na toxicity and K deficiency in plants when low K level available in root medium (Zörb et al. 2005). Cell growth may also reduce due to leakage of K out of the cell under stress and lead to cell death. Salt stress limits essential nutrient transport from root to shoot (Termaat and Munus 1986) and lowers the net transport of different nutrients such as Ca^{2+}, NH_4^+, Mg^{2+} and mainly K^+ in plants. The plant tolerance mechanism is the selectivity of K ions over Na^+ (Cuartero et al. 1992) and also partially linked with a Na^+ concentration in leaf (Taleisnik and Grunberg 1994). Plants have the ability to adopt different mechanism to avoid entry of Na^+ reaching to leaves: controlling influx of Na^+ through root cells to plasmalemma (Jacoby and Hanson 1985); Na^+ sequestration into lower parts of the stem and root parenchyma cells and Na^+ removing from xylem streams (Johanson and Cheesman 1983) and Na^+ re-translocation via phloem from shoots to roots (Jacoby 1979). Decreased K supply in plants reduced the photosynthesis sharply due to disturbance in photosystem II (Chow et al. 1990). The higher requirement of K under 250 mM NaCl salt stress than lower salt stress level (50 mM NaCl) may not be ignored. Addition to K supply to roots at increased salt stress can ameliorate reductions in plant and shoot biomass and overwhelmed Na ion toxicity.

(c) Crop Lodging

Lodging is the stems displacement from upright position. Lodging may be partially disturbance or permanent, depends on extent of bending level of stem. Plants affected by two type of lodging: (a) shoot/stem lodging, lower culm internodes bending, (b) root lodging-root system disturbance cause crown leaning (Pinthus 1973) mainly occurs in cereals. Two stages are more critical for lodging, early grain development and heading stage for yield losses due to lodging. This may be due to the interactive factors of plant type, soil and environmental conditions and nutrient management strategies. Plant becomes susceptible for lodging because stem diameter was decreased in K-deficient plants. In K-deficient stem woody parenchyma and sclerenchyma fiber cell form poorly thin lignified cell walls which reduced its diameter (Mulder 1954; Wakhloo 1975). Optimum application of K fertilization increased sclerenchyma tissue layers thickness in rice (Vaithilingam and Balasubramanian 1976). Wheat internode cross section (Fig. 4a) shows that optimum K fertilization has improved thick stalk walls of plants. Improved stem thickness and stability associated with low senescing pith parenschyma and highly active plant defense mechanism with optimum K nutrition (Abney and Foley 1971). Addition of K fertilizer improves the growth of root length which ultimately supports the plant stand strengthening (Fig. 4).

K increased resistance against lodging in different crop species especially in rice (Datta and Mikkelsen 1985), corn (Welch and Flannery 1985; Csatho 1991), oilseed rape (Sharma and Kolte 1994) and wheat (Beaton and Sekhon 1985; Khurana and Bhaya 1990). Recently, Zaman et al. (2015) found that application of K fertilizer reduced the basal node length of rice plants, which is very supportive parameter to conclude the role of K fertilization in strengthening the stem strength.

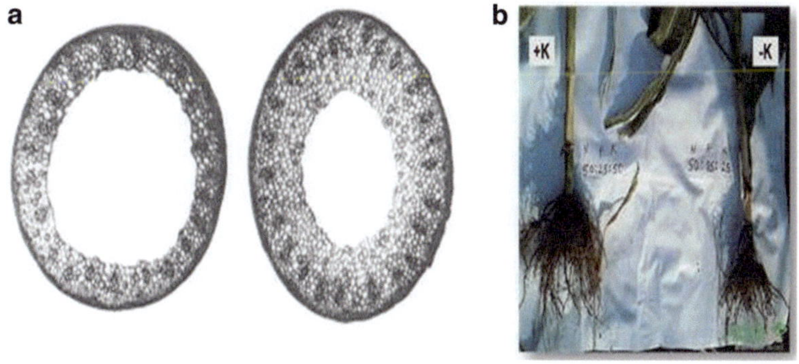

Fig. 4 Cross-section of 3rd internode of wheat plant with low (*left*) and high (*right*) K^+ nutrition (**a**), application of potash develop more intensive roots which may strengthen plant stand (**b**) (Source: Melis and Farina (1984) and International Potash Institute Switzerland)

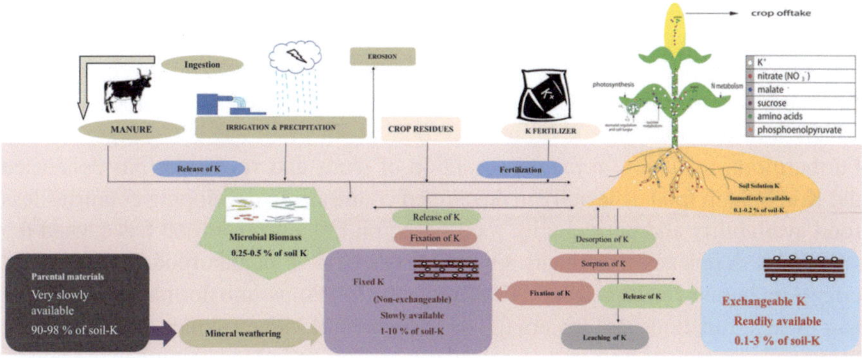

Fig. 5 Distribution of different forms of potassium in an agricultural system (modified from International Potash Institute Switzerland)

2.4 *Potassium for Sustainable Soil Fertility*

Potassium is present in soil in four distinctly different pools from where plants accessibility is differently and this should be considered while understanding the soil K availability to plants (Syers 2003). The conceptual understanding of K cycle in soil is illustrated in Fig. 5. Main pathway for K movement towards plants is via soil solution and soil solution K contributes only 5 % of total crop demand. In soil, it is about 0.1–0.2 % of the total soil K, however, this fraction is immediately available to the plants and replenished quicker than the exchangeable and non-exchangeable K pools. Exchangeable and non-exchangeable K make up about 1–2 % and 1–10 % of the total K respectively, and are the main contributors to K uptake by plants. Non-exchangeable form of K is slowly available to plants and is released from wedge sites of lattice weathered micaceous clay minerals (Mengel and Kirkby 2001). The main pool stock holds the bulk of K i.e. 90–98 % of the total soil K and release slowly because it is held in primary K bearing minerals structure. In Fig. 5 all possible sources of K have also been shown with their contribution to

K cycle. Although, the soils are considered very rich in K, however the available fraction is very low which is also decreasing very rapidly due to cultivation of high yielding crops. Potassium fertilizers need to be applied to soils to replenish the exchangeable K and non-exchangeable K for sustainable soil fertility, otherwise severe deficiency of K may occur which will not be easily cured because of strong interaction of K with soil minerals and huge quantity may be required to get response on crop growth (Wakeel et al. 2013).

3 Potassium and Human Health

Potassium is one of the electrolytes (chloride, calcium, magnesium, sodium and phosphorus) which play a very important role in human body. It is vital for optimum functioning of body cells, tissues and organs. Therefore, an optimum dose of K is required for human body on daily basis. According to the American National Academies of Food and Nutrition, an adequate intake of K is 4.7 g per person per day. Food enriched with K was always consumed by humans; however K mining from soils due to high yielding verities/hybrids and increased use of processed food which has removed the K from the food chain (He and MacGregor 2008). Furthermore, reduction in the consumption of fruits and vegetables has decreased the K in human food especially in developed countries. In developing countries less food availability and poor quality food based on less nutritive food has caused the issue of K deficiency in humans. Optimum use of K in human food has beneficial effects on human health (He and MacGregor 2008). Potassium normalizes the nerve and muscle functions and lowers the development of kidney stones due to reduced urinary Ca excretion. As K reduces the Ca excretion, therefore strengthen the bones. Diet containing sufficient K lowers the blood pressure decreasing the risk of cardio vascular diseases. Hypo serum K is causing glucose intolerance, and increasing K intake may prevent the development of diabetes (He and MacGregor 2008).

Peoples of developing/poor countries who have very less food diversity and low fruit and vegetables intake are even more susceptible to K deficiency. Use of fruits and vegetables rich in K can eliminate its deficiency in humans (Fig. 6). As also mentioned earlier, a very significant reason for such deficiency is also K depletion from soils due to high yielding intensive cropping without addition of K-fertilizers. Promoting biofortification of agricultural products through application of K fertilizers is essentially required for better human nutrition and its impact on human health.

4 Potassium Evaluation in Soils

Potassium evaluation in soils is like other nutrients and three standard methods are used which are based on time and resources available. These three evaluation processes includes visual observations, soil analysis and plant tissue analysis.

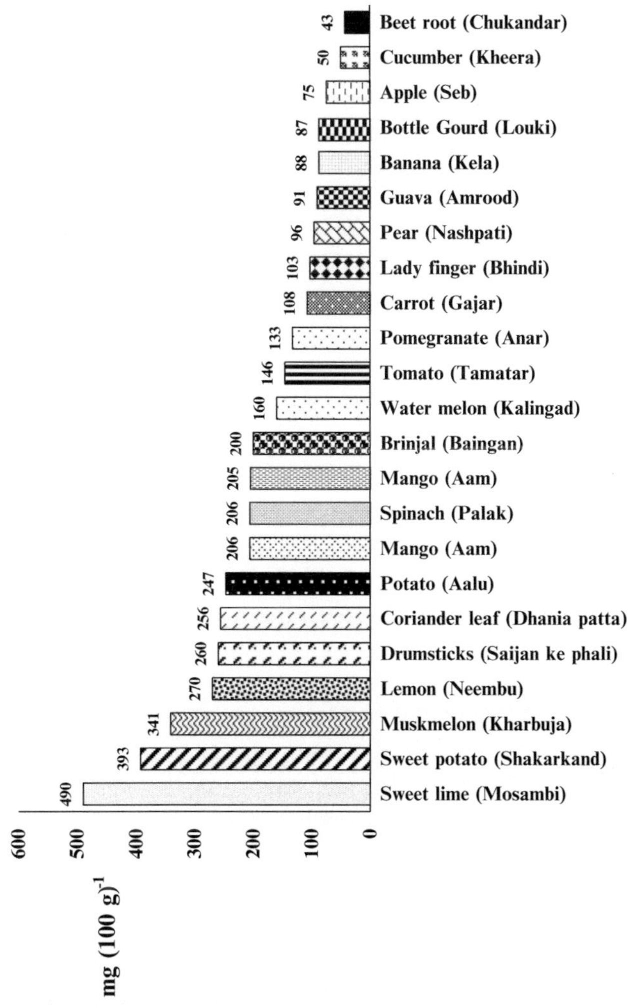

Fig. 6 Potassium concentration in fruits and vegetables (Source: National Institute of Nutrition, Hyderabad, India)

4.1 *Visual Observations of Plants*

Higher and lower concentrations of nutrients in soils lead to nutrient deficiencies and toxicities in plant which show unusual appearance which can be observed on plant leaves, stems or fruits. Generally, deficiency symptoms appears as stunted growth, interveinal chlorosis, chlorosis, red or purple discoloration and necrosis. However, to determine the deficiency or toxicity symptoms, it is necessary to understand essential role of nutrient and its mode of mobility in plant. Mobile nutrients deficiency symptoms first appear in older, lower leaves, whereas immobile nutrients deficiencies symptoms appear in younger, and upper leaves. Potassium is very mobile in plants, therefore its specific symptoms appears in older leaves first.

Plate 1 Chlorosis at old leaf margins of maize due to potassium deficiency (Source: International potash institute Switzerland)

Plate 2 Pottassium-deficiency symptoms due to sucrose accumulation in leaves of *Phaseolus vulgans* (Source: Cramer et al. 1994)

Chlorosis at the margins of older leaves appears firstly which leads to necrosis and ultimately cause plant tissue death (Plate 1).

As K is also involved in transport of sugar produced in leaves, therefore its deficiency may cause sucrose accumulation in leaves and typical symptoms may appear (Plate 2).

Deficiency symptoms of K are usually observed in early growth stages. In maize plant, at vegetative stage, K deficiency symptoms appear as brown or yellow leaf margins, however deficiency symptoms of K are difficult to recognize at harvest stages. At harvesting stages, maize shows some visual hints of K deficiency such as broken stalk, smaller cobs and unfilled cobs tips, which can be used to plan fertilizers for coming season.

Visual observation is diagnostic tool but not an authenticated method and can be limited by various factors, including pseudo deficiencies and hidden hunger. Therefore, plant and soil analytical testing is crucial to verify soil nutrient status and nutrient stress symptoms. Visual symptoms evaluation in field is quick and an inexpensive method to identify or detect deficiency or toxicity of nutrients in crops.

4.2 *Soil Testing*

Soil testing is a quantitative analysis of soil for its physicochemical properties. Soil analysis gives the precise amount of all nutrients; however the methods of analysis may be different for different elemental analysis. Good soil test report is much dependent on a truly representative soil sample. If soil seems homogeneous, one representative sample may be sufficient, but when abnormal growth areas present, then one sample must be taken from the normal area and from the problem area, each. Potassium is extracted from soils generally by NH_4-acetate method which includes the water soluble and NH_4 replaceable K in the soil, which are considered as readily available forms of K. The other method used is ammonium bicarbonate diethylene triamine penta acetic acid (AB-DTPA) method which also includes exchangeable and water soluble K; however its value is relatively lower than NH_4-acetate exchangeable K.

Usually, when K is less than 150 mg Kg^{-1}, it is classified as low K and K concentration in range between 250 and 800 ppm is considered high to excessive. However, soils are much diverse and crop response to K fertilizers is variable even at optimum K levels because of a close relation of K to soil minerals plants can reveal deficiencies. Along with soil-K concentration, K-fixing capacity of soils and soil mineralogy should also be considered while recommending potassic fertilizers.

4.3 *Plant-Tissue Analysis*

In the nineteenth century, several elements were recognized as essential nutrients for plant growth by different scientists. Justus von Liebig and later on others scientist started to analyze mineral contents in plants. The idea was conceived by Weinhold (1862) to analyze plant to determine the available nutrient supply. Goodall and Gregory (1947) concluded that much of the work done prior to 1947 could be classified into one of four categories in terms of research as well as plant analysis utility in crop production. These are categorized as: (a) To detect nutritional disorder caused by certain deficiency symptoms, (b) Field trials results interpretation, (c) Use of advanced testing methods for advisory work, (d) Plant analysis for interpretation of nutritional survey.

Plant analysis of different parts is a valuable service for crop management practices. Alone, this makes proper fertilizer recommendation used for certain crops, fruit trees and also for grapes. Plant tissue analysis and soil testing in combination is a best approach for detecting nutrient deficiency symptoms and fertilizer recommendations. Plant analysis alone is not a substitute of soil testing method but plant tissue analysis is more effective in conjunction with soil analysis. Several environmental and soil characteristics contribute soil fertility and influence nutrient uptake in plant. These factors herbicide damage, soil pH, drought, waterlogging, air temperatures, diseases, insects, soil compaction etc. can reduce the nutrient uptake by plants, therefore only addition of deficient nutrient in soil may not tell the true availability of nutrients to plants. Plant tissue analysis represent the actual availability of nutrients to plants. Plant analysis is also useful tool when small field area is discolored or show stunted growth.

K concentration in plant tissues can be determined from fresh material, (i.e. plant tissue analysis), and also from total dry plant matter. This test is basically qualitative method and appropriately quick for growing crops. Two methods for tissue analysis are used that are wet digestion and dry-ashing. Dry-ashing is simple digestion method, less expensive and non-hazardous as compared to wet digestion with HNO_3-$HClO_4$. Plant analysis with dry ashing for Ca, Mg, Na, P and K. Different micronutrients cations (Zn, Mn, Cu and Fe) can also be determined by dry-ashing method but sophisticated for low silica contents crops (legumes). The wet digestion with HNO_3-$HClO_4$ is required for micronutrient full recovery method in that plant tissue which has high silica contents (like barley, wheat, sugarcane and rice etc.).

After digesting the plant sample, K concentration can easily be detected by flame photometer. The general optimum concentration in plant sample is >1.5 % dry plant materials, however the actual optimum concentration varies with plant species. The optimum concentration in different plant species is given in Table 4.

5 Conclusion and Recommendations

Role of K in crop growth and yield development is clear; however the K fertilization may not be required at all agricultural sites because deficiency of K may not be observed at all agricultural sites. However, there is a need to ensure the site-specific balanced fertilization based on soil analysis. Application of K fertilizers should be dealt more aggressively because in addition to direct crop-growth promoting effects, it also improves the nutrient use efficiency of plants and better N use efficiency has been observed at K fertilized agricultural sites. As K has ameliorating effects under various environmental stresses, such as drought, salinity and lodging, therefore this aspect should be more highlighted. Soils are being mined for K due to its higher concentrations in harvested crop parts, the removal of available K must be replenished for agricultural sustainability. In human and animal nutrition K has significant role and its deficiency in humans can cause cardiovascular and nervous diseases. The ultimate source of K in human nutrition is the soil which provides K to plants

Table 4 Sufficient range of K concentration in plant parts for optimum plant growth

Plant species	Concentration range (mg K/g DM)
Cereals, young shoot 5–8 cm above soil surface	
Wheat (*Triticum aestivum*)	35–55
Rice (*Oryza sativa*)[a] before anthesis	20–30
Oat (*Avena sativa*)	45–58
Barley (*Hordeum vulgare*)	35–55
Maize (*Zea mays*)[a]	20–35
Rye (*Secale cereale*)	28–45
Dicotyledonous field crops	
Forage and sugar beets (*Beta vulgaris*)	35–60
Sunflower (*Helianthus annuus*)[a] at anthesis	30–45
Potato (Solanum tuberosum)[a] at flowering	50–66
Soya bean (*Glycine max*)	25–37
Phaseolus beans (*Phaseolus vulgaris*)	20–30
Cotton (*Gossypium*) anthesis to fruit setting	17–35
Rape (*Brassica napus*)[a]	28–50
Forage Crops	
Total shoot at flowering 5 cm above soil surface, Dactylis glomerata, Poa pratensis, Phleum pratensis, Lolium perenne, Festuca pratensis	25–35
Vegetables	
Lettuce (*Lactuca sativa*)[a]	42–60
Spinach (*Spinacia oleracea*)	35–53
Pepper (*Capsicum annuum*)[a]	40–54
Watermelon (*Cirulus vulgaris*)[a]	25–35
Tomatoes (*Lycopersicon esculentum*)[a] at first fruit setting	30–40
Onions (*Allium cepa*) at mid vegetation stage	25–30
Carrot (*Daucus carota sativus*)[a]	27–40
Asparagus (*Asparagus officinalis*) fully developed shoot	15–24
Berry fruits[a]	
From anthesis until fruit maturation *Fragaria ananassa, Rubus idaeus, Ribes rubrum, Ribes nigrum, Ribes grossularia*	18–25
Miscellaneous crops	
Tea (*Camellia sinensis*)[a] at the mid of the vegetation season	16–23
Tobacco (*Nicotiana tabacum*)[a] at the mid of the vegetation season	25–45
[a]Young fully developed leaf	

Source: W. Bermann, *Nutritional Disorders of Crop Plants*. VEB Gustav Fischer Verlag,. Jena 1986

and is then used by humans and animals. In a developing world, it is required to optimize the K concentration in staple crops to fulfill the nutritional needs. Precise recommendation of K is very critical as K has strong interactions with clay minerals, therefore, while recommending K fertilizers soil minerals should be considered for better agricultural returns; otherwise plants may not respond to K fertilizers due to its unavailability to plants.

References

Abd-Alla MH, Wahab AMA (1995) Response of nitrogen fixation, nodule activities, and growth to K supply in water-stressed broad bean. J Plant Nutr 18:1391–1402

Abd-El-Motagally F (2004) Evaluation of two sugar beet cultivars (*Beta vulgaris* L.) for growth and yield under drought and heat conditions. PhD thesis. Faculty of Agricultural and Nutritional Sciences, Home Economics and Environmental Management, Justus Liebig University, Giessen

Abdi M, Nour-Mohamedi G, Golchin A (2002) The influence of foliar nutrition of urea and potassium chloride on grain yield, grain protein content, yield components and leaf relative water content of sardari wheat under rain-fed conditions. J Agric Sci Islamic Azad Univ 8:29–38

Abney TS, Foley DC (1971) Influence of nutrition on stalk rot development of *Zea mays*. Phytopathology 61:1125–1129

Ache P, Becker D, DeekenR DI, Weber H, Fromm J, Hedrich J (2001) VFK1, a Vicia faba K^+ channel involved in phloem unloading. Plant J 27:571–580

Ahmad P, Ashraf M, Hakeem KR, Azooz MM, Rasool S, Akram NA (2013) Potassium starvation-induced oxidative stress and antioxidant defense responses in *Brassica juncea*. J Plant 9(1):1–9

Allen VB, David J (2007) Handbook of plant nutrition. Pilbeam (eds). CRC Press, 600 Broken Sound Parkway NW, Suite 300, Boca Raton, FL 33487. 613 p, $139.95 hardcover. ISBN 0-8247-5904-4

Amtmann A, Sanders D (1999) Mechanism of Na^+ uptake by plant cells. Adv Bot Res 29:75–112

Amtmann A, Troufflard S, Armengaud P (2008) The effect of potassium nutrition on pest and disease resistance in plants. Physiol Planta 133:682–691

ArabiMIE MN, Jawhar M (2002) Effect of foliar and soil potassium fertilization on wheat yield and severity of septoria tritici blotch. Aust Plant Pathol 31:359–362

Beaton JD, Sekhon GS (1985) Potassium nutrition of wheat and other small grains. In: Munson RD (ed). Potassium in agriculture ASA, Madison

Bergmann E, Bergmann HW (1985) Comparing diagrams of plant/leaf analysis presenting by rapid inspection the mineral nutrient element status of agricultural crop plants. Potash Rev. Sub. 5 No. 2/1985 pp 1–10

Berkowitz GA, Kroll KS (1988) Acclimation of photosynthesis in *Zea mays* to low water potentials involves altered degree of protoplast volume reduction. Planta 175:374–379

Berkowitz GA, Peters JS (1993) Chloroplast inner-envelope ATPase acts as a primary H^+ pump. Plant Physiol 102:261–267

Bhatti AU, Wadud A, Shah GS (1981) Response of some major cereals of NWFP to nitrophos and nitrophoska complex fertilizers. Rice J Agric Res Pak 19:163–167

Bohn HL, Mcneal LB, O'connor AG (2001) Soil chemistry, 3rd edn. Wiley, New York/Chi Chester/Toronto/Singapore, p 120

Bohra JS, Doerffling K (1993) Potassium nutrition of rice (*Oryza sativa* L.) varieties under NaCl salinity. Plant Soil 152:299–303

Bolwell GP, Wojtaszek P (1997) Mechanisms for the generation of reactive oxygen species in plant defence: a broad perspective. Physiol Mol Plant Pathol 51:347–366

Botella MA, Martinez V, Pardines J, Cerda A (1997) Salinity induced potassium deficiency in maize plants. J Plant Physiol 150:200–205

Cakmak I (1994) Activity of ascorbate-dependent H_2O_2-scavenging enzymes and leaf chlorosis are enhanced in magnesium and potassium deficient leaves, but not in phosphorus deficient leaves. J Exp Bot 45:1259–1266

Chow WS, Marylin CB, Anderson JM (1990) Growth and photosynthetic responses of spinach to salinity: implications of K^+ nutrition for salt tolerance. Aust J Plant Physiol 17:563–578

Cox AE, Joern BC, Brounder SM, Gao D (1999) Plant-available potassium assessment with a modified sodium tetraphenylboron method. Soil Sci Soc Am J 63:902–911

Cramer GR, Lauchli A, Polito VS (1985) Displacement of Ca^{2+} from the plasmalemma of root cells. Plant Physiol 79:211–297

Cramer GR, Lynch J, Lauchli A, Epstein E (1987) Influx of Na^+, K^+ and Ca^{2+} into roots of salt-stressed cotton seedlings. Plant Physiol 83:510–516

Cramer GR, Alberico GJ, Schmidt C (1994) Salt tolerance is not associated with the sodium accumulation of two maize hybrids. Aust J Plant Physiol 21:675–692

Csatho P (1991) Effect of NPK-fertilization and split application of nitrogen on lodging due to windstorm and harvestable grain yield of maize. Acta Agron Hung 40:281–294

Cuartero J, Yeo AR, Flowers TJ (1992) Selection of donors for salt tolerance in tomato using physiological traits. New Phytol 121:63–69

Datta SK, Mikkelsen DS (1985) Potassium nutrition of rice. In: Munson RD (ed) Potassium in agriculture. ASA, Madison

Davis JG, Walker ME, Parker MB, Mullinix B (1996) Long term phosphorus and potassium application to corn on coastal plain soils. J Prod Agric 9:88–94

Doll EC, Lucas RE (1973) Testing soils for potassium, calcium, and magnesium. p 133–151. In: Walsh LM, Beaton JD (eds). Soil testing and plant analysis. Soil Science Society of America, Madison, Wisconsin

Dreyer I, Uozumi N (2011) Potassium channels in plant cells. FEBS J 278:4293–4303

Egilla JN, Davies FTJ, Drew MC (2001) Effect of potassium on drought resistance of Hibiscus rosa-sinensis cv. Leprechaun: plant growth, leaf macro and micronutrient content and root longevity. Plant Soil 229:213–224

e-ifc No. 13. 2007. Improving the efficiency of nitrogen use with potassium: De-Bottlenecking nitrogen use efficiency through balanced fertilization and adequate supply of potassium. International Potash Institute, Switzerland. http://www.ipipotash.org/eifc/2007/13/2/

Engels C, Marschner H (1992) Adaptation of potassium translocation into the shoot of maize (*Zea mays* L.) to shoot demand: evidence for xylem loading as a regulating step. Physiol Planat 86:263–268

Fernando M, Mehroke J, Glass DM (1992) De novo synthesis of plasma membrane and tonoplast polypeptides of barley roots during short-term K^+ deprivation. In search of the high-affinity K^+ transport system. Plant Physiol 100:1269–1276

Flowers T, Troke PF, Yeo AR (1977) The mechanisms of salt tolerance in halophytes. Annu Rev Plant Physiol 28:89–121

Ford CW, Wilson JR (1981) Changes in levels of solutes during osmotic adjustment to water stress in leaves of four tropical pasture species. Aust J Plant Physiol 8:77–91

Gajdanowicz P, Michard E, Sandmann M, Rocha M, Correa LGG, Ramìrez-Aguilar SJ (2011) Potassium (K^+) gradients serve as a mobile energy source in plant vascular tissues. Proc Natl Acad Sci U S A 108:864–869

Gerendas J, Heuser F, Sattelmacher B (2007) Influence of Nitrogen and Potassium Supply on Contents of Acrylamide Precursors in Potato Tubers and on Acrylamide Accumulation in French Fries. J Plant Nutr 30:1499–1516

Gething PA (1990) Potassium and water relationships. In: Potash facts. IPI, Bern

Goodall DW, Gregory PG (1947) Chemical composition of plants as an index of their nutritional status. Imperial Bureau Horticultural Plantation Crops. Technical Communication No. 17. Ministry of Agriculture, London, England

Gregory PJ, Crawford DV, McGowan M (1979) Nutrient relations of winter wheat. 1. Accumulation and distribution of Na, K, Ca, Mg, P, S and N. J Agric Sci Camb 93:485–494

Havlin JL, Beaton JD, Tisdale SL, Nelson WL (1999) Soil Fertility and Fertilizers, 6th edn. Prentice Hall, Upper Saddle River, 499 p

He FJ, MacGregor GA (2008) Beneficial effects of potassium on human health. Physiol Plant 133:725–735

Hsiao TC, Lauchli A (1986) Role of potassium in plant-water relations. In: Tinker B, Lauchli A (eds) Advances in Plant Nutrition, vol 2. Praeger Scientific, New York, pp 281–312

Humble GD, Raschke K (1971) Stomatal opening quantitatively related to potassium transport. Plant Physiol 48:447–453

Hussain F, Yasin M (2003) Soil Fertility monitoring and management in rice-wheat system. Ann. Report LRRP, NARC, Islamabad. pp 1–16

Int'l Potash Instt, 1986. Effect of potash in long term experiments. Int'l Fertilizer Correspondent 27:3–4
Itoh R, Kumara A (1987) Acclimation of soybean plants to water deficit. V. Contribution of potassium and sugar to osmotic concentration in leaves. Jpn J Crop Sci 56:678–684
Jacoby B (1979) Sodium recirculation and loss from Phaseolus vulgaris L. Ann Bot 43:741–744
Jacoby B, Hanson JB (1985) Controls on $^{22}Na^{+}$ influx in corn roots. Plant Physiol 77:930–934
Johanson JG, Cheesman JM (1983) Uptake and distribution of sodium and potassium by corn seedlings. I. Role of the mesocotyl in "sodium exclusion". Plant Physiol 73:153–158
Jones RG, Pollard A (1983) Proteins, enzymes and inorganic ions. In: Läuchli A, Bieleski RL, (ed) Inorganic plant nutrition. Springer; New York, pp 528–562
Kafkafi U (1990) The functions of plant K in overcoming environmental stress situations. In: Proceedings of the 22nd colloquium of IPI, pp 81–93, held in Soligorsk, USSR, IPI, Bern
Kafkafi U (1991) Root growth under stress. In: Waisel Y, Eshel A, Kafkafi U (eds). Plant roots, the hidden half. Marcel Dekker, Inc. New York, pp 375–391
Khattak JK, Bhatti AU (1986) Coop. Res. Prog, National out reach Res. Project, Soil. Fertility and Fertilizer use in Pakistan. Deptt. of Soil Sci. N.W.F.P. Agric. Univ. Peshawar. pp 60–69.
Khurana GP, Bhaya GP (1990) Effect of potash on wheat irrigated with nitrate waters. Ind J Agron 35:429–431
Krishnan KS (1977) Response of high yielding varieties of wheat to fertilizer application with special reference to potassium. Exp Agric 13:337–340
Kuchenbuch R, Jungk A (1984) Z Pflanzenernahr Bodenkd 147:435–448
Lamb C, Dixon RA (1997) The oxidative burst in plant disease resistance. Annu Rev Plant Physiol Plant Mol Biol 48:251–275
Leigh RA (2001) Potassium homeostasis and membrane transport. J Plant Nutr Soil Sci 164:193–198
Leigh RA, Jones RGW (1984) A hypothesis relating critical potassium concentrations forgrowth to the distribution and functions of this ion in the plant cell. New Phytol 97:1–13
Leigh RA, Jones RGW (1986) Cellular compartmentation in plant nutrition: the selective cytoplasm and the promiscuous vacuole. In: Tinker B, Lauchli A (eds) Advances in plant nutrition 2. Praeger Scientific, New York, pp 249–279
MacRobbie EAC (1977) Functions of ion transport in plant cells and tissues. MTP Int Rev Sci 13:211–247
Marschner H (1995) Mineral Nutrition of Higher Plants: functions of macronutrients. San Diego, USA
Marschner P (2012) Marschner's mineral nutrition of higher plants. 3rd ed. Academic Press, London, pp 178–189
Marschner H, Krauss A (1980) Correlations between potassium content and quality of potatoes. Kartoffelbau 131:65–67
Melis M, Farina MPW (1984) Potassium effects on stalk strength, premature death and lodging of maize (Zea mays L.). S Afr J Plant Soil 1:122–124
Mengel K (2001) Principles of plant nutrition. 5th edn. Kluwer Academic Publishers; Dordrecht, the Netherlands, pp 481–509
Mengel K, Haeder HE (1977) Effect of potassium supply on the rate of phloem sap exudation and the composition of phloem sap of *Ricinus communis*. Plant Physiol 59:282–284
Mengel K, Kirkby EA (2001) Principles of plant nutrition. Kluwer Academic Publishers, Dordrecht
Mesbah EAE (2009) Effect of irrigation regimes and foliar spraying of potassium on yield, yield components and water use efficiency of wheat (*Triticum aestivum* L.) in sandy soils. World J Agric Sci 5:662–669
Miller C (1993) Potassium selectivity in proteins - oxygen cage or pi in the face. Science 261:1692–1693
Moran N, Ehrenstein G, Iwasa K, Mischke C, Bare C, Satter RL (1988) Potassium channels in motor cells of Samanea saman. A patch clamp study. Plant Physiol 88:643–648

Mpelasoka BS, Schachman DP, Treeby MT, Thomas MR (2003) A review of potassium nutrition in grapevines with special emphasis on berry accumulation. Aust J Grape Wine Res 9:154–168

Mulder EG (1954) Effect of mineral nutrition on lodging of cereals. Plant Soil 5:245–306

Nandwal AS, Hooda A, Datta D (1998) Effect of substrate moisture and potassium on water relations and C, N and K distribution in *Vigna radiata*. Biol Planata 41:149–153

Nunes MA, Correia MM, Lucas MD (1983) NaCl-stimulated proton efflux and cell expansion in sugar beet leaf discs. Planta 158:103–107

Perrenoud S (1990) Potassium and plant health. 2nd edn. International Potash Institute; Bern, Switzerland: 1990. pp 8–10

Pettigrew WT (2008) Potassium influences on yield and quality production for maize, wheat, soybean and cotton. Physiol Plant 133:670–681

Pier PA, Berkowitz GA (1987) Modulation of water stress effects on photosynthesis by altered leaf K^+. Plant Physiol 85:655–661

Pinthus MJ (1973) Lodging in wheat, barley and oats: the phenomenon, its causes and preventive measures. Adv Agron 25:209–263

Prabhu AS, Fageria NK, Huber DM, Rodrigues FA (2007) Potassium and Plant Disease. In: Datnoff LE, Elmer WH, Huber DM (eds) Mineral Nutrition and Plant Disease. American Phytopathological Society, St. Paul, pp 57–78

Premachandra GS, Saneoka H, Ogata S (1991) Cell membrane stability and leaf water relations as affected by potassium nutrition of water-stressed maize. J Exp Bot 42:739–745

Rehman H, Bhatti AU, Gurmani AH (1982) Fertilizer experiments on cereal crops in D.I.Khan district. Soil. Sci. Div. Agric. Res. Instt. Tarnab, Peshawar. 36 p

Roberts SK, Tester M (1997) Permeation of Ca^{2+} and monovalent cations through an outwardly rectifying channel in maize root stelar cells. J Exp Bot 48:839–846

Rus A, Yokoi S, Sharkhuu A, Reddy M, Lee BH, Matsumoto TK, Koiwa H, Zhu JK, Bressan RA, Hasegawa PM (2001) AtHKT1 is a salt tolerance determinant that controls Na^+ entry into plant roots. Proc Natl Acad Sci U S A 98:14150–14155

Schubert S (2006) Pflanzenerri¨ahrung. Verlag Eugen Ulmer, Stuttgart

Scott AD, Smith SJ (1987) Sources, amount and forms of alkali elements in the soil. Adv Soil Sci 6:101–147

Sen-Gupta A, Berkowitz GA, Pier PA (1989) Maintenance of photosynthesis at low leaf water potential in wheat. Plant Physiol 89:1358–1365

Shabala S (2003) Regulation of potassium transport in leaves: from molecular to tissue level. Ann Bot 92:627–634

Shabala S, Babourina O, Rengel Z, Nemchinov LG (2010) Non-invasive microelectrode potassium flux measurements as a potential tool for early recognition of virus–host compatibility in plants. Planta 232:807–815

Sharma SR, Kolte SJ (1994) Effect of soil-applied NPK fertilizers on severity of black spot disease (*Alternaria brassicae*) and yield of oilseed rape. Plant Soil 167:313–320

Sharma KD, Nandwal AS, Kuhad MS (1996) Potassium effects on CO_2 exchange, ARA and yield of cluster bean cultivars under water stress. J Pot Res 12:412–423

Siddique T, Aslam J, Alyas M (1997) Response of wheat to potassium application in different soil series of Punjab. Pak J Soil Sci 13:92–96

Smith JAC, Milburn JA (1980) Phloem turgor and regulation of sucrose loading in *Ricinus communis* L. Planta 148:42–48

Somers TC (1977) A connection between potassium levels in the harvest and relative quality in Australian red wines. Aust Wine Brew. Spirit Rev 96:32–34

Syers JK (2003) Potassium in soils: current concepts. In: Johnston AE (ed) Proceedings of the IPI Golden Jubilee congress 1952–2002 held in Basel, Switzerland 8–10 October 2002. Feed the Soil to Feed the People. The Role of Potash in Sustainable Agriculture. International Potash Institute, Switzerland. p 301–310

Taiz L, Zeiger (2006) Plant physiology, Fourth Edition. Sinauer Associates. Sunderland, MA. 764 pages

Taleisnik E, Grunberg K (1994) Ion balance in tomato cultivars differening in salt tolerence. I. Sodium and potassium accumulation and fluxes under moderate salinity. Physiol Planat 92:528–534

Termaat A, Munus R (1986) Use of concentrated macronutrient solutions to separate osmotic from NaCl-specific effects on plant growth. Aust J Plant Physiol 13:509–522

Tisdale S, Nelson WL, Beaton JD, Havlin JL (1993) Soil Fertility and Fertilizers, 5th edn. MacMillan Publishing Company, New York

Vaithilingam C, Balasubramanian M (1976) Effect of potash on sclerenchyma thickness and silica content in rice. Indian Potash J 1:17–23

Wakeel A, Hanstein S, Pitann B, Schubert S (2010) Hydrolytic and pumping activity of H^+-ATPase from leaves of sugar beet (*Beta Vulgaris* L.) as affected by salt stress. J Plant Physiol 167:725–731

Wakeel A, Gul M, Sanaullah M (2013) Potassium dynamics in three alluvial soils differing in clay contents. Emir J Food Agric 25:39–44

Wakhloo JL (1975) Studies on the growth, flowering and production of female sterile flowers as affected by different levels of foliar potassium in *Solanum sisymbrifolium* Lam. J Exp Bot 26:425–450

Weinhold A (1862) Analyze von unkiauterm des bodens der versuchsstation chemnitz. Landeo Vers Sta 4(188–1):93

Welch LF, Flannery RL (1985) Potassium nutrition of corn. In: Munson RD (ed) Potassium in agriculture. ASA, Madison

Yadav DS, Goyal AK, Vats BK (1999) Effect of potassium in *Eleusine coracana* (L.) Gaertn. under moisture stress conditions. J Pot Res 15:131–134

Zaman U, Ahmad Z, Farooq M, Saeed S, Ahmad M, Wakeel A (2015) Potassium fertilization may improve stem strength and yield of basmati rice grown on nitrogen-fertilized soils. Pak J Agric Sci 52:439–445

Zhang YH, Cheng H, Alexeenko VA, Dempsey CE, Hancox (2010) Characterization of recombinant hERG K^+ channel inhibition by the active metabolite of amiodarone desethyl-amiodarone. J Electrocardiol 43:440–448

Zörb C, Noll A, Karl S, Leib K, Yan F, Schubert S (2005) Molecular characterization of Na^+/H^+ antiporters (ZmNHX) of maize (*Zea mays* L.) and their expression under salt stress. J Plant Physiol 162:55–66

Zörb C, Senbayram M, Peiter E (2014) Potassium in agriculture, Status and perspectives. J Plant Physiol 171:656–669

Weathering and Approaches to Evaluation of Weathering Indices for Soil Profile Studies – An Overview

Sartaj A. Wani, G.R. Najar, J.A. Wani, Mohmad Ramzan, and Khalid Rehman Hakeem

Contents

Abstract Weathering involves a gradual progressive physico-chemical and hydrothermal alterations of the original parent material and transformation to more stable minerals. Chemical weathering indices are commonly used for characterising weathering profiles by incorporating bulk major element oxide chemistry into a single metric for each element. The principal assumption in formulating chemical weathering indices is that the behavior of chemical elements is controlled solely by the degree of weathering. A comprehensive review of a great number of geochemi-

S.A. Wani (✉) • G.R. Najar • J.A. Wani
Department of Soil Science, Sher-e-Kashmir University of Agricultural Sciences and Technology-Shalimar, Srinagar, Jammu and Kashmir 190001, India
e-mail: sartajmess@gmail.com

M. Ramzan
Department of Botany, Aligarh Muslim University, Aligarh 202002, Uttar Pradesh, India

K.R. Hakeem (✉)
Faculty of Forestry, Universiti Putra Malaysia, 43400 Serdang, Selangor, Malaysia
e-mail: kur.hakeem@gmail.com

K.R. Hakeem et al. (eds.), *Soil Science: Agricultural and Environmental Prospectives*, DOI 10.1007/978-3-319-34451-5_8

cal qualitative and quantitative weathering index approaches have been devised to quantify the changes most effectively to indicate mineralogical composition and mobilisation in the index properties of rock materials and evaluating through various related statistical analytical approaches. Understanding of the nature and indices of weathering within weathered profiles is of a paramount importance. Past conditions of physical and chemical weathering can be reliably inferred if application of the Chemical Index of Weathering is combined with a comprehensive facies analysis. The criteria applied in evaluating the utility of weathering indices on parent materials and weathering profiles and its applications for characterizing weathering profiles, soil fertility and understanding salinity and groundwater management determines its degree of scope.

Keywords Chemical weathering elements • Minerals • Profiles • Weathering indices

1 Introduction

Weathering the process of alteration and breakdown of rocks at or near the surface of earth by physical and chemical effects either natural or athropogenic leads to a number of physical and chemical changes (Miscevic and Vlastelica 2009) in the rocks. Weathering is the process of alteration and breakdown of rock and soil materials at near the Earth's surface by chemical decomposition and physical disintegration (Selby 1993). According to some researchers, weathering is the irreversible response of soil and rock materials and masses to their natural or artificial exposure to the near surface geomorphological or engineering environment. Weathering of rocks to form minerals results in the gradual change in the amount of elements constituting the rocks. Exposed rocks are affected to variable degrees by a combination of chemical and physical weathering (Bland and Rolls 1998 and Etienne et al. 2008). Weathering process and soil development are classically considered as two aspects of a single phenomenon which is the transformation of rocks under climatic conditions (meteroization). When a soil develops on rock, a soil profile develops. The different layers or horizons of the profile are not the same as beds formed by sedimentation; instead each of the horizons forms and grows in place by weathering and the addition of organic material from decaying plants and plant roots.

2 Physical Weathering

Physical weathering causes disintegration and separation of rocks without changing their chemical composition. Typical examples of these processes are the splitting of rocks by the daily warming of the sun and cooling during the night (typical of desert environments) or by the repeated freezing and thawing of water (when water freezes, its volume increases by 10 %, causing tremendous pressures if it occurs in confined spaces such as crevices in rocks). A layer of loose material produced during physical weathering which covers the underlying solid rock. This material also known as regolith can vary from a few millimeters to tens of meters in thickness. Regolith layers in some parts of West Africa have been found to be more than 150 m thick. There is often a sharp boundary or line of divide between the bottom of the regolith and the bedrock. This narrow zone is known as the weathering front and is considered as most active weathering zone.

Mechanical weathering and physical erosion are driven by stress gradients that break or move apart the detached material. These stress gradients may arise from gravitational stress, phase transformations (such as ice growth, salt precipitation, oxidation or hydration of minerals) or biological agents (roots, burrowing animals, worms, and the like). Physical weathering leads to the degradation or break down of rocks to smaller grain sizes, ideally without causing geochemical and mineralogical changes during the weathering process. If physical weathering and the grinding action of moving ice lodged debris in particular, degrades a source rock into a clay-sized deposit, it should essentially preserve the mineralogical and geochemical composition of the original rock (Nesbitt and Young 1996).

3 Chemical Weathering

The chemical weathering of rock proceeds by water-rock interaction. The interactions between minerals present in soils and water play an important role in geochemical processes, i.e. soil formation, elemental mobility, bio-mineralization, nutrient availability, etc (Lin 2010 and Putins and Ruiz-Agudo 2013). Hydrolytic weathering in general, causes a progressive transformation of affected components into clay minerals, ultimately kaolinites (Miscevic and Vlastelica 2011). During chemical weathering, some alkali and alkaline earth mineral elements are easily leached from rock and the residual elements are redistributed to secondary minerals (Loughnan 1969). This is the fundamental system on the chemical weathering of rocks. Progressive chemical weathering of labile minerals like feldspar leads to the loss of Ca^{2+}, K^+ and Na^+ and thereby transformation to minerals more stable under surface conditions (Fedo et al. 1995). It results in the formation of shales rich in clay minerals like illite and kaolinite and Fe-oxyhydrates like goethite. Consequently, the character of the climate framework and the way it governs weathering conditions and reactions is reflected by the mineralogical and mobile element geochemical composition of the resulting deposits. Chemical weathering of silicate rocks ultimately leads through hydrolysis to an exchange of the cations like Na^+, K^+, Ca^{2+} and Mg^{2+} for H^+, and may be a loss of Si^{4+} (Kramer 1968). Na^+, K^+ and Ca^{2+} are commonly supplied by the weathering of feldspar and volcanic glass, together accounting for 58 % Ca^{2+} of the exposed crust. Mg^{2+} derived from glasses, sheet silicates etc. resides in chloritic and smectitic clays (Nesbitt and Young 1984; Pettijohn et al. 1987). It has been observed generally that mica minerals, particularly biotite, are more prone to chemical weathering and release large amounts of major and trace elements to be moved into solution (Price et al. 2014). Marl is significantly influenced by weathering around the sandstone layer. It disintegrates quickly and is gradually removed by the action of gravity and precipitation (Adamassu et al. 2004). The different weathering processes occurring under natural conditions are shown in Fig. 1.

4 Relationship Between Physical and Chemical Weathering

Understanding the interplay between physical and chemical weathering processes and their relationship to erosion rates has become increasingly important in understanding geochemical cycles (Stallard 1995). The potential feedbacks between erosion and chemical weathering were brought to prominence by the erosion-driven

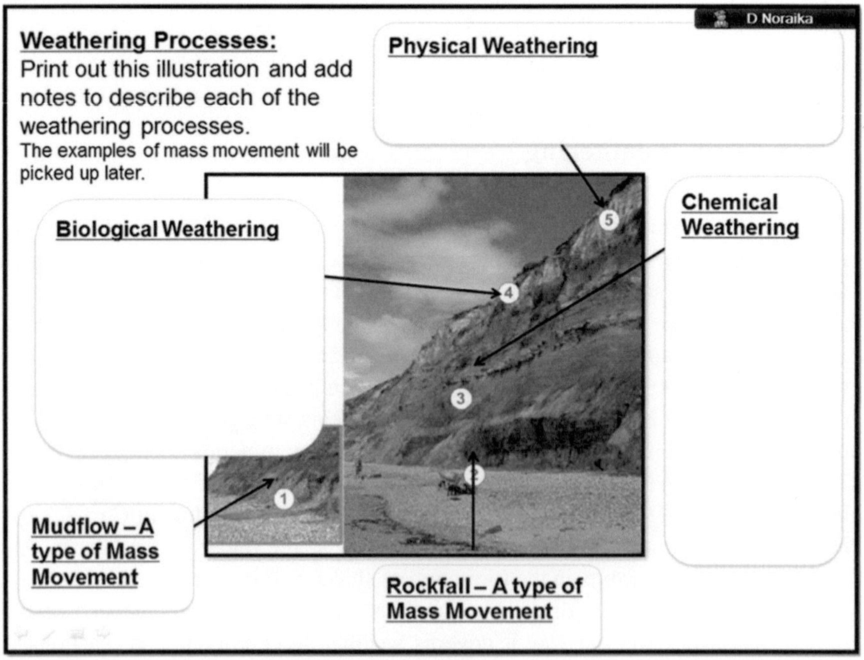

Fig. 1 Showing different weathering phenomenon

climate change hypothesis (Raymo et al. 1988; Raymo and Ruddiman 1992). Given that global silicate weathering rates exert an important control on the long-term C dioxide content of the atmosphere (Holland 1978; Berner 1990), if silicate chemical weathering is linked to physical denudation and tectonic uplift (Ruddiman and Prell 1997), then weathering and erosion may drive climate change. A number of workers have stressed the significant role played by physical erosion in enhancing chemical-weathering rates (Gardner and Walsh 1996; Gaillardet et al. 1999; Riebe et al. 2001), whereas White and Blum (1995) observed little correlation between physical erosion rates and chemical weathering. The relationship between physical and chemical weathering processes has been accomplished in a small catchment, where there is more control on rock type, vegetation and local climate. Long-term chemical weathering rates have typically been measured using soil mass-balance techniques (April et al. 1986), which quantify the total mass that is removed when a specified volume of unweathered parent material is converted to soil (Brimhall and Dietrich 1987). If physical weathering process (erosion) has been negligible, then mass loss from a soil can be attributed to chemical weathering alone. In that case, the mass loss of soil can be interpreted as a chemical weathering rate, using the soil's age (if known) to average the loss over the time since the soil began to form (Bain et al. 1993; Taylor and Blum 1995). But non-eroding soils of known age are rare and by their nature preclude comparisons of physical erosion and chemical weathering. Fletcher et al. (2006) explored complex coupling between chemical and mechanical processes that advance the weathering front through spheroidal weathering of granite. Increased volume from oxidation of iron in biotite exerts stresses, ultimately fracturing the rock, forming multiple, onion-skin layers separated by cracks. Although diffusion of oxygen controls initial chemical weathering, the flow paths created by fractures allow rapid transport of oxygen and water deeper into the unweathered core. The rate of advance of the weathering front is linear with time, rather than a function of the square root of time, as expected for true diffusion.

5 Quantification of Weathering

Quantifying long-term rates of chemical weathering (Heidari et al. 2010; Khanlari et al. 2012) and physical erosion (weathering) is important for understanding the long-term evolution of soils, landscapes and earth's climate. Weathering indices have been devised to quantify changes in the index properties of rock materials; some of which are relevant to engineering properties. Weathering indices are typically applied by plotting some specific index versus depth in the weathering profile, providing a visual representation of changes in bulk chemistry with presumed increasing (or decreasing) weathering of the parent rock. The role of chemical weathering indices is essentially to quantify the degree of depletion of mobile components relative to immobile components during weathering (Harnois 1988). These indices can then be applied to standard weathering grades of material set up by specific weathering classifications systems which in turn may be correlated with engineering behaviour. In a qualitative way, mineralogical changes can be detected in shales and sandstones using X-ray diffractometry. Optical analysis is a quantitative option in sandstones only. Here, the mineralogical index of alteration (MIA) permits an assessment of weathering effects (Johnsson et al. 1988).

Quantitative estimations of weathering in coarse-grained and fine-grained rocks are relatively easily achieved by using whole rock geochemical data to calculate geochemical weathering proxies. A comprehensive review of a number of geochemical weathering indices was presented by Duzgoren-Aydin et al. (2002). Recently, von Eynatten et al. (2003 and Von Eynatten 2004) presented a quantitative approach, the t-index, to statistically model linear compositional and weathering trends. If corroborated by further studies the t-index will likely lead to more quantitative definitions of weathering trends. A related statistical approach has been taken by Ohta and Arai (2007) whose weathering index (WI) is based on principal component analysis of 8 major oxides. However, as yet it has been applied only to igneous rocks.

5.1 Weathering Indices

Chemical changes during weathering and hydrothermal alterations are quantified in several ways including the ratios of elements to elements, measurement and calculation of loss and gain of weight based on immobile elements and chemical weathering indices. Geochemical estimations of weathering effects need to be considered carefully because the major cations of $Na^{+,}$ K^+ and Ca^{2+} may be mobile also under diagenetic conditions (Wintsch and Kvale 1994). A very simple proxy is the Ruxton Ratio R (Ruxton 1968) given by the SiO_2/Al_2O_3 ratio. This assumes that Al_2O_3 remains immobile during weathering so that changes in R reflect silica loss as a proxy for total element loss. The Ruxton Ratio R may be useful when weathering profiles on rocks of felsic and intermediate composition are considered, but was found to be poorly correlated to the actual weathering grade of a silicate rock (Duzgoren-Aydin et al. 2002).

5.1.1 Weathering Index of Parker (WIP)

Since the transformation of feldspar to clay minerals and the coincident mobility of the main cations is a major process of chemical (hydrolytical) weathering, Parker (1970) considered it more useful to monitor changes in Na^+, K^+, Ca^{2+} and Mg^{2+} and created a weathering index (WIP, Weathering Index of Parker) given by

$$\mathrm{WIP} = (\mathrm{K}^* / 0.25 + \mathrm{Na}^* / 0.35 + \mathrm{Ca}^* / 0.69 + \mathrm{Mg}^* / 0.98)100$$

Where the cations* represent the atomic percentage of an element divided by the atomic weight.

Parker (1970) also considered the susceptibility of these elements to weathering by including in the denominator Nicholls' values of bond strength as a measure of the energy necessary to break the cation-to-oxygen bonds of the respective oxides. These different values are considered to reflect the probability of an element to be mobilized during the weathering process. Values of WIP are commonly between $\geq$100 and 0 with the least weathered rocks having the highest values. The WIP simply assumes that all Ca^{2+} in a silicate rock is contained in silicate minerals. This simplification is a source of imprecision of the index, particularly if larger amounts of carbonate detritus or cements are present in the rock.

5.1.2 Chemical Weathering Indices (CWI)

Chemical weathering indices (CWI) as like other weathering indices (physical, micro structural, mechanical) were introduced as a result of dissatisfaction with and the subjectivity of the sixfold weathering classification (Moye 1995) and its improved versions (GSL 1995). More than 30 different CWI, mostly proposed for felsic igneous weathered materials were developed under tropical and subtropical environments. These indices were expressed mostly as molecular ratios of major elements. Chemical weathering index, sometimes referred to as Chemical Index of Alternation (CIA), is the most accepted available weathering index for estimating weathering profile (Price et al. 2003). The principal assumption in formulating these indices is that behaviour of chemical elements involved principally controlled by the degree of weathering. Chemical weathering indices incorporate bulk major element oxide chemistry into a single value for each sample. The WIP (Weathering Index of Parker) in the Chemical Index of Alteration (CIA) uses whole rock geochemical data of major element oxides (Nesbitt and Young 1982). The index is essentially based on the same considerations which led Kramer (1968) to conclude that monitoring the hydrolysis of feldspar and volcanic glass and the respective changes in the content of the major cations offers the best quantitative measure of chemical weathering. It represents a ratio of predominantly immobile Al_2O_3 to the mobile cations Na^+, K^+ and Ca^{2+} given as oxides. The CIA is defined as

$$\mathrm{CIA} = \frac{(\mathrm{Al_2O_3})}{\mathrm{Al_2O_3} + \mathrm{Na_2O} + \mathrm{K_2O} + \mathrm{CaO}^*} \times 100$$

Where the major element oxides are given in molecular proportions. CaO* represents the CaO content of silicate minerals only (Fedo et al. 1995) and thus eliminates one of the disadvantages of the WIP. Kaolinite with CIA value of 100 represents the highest degree of weathering followed by Illite, muscovite, feldspars, Fresh basalts, fresh granites and granodiorites with their CIA values shown in Table 1 (Bahlburg 2009, and Dhannoun et al. 2010). The CIA measures weathering intensity on a scale from 1 to 100, where a higher value represents more intense chemical weathering of the material, indicating Kaolinite the most and fresh basalt as least weatherable clay minerals.

Table 1 Approximate range of values of CIA in various rocks and clay minerals

Rock/Clay	CIA-value
Kaolinite	~100
Illite	~75–90
Muscovite	~75
Feldspar	~50
Fresh Granite/ Granodiorite	~45–55
Fresh Basalt	~30–45

Bahlburg (2009) and Dhannoun et al. (2010)

The values of weathering index (CIA) seem to demonstrate that both the glacial and non-glacial Neoproterozoic sedimentary rocks in South China consist of a mixture of detritus of glacial and non-glacial weathering provenance, albeit in different proportions (Fig. 2). The initial incorporation of weathered older detritus into the

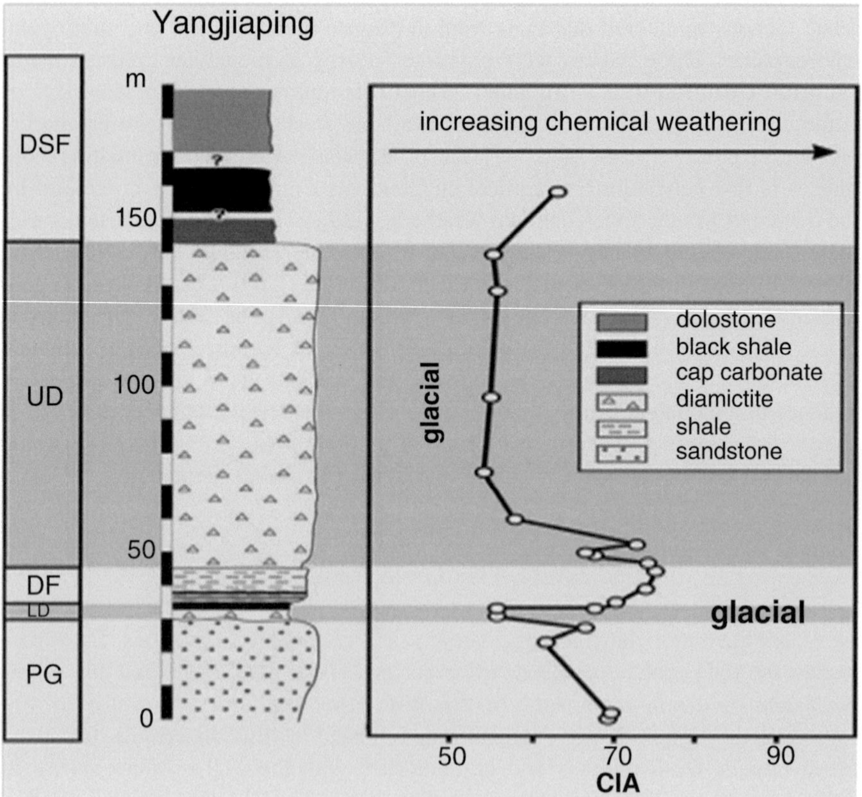

Fig. 2 The Yangjiaping section in South China and its Neoproterozoic climate record indicated by *CIA* values. *PG* preglacial units, *LD* lower diamictite, *DF* Datangpo Formation, *UD* upper diamictite, *DSF* Doushantuo Formation (Modified from Dobrzinski et al. (2004))

Neoproterozoic glacial deposits was demonstrated by Dobrzinski et al. (2004) by upward trends to lower CIA values in several sections in South China.

The CIA values alone are almost meaningless as climate indicators when not considered in the context of the stratigraphic framework and facies of the analysed sedimentary rock. This is demonstrated very well in a study of the Fiq Formation in Oman by Rieu et al. (2007) as shown in Fig. 3, an association of alternating glacial diamictites, debris-flow deposits, turbidite sandstones, hemipelagic shales, and wave-rippled shoreface deposits.

5.1.3 Chemical Proxy of Alteration (CPA)

Another weathering index, the Chemical Proxy of Alteration (CPA) was used as a complement to the CIA being insensitive to bias effects caused by CaO from minerals other than silicates. The CPA is calculated as follows:

$$\mathrm{CPA} = \frac{(\mathrm{Al_2O_3})}{(\mathrm{Al_2O_3} + \mathrm{Na_2O})} \times 100$$

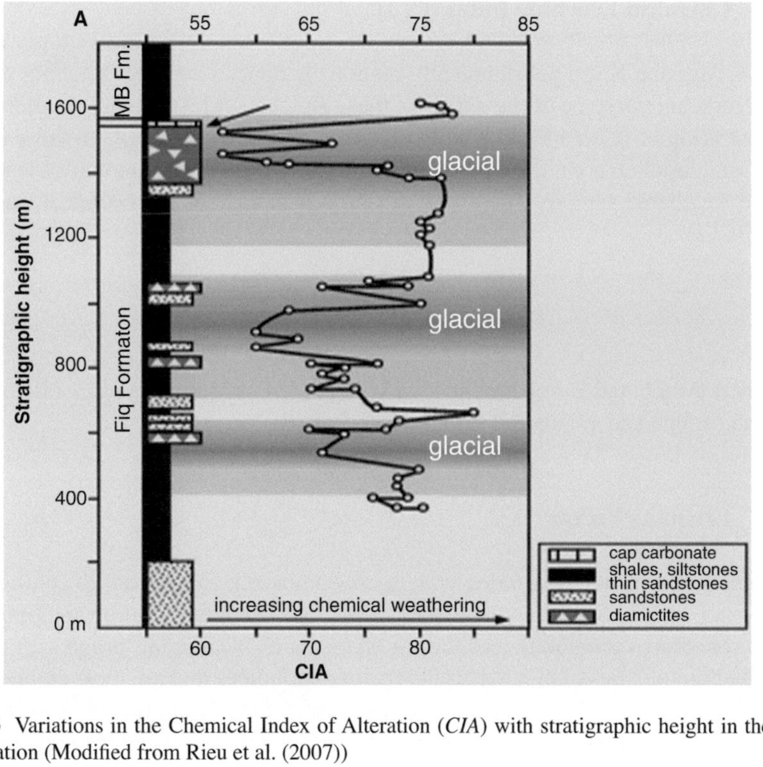

Fig. 3 Variations in the Chemical Index of Alteration (*CIA*) with stratigraphic height in the Fiq Formation (Modified from Rieu et al. (2007))

Table 2 Summary of weathering indices

Index	Formula	Optimum value	Fresh optimum weathered value	References
CIA	$[(Al_2O_3/Al_2O_3+CaO+Na_2O+K_2O)]\times 100$	≤50	100	Nesbitt and Young 1982
CIW	$[(Al_2O_3/Al_2O_3+CaO+Na_2O)]\times 100$	≤50	100	Harnois and Moore (1988)
WIP	$[(2Na_2O/0.35)+(MgO/0.9)+(2K_2O/0.25)+(CaO/0.7)]$	>100	0	Parker 1970
PWI	$[(SiO_2/(TiO_2+Fe_2O_3+SiO_2+Al_2O_3)]\times 100$	>50	0	Souri et al. (2006)

According to Buggle et al. (2011), this may be the most appropriate geochemical proxy for silicate weathering in loess-paleosol sequences. Like the CIA, the CPA has values in a range of 1–100, where a higher value indicates more intense weathering. Assessing that geochemical component and weathering indices can improve our understanding of soil formation and process. Among the indices proposed by the researcher workers (Lee et al. 2008; Heidari et al. 2011; Khanlari et al. 2012), we used CIA, CIW, PWI and WIP as commonly used calculated indices (Table 2).

5.1.4 Chemical Leaching Index (CLI)

Ca, Na, Mg, and K are geochemically mobile elements. Chemical leaching results in a significant decrease of the oxides of these elements. The ratio of the volumetric concentration of ($CaO+MgO+Na_2O+K_2O$) in a weathered sample to those in the fresh sample taken from the same weathering profile gives the amount of leaching for the weathered sample. Therefore, this ratio is defined as the chemical leaching index (CLI).

$$\mathrm{CLI} = \frac{100\,(\mathrm{Amob}-\mathrm{Bmob})}{\mathrm{Amob}}$$

Where Amob and Bmob are the total volumetric concentration of mobile oxides in fresh sample and weathered sample, respectively.

5.1.5 Leaching Factor

Changes in the weathering index with depth commonly are gradual or continuous, steady and systematic for homogeneous parent rocks (granite) (Sutton and Maynard 1992) reflecting continuous leaching of elements as weathering progresses on an initially homogenous parent material. Weathering indices that are most mobile during weathering (i.e., alkali and alkaline earth metals) are most effective in quantifying the effects of chemical weathering relative to the unweathered bedrock and are best suited for application to heterogeneous weathering residual. The oxides of Al, Fe and Ti i.e., Al_2O_3, Fe_2O_3, and TiO_2 respectively gradually increase with increase in weathering while all other elements are gradually reduced. This is the basis of calculating the weathering indices of soil which actually indicate the degree of weathering undergone by rocks. One of the earliest weathering indexes is the ratio of the sum of oxides of Potassium K, and Sodium Na, to Silica $SiO_{2,}$ in the weathered horizon and in the parent material. This has been designated as the leaching factor (LF) by Jenny (1941) as

$$\text{Leaching factor} = \frac{(K_2O + Na_2O / SiO_2)\text{ of weathering horizon}}{(K_2O + Na_2O / SiO_2)\text{ of the parent material}}$$

Molar ratios have also been considered to be the good indices of weathering. Birkland (1999) have quoted the undermined ratios:-

Silica : Alumina ratio	...	SiO_2 / Al_2O_3
Silica : Iron oxide ratio	...	SiO_2 / Fe_2O_3
Silica : Sesquioxide ratio	...	$Sio_2 / SiO_2 + Fe_2O_3$
Silica : R_2O_3 ratio	...	$SiO_2 / Fe_2O_3 + Al_2O_3 + TiO_2$
Bases : Alumina ratio	...	$K_2O + Na_2O + CaO + MgO / Al_2O_3$
Bases : R_2O_3 ratio	...	$K_2O + NaO + CaO + MgO / Fe_2O + Al_2O + TiO_2$
Parker weathering index	...	$100\ (K_2O / 0.25 + Na_2O / 0.35 + CaO / 0.69 + MgO / 0.98$
Reiches Weathering Potential index	...	$100\ (\text{Sum of bases} - H_2O) / \text{sum of bases} + SiO_2 + R_2O_3)$
Reichmann product index	...	$100\,SiO_2 / SiO_2 + R_2O_3$

The first six ratios should decrease in a linear environment i.e., high rainfall (Birkland 1999). The weathering indices given by Bear (1964) of some clay sized minerals in soils are shown in Table 3.

5.1.6 Chemical Weathering Product Index (CWPI)

Some elements like Al, Fe and Ti are less affected by chemical leaching than alkali and alkali-earth elements, but tend to concentrate in weathering products (Loughnan 1969). Depending on development of drainage, the Si mobility is controlled in the weathering products. The ratio of the total amount of these oxides in weathering product to those in the respective sample yields the amount of weathering products. Therefore, the chemical weathering product index (CWPI) is defined through the following equation:

Table 3 Weathering indices of some clay sized minerals in soils

S.No.	Typical Clay-size minerals	Weathering index
I	Gypsum	1
II	Calcite, dolomite, aragonite, apatite	2
III	Olivine-hornblende, pyroxenes, diopsides	2
IV	Biotite, glauconite, nontronite	3
V	Albite, microcline	4
VI	Quartz, cristobalite	5
VII	10 A Mica, muscovite, sericite	6
VIII	Vermiculite, 2:1:1minerals, where brucite layer is Fe/Mg	7
IX	Montmorillonite, beidellite	8
X	Pedogenic chlorite (2:1:1 minerals)	9
XI	Kaolinite, halloysite	9
XII	Gibbsite, boehmite,	10
XIII	Allophane	11
XIV	Hematite	11
XV	Anatase	12

Source: Bear (1964)

$$\mathrm{CWPI} = \frac{100\,(\mathrm{Bimmob} - \mathrm{Cimmob})}{\mathrm{Bimmob}}$$

Where Bimmob and Cimmob are the total volumetric concentration of immobile oxides in the whole sample and unaltered portion of the sample, respectively.

5.1.7 Total Chemical Weathering Index (TCWI)

Considering the definition of Loughnan (1969), total chemical weathering index can be defined as the sum of CWPI and CLI. Since the rock material can be weathered 100 % at most, TCWI value should be also at most 100. Therefore, (CWPI + CLI) value has been divided by 2 in order to get TCWI given by the following equation

$$\mathrm{TCWI} = \frac{(\mathrm{CWPI} + \mathrm{CLI})}{2}$$

6 Criteria Applied in Evaluating the Utility of Weathering Indices

A useful chemical weathering index should provide values that do more than simply vary relative to one another on a given weathering profile. Ideally, chemical weathering index should permit comparison between studies performed at different localities, on different parent materials and on weathering profiles of different ages.

Based on review of literature the following criteria are applied in evaluating the utility of weathering indices.

1. A chemical weathering index should be easy to use, involve chemical elements common in soil analysis (Harnois 1988). This is desirable for wide applicability and maximum comparison with other studies.
2. A weathering index should incorporate with a range of mobility in the weathering environment (Harnois 1988).
3. A useful weathering index should exhibit chemically appropriate trends with increased trends and should change greatly with increasing weathering. This is especially important in weathering profiles developed from meta sedimentary parent rocks which are commonly compositionally layered at scales much smaller than overall dimensions of the weathering profile.
4. A useful weathering index should be applicable to wide range of rocks and rock types and should yield values for unweathered parent material, regarding of rock type.
5. Ideally a chemical weathering index should not assume that any element is immobile. This is difficult to apply in practice. At best, those elements that are least mobile are used.

7 Applications of Weathering Indices

1. Chemical weathering indices are commonly used for characterizing weathering profiles (Kirschbaum et al. 2005) by incorporating bulk major element oxide chemistry into a single metric for each sample. Weathering indices are typically applied by plotting some specific index vs. depth in the weathering profiles, providing representation of the changes in bulk chemistry with presumed increasing or decreasing weathering of parent material.
2. Weathering indices are commonly used in evaluating soil fertility and development (Delvaux et al. 1989), demonstrating the impact of climate on bedrock weathering characterizing alteration (Souri et al. 2006) thereby quantifying engineering properties of regolith.
3. Weathering indices of parker (WIP) is most appropriate for application to weathering profiles on homogenous and heterogeneous parent rocks because weathering of parker includes only the highly mobile alkali and alkaline earth elements in its formation, it yields values that differ greatly from those of parent rock. WIP allows for minimum mobility and thus WIP is most appropriate for studying the weathering of heterogeneous sedimentary rocks.
4. The weathering intensity index has broad application in understanding weathering and geomorphic processes across a range of spatial and temporal landscape scales (Wilford 2011). The index can indicate highly weathered paleo-surfaces, assess chemical and physical denudation processes and map the relative rates of regolith formation and its removal through erosion across different landscapes.

5. The index also has the potential to be used in combination with other environmental covariates in a range of soil property predictions including texture, chemical composition, depth, fertility, pH, porosity and permeability. The latter two properties are important for understanding the way in which solutes move through groundwater and the interflow pathways within the regolith and the consequent hydro-pedological processes and characteristics within different landscapes.
6. The weathering intensity index is currently being integrated with other datasets to develop an improved hydro-geological framework to assist in improved salinity and groundwater management. Correlations are expected when using the weathering index for broad-scale ecological studies where biological processes are underpinned by soil fertility and water availability. The index is therefore likely to be useful in mapping and modelling plant types and/or for predicting the distribution of plant communities as well as assisting a more general understanding of the interrelationships between regolith, climate (present and paleo) and vegetation at local, regional or even continental scale.

References

Adamassu Y, Shakoor A, Wells NA (2004) Evauating selected factors affecting the depth of undercutting in rocks subjected to differential weathering. Eng Geol 124:1–11

April R, Newton R, Coles LT (1986) Chemical weathering in two Adirondack watersheds: past and present day rates. Geol Soc Am Bull 97:1232–1238

Bahlburg H (2009) A review of the Chemical Index of Alteration (CIA) and its application to the study of Neoproterozoic glacial deposits and climate transitions. In: Arnaud E, Halverson GP, Shields GA (eds) The geological record of Neoproterozoic glaciations. Geological Society, Memoir, London

Bain DC, Mellor A, Robertson Rintoul MSE, Buckland ST (1993) Variations in weathering processes and rates with time in a chronosequence of soils from Glen Freshie, Scotland. Geoderma 57:275–293

Bear FE (ed) (1964) Chemistry of the soil. van Norshtand Reinhold, New York, 515 p

Berner RA (1990) Atmospheric carbon dioxide levels over Phanerozoic time. Science 249:1382–1386

Birkland PW (1999) Soils and geomorphology, 3rd edn. Oxford University Press, New York

Bland W, Rolls D (1998) Weathering an introduction to the scientific principles. Arnold Publishers, London, p 271

Brimhall GH, Dietrich WE (1987) Constitutive mass balance relations between chemical-composition, volume, density, porosity and strain in metasomatic hydrochemical systems-Results on weathering and pedogenesis. Geochim Cosmochim Acta 51:567–587

Buggle B, Glaser B, Hambach U, Gerasimenko N, Markovi S (2011) An evaluation of geochemical weathering indices in loess-paleosol studies. Quat Int 240:12–21

Delvaux B, Herbillon AJ, Vielvoye L (1989) Characterization of a weathering sequence of soils derived from volcanic ash in Cameroon. Taxonomic, mineralogical and agronomic implications. Geoderma 45:375–388

Dhannoun HY, Othman SM, Al-Dabbagh SM (2010) The relationship between chemical index of alteration and some major and trace elements content in Rocks of Injana Formation of Northern Iraq. Iraqi Nat J Earth Sci 11(1):1–22

Dobrzinski N, Bahlburg H, Strauss H, Zhang QR (2004) Geochemical climate proxies applied to the Neoproterozoic glacial succession on the Yangtze Platform, South China. In: JENKINS G, Mc MEMENAMIN M, MACKAY, CP, and SOHL L, (eds.) The Extreme Petrozoic: Geology, Geochemistry and Climate, American Geophysical Union, Monograph Series 146:13–32

Duzgoren-Aydin NS, Aydin A, Malpas J (2002) Re-assessment of chemical weathering indices: case study of pyroclastic rocks of Hong Kong. Eng Geol 63:99–119

Etienne JL, Allen PA, Rieu R, Le Gerroue E (2008) Neoproterozoic glaciated basins: a critical review of the Snowball Earth hypothesis by comparison with Phanerozoic glaciations. Special Publication of the International Association of Sedimentologists Special Publication 39

Fedo CM, Nesbitt HW, Young GM (1995) Unraveling the effects of potassium metasomatism in sedimentary rocks and paleosols, with implications for paleoweathering conditions and provenance. Geology 23:921–924

Fletcher RC, Buss HL, Brantley SL (2006) A spheroidal weathering model coupling porewater chemistry to soil thicknesses during steady-state denudation. Earth Planet Sci Lett 244:444–457

Gaillardet J, Dupre B, Louvat P, Allegre CJ (1999) Global silicate weathering and CO_2 consumption rates deduced from the chemistry of large rivers. Chem Geol 159:3–30

Gardner R, Walsh N (1996) Chemical weathering of metamorphic rocks from low elevations in Southern Himalayas. Chem Geol 127:161–176

Geological Society of London GSL (1995) The description and classification of weathered rocks for engineering purposes. Geological Society Engineering Group Working Party Report. Q J Eng Geol 28:207–242a

Harnois L (1988) The CIW of weathering: a new chemical index of weathering. Sediment Geol 55:319–322

Harnois L, Moore JM (1988) Geochemistry and origin of the Ore Chimney formation: transported Paleoregolith in the Grenville Province of Southeastern Ontario, Canada. Chem Geol 69:267–289

Heidari M, Khanlari GR, Momeni AA (2010) Weathering indices and its relation to uniaxial compressive strength of Hamedan Hololeucogranite Rocks in West of Iran. World Appl Sci J 11:142–150

Heidari M, Khanlari GR, Momeni AA, Jafargholizadeh H (2011) The relationship between geomechanical properties and weathering indices of granitic rocks, Hamedan, Iran. Geomech Geoeng 6(1):59–68

Holland HD (1978) The chemistry of the atmosphere and oceans. Wiley, New York, 351 p

Jenny H (1941) Factors in soil formation. McGraw Hill, New York, 271

Johnsson MJ, Stallard RF, Meade RH (1988) First-cycle quartz arenites in the Orinoco River Basin, Venezuela and Colombia. J Geol 96:263–277

Khanlari GR, Heidari M, Momeni AA (2012) Assessment of weathering processes effect on engineering properties of Alvand granitic rocks (west of Iran), based on weathering indices. Environ Earth Sci 67(3):713–725

Kirschbaum A, Martinez E, Pettinari G, Herrero S (2005) Weathering profiles in granites, Sierra Notre (Cordoba, Argentina). J South Am Earth Sci 19:479–493

Kramer JR (1968) Mineral-water equilibria in silicate weathering. International geological congress, 23rd session, Section 6:149–160

Lee SY, Kim SJ, Baik MH (2008) Chemical weathering of granite under acid rainfall environment, Korea. Environ Geol 55(4):853–862

Lin H (2010) Earth's Critical Zone and hydropedology: concepts, characteristics, and advances. Hydrol Earth Syst Sci 14

Loughnan FC (1969) Chemical weathering of the silicate minerals. Elsevier, London

Miscevic P, Vlastelica G (2009) Shear strength of weathered soft rock e proposal of test method additions. In: Vrkljan I (ed) Proceedings of regional symposium of ISRMe EUROCK, Rock engineering in difficult conditions e Soft Rock and Karst. CRC Press, pp 303–308

Miscevic P, Vlastelica G (2011) Durability characterization of marls from the region of Dalmatia, Croatia. Geotech Geol Eng 29(5):771–781

Moye DG (1995) Engineering geology of Snowy Mountains scheme. J Inst Eng Aust 27:287–298
Nesbitt HW, Young GM (1982) Early Proterozoic climates and plate motions inferred from major element chemistry of lutites. Nature 299:715–717
Nesbitt HW, Young GM (1984) Prediction of some weathering trends of plutonic and volcanic rocks based on thermodynamic and kinetic considerations. Geochim Cosmochim Acta 48:1523–1534
Nesbitt HW, Young GM (1996) Petrogenesis of sediments in the absence of chemical weathering: effects of abrasion and sorting on bulk composition and mineralogy. Sedimentology 42:341–358
Ohta T, Arai H (2007) Statistical empirical index of chemical weathering in igneous rocks: a new tool for evaluating the degree of weathering. Chem Geol 240:280–297
Parker A (1970) An index of weathering for silicate rocks. Geol Mag 107:501–504
Pettijohn FJ, Potter PE, Siever R (1987) Sand and Sandstone. Springer, New York, 553 pp
Price R, Velbel J, Michael A (2003) Chemical weathering indices applied to weathering profiles developed on heterogenous felsic-metamorphic parent rocks. ELSEVIER Chem Geol 202:397–416
Price R, Velbel J, Michael A (2014) Rates of Biotite weathering, and clay mineral transformation and neoformation, determined from Watershed geochemical mass-balance methods for the Coweeta Hydrologic Laboratory, Southern Blue Ridge Mountains, North Carolina, USA. Aquat Geochem 20:203–224
Putins CV, Ruiz-Agudo E (2013) Coupled dissolution and precipitation at mineral–fluid interfaces. Elements 9:177–182
Raymo ME, Ruddiman WF (1992) Tectonic forcing of late Cenozoic climate. Nature 359:117–122
Raymo ME, Ruddiman WF, Froelich PN (1988) Influence of late Cenozoic mountain building on ocean geochemical cycles. Geology 16:649–653
Riebe CS, Kirchner JW, Granger DE, Finkel RC (2001) Strong tectonic and weak climatic control of long-term chemical weathering rates. Geology 29:511–514
Rieu R, Allen PA, Plotze M, Pettke T (2007) Compositional and mineralogical variations in a Neoproterozoic glacially influenced succession, Mirbat area, south Oman: implications for paleo-weathering conditions. Precambrian Res 154:248–265
Ruddiman WF, Prell WL (1997) Introduction to the uplift-climate connection. In: Ruddiman WF (ed) Tectonic uplift and climate change. Plenum Press, New York, pp 3–15
Ruxton BP (1968) Measures of the degree of chemical weathering of rocks. J Geol 76:518–527
Selby MJ (1993) Hillslope materials and processes, 2nd edn. Oxford University Press, Oxford
Souri T, Watanabe M, Sakagami K (2006) Contribution of parker and product indexes to evaluate weathering condition of yellow brown forest soils in Japan. Geoderma 130:346–355
Stallard RF (1995) Tectonic, environmental, and human aspects of weathering and erosion: a global review using a steady-state perspective. Annu Rev Earth Planet Sci 23:11–39
Sutton SJ, Maynard JB (1992) Multiple alteration events in the history of a sub-Huronian regolith at Lauzon Bay, Ontario. Can J Earth Sci 29:432–445
Taylor A, Blum JD (1995) Relation between soil age and silicate weathering rates determined from the chemical evolution of a glacial chronosequence. Geology 23:979–982
von Eynatten H (2004) Statistical modelling of compositional trends in sediments. Sediment Geol 171:79–89
Von Eynatten H, Barcelo-Vidal C, Pawlowsky-Glahn V (2003) Modelling compositional change: the example of chemical weathering of granitoid rocks. Math Geol 35:231–251
White AF, Blum AE (1995) Effects of climate on chemical weathering in watersheds. Geochimica et Cosmochimica Acta 59:1729–1747
Wilford J (2011) in press. A weathering intensity index for the Australian continent using airborne gamma-ray spectrometry and digital terrain analysis
Wintsch RP, Kvale CM (1994) Differential mobility of elements in burial diagenesis of siliciclastic rocks. J Sediment Res 64:349–361

Pesticides Pollution in Agricultural Soils of Pakistan

Muhammad Shahid, Ashfaq Ahmad, Sana Khalid, Hafiz Faiq Siddique, Muhammad Farhan Saeed, Muhammad Rizwan Ashraf, Muhammad Sabir, Nabeel Khan Niazi, Muhammad Bilal, Syed Tatheer Alam Naqvi, Irshad Bibi, and Eric Pinelli

Contents

M. Shahid (✉) • A. Ahmad • S. Khalid • H.F. Siddique • M.F. Saeed • M.R. Ashraf
Department of Environmental Sciences, COMSATS Institute of Information Technology, Vehari 61100, Pakistan
e-mail: muhammadshahid@ciitvehari.edu.pk

M. Sabir • N.K. Niazi • I. Bibi
Institute of Soil and Environmental Sciences, University of Agriculture Faisalabad, Faisalabad 38040, Pakistan

M. Bilal • S.T.A. Naqvi
Department of Environmental Sciences, COMSATS Institute of Information Technology, Abbottabad, Pakistan

E. Pinelli
Université de Toulouse; INP, UPS; EcoLab (Laboratoire Ecologie Fonctionnelle et Environnement); ENSAT, Avenue de l'Agrobiopole, 31326 Castanet Tolosan, France

CNRS; EcoLab, 31326 Castanet Tolosan, France

K.R. Hakeem et al. (eds.), *Soil Science: Agricultural and Environmental Prospectives*, DOI 10.1007/978-3-319-34451-5_9

Abstract Pesticides are widely used worldwide to control a range of pests infesting the agricultural crops. Increased use of pesticides has threaten human and viroenmnetal health. In this book chapter, we have compiled data regarding pesticide use, soil and water contamination, and human poining in Pakistan. Pesticide utilization in Pakistan started in 1954 and is currently on the rise. Of the total pesticides used in Pakistan, insecticides shared major portion, followed by herbicides, acaricides, and fumigants. High percentage of pesticides is being applied in the Punjab province, followed by Sindh, Khyber Pakhtoonkhaw and Balochistan. In Pakistan, the pesticide uses are mostly focused on cotton crop (almost 70–85 % of total pesticides use) and other crops such as wheat, sugarcane, maize, rice and tobacco as well as for vegetables and fruits. Different groups of pesticides, especially the residues of organochlorine, have been reported in soils and waters in different areas of Pakistan. The fate and charactiristics of pesticides in soils and percolation to deep soil depends vary with soil physico-chemical properties. Over 500,000 Pakistanis suffered annually from poisoning due to agro-chemicals, out of which 10,000 died. Regulations have also been developed for safe use of pesticides in Pakistan such as Farmer Field School led Integrated Pest Management model.

Keywords Pesticide pollution • Soil health • Pakistan • Integrated pest management • Pesticide use

1 Introduction

Increased crop production has appeared an inevitable component of modern agricultural system in order to support the increasing population pressure (Omer et al. 2010; Sabir et al. 2013; Pierart et al. 2015; Hakeem 2015). Pesticides are widely used to control a range of pests infesting the agricultural crops. In the context of current farming practices, if pesticide use is banned, the crop production will decrease significantly and food prices would soar drastically. Under such circumstances, it would become impossible to feed the ever increasing population of the world on sustainable basis. Therefore, use of pesticides has become imminent for increased crop protection and production. Application of pesticides is considered a cheap and efficient defence against the attack of pests which contributes to increased crop productivity. The use and development of pesticides to protect crops against insects, pests and diseases increased steadily during the last three to four decades. Nowadays, the development in pesticide sector has given rise to entirely new ways of production and protection of crops, as well as preservation of stored grains and their products in the godowns and warehouses.

The risks associated with the pesticides use are serious as well (Kouser and Qaim 2011; Lee et al. 2014). From environmental and health perspectives, the use of pesticides in contemporary agriculture has enhanced the impact of these chemicals on environment (Debenest et al. 2009, 2010; Kouser and Qaim 2011). Pesticide-induced chronic poisoning causes health hazards to farming community who are at high risk of being poisoned upon pesticides exposure (Lekei et al. 2014). Almost three million people have been poisoned and over 200,000 are reported to die each year around the globe as a result of pesticide poisoning (Sheikh 2011). The situation might be even worse in Pakistan, but for the moment there is very rare data available regarding pesticide contamination of air, water, soil, plants, animals and human beings.

In addition to human poisoning, use of pesticide also contaminate the different parts of an environment i.e., soil, water and air. Soil being the basic and most essential part of the ecological system is heavily contaminated by organic and inorganic pollutants including the pesticides throughout the world (Farlin et al. 2013; Shahid et al. 2013, 2015; Sultana et al. 2014; Mombo et al. 2015). Soil contamination with pesticides illustrates great attention owing to their potential risk to environmental and food safety. Pesticides pollution in agricultural soils may lead to the functional disorder of soil (Niemi et al. 2009; Karpouzas et al. 2014). For example, pesticides disturb enzymatic activity, which is considered as an indicator of soil tolerance to hazardous pollutants (Niemi et al. 2009). Pesticides interfere with soil properties and nutrient behaviour (Hussain et al. 2009; Hakeem 2015). Soil ecosystem may be disturbed by extensive use of pesticides by by harming soil microorganisms. Several previous studies have reported the harmful effects of pesticides on soil microbial diversity and activities (Hussain et al. 2009; Karpouzas et al. 2014). Pesticides also influence the microbial assisted breakdown of organic matter and nutrient dynamics and availability in soil (Kinney et al. 2005). However, the environmental and health hazards of pesticides depend on their persistence in the soil environment. Several pesticides have been banned worldwide owing to their long environmental persistence and acute toxicity.

Inside the soil, pesticides have several fates including retention or degradation, uptake by animals, plants (bioavailability) and leaching to groundwater (Fenoll et al. 2014; Paszko 2014a, b). The fate and behaviour of pesticides in soils and percolation to deep soil depends on a variety of complex biological, chemical and physical processes, including: volatilization, biochemical degradation, sorption–desorption, uptake by plants and leaching (Arias-Estévez et al. 2008; Fenoll et al. 2014). The relative importance of these processes varies with molecules properties ($LogK_{ow}$, K_{oc}, K_d, solubility, pKa,…), soil physico-chemical characteristics (pH, moisture content, cation exchange capacity, soil mineralogy, biological and microbial conditions) and soil organic matter contents (Gai et al. 2014; Cabrera et al. 2014). All these factors separately or in combination with each other affect pesticides behaviour and fate in soil system.

Pakistan is primarily an agricultural country where pesticides use is an integrated component of crop production. The use of pesticide in agriculture is on the rise in Pakistan (Economic Survey of Pakistan 2010–2011). In Pakistan, crop production and yield per acre have increased over the past 40 years (Economic Survey of Pakistan 2010–2011). However, the crop production in Pakistan is not constant, but fluctuates from year to year due to damage caused by insects, pests and diseases. Currently several types of fungicides, insecticides, herbicides, rodenticides and acaricides are being applied in Pakistan for crop protection (Khan et al. 2010). Several previous studies reported pesticide residues in soils and groundwater of Pakistan, however, there is a lack of comprehensive report regarding pesticide residues/persistence in the agroultural soils of Pakistan. The objective of this chapter is to summaries the use of pesticide in agricultural sector of Pakistan and the persistence of pesticides in soil.

2 Pesticide Use and Their Classification

Pesticides constitute a heterogeneous category of chemical substances, either natural or synthetic, specifically designed for the control of pests, weeds or plant diseases. There exist more than 50,000 species of plant pathogens, 9000 species of mites and insects, and 8000 species of weeds, which can damage the crops. Pests may cause annually up to 30 % loss to crop production (Saleem and Ashfaq 2004). According to the Food and Agriculture Organization of the United Nations (UN FAO), the potential human food loss worldwide is about 55 % that include 35 % preharvest and 20 % post-harvest loss. It is reported that approximately 14 % of loss is caused by insect pests, 13 % by plant pathogens, and 13 % by weeds (Pimentel 2009). Without pesticide use, the losses of vegetables, fruits and cereals from pest injury would reach respectively 54 %, 78 %, and 32 % (Cai 2008). Pesticide is so indispensable in agricultural production that approximately 1/3rd of the agricultural products are produced by using pesticides (Liu et al. 2002).

In order to control crop losses by pests, insects and diseases, thousands of different kinds of chemicals are being used worldwide. Some of the pesticide contain several active substances to combat two or more group of parasites, make their classification a real challenging task. According to the status list of active substances available commercially in the European Union (EU), >1100 pesticide substances are currently registered (Húšková et al. 2008). These active substances are classified based on (i) molecular structure of the active molecule, (ii) the formulation and (ii) the biological target. In Pakistan active substances are generally known by their brand name and function in the market and among the farming community.

Based on molecular structure of the active molecule, the pesticides are classified into; organic/botanical and inorganic. Organic pesticides are generally the products of living organisms. Organic pesticides used in Pakistan are nicotine, sabadilla, rotenone, pyrethrum, ryania and neem. Inorganic pesticides include the pesticides that are mined from the earth and ground into a fine powder. Inorganic pesticides used in Pakistan include cryptile, borates and borax. Biorational pesticides refer to synthetic, organic, or inorganic pesticides having low toxicity and little impact on the environment, which are not commonly used in Pakistan.

Taking into account the formulation and the active substances used in the chemical formula, pesticides are classified as (i) organochlorine (commonly used in Pakistan include endosulfan, dichlorodiphenyle trichloroethane (DDT), dieldrin, aldrin, heptachlor, chlordane), (ii) organophosphates (commonly used in Pakistan include malathion, parathion, chlorpyrifos), (iii) phenoxyacetic acid herbicides (2.4-D, MCPA), (iv) carbamate (carbaryl, aldicarb, carbofuran, carbaryl), (v) pyridinium herbicides: (commonly used in Pakistan include picloram, paraquat, diquat), (vi) triazine herbicides: (cyanazine, simazine, trietazine, atrazine are commonly used in Pakistan), and (vii) substitute urea (chlorotoluron, isoproturon). In Pakistan, about 145 active substances have been registered, with pyrethroids having the greatest share (45 %), followed by organophosphates (39 %), organochlorine (9 %) and carbamates (4 %). Organochlorine and organophosphates are formulated and manufactured locally in Pakistan and are used commonly for cotton protection against insects.

Based on their biological target and function, pesticides are generally classified into three families: herbicides, fungicides and insecticides. In Pakistan, currently more than 39 types of herbicides, 30 types of fungicides and 108 types of insecticides are used in Pakistan for crop protection (Khan et al. 2010). Herbicides are reported to inhibit: cell division (pendimethalin, trifluralin), photosynthesis (isoproturon, atrazine), cellulose synthesis (chlorotiamide), amino acid synthesis (glyphosate) and lipid synthesis (cycloxydime) (Dayan and Watson 2011; Dayan et al. 2012). Fungicides fight against the disease of plants caused by fungus. Fungicides are also reported to function in different ways as they can inhibit cell division and respiration of target organism (benomyl) and disturb the metabolism of carbohydrates and biosynthesis of amino acids or proteins (cyprodinil) (Fairbanks et al. 2002; Dane and Dalgiç 2005). Insecticides are employed worldwide against harmful insects to eliminate and prevent them from reproducing. Insecticides can be growth regulators (teflubenzuron), neurotoxins (indoxacarbe) and cell respiration inhibitors (cyhexatin) (James et al. 2008).

3 Pesticide Use History

Use of pesticide by human to protect crops against pest, insects and diseases dated back to 2000 BC. The wave of chemicals introduced by humans commenced with simple elements and plant derivatives (Tierney et al. 2014). Sumerians had been reported to use a sulfur containing pesticide over 4500 years ago. Chinese used sulfur as an antibacterial and fungicide at least 3000 years ago. In the Middle East, arsenic (As) was used for a variety of pests control purposes over 2000 years ago (Bentley and Chasteen 2002). Different metals {(lead (Pb), mercury (Hg) and As)} were commonly used for crop production by the fifteenth century. At some point in antiquity, humans discovered the power of plant-based chemicals for insect and/or pest control. Tobacco extract (nicotine sulfate) was used as an insecticide early in the seventeenth century. Later on rotenone (a derivative of roots of tropical vegetables) was also introduced for crop protection. Until the 1950s, pesticides containing As were dominant, such as: arsenical pesticides were widely used to control ticks in cattle, as well as to preserve the wood timber using a chromium-copper-arsenate (CCA) containing pesticide in many countries e.g., Australia, New Zealand, the USA, China (Niazi et al. 2012).

The mid-twenty century is known as a revolution in pesticide use and development. Use of herbicides was very common in the 1960s. The modern synthetic pesticides were developed around World War II when the insecticidal potential of DDT was discovered in Switzerland. Use of organochlorines, organophosphates and carbamates became common in the late 1970s. DDT and pyrethrin were dominant chemical substances in use in the whole world during 1970–1980. The discovery of DDT and analogues was thought to be miracle and a permanent solution to pest problem.

4 Worldwide Use of Pesticides

Pesticides are used widely and globally in agriculture for crop production, and have become an enduring feature of modern time. It is estimated that almost 3×10^9 kg of pesticides is used annually throughout the world with a price value of nearly $40 billion (Hussain et al. 2009). World pesticide use was more than 33390 billion dollars in 2006 and 2007 (EPA 2007). The contribution of each continent was: Europe ($10568 billion), Asia ($7815 billion), North America ($7507 billion), Latin America ($6170 billion) and Africa ($1330 billion). The quantity of herbicides accounted for the largest part ($16115 billion), followed by fungicides ($8105 billion), insecticides ($8016 billion) and other pesticides ($1154 billion). Of the total quantity of pesticides used worldwide, herbicides shared about 40 % followed by insecticides (18 %) and fungicides (10 %). It is reported that almost 80 % of the total world pesticides produced are consumed in industrialized countries and remaining 20 % in developing countries (EPA 2007). Currently, Europe is known to be the largest consumer of pesticides in the world, followed by Asia. In term of countries, China, United States, France, Brazil and Japan are the largest pesticide producers, consumers or traders in the world. However, the per acre use of pesticide is high in Costa Rica and Colombia (Table 1).

Table 1 World ranking of pesticide use per acre

Sr #	Country	Ranking	Pesticide use per acre
1	Costa Rica	1	51.2 kg
2	Colombia	2	16.7 kg
3	Netherlands	3	9.4 kg
4	Ecuador	4	6 kg
5	Portugal	5	5.3 kg
6	France	6	4.6 kg
7	Greece	7	2.8 kg
8	Uruguay	8	2.7 kg
9	Suriname	9	2.6 kg
10	Honduras	10	2.5 kg
11	Germany	10	2.5 kg
12	Austria	12	2.4 kg
13	Dominican Republic	13	2.1 kg
14	Ireland	14	1.8 kg
15	Slovakia	14	1.8 kg
16	Paraguay	16	1.5 kg
17	Denmark	17	1.4 kg
18	Jordan	17	1.4 kg
19	Czech Republic	19	1.3 kg
20	*Pakistan*	*19*	*1.3 kg*
21	Turkey	19	1.3 kg

Source: Nation Master (Year 2000), derived June 2015

5 Pesticides Use in Agriculture Sector of Pakistan

Pakistan is an agricultural country where an area of about 22.2 million ha is used for crop production (FAO 2006). Agriculture is considered as the mainstay of the Pakistan's economy because it gives jobs to more than 40 % of the population (Economic Survey of Pakistan 2010–2011). Agriculture is the sole key segment and the major source of income for about 66 % of the country's inhabitants. Agriculture sector adds 21 % of the GDP (Economic Survey of Pakistan 2010–2011). In agriculture sector, Pakistan is ranked 3rd, 7th, 9th, 11th, 19th and 20th, respectively for cotton production, farm workers, arable land, gross value added, agricultural machinery and pesticide use in the world (Nation Master 2013). The use of pesticide in agriculture is on the rise in Pakistan (Economic Survey of Pakistan 2010–2011).

The use of different substances for pest control and enhanced crop production in agricultural sector is being practised in Pakistani region since centuries. However, pesticide utilization in Pakistan started in 1954 with an import of 254 metric tons of formulated product. Pesticide consumption in Pakistan increased from 7000 tons per annum by mid-1960s to 906 metric tons in 1980. The use of synthetic pyrethroids started in Pakistan in 1980 when fenvalerate, deltamethrin and permethrin

Table 2 Pesticide imported during 1960–2010

Year	Rs (Millions)	Quantity
1960	20	4979
1970	42	2248
1980	225	7105
1990	1489	13030
2000	3477	21255
2005	8281	41561
2008	6330	27814
2009	5498	16495
2010	8741	27995

Source: Economic Survey of Pakistan

were commercially introduced. More than a dozen brands of pyrethroids were launched during 1980–1985, which made synthetic pyrethroids a major shareholder (70 %) of the total pesticides market in Pakistan. Tariq (2002) reported that during last two decades, use of pesticides increased by about 70 times in Pakistan. According to Economic Survey of Pakistan (2002–2006a, b), the pesticide import increased from 4979 to 27995 tons between 1960 and 2010, and pesticides use increased from 42732 to 46951 tons between 1998 and 2006 in Pakistan (Table 2). One of the key reasons behind enhanced use of pesticides in Pakistan is associated with very soft legislation regarding the registration and import of pesticides at that time, according to which any generic compound registered elsewhere can be imported to country without any field testing. Therefore, a major portion of pesticides applied in Pakistan are imported from Europe, US, China and India (Table 3). Of the total pesticides used in Pakistan, insecticides shared major portion (74 %), followed by herbicides (14 %), fungicides (9 %), acaricides (2 %), and fumigants (1 %) (Fig. 1) (Khooharo et al. 2008). High percentage of pesticides is being applied in the Punjab province (89 %), followed by Sindh (8 %), Khyber Pakhtoonkhaw (2 %) and Balochistan (1 %) (Khan 2000).

In Pakistan, pesticide legislation was not in practice by 1971, and the Federal Government Department of Plant Protection (DPP) was responsible for standardization of pesticides import. The agricultural pesticides ordinance (APO) and agricultural pesticide rules were formulated respectively in 1971 and 1973 under the guidelines of Food and Agriculture Organization (FAO), which governed the formulation, production, import, use, distribution and sale in the country (Mazari 2005). The ordinance also implemented registration scheme for pesticide companies, which was amended later in favour of importers (Ali et al. 2014a, b). Food and agricultural regulation was promulgated in 1965 for controlling the pesticide residues in food samples, and is not amended up-till now. By 1989, the marketing, distribution and sale of pesticides were governed by public sector, which has been transferred to the private sector since 1989. In 1999, Pakistan became the 67th member country of the Convention at United Nations in New York by signing the Rotterdam Convention. Pakistan also endorsed the Stockholm Convention since July 2000.

Table 3 Main products and area of origin of major pesticide companies working in Pakistan

Sr #	Company name	Major products	Origion
1	Syngenta	Polo, DualGold, Poytrin-C	Switzerland
2	FMC (United) Pvt.Ltd	Talstar, Furadon, Commando, Acelan	USA, China
3	Target-Ali Akbar Group	Timer, Capital+, Track, Furon, Pilot	USA, China
4	DJC-Ali Akbar Group	Shark, Shooter, Stater, Fusion, Launcher	China
5	KanzoAg	Referry, Persept, Momentu, Priorty, TopMax	China, India
6	Bayer Crop Sciences	Movento, Oberan, Lasenta, Intracal	Germany
7	Arysta Life Sciences	Padan, Radient, Tracer, Topsun, Susses	USA, China
8	Suncrop Chemicals	Leumax Extra, Trunk	China
9	SunGrow Chemicals	Mauzer, Hiflow, Spartan Super	China
10	Welcone Chemicals Pvt.Ltd	Bruce, Jassper, Jailer	China, India
11	Warble Chemicals Pvt.Ltd	Active, Bruce, Prifle, Hilt, Treasue	China, India
12	Auriga Chemicals Pvt.Ltd	Gangvi, Zarkon, Fokker	China, India
13	Sayban International	Jerk, Grow Up, Pythen	China, India

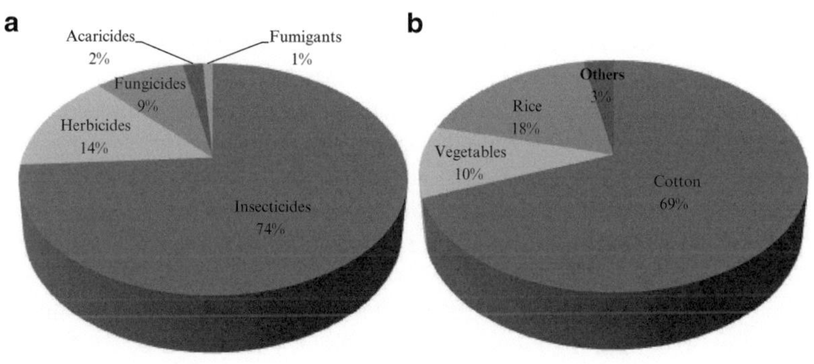

Fig. 1 Percent share of different pesticides (**a**) and their percent use for major crops (**b**) in Pakistan (2007–2012)

6 Major Crops in Pakistan and Pesticides Use

Nature has endowed Pakistan with four seasons (summer, autumn, winter and spring) and versatile land (arid, semi-arid, rainfed and mountains). Therefore, a great variety of crops, vegetables, fruits are cultivated in Pakistan. There are two major cropping seasons in Pakistan i.e., Kharif (hot summer season) and Rabi (cool winter season). Cotton, rice, sugarcane, maize, sorghum and millet are main Kharif crops while wheat, lentil, tobacco, mustard, barley, gram and rapeseed are main Rabi crops in Pakistan. To a great extent, agriculture in Pakistan is supported by major crops such as, wheat, rice, cotton and sugarcane. These major crops are reported to share almost 90 % of the value added in the major crops, which accounts for 31 % of the value added in the agriculture and 5.9 % to GDP (Economic Survey of Pakistan 2008–2009). The minor crops account for 10 % of the value added in overall agriculture. The introduction of Green Revolution technologies (use of hybrid seeds, synthetic fertilizers and pesticides) in Pakistan resulted in significant increase of crop production. For example, wheat production enhanced from 3.3 m tonnes during 1950–1951 to 22.4 m tonnes during 2008–2009. Similarly, rice production rose during the same period, from 0.86 m tonnes to 6.95 m tonnes. The production of cotton reached 11.81 million bales during 2008–2009, which was maximum (14.26 million bales) in 2004–2005. This increase in crop production is mainly attributed to increased use of agrochemical especially pesticides (Tables 4a, 4b, 5, and 6).

Among the major crops, cotton, known as white gold for Pakistan, is the backbone for the economy which contributes 1.4 % of GDP and 7 % of value added in agriculture (Economic Survey of Pakistan 2010–2011). Almost 26 % of all farmers in Pakistan grow cotton, and more than 15 % of the total cultivated area in Pakistan is devoted to this crop (Sabir et al. 2011). In Pakistan, the pesticide use is mostly focused on cotton-growing areas (almost 70–85 % of total pesticides use) (Fig. 1) (Khan et al. 2011), as cotton is the most vulnerable to pest attacks. The spray frequency in Pakistan is 10 sprays per crop season (73 % farmers), which can be upto 16 sprays in one season, especially for Bt cotton (*Bacillus thuringiensis*) (Khan et al. 2011). Most of the pesticides used in Pakistan are insecticides. Whitefly, a vector of cotton leaf curl virus, is considered among the most serious cotton pests. It is reported that during 1993, whitefly caused a loss of about three million bales of cotton, equivalent to almost 25 % of total production. Therefore, a major portion of pesticides used was consumed to control this insect (Khan 1998; Khooharo et al. 2008). However, recently a decreasing trend in cotton cultivation has been observed in Pakistan, particularly in Punjab (Sahiwal, Khanewal, Chichwatni and Vehari), which has resulted in decreased use of pesticide for cotton crop in Pakistan.

In addition to cotton, high quantities of pesticides are also used for other major and minor crops such as wheat, sugarcane, maize, rice and tobacco as well as for vegetables and fruits (Tables 4a, 4b, 5, and 6) (10–20 %) (Eavy et al. 1995). In Khyber Pukhtoonkhwa province of Pakistan, high doses of pesticides are applied on sugarcane, tobacco and maize (Ahad et al. 2000). Spray frequency is sometime more than 10 sprays per crop of tobacco. The proportion of pesticide used for these major crops has increased recently owing to increased cultivation of these and other major and minor crops. For example, during 2010–2011, the production of maize, millet, mustard and tobacco increased by 2.4 %, 18 %, 11 % and 13 %, respectively, in Pakistan. Consequently, the use of pesticides has also increased these days in Pakistan to control insect, pests and diseases of these major and minor crops. Similarly, the fruits (citrus, mangoes, apples, bananas, guavas, grapes) and vegetables (cabbage, okra, carrot, chilies, onion, potato, radish, spinach and tomatoes) are important horticultural crops in Pakistan, and account minor portion of pesticide use (Tables 4a, 4b, 5, and 6). Khan et al (2010) reported that almost 12 % pesticides are being used on vegetables and fruit crops in Pakistan. Indiscriminate and enormous use of pesticides on vegetables has been observed in Pakistan (Khan 2004).

Table 4a Pesticides used against insects, pests and diseases of major crops in Pakistan

Crops	Diseases, insects/pests	Active ingredient	Applied dose	Stage of application
Cotton	Narrow & Broad Leaf	Acetachlore	500–650 ml/acre	With in 24 h of sowing
	Jassid, Thrips, W.Fly	Imadacloprid	8–10 g/kg seed	Seed dressing
	W.Fly, Thrips, Jassid	Imadacloprid	250 ml/acre	3/leaf, 5/leaf, 1.0/plant
	Thrips	Acephate	400–500 g/acre	3/leaf
	Whitefly	Diafenthiuron	200 ml/acre	3/leaf
	Spotted BW, Pink BW, mites	Bifenthrin	250 ml/acre	Apprence of bud
	American BW, Spotted BW, Army BW	Emmamectin Benzoate	200 ml/acre	Apprence of pest
	Armyworm	Leufenoran	100 ml/acre	Apprence of pest
	Spotted BW, Pink BW	Lambada Cyhalothrin	330 ml/acre	Start of squres
	Spotted BW, Pink BW	Profenofos + Lamda Cyhalothrin	500 ml/acre	Start of squres
	Spotted BW, Pink BW	Betacyflothrin + triazohos	500 ml/acre	Apprence of bud
	Spotted BW, Pink BW	Deltamethrin	250–300 ml/acre	Appreance of bud
	W.Fly, Thrips, Jassid, Aphid	Acetameprid	125 ml/acre	5/leaf, 8–10/leaf, 1.0/leaf, 15/leaf
	Spotted BW, Pink BW	Triazophos	1000 ml/acre	Appreance of roset flower
	W.Fly Nymph	Buprofezin	600 g/acre	5/leaf
	Two Spotted, Red Mites	Pyridaben	500 ml/acre	At appreance
	Growth of Plant; escape from virus	Chelated Zinc	500 ml/acre	30–60 days of crop
Wheat	American worm & fruit borer	Leufenoran	100 ml/acre	Friut setting till maturity
	Whitefly, aphid	Acetameprid	125 ml/acre	At appreance
	Smut & Rust, karnal bunt	Thiophenate Methyl	2 g/kg seed	Seed treatment
	Narrow leaf Weeds; Bumbi sitti and jugli jaii	Clodinofop	136 g/acre	40–50 days after sowing
	Dumbi sitte & Broad leaf weeds	Metribuzin	100 g/acre	After 1st irrigation when weeds are at 2–4 leaf stage

(continued)

Table 4a (continued)

Crops	Diseases, insects/pests	Active ingredient	Applied dose	Stage of application
	Narrow Leaf Weeds	Phenoxaprop P-Ethyl	400 ml/acre	40–50 days after sowing
	Narrow leaf weed	Isoproturan	800 g/acre	After first irrigation
	Broad leaf weeds; Including "Matri"	Fluroxypyr Methyl + MCPA	150 + 250 ml/acre	After first irrigation
	Broad Leaf Weeds	Bromoxynil + MCPA	500 ml/acre	After 1st irrigation when weeds are at 2–4 leaf stage
Sugarcane	Whipe smut and Red Rot of Sugarcane	Mencozeb + Cymoxynil	250 g/400 lit water	Soak sugarcane sets in solution for 5–10 min
	Narrow & Broad leaf	Atrazine + Ametryn	1.0–1.5 kg/acre	35–40 days after sowing
	Root, stem, top, Gurdaspur borer	Carbofuron	8–14 Kg/acre	1st at sowing, 2nd at completion of hoeing
	Pyrilla	Lambda Cyhalothrin	330 ml/acre	At appreance
	Black bug	Acetameprid	125 ml/acre	At appreence

BW Bollworm, *W.Fly* White fly

Table 4b Pesticides used against insects, pests and diseases of major crops in Pakistan

Crops	Diseases, insects/pests	Active ingredient	Dose	Stage for application
Rice	Narrow & Broad leaf weeds	Butachlore	800 ml/acre	In soils where water can not stand for 4–5 days
	Narrow & some Broad	Acetachlore	100 ml/acre	In soils where water can not stand for 4–5 days
	Stem borer, leaf roller	Cartap	9 kg/acre	Two application 35 days of crop and 55 days of crop
	Leaf roller	Lambada Cyhalothrin	250 ml/acre	Before attack
	Adult and nymph of White backed Hopper	Acetameprid	125 ml/acre	Application at 45 days crop
	Nymph of White backed Hopper	Buprofezin	300 g/acre	Application at 45 days crop

(continued)

Table 4b (continued)

Crops	Diseases, insects/pests	Active ingredient	Dose	Stage for application
	Grass hopper	Carbryl	1.0 kg/acre	9 days after nursery sowing
	Bakaenee, brown leaf spot	Thiophenate Methyl	2.5 g/kg Seed (seed treatment)	15 days before sowing in dry method
	Bacterial Blight	Copper oxychloride	500–1000 g/acre	500 g for small and 1000 g for large crop
	Rice Blast or "bhabka" and brown leaf spot	Tebuconazol	250 ml/acre	Two application; 5–7 before and after penical apprence
	Rice Blast or "bhabka"	Thiophenate Methyl	500 g/acre	1st at leaves, 2nd 3–4 before penicle, 3rd 6–7 days after penicle emergence
Maize	Narrow leaf weeds	Acetochlor	500 ml/acre	With in 24 h after sowing
	Broad leaf weeds	Atrazine	500 g/acre	After sowing in tar wattar
	Narrow & Broad leaf weeds	Atrazine + Propisochlore	650 ml/acre	After sowing in tar wattar
	Shoot fly, W.Fly	Imadacloprid	5–7 g/kg Seed	Seed dressing
	Maize Borer, Nematods	Carbofuron	8 kg/acre	1st in whorles at 1–1.5 ft height of crop, 2nd at 2.5–3.0 ft height of crop
Tobacco	Termite	Chlorpyriphos	1500–2000 Ml/acre	With irrigation water
	Bacterial wilt, Black shink angular leaf spot	Chlorthalonil	200 g/acre	Early stage of crop
	Bacterial wilt, Black shink, angular leaf spot	Mancozib + metalyxal	250 g/acre	Mid crop stage
	Bacterial wilt, Black shink, angular leaf spot	Mencozeb + Cymoxynil	250–300 g/acre	Late crop stage
	Army worm, bud worm	Leufenoran	100 ml/acre	Apprence
	Tobacco bud worm, Armyworm	Emmamectin Benzoate	150 ml/acre	Apprence

ml/acre indicates active ingredients in powder or granular form, ml/acre indicates active ingredients in liquid form

Table 5 Pesticides used against insects, pests and diseases of major fruits in Pakistan

Fruits	Diseases, insects/ pests	Active ingredient	Dose	Stage for application
Apple	Aphid, Seen Jozz Scale	Imadacloprid	100 ml/acre	Bud burst
	Codding Moth, American worm	Emmamectin Benzoate	20–25 ml/100 l water	Bud burst to fruit setting
	Spider Mites	Pyridaben	100 ml/100 l water	At symptoms
	Spider Mites	Propargite	100 ml/100 l water	At symptoms
	Powdery mildew	Tebuconazol	50 ml/100 l water	Bud burst to fruit setting
	Powdery mildew	Thiophenate Methyl	200 g/100 l water	Bud burst to fruit setting
	Apple Scap	Mencozeb + Cymoxynil	100 g/100 l water	Bud formation to fruit setting
Banana	Fingure tip rot (rotting of small comb)	Copper oxychloride	1.0 kg/400 lit water	Preventive
	Brown leaf spot (small brown spots, leaves dry and broken)	Copper oxychloride	1.0 kg/200 lit water	Preventive
	Nematodes	Carbofuron	5–10 g/plant	Preventive
	Bunchy top disease	Carbofuron	5–10 g/plant	Before transplanting the young suckers
Citrus	W.Fly, ahid	Imadacloprid	100 ml/100 literer water	Flower stage
	Leaf Minor, Citrus Psylla, lemon butterfly	Bifenthrin	50–60 ml/acre	Flower stage
	Gomosis & die back	Tebuconazol	125 ml/acre	Preventive
	Nematodes	Carbofuron	250 g/plant	August
	Die Back, Anthracnose, Gamosis	Thiophenate Methyl	200 g/100 l water	April
	Sudden death	Thiophenate Methyl	500 g/plant	April
	Fruit Fly	Trichlorofon	100 g/100 l water	May–June
	Citrus canker	Copper oxychloride	500 g/100 l water	July–september

(continued)

Table 5 (continued)

Fruits	Diseases, insects/ pests	Active ingredient	Dose	Stage for application
Mango	Mango Hopper	Imadacloprid	100 ml/100 l water	Preventive before flowing
	Scale	Imadaclopride	100 ml/100 l water	Preventive before flowering
	Fruit Fly	Tricholorofon	100–150 g/100 l water	April at 15 days interval
	Powder mildew/ blosum blight	Tebuconazol	125–150 ml/100 l water	At 30 % flowering
	Antracnose/die back	Copper oxychloride	500 g/100 l of water	Preventive in April
	Sudden death	Thiophenate Methyl	500 g/plant	Canopy of plat with irrigation, preventive

For orchards pesticides are dissolved in big water tanks (100 ml or 100 ml) and sprayed on fruit plants

Table 6 Pesticides used against insects, pests and diseases of some vegetables in Pakistan

Vegetables	Diseases, insects/ pests	Active ingredient	Dose	Stage for application
Onion	Stem rot	Thiophenate Methyl	2.5 g/l of water	Preventive
	Mites	Pyridaben	500 ml/acre	At appreance
	Purple bloch and downy mildew	Chlorthalonil	200 g/acre	Early stage of crop
	Purple bloch and downy mildew	Chlorthalonil	200 g/acre	Early stage of crop
	Purple bloch and downy mildew	Mancozib + metalyxal	250 g/acre	Mid crop stage
	Purple bloch and downy mildew	Mencozeb + Cymoxynil	250–300 g/ acre	Late crop stage
	Blast	Tebuconazol	250 ml/acre	Preventive
	American worm	Leufenoran	100 ml/acre	Apprence
Potato	Jassid	Imadacloprid	80 ml/acre	15–30 days after sowing
	Armyworm & american aorm	Leufenoran	100 ml/acre	Apprence
	Americn worm + jassid	Bifenthrin	150 ml/100 lit water	Preventive
	American, Armyworm	Emmamectin Benzoate	150 ml/acre	Preventive
	Purple bloch and downy mildew	Chlorthalonil	200 g/acre	Early stage of crop
	Purple bloch and downy mildew	Mancozib + metalyxal	250 g/acre	Mid crop stage
	Early & Late Blight	Mencozeb + Cymoxynil	250 g/acre	Later stage of crop

(continued)

Table 6 (continued)

Vegetables	Diseases, insects/ pests	Active ingredient	Dose	Stage for application
Tomato	Cut worm	chlorpyrifos	1500–2000 ml/acre	Fertigation
	After transplanting within 24 h	Acetochlor	500 ml/acre	Tansplanting
	Cut worm	chlorpyrifos	1500–2000 ml/acre	Fertigation
	After transplanting within 24 h	Acetochlor	500 ml/acre	Tansplanting
	Early blight	Chlorthalonil	200 g/acre	At early stage of crop
	Early and Late blight	Mancozib + Metalyxal	250 g/acre	At early stage of crop
	Fruit budworm	Emmamectin Benzoate	150 ml/acre	Leaves and fruit stage
Chilies	Narrow leaf	Acetochlor	500 ml/acre	24 h after transplantion
	Cut worms	Chlorpyriphos	1500–2000 ml/acre	Fertigation
	Colar rot	Mencozeb + Cymoxynil	250 g/acre	Preventive
	Leaf blight	Copper exychloride	500 g/acre	Preventive
	Bud Mites	Pyridaben	500 ml	At symptoms

7 Pesticide Occurrence in Agricultural Soils of Pakistan

Soil being the vital and most important part of the ecosystem is critically contaminated by persistent organic and inorganic pollutants throughout the world (Pourrut et al. 2013; Foucault et al. 2013; Sultana et al. 2014; Austruy et al. 2014; Shahid et al. 2014a, b). Soil acts as a buffer and filter regarding storage of pollutant (Burauel and Baßmann 2005; Pourrut et al. 2011; Xiong et al. 2014). It has been estimated that less than 1 % of sprayed pesticides reach the target organism while the remaining bulk flows into the soil, water and air environments (Carriger et al. 2006). It is recognized that the soil is also a potential pathway of pesticide transport to contaminate water, air, plants, food and humans. Several previous studies in Pakistan have highlighted the pollution of pesticide residues in soils (Jabbar et al. 1993; Tehseen et al. 1994; Tariq et al. 2007; Anwar 2009; Anwar et al. 2012; Syed et al. 2013; Sultana et al. 2014).

Different groups of pesticides, especially the residues of organochlorine, have been reported in soils and waters in different areas of Pakistan. Organochloride pesticides are organic compounds which contain at least one covalently bonded chlorine atom (Jiang et al. 2009). Organochloride pesticides have been used extensively in Pakistan to control crop pests and against malaria, and are detected in various environmental compartments (Bano and Siddique 1991; Tehseen et al. 1994; Tariq et al. 2007; Eqani et al. 2011; Malik et al. 2010; Syed and Malik 2011). Pakistan does not manufacture organochlorien pesticides (OCPs) and most of the chemicals are generally imported to meet the pesticides demand (Syed and Malik 2011). Organochlorine has been reported to persist in soil for long time and therefore the use of most of the organochlorine pesticides has been banned in Pakistan.

Alamdar et al. (2014) reported surface soil and air pollution of dichlorodiphenyltrichloroethane (DDTs), chlordane, heptachlor, hexachlorocyclohexane (HCH) and hexachlorobenzene in Hyderabad city, Pakistan. Mahmood et al (2014) reported contamination of DDT (320 ng/g) in agricultural soil of Gujranwala, Pakistan. Sultana et al. (2014) showed soil pollution by organochlorine pesticides especially DDT and HCHs at downstream agricultural sites, particularly at Head Panjnad. Dichlorvos soil pollution was reported in Bahawalpur district by Anwar (2009) and later in Lodhran district by Anwar et al. (2012). The Federal Pesticide Laboratories, Karachi continually monitor the pesticides level in food samples and the environment, and reported that upto 0.2–0.59 ppm of DDT was present in the sugarcane and tobacco fields of Khyber Pakhtunkhwa (KPK) and in the orchards of Bhalwal, upto 0.6 ppm in the rice fields around Kala Shah Kaku and upto 6 ppm in cotton growing areas of Multan. Hussain et al (1988) from Nuclear Institute for Agriculture and Biology, Faisalabad reported major portion of the applied DDT in the top 5 cm layer of sandy loam soils, with a very low concentration in deep soil layers. They also reported that the half life dissipation of DDT was 890 days in the laboratory, but 110 days in irrigated and 112 days in rainfed soils under field conditions. Many other researchers also reported that organochlorines (DDT) decompose very slowly and may persist for years, and are retained by soil due to their insolubility in water (Alamdar et al. 2014; Mahmood et al. 2014).

8 Groundwater and Surface Water Pollution by Pesticides in Pakistan

Leaching is a fundamental soil process whereby pesticides and other pollutants move down the soil profile by percolating water. Pesticide leaching is a key factor of groundwater pesticide pollution. The problem has become more prominent in areas where water table is high and the groundwater is the main drinking water resource (Tariq et al. 2004a; Sankararamakrishnan, et al. 2005). In the 1980s, different studies in Pakistan reported traces of pesticides in shallow drinking water wells (Ahad et al. 2001; Tariq et al. 2004a) as well as in the surface waters (Ahad et al. 2006). Parveen and Masud (1988) reported pesticides residues in cattle drinking water in Karachi. Groundwater and surface water in areas of the cotton belt in the south-eastern Punjab and Sindh plains are also reported to be polluted with pesticides (Tariq et al. 2004b). Variable concentrations (0.017–1.06 ng/mL) of DDT were detected in surface water and groundwater of different districts of Punjab, Pakistan (Asi et al. 2008). In Mianchannu, district Khanewal, the concentration of DDT (1.06 ng/mL) was 10 times more than the maximum admissible limits set by the European Community (Asi et al. 2008). Twelve groundwater samples taken from 6 different sites in Multan showed pesticide contamination, with 33 % samples exceeding the maximum threshold levels (Ahad et al. 2001). Jan et al. (2008) reported 0.07–0.40 μg/mL of DDT in and around a former DDT-producing factory in Aman Gharh, Nowshera. Jabbar et al. (1993) analysed shallow groundwater samples in Faisalabad and reported the residues of cyhalothrin (traces to 0.2 μg/L), monocrotophos (40–60 μg/L) and endrine (0. 1–0.2 μg/L).

Groundwater analysis carried out by Tariq et al. (2004b) in four cotton growing districts (Dera Ghazi Khan, Muzafargarh, Bahawalnagar and Rajan Pur) of Punjab province showed that 6 out of 8 pesticides were present in water samples at various levels but they did not exceed the maximum contaminant levels (MCLs) for drinking water set by the United States Environment Protection Agency (USEPA). Similarly, another groundwater study at twelve different sites in Mardan, reported pesticide contamination of all the samples (Ahad et al. 2000). In this study water samples from Madras Kalay (0.64 μg/L), Swabi (0.50 μg/L) and Amber (0.82 μg/L), exceeded the maximum acceptance concentration (MAC) established by the European community. Recently, press news regarding fish killing disaster in the Rawal Lake situated near federal capital (Islamabad) of Pakistan highlighted water contamination by pesticides. It was reported that the resisdues of pyrethroids pesticide residue were almost 4 times higher than Environment – European Commission (EEC) standards for drinking water (Ahad et al. 2006).

In Pakistan, pesticides do not actually present the pollution potential yet they show environmental pollution due to other factors such as shallow water tables, soil characteristics and intensive spraying (Tariq et al. 2006). Moreover, ponding irrigation may also cause pesticide contamination of the groundwater due to faster water flow infiltration (Flury et al. 1994). Other possible reasons of groundwater contamination can be leaching, careless disposal of empty containers and direct runoff (Ahad et al. 2001; Tariq et al. 2006).

9 Fate of Pesticides in Soil

Once pesticides enter the soil environment, they have several fates in soil, including retention or degradation, uptake by animals, plants and leaching to groundwater (Fenoll et al. 2014; Paszko 2014a, b). The processes relating to transformation, retention and transport reflect the fate of pesticides in soil. The fate and charactiristics of pesticides in soils and percolation to deep soil depends on a variety of complex dynamic physical, chemical and biological processes. The relative importance of these processes varies with soil physico-chemical properties (Cabrera et al. 2014).

9.1 *Effect of Soil pH on Pesticide Retention in Soil*

Soil pH is an important parameter, which affects the fate of pesticide in soil. The half life of pesticides in soil/water is pH dependent (Awasthi et al. 2000; Kumar and Philip 2006). Soil pH affects the adsorption/desorption reactions and chemical forms of pesticides in soils. Soil pH alters the electrical charge of certain pesticides which in turns influences the degree and the type of pesticide adsorption and degradation. The adsorption of pesticides may increase or decrease with pH depending on the pKa of the pesticide (Mamy and Barriuso 2005). Under alkaline soils conditions, endosulfan, particularly the α-isomer, is subjected to significant degradation within a week. Soil pH can influence the pathway of pesticide degradation either by affecting the adsorption of pesticides on soil components or by influencing the microbial activity (Ali et al. 2014a, b; Yenisoy-Karakaş 2006). The effect of soil pH on pesticide adsorption to soil particles is more prominent for less basic pesticides. These basic pesticides become cationic depending on their basicity and soil pH. Generally the pH is alkaline (7.5–10) in most agricultural areas of Pakistan that include arid and semi-arid regions. Under alkaline conditions, the alkaline hydrolysis occurs which breaks down the pesticide. Generally, insecticides (particularly carbamates and organophosphates) are highly susceptible to alkaline hydrolysis. The effects of pH on pesticides degradation vary with the type of pesticides and by the proportion of buffering solutions contained in the pesticide formulation. There is generally very low effect of soil pH on the degradation of pesticides having high proportion of buffering solutions in their formulation.

9.2 *Effect of Soil Texture on Pesticide Retention in Soil*

Soil texture is an important parameter controlling pesticide retention, leaching or runoff phenomenon. Compared to other soil components, clay is reported to be actively involved in pesticide retention in soil (Rauf et al. 2012). Generally, pesticide retention is the highest in fine grained soils (clay loam or clay soils) compared to coarse grained soils (sand or loam) (Atasoy et al. 2009; Grondona et al. 2014). The high specific surface area of clays makes them capable to bind higher amount of pesticides in soil. Indeed, the soils with high clay contents have large amounts of active surface sites such as Fe- and Mn-oxyhydroxides, clay minerals and humic acids (Owojori et al. 2010), which retain higher amounts of pesticide than coarse textured soils. Richardson and Epstein (1971) reported that endosulfan was primarily adsorbed by soil clay particles when organic matter was removed. Zhou and Zhu (2007) reported that Triton X-100, a non-ionic surfactant, was more efficient in desorbing polycyclic aromatic hydrocarbons (PAHs) pesticides from soils due to low clay content. In Pakistan, soil texture of agricultural areas ranges from sand to sandy loam. Less clay contents in Pakistani soil make it vulnerable to pesticide leaching to groundwater. Several studies in Pakistan have highlighted groundwater contamination by pesticides, which can be due to less adsorption and more leaching of these pesticides in soil.

9.3 *Effect of Soil Organic Matter (SOM) on Pesticide Retention in Soil*

Soil organic matter is one of the most dynamic and largest reservoirs of carbon on Earth (Cerli et al. 2012). Soil organic matter contents play an important role in controlling pesticide mobility and adsorption in soil (Atasoy et al. 2009; Shahid and Hussain 2011). Soil organic matter is one of the most important factors that affect pesticide adsorption, mobility and leaching in soil (Shahid and Hussain 2011). It is well established that the pesticide adsorption/retention in soil is strongly correlated with SOM content (Shahid and Hussain 2011). SOM increases pesticides retention in upper layers of soil by binding pesticides. A soil with higher SOM content is reported to have higher ability for pesticide retention owing to high chemical reactivity of SOM for organic molecules. The sorption/adsorption abilities of SOM vary with their size and chemical composition (Shahid et al. 2012a). The stability, mobility and leaching of pesticide complexes with SOM vary greatly due to variation in size and chemical composition of SOM (Rodríguez-Liébana et al. 2011). This is due to the fact that pesticides binding to SOM may take place by electrostatic interactions (charge transfer, ion exchange or ligand exchange), sorption (hydrogen bonding, hydrophobic bonding and Van der Waal's forces,), covalent bonding or combinations of these reactions (Shahid et al. 2012b). The strength of pesticide binding with SOM depends on the type of bond. Therefore, amount and type of SOM plays a key role in determining pesticide pretension in soil and percolation to deep layers. Recently, several studies reported the application of organic amendments to soil to decrease pesticide mobility in the soil profile (Shahid and Hussain 2011; Rojas et al. 2013). A number of low-cost and locally available organic adsorbents such as straw, peat mix, cow manure and charcoal have been reported by several researchers for pesticide adsorption and removal (Rojas et al. 2013). Agricultural soil in Pakistan has very less amount of organic matter (average <1 % and maximum 1.5 %), which further decreases with soil depth. Therefore, there is low retention of pesticides in Pakistani soils, resulting in maximum leaching to groundwater. SOM is also reported to affect persistence and retention of pesticides by influencing their rate of degradation or half-life in soils. The pesticides are considered persistent when DT-50 is >100 days, slightly persistent when DT-50 is 30–100 days and non-persistent when DT-50 is <30 days.

The pesticides adsorption/desorption in soil and possible leaching to groundwater is calculated using organic sorption coefficient (K_{oc}), which is the ratio of solution-phase and adsorbed-phase pesticides normalized in term of SOM contents. Pesticides are sorbed strongly in soils with high K_{oc} values. Ahmad et al. (2001) reported a positive correlation of aromaticity of SOM and K_{oc} values for several areas of Australia and Pakistan. The survey for soil and groundwater pollution by pesticides was carried out by Tariq et al. (2004b) to determine the degradation and buffering potentials of different soil series of Pakistan on the basis of pesticide half lives and observed Koc. Ahmad et al. (2001a) used solid-state Cross-Polarization Magic Angle Spinning Carbon-13 Nuclear Magnetic Resonance (CPMAS ^{13}C

NMR) to evaluate the relationships between the nature of organic matter and Koc of different pesticides by determining the structural composition of SOM in twenty seven soils of Australia and Pakistan. They reported highly significant positive correlations of Koc values and SOM aromaticity, and revealed that the aromatic component of SOM is a good indicator of a soil's potential to bind pesticides. Later on, Ahmad et al. (2001b) determined the sorption affinities of phosalone and carbaryl pesticides on forty eight different soils from the United Kingdom, Pakistan and Australia, which confirmed their initial findings (Ahmad et al. 2001a). Similarly, Tariq et al. (2004b, 2006) reported that persistence and hydrophobicity are the key properties of pesticides which control their accumulation in different soil series of Pakistan.

10 Toxic Effects of Pesticides in Soil

From environmental and health perspectives, high use of pesticides in contemporary agriculture has enhanced the impact of these chemicals on environment (Baxter and Cummings 2008; Kouser and Qasim 2011). Despite a number of benefits, pesticides may have toxic side effects, causing potential environmental and health risks. Therefore, the use of agro-chemicals, especially pesticides has become controversial owing to their environmental concerns (Henry et al. 2012; Popp et al. 2013). Soil contamination with pesticides draws great attention because of their potential threat to food safety and detrimental effects on the ecosystem. Application of pesticides in agriculture has significantly decreased the biodiversity of stream invertebrates in Australia and Europe, which resulted in a loss of 42 % species pools (Beketov et al. 2013). Pesticides pollution in agricultural soils may lead to the functional disorder of soil that interferes with soil properties and nutrient behaviour (Niemi et al. 2009; Karpouzas et al. 2014; Rodríguez-Liébana et al. 2014). Pesticides may disturb the soil ecosystem by harming soil microorganisms. Several previous studies have reported the harmful effects of pesticides on soil microbial diversity and activities (Littlefield-Wyer et al. 2008; Karpouzas et al. 2014). It is reported that pesticides can adversely influence the proliferation and associated biotransformation of beneficial soil microorganisms in the soil. Pesticides can inactivate phosphorus-solubilizing and nitrogen-fixing microorganisms in soils, and consequently affect the vital processes of biological nitrogen fixation and phosphorous solubilzation in soil.

Pesticides also affect the microbial assisted mineralization of organic matter and nutrient dynamics and bioavailability in soil (Mahía et al. 2007; Hussain et al. 2009). Soil pesticides pollution may cause functional disturbance of soil resulting in reduced soil and crop productivity (Tariq et al. 2007; Hussain et al. 2009; Tarcau et al. 2013).

Soil contains enzymes within microbial cells, immobilized extracellular enzymes and free enzymes (Mayanglambam et al. 2005; Hussain et al. 2009). Pesticides can disturb enzymatic activity of soil which is considered as useful integrative indicators of soil health (Niemi et al. 2009; Hussain et al. 2009). Soil enzymes play an important role in nutrient mineralization (urease, amidase, sulfates, phosphatase) and the breakdown of organic matter (hydrolase, glucosidase). Pesticides influences soil biochemical processes driven by microbial and enzymatic reactions (Kinney et al. 2005; Mahía et al. 2007). Negative impact of pesticides on soil enzymes such as dehydrogenase, oxidoreductases and hydrolases activities has been extensively documented previously (Menon et al. 2005).

Generally, pesticides do not affect soil enzymes activities when applied at normal/recommended doses. On the contrary, considerable effects on soil enzymatic activity have been reported when pesticides are sprayed for long periods or at higher than recommended doses. For example, Voets et al (1974) reported significant reduction in the activity of β- invertase, glucosidase, urease and phosphatase in soils after long-term atrazine applications. Pozo et al. (2011) showed a temporary reduction in phosphatase and dehydrogenase activity under chlorpyrifos application. Other similar reports include a reduction in the activity of phosphatase after long-term glyphosate applications (Sannino and Gianfreda 2001), a decrease in arylsulfatase and dehydrogenase activity following long-term atrazine applications (Megharaj 2002), and a significant reduction in the activities of urease and dehydrogenase following 15 years application of 2,4-D (isoctyl ester formulation) (Rai 1992). Some current reports have indicated the increase in development of resistance in insects and pests against pesticides. Azeem et al (2002) stated that the condition of the environment and agricultural sustainability in cotton growing areas of Punjab are going steeply downhill. Despite tremendous increase in pesticide use, cotton crops cannot be properly protected from pest's damage.

11 Risks Associated with Pesticides Use

Concerns regarding environmental and human health of pesticides have increased over the last three decades (Lekei et al. 2014). Pesticide poisoning can cause severe health hazards to agricultural workers who are at elevated risk of being poisoned (Lekei et al. 2014). Prolonged exposure to multiple pesticides may cause cytotoxic changes and negatively affect the regular functioning of organs like kidney and liver (Azmi et al. 2006; Khan et al. 2008). Pesticide poisoning results in allergic reactions and peripheral neuropathies (Corsini et al. 2013). Pesticide-induced chronic toxicity may vary from skin irritation to dysfunctioning of essential organs resulting in death, and found to be the major cause of cancer in farming community (Horrigan et al. 2002). Besides, pesticides poisoning can adversely influence the endocrine systems of humans, which may result in hormonal dysfunctioning (Ejaz et al. 2004).

The presence of pesticides in food items and their accumulation in tissues has direct toxic effects on humans and other non-target organisms. The organochlorine pesticides present in human and cow's milk are transferred to the infants. Due to their lipophilic nature, organochlorine insecticides accumulate in fat tissues of animals and are released in situations of fasting or pregnancy. Numerous previous studies in Pakistan have showed pesticide residues in fat samples and blood serum in residents of cotton-growing areas in the Punjab, Sindh and Balochistan provinces (Naqvi and Jahan 1996; Parveen et al. 2004). Hayat et al. (2010) reported pesticide in blood samples of cotton cultivating farmers. Unsafe use of pesticides is damaging the health of the farmers and the community in Pakistan. According to the UN's 1998 report, over 500,000 Pakistanis suffered annually from poisoning due to agrochemicals, out of which 10,000 died (DAWN 2004). According to some reports, annually 10,000 farmers and field worker get poisoned by pesticides in Pakistan while unintentional acute pesticide poisoning cases are observed due to occupational exposure (Hashmi and Khan 2011; Tahir and Anwar 2012). This presents an alarming situation and serves food for thought for all those who are interested in ameliorating the plight of farming community. The situation is very much similar in other developing Asian countries. Jayaratnum et al. (1987) carried out a detailed survey of acute poisoning among farming community in four Asian countries. They reported 69 % pesticide poisoning (out of total poisoning cases) in Sri Lanka, 54 % in Malaysia, 27 % in Thailand and 23 % in Indonesia.

12 Management of Pesticide Use and Integrated Pest Management in Pakistan

Integrated Pest Management (IPM) is economical, effective and environmentally sensitive approach that combines different management practices and strategies to cultivate healthy crops with minimum use of pesticides. FAO endorses IPM as the best approach for crop protection and consider it as a mainstay of environmentally sustainable crop production. During early 70s through Agricultural Pesticide Ordinance (APO 1971), the Government of Pakistan tried to regulate production and consumption of pesticides. The legislation regarding specifications of pesticide exists in the Agricultural Pesticide Rules 1973. Regulations have also been developed for safe use of pesticides (Rasheed 2007). Recognizing and realizing health and environmental hazards attached to pesticide use, reliance on IPM has been stressed in the National Agricultural Policies. A Farmer Field School led Integrated Pest Management model popularly known as "Vehari Model", was implemented in Pakistan during 1996, which clearly showed that IPM technique can be practiced at the farm scale level. It was concluded that pesticide led control of pests and insects has actually further enhanced the pest problems, by disturbing the agroecosystem and killing the environment friendly and non-targeted organisms such as predators, parasitoids and birds. The results of this model led to the establishment of National IPM Programme of Pakistan in December 2000. This programme helps to create awareness among the farming community of the worth of biodiversity. Under this programme, Farmers Field School (FFS) activities and Training of Facilitators (TOF) were organised during 2001 in Punjab, Sindh and Balochistan provinces, in order to enhance capacity building of the farmers, through participatory learning processes. The monetary reliability of IPM has already been well established in Pakistan by the research trials in the cotton growing area of the Punjab during 1995–1996, which showed that pesticide use can be decreased upto 50 % without any major decrease in crop yield. After the successful completion of National IPM Project, currently, the National IPM Programme of Pakistan has implemented two projects i.e. "Management of cotton leaf curl virus (CLCV) disease through IPM technique by adopting Farmers Field School (FFS)" supported by International Center for Agricultural Research in the Dry Area (ICARDA) and "Integrated Crop Management Practices to enhance value chain outcome for Mango industry in Pakistan and Australia" supported by Australia Pakistan Agriculture Sector (ASLP). Despite wide acceptance and institutionalization of IPM through National IPM Programme, concerted approaches are still necessary by the government to educate the farming community on a large scale. Moreover, future research on the risk assessment, cost-benefit and feasibility of various active ingredients and other alternatives is essential to develop an effective pesticide use strategy.

13 Conclusion

In Pakistan food demand has increased due to rapid increase in population. This population flux enforced the farmers to increase the crop yield by using the agrochemicals especially the pesticides. Use of pesticides in Pakistan started in 1954, which increased rapidly with time especially during 1980–2000 owing to very soft legislation regarding pesticide registration and import. Pesticide usage is not properly regulated due to lack of awareness, ineffective legislation and technical knowhow among the farming community in Pakistan. Currently, about 145 active substances have been registered in Pakistan. Insecticides are main pesticides utilized in Pakistan especially in Punjab province. Cotton shares about 80 % of pesticide use in Pakistan, therefore maximum soil and water contamination is observed in cotton zone. Pesticide residues especially organochlorines have been reported in groundwater, surface water, wells and soils of Pakistan particularly in cotton growing areas of Sindh and Punjab. The soil of Pakistan generally contains less clay, low organic matter contents with alkaline pH. Therefore, pesticides do not persist for long time in soil of Pakistan. Pesticide leaching to groundwater is common in Pakistan. Pesticides poisoning has been observed in farming community, which is causing sever health issues in Pakistan. Despite several environmental issues related to pesticide use, still the use of chemical pesticides is the only possible and feasiable way of crop protection in Pakistan and there is no shift away from it. The government of Pakistan has launched National IPM Programme in December 2000 to educate the farming community regarding minimize use of Pesticides. Still there is a dire need of rigorous approaches at national level to educate the farming community regarding pesticide use and poisoning.

References

Ahad K, Anwar T, Ahmad I, Mohammad A, Tahir S, Aziz S, Baloch UK (2000) Determination of insecticide residues in groundwater of Mardan Division, NWFP, Pakistan: a case study. Water SA 26(3):409–412

Ahad K, Hayat Y, Ahmad I, Soomro MH (2001) Capillary chromatographic determination of pesticides residues in ground water of Multan division. Nucleous 38:145–149

Ahad K, Mohammad A, Mehboob F, Sattar A, Ahmad I (2006) Pesticide residues in Rawal Lake, Islamabad, Pakistan. Bull Environ Contam Toxicol 76:463–470

Ahmad R, Kookana RS, Alston AM, Bromilow RH (2001a) Differences in sorption behaviour of carbaryl and phosalone in soils from Australia, Pakistan, and the United Kingdom. Aust J Soil Res 39:893–908

Ahmad R, Kookana RS, Alston AM, Skjemstad JO (2001b) The nature of soil organic matter affects sorption of pesticides. Relationships with carbon chemistry as determined by ^{13}C CPMAS NMR spectroscopy. Environ Sci Technol 35:878–884

Alamdar A, Syed JH, Malik RN, Katsoyiannis A, Liu J, Li J, Zhang G, Jones KC (2014) Organochlorine pesticides in surface soils from obsolete pesticide dumping ground in Hyderabad City, Pakistan: contamination levels and their potential for air–soil exchange. Sci Total Environ 470:733–741

Ali M, Kazmi AA, Ahmed N (2014a) Study on effects of temperature, moisture and pH in degradation and degradation kinetics of aldrin, endosulfan, lindane pesticides during full-scale continuous rotary drum composting. Chemosphere 102:68–75

Ali U, Syed JH, Malik RN, Katsoyiannis A, Li J, Zhang G, Jones KC (2014b) Organochlorine pesticides (OCPs) in South Asian region: a review. Sci Total Environ 476–477:705–717

Anwar T (2009) Pesticides residues in water, soil, fruits and vegetables in cotton growing areas of Sindh and lower Punjab. Ph.D. thesis, Department of Zoology, University of Karachi

Anwar T, Ahmad I, Tahir S (2012) Determination of pesticide residues in soil of Nawabshah District, Sindh, Pakistan. Pak J Zool 44(1):87–93

APO (1971) Agricultural Pesticides Ordinance, 1971 (II OF 1971)

Arias-Estévez M, López-Periago E, Martínez-Carballo E, Simal-Gándara J, Mejuto J-C, García-Río L (2008) The mobility and degradation of pesticides in soils and the pollution of groundwater resources. Agric Ecosyst Environ 123:247–260

Asi MR, Hussain A, Muhmood ST (2008) Solid phase extraction of pesticide residues in water samples: DDT and its metabolites. Int J Environ Res 2:43–48

Atasoy AD, Mermut AR, Kumbur H, Ince F, Arslan H, Avci ED (2009) Sorption of alpha and beta hydrophobic endosulfan in a Vertisol from southeast region of Turkey. Chemosphere 74:1450–1456

Austruy A, Shahid M, Xiong T, Castrec M, Payre V, Niazi NK, Sabir M, Dumat C (2014) Mechanisms of metal-phosphates formation in the rhizosphere soils of pea and tomato: environmental and sanitary consequences. J Soils Sediments 14:666–678

Awasthi N, Ahuja R, Kumar A (2000) Factors influencing the degradation of soil-applied endosulfan isomers. Soil Biol Biochem 32:1697–1705

Azeem M, Iqbal M, Ahmad I, Soomro MH (2002) Economic evaluation of pesticide use externalities in the cotton zones of Punjab, Pakistan. Pak Dev Rev 41:4 part 2, 683–698

Azmi MA, Naqvi SNH, Azmi MA, Aslam M (2006) Effect of pesticide residues on health and different enzyme levels in the blood of farm workers from Gadap (rural area) Karachi-Pakistan. Chemosphere 64:1739–1744

Bano A, Siddique SA (1991) Chlorinated hydrocarbons in the sediments from the coastal waters of Karachi (Pakistan). Pak J Sci Ind Res 34:70–74

Baxter J, Sp C (2008) The degradation of the herbicide bromoxynil and its impact on bacterial diversity in a top soil. J Appl Microbiol 104:1605–1616

Beketov MA, Kefford BJ, Schäfer RB, Liess M (2013) Pesticides reduce regional biodiversity of stream invertebrates. Proc Natl Acad Sci U S A 110:11039–11043

Bentley R, Chasteen TG (2002) Microbial methylation of metalloids: arsenic, antimony, and bismuth. Microbiol Mol Biol Rev 66:250–271
Burauel P, Baßmann F (2005) Soils as filter and buffer for pesticides—experimental concepts to understand soil functions. Environ Pollut 133:11–16
Cabrera A, Cox L, Spokas K, Hermosín MC, Cornejo J, Koskinen WC (2014) Influence of biochar amendments on the sorption–desorption of aminocyclopyrachlor, bentazone and pyraclostrobin pesticides to an agricultural soil. Sci Total Environ 470–471:438–443
Cai DW (2008) Understand the role of chemical pesticides and prevent misuses of pesticides. J Agric Sci Technol 1:36–38
Carriger JF, Rand GM, Gardinali PR, Perry WB, Tompkins MS, Fernandez AM (2006) Pesticides of potential ecological concern in sediment from South Florida canals: an ecological risk prioritization for aquatic arthropods. Soil Sediment Contam 15:21–45
Cerli C, Celi L, Kalbitz K, Guggenberger G, Kaiser K (2012) Separation of light and heavy organic matter fractions in soil —Testing for proper density cut-off and dispersion level. Geoderma 170:403–416
Corsini E, Sokooti M, Galli CL, Moretto A, Colosio C (2013) Pesticide induced immunotoxicity in humans: a comprehensive review of the existing evidence. Toxicology 307:123–135
Dane F, Dalgiç Ö (2005) The effects of fungicide benomyl (benlate) on growth and mitosis in onion (*Allium cepa* L.) root apical meristem. Acta Biol Hung 56:119–128
DAWN (2004) Deaths from poisoning of agro-chemicals on the rise. The Daily Dawn, Internet Edition, July 29
Dayan FE, Watson SB (2011) Plant cell membrane as a marker for light-dependent and light-independent herbicide mechanisms of action. Pest Biochem Phys 101:182–190
Dayan FE, Owens DK, Duke SO (2012) Rationale for a natural products approach to herbicide discovery. Pest Manag Sci 68:519–528
Debenest T, Pinelli E, Coste M, Silvestre J, Mazzella N, Madigou C, Delmas F (2009) Sensitivity of freshwater periphytic diatoms to agricultural herbicides. Aquat Toxicol 93:11–17
Debenest T, Silvestre J, Coste M, Pinelli E (2010) Effects of pesticides on freshwater Diatoms. Rev Environ Contam Toxicol 203:87–103
Eavy AL, Ahmed F, Buriro AS (1995) Final report on integrated pest/production/plant management (IPM). Development program ARP-II Sindh submitted by Winrock international institute for agricultural development. Dir Gen Agric Res 2:1–27
Economic Survey of Pakistan 2002–06a. Ministry of Finance, Government of Pakistan. http://www.finance.gov.pk/survey_1112.html
Economic Survey of Pakistan 2002–06b. Ministry of Finance, Government of Pakistan. http://www.finance.gov.pk/survey_0607.html
Economic Survey of Pakistan 2008–09. Ministry of Finance, Government of Pakistan. http://www.finance.gov.pk/survey_0809.html
Economic Survey of Pakistan 2010–11. Ministry of Finance, Government of Pakistan. http://www.finance.gov.pk/survey_1112.html
Ejaz S, Akram W, Lim CW, Lee JJ, Hussain I (2004) Endocrine disrupting pesticides: a leading cause of cancer among rural people in Pakistan. Exp Oncol 26:98–105
EPA (2007) Pesticides industry sales and usage 2006-2007 market estimates. Environmental Protection Agency, Washington, DC
Eqani SA, MAS, Malik RN, Mohammad A (2011) The level and distribution of selected organochlorine pesticides in sediments from River Chenab, Pakistan. Environ Geochem Health 33:33–47
Fairbanks MM, Hardy GESJ, McComb JA (2002) Mitosis and meiosis in plants are affected by the fungicide phosphite. Aust Plant Pathol 31:281–289
FAO (2006) Global forest resources assessment 2005: progress towards sustainable forest management. Forestry paper 147
Farlin J, Gallé T, Bayerle M, Pittois D, Braun C, El Khabbaz H, Lallement C, Leopold U, Vanderborght J, Weihermueller L (2013) Using the long-term memory effect of pesticide and

metabolite soil residues to estimate field degradation half-life and test leaching predictions. Geoderma 207–208:15–24

Fenoll J, Vela N, Navarro G, Pérez-Lucas G, Navarro S (2014) Assessment of agro-industrial and composted organic wastes for reducing the potential leaching of triazine herbicide residues through the soil. Sci Total Environ 493:124–132

Flury M, Flühler H, Jury WA, Leuenberger J (1994) Susceptibility of soils to preferential flow of water: a field study. Water Resour Res 30:1945–1954

Foucault Y, Lévêque T, Xiong T, Schreck E, Austruy A, Shahid M, Dumat C (2013) Green manure plants for remediation of soils polluted by metals and metalloids: ecotoxicity and human bioavailability assessment. Chemosphere 93:1430–1435

Gai N, Pan J, Tang H, Chen S, Chen D, Zhu X, Lu G, Yang Y (2014) Organochlorine pesticides and polychlorinated biphenyls in surface soils from Ruoergai high altitude prairie, east edge of Qinghai-Tibet Plateau. Sci Total Environ 478:90–97

Grondona SI, Gonzalez M, Martínez DE, Massone HE, Miglioranza KSB (2014) Endosulfan leaching from Typic Argiudolls in soybean tillage areas and groundwater pollution implications. Sci Total Environ 484:146–153

Hakeem KR (2015) Crop production and global environmental issues. Springer International Publishing AG, Gewerbestrasse 11, 6330 Cham, Switzerland, 598 pp

Hashmi I, Khan AD (2011) Adverse health effects of pesticide exposure in agricultural and industrial workers of developing country. Pesticides—The Impacts of Pesticides Exposure

Hayat R, Ali S, Amara U, Khalid R, Ahmed I (2010) Soil beneficial bacteria and their role in plant growth promotion: a review. Ann Microbiol 60:579–598

Henry M, Béguin M, Requier F, Rollin O, Odoux JF, Aupinel P, Aptel J, Tchamitchian S, Decourtye A (2012) A common pesticide decreases foraging success and survival in honey bees. Science 336:348–350

Horrigan L, Lawrence RS, Walker P (2002) How sustainable agriculture can address the environmental and human health harms of industrial agriculture. Environ Health Perspect 110:445–456

Húšková R, Matisová E, Kirchner M (2008) Fast GC–MS pesticide multiresidue analysis of apples. Chroma 68:49–55

Hussain AH, Tirmizi, Babar ZK (1988) Studies on the degradation dissipation and persistence of DDT in sandy loam soil under laboratory and field conditions. In: Proceedings of the third (Final) FAO/IAEA research coordination meeting on the fate of persistent pesticides in the tropics, using isotope technique, January 25–29 1998: Bangkok, Thailand

Hussain S, Siddique T, Saleem M, Arshad M, Khalid A (2009) Chapter 5 Impact of pesticides on soil microbial diversity, enzymes, and biochemical reactions. Adv Agron 102:159–200

Jabbar A, Masud SZ, Parveen Z, Ali M (1993) Pesticide residues in cropland soils and shallow groundwater in Punjab Pakistan. Bull Environ Contam Toxicol 51:268–273

James PJ, Cramp AP, Hook SE (2008) Resistance to insect growth regulator insecticides in populations of sheep lice as assessed by a moulting disruption assay. Med Vet Entomol 22:326–330

Jan MR, Shah J, Khawaja MA, Gul K (2008) DDT residue in soil and water in and around abandoned DDT manufacturing factory. Environ Monit Assess 155:31–38

Jiang YF, Wang XT, Jia Y, Wang F, Wu MH, Sheng GY, Fu JM (2009) Occurrence, distribution and possible sources of organochlorine pesticides in agricultural soil of Shanghai, China. J Hazard Mater 170:989–997

Karpouzas DG, Papadopoulou E, Ipsilantis I, Friedel I, Petric I, Udikovic-Kolic N, Djuric S, Kandeler E, Menkissoglu-Spiroudi U, Martin-Laurent F (2014) Effects of nicosulfuron on the abundance and diversity of arbuscular mycorrhizal fungi used as indicators of pesticide soil microbial toxicity. Ecol Indic 39:44–53

Khan MS (1998) Pakistan crop protection market. PAPA Bull 9:7–9

Khan AM (2000) Pakistan agricultural pesticides association. Internet www page, at URL: http://www.cpp.org.pk/Assoc/associations.html

Khan AF (2004) Hazards of pesticide spray on vegetables. The Daily Dawn Pakistan, Internet Edition, June 12
Khan DA, Bhatti MM, Khan FA, Naqvi ST, Karam A (2008) Adverse effects of pesticides residues on biochemical markers in Pakistani tobacco farmers. Int J Clin Exp Med 1:274–282
Khan DA, Hashmi I, Mahjabeen W, Naqvi TA (2010) Monitoring health implications of pesticide exposure in factory workers in Pakistan. Environ Monit Assess 168:231–240
Khan HAA, Akram W, Shehzad K, Shaalan EA (2011) First report of field evolved resistance to agrochemicals in dengue mosquito, Aedes albopictus (Diptera: Culicidae), from Pakistan. Parasit Vectors 4:146
Khooharo AA, Memon RA, Mallah MU (2008) An empirical analysis of pesticide marketing in Pakistan. Pakistan economic and social review: incorporating the Punjab University economist; bi-annual journal of the Department of Economics, University of the Punjab. – Lahore 46, 57–74
Kinney CA, Mandernack KW, Mosier AR (2005) Laboratory investigations into the effects of the pesticides mancozeb, chlorothalonil, and prosulfuron on nitrous oxide and nitric oxide production in fertilized soil. Soil Biol Biochem 37:837–850
Kouser S, Qaim M (2011) Impact of Bt cotton on pesticide poisoning in smallholder agriculture:A panel data analysis. Ecol Econ 70:2105–2113
Kumar M, Philip L (2006) Bioremediation of endosulfan contaminated soil and water—Optimization of operating conditions in laboratory scale reactors. J Hazard Mater 136:354–364
Lee KJ, Shin JW, Moon J, Lim JA, Byun JI, Kim TJ, Shin YW, Lee ST, Jung KH, Lee SK et al (2014) An illustrative case of mixed pesticide poisoning with remarkable improvement:A case report. J Neurol Sci 344:232–233
Lekei E, Ngowi AV, London L (2014) Hospital-based surveillance for acute pesticide poisoning caused by neurotoxic and other pesticides in Tanzania. Neuro Toxicol 45:318–326
Littlefield-Wyer JG, Brooks P, Katouli M (2008) Application of biochemical fingerprinting and fatty acid methyl ester profiling to assess the effect of the pesticide Atradex on aquatic microbial communities. Environ Pollut 153:393–400
Liu B, McConnell LL, Torrents A (2002) Herbicide and insecticide loadings from the Susquehanna river to the Northern Chesapeake Bay. J Agric Food Chem 50:4385–4392
Mahía J, Martín A, Carballas T, Díaz-Raviña M (2007) Atrazine degradation and enzyme activities in an agricultural soil under two tillage systems. Sci Total Environ 378:187–194
Mahmood A, Malik RN, Li J, Zhang G (2014) Human health risk assessment and dietary intake of organochlorine pesticides through air, soil and food crops (wheat and rice) along two tributaries of river Chenab, Pakistan. Food Chem Toxicol 71:17–25
Malik RN, Rauf S, Mohammad A, Eqani SAMAS, Ahad K (2010) Organochlorine residual concentrations in cattle egret from the Punjab Province, Pakistan. Environ Monit Assess 173:325–341
Mamy L, Barriuso E (2005) Glyphosate adsorption in soils compared to herbicides replaced with the introduction of glyphosate resistant crops. Chemosphere 61:844–855
Mayanglambam T, Vig K, Singh DK (2005) Quinalphos persistence and leaching under field conditions and effects of residues on dehydrogenase and alkaline phosphomonoesterases activities in soil. Bull Environ Contam Toxicol 75:1067–1076
Mazari RB (2005) Country report on international code of conduct on the distribution and use of pesticides. Department of Plant Protection, Ministry of Food, Agriculture & Livestock Government of Pakistan. 26th July; 2005
Megharaj M (2002) Heavy pesticide use lowers soil health. Kondinin Landc Group Magaz Farm Ahead 121:37–38
Menon M, Hermle S, Abbaspour KC, Günthardt-Goerg MS, Oswald SE, Schulin R (2005) Water regime of metal-contaminated soil under juvenile forest vegetation. Plant Soil 271:227–241

Mombo S, Foucault Y, Deola F, Gaillard I, Goix S, Shahid M, Schreck E, Pierart A, Dumat C (2015) Management of human health risk in the context of kitchen gardens polluted by lead and cadmium near a lead recycling company. J Soils Sediments. doi:10.1007/s11368-015-1069-7

Naqvi SNH, Jahan M (1996) Pesticide residues in random blood samples of human population in Karachi. J Coll Physicians Surg Pak 6(3):151–153

Nation Master (2013). http://www.nationmaster.com/country-info/profiles/Pakistan/Agriculture. Accessed 30 June 2015

Niazi NK, Singh B, Zwieten LV, Kachenko AG (2012) Phytoremediation of an arsenic-contaminated site using *Pteris vittata* L. and *Pityrogramma calomelanos* var. austroamericana:a long-term study. Environ Sci Pollut Res 19:3506–3515

Niemi RM, Heiskanen I, Ahtiainen JH, Rahkonen A, Mäntykoski K, Welling L, Laitinen P, Ruuttunen P (2009) Microbial toxicity and impacts on soil enzyme activities of pesticides used in potato cultivation. Appl Soil Ecol 41:293–304

Omer A, Pascual U, Russell N (2010) A theoretical model of agrobiodiversity as a supporting service for sustainable agricultural intensification. Ecol Econ 69:1926–1933

Owojori OJ, Reinecke AJ, Rozanov AB (2010) Influence of clay content on bioavailability of copper in the earthworm Eisenia fetida. Ecotoxicol Environ Saf 73:407–414

Parveen Z, Masud SZ (1988) Organochlorine pesticide residues in cattle drinking water. Pak J Sci Ind Res 31:53–56

Parveen Z, Khuhro MI, Rafiq N, Kausar N (2004) Evaluation of multiple pesticide residues in apple and citrus fruits, 1999–2001. Bull Environ Contam Toxicol 73:312–318

Paszko T (2014a) Adsorption, degradation and mobility of carbendazim in profiles of Polish mineral soils. Geoderma 226–227:160–169

Paszko T (2014b) Modeling of pH-dependent adsorption and leaching of MCPA in profiles of Polish mineral soils. Sci Total Environ 494–495:229–240

Pierart A, Shahid M, Séjalon-Delmas N, Dumat C (2015) Antimony bioavailability: knowledge and research perspectives for sustainable agricultures. J Hazard Mater 289:219–234

Pimentel D (2009) Environmental and economic costs of the application of pesticides primarily in the United States. In: D.R. Peshin, A.K. Dhawan (eds) Integrated pest management: innovation-development process. Springer, Dordrecht, pp 89–111

Popp J, Pető K, Nagy J (2013) Pesticide productivity and food security. A review. Agron Sustain Dev 33:243–255

Pourrut B, Shahid M, Dumat C, Winterton P, Pinelli E (2011) Lead uptake, toxicity, and detoxification in plants. Rev Environ Contam Toxicol 213:113–136

Pourrut B, Shahid M, Douay F, Dumat C, Pinelli E (2013) Molecular mechanisms involved in lead uptake, toxicity and detoxification in higher plants. In: Heavy Metal Stress in Plants 1:121–147. Springer Berlin Heidelberg

Pozo K, Harner T, Lee SC, Sinha RK, Sengupta B, Loewen M et al (2011) Assessing seasonal and spatial trends of persistent organic pollutants (POPs) in Indian agricultural regions using PUF disk passive air samplers. Environ Pollut 159:646–653

Rai JPN (1992) Effects of long-term 2,4-D application on microbial populations and biochemical processes in cultivated soil. Biol Fertil Soils 13:187–191

Rasheed BM (2007) Country report on international code of conduct on the distribution and use of pesticide. Department of Plant Protection, Ministry of Food. Agriculture and Livestock, Government of Pakistan

Rauf N, Tahir SS, Kang J-H, Chang Y-S (2012) Equilibrium, thermodynamics and kinetics studies for the removal of alpha and beta endosulfan by adsorption onto bentonite clay. Chem Eng J 192:369–376

Richardson EM, Epstein E (1971) Retention of three insecticides on different size soil particles suspended in water1. Soil Sci Soc Am J 35:884

Rodríguez-Liébana JA, Mingorance MD, Peña A (2011) Sorption of hydrophobic pesticides on a Mediterranean soil affected by wastewater, dissolved organic matter and salts. J Environ Manage 92:650–654

Rodríguez-Liébana JA, Mingorance MD, Peña A (2014) Pesticide mobility and leachate toxicity in two abandoned mine soils. Effect of organic amendments. Sci Total Environ 497–498:561–569

Rojas R, Morillo J, Usero J, Delgado-Moreno L, Gan J (2013) Enhancing soil sorption capacity of an agricultural soil by addition of three different organic wastes. Sci Total Environ 458–460:614–623

Sabir HM, Tahir SH, Khan MB (2011) BT Cotton and its impact on cropping pattern in Punjab. PJSS 31:127–134

Sabir M, Hanafi MM, Malik MT, Aziz T, Zia-ur-Rehman M, Ahmad HR, Hakeem KR, Shahid M (2013) Differential effect of nitrogen forms on physiological parameters and micronutrient concentration in maize (*Zea mays* L.). Aust J Crop Sci 7:1836–1842

Saleem MA, Ashfaq M (2004) Environmental pollution and agriculture, B.Z. University Press, Multan. Effects of pesticides used in sugarcane cropping systems on soil organisms and biological functions associated with soil health

Sankararamakrishnan N, Kumar Sharma A, Sanghi R (2005) Organochlorine and organophosphorous pesticide residues in ground water and surface waters of Kanpur, Uttar Pradesh, India. Environ Int 31:113–120

Sannino F, Gianfreda L (2001) Pesticide influence on soil enzymatic activities. Chemosphere 45:417–425

Shahid M, Hussain I (2011) Effect of soil organic matter on pesticides retention in soil and leaching to ground water. Int Post J Sci Technol 1:12

Shahid M, Dumat C, Silvestre J, Pinelli E (2012a) Effect of fulvic acids on lead-induced oxidative stress to metal sensitive *Vicia faba* L. plant. Biol Fertil Soils 48:689–697

Shahid M, Pinelli E, Dumat C (2012b) Review of Pb availability and toxicity to plants in relation with metal speciation; role of synthetic and natural organic ligands. J Hazard Mater 219–220:1–12

Shahid M, Xiong T, Castrec-Rouelle M, Leveque T, Dumat C (2013) Water extraction kinetics of metals, arsenic and dissolved organic carbon from industrial contaminated poplar leaves. J Environ Sci 25:2451–2459

Shahid M, Pinelli E, Pourrut B, Dumat C (2014a) Effect of organic ligands on lead-induced oxidative damage and enhanced antioxidant defense in the leaves of *Vicia faba* plants. J Geochem Explor 144:282–289

Shahid M, Xiong T, Masood N, Leveque T, Quenea K, Austruy A, Foucault Y, Dumat C (2014b) Influence of plant species and phosphorus amendments on metal speciation and bioavailability in a smelter impacted soil: a case study of food-chain contamination. J Soils Sediments 14:655–665

Shahid M, Dumat C, Pourrut B, Abbas G, Shahid N, Pinelli E (2015) Role of metal speciation in lead-induced oxidative stress to Vicia faba roots. Russ J Plant Physiol 62:448–454

Sheikh MA (2011) Mortality in patients presenting with organophosphorus poisoning at Liaquat University of Medical and Health Sciences. Pak J Med Sci http://pjms.com.pk/index.php/pjms/article/view/1176

Sultana J, Syed JH, Mahmood A, Ali U, Rehman MYA, Malik RN, Li J, Zhang G (2014) Investigation of organochlorine pesticides from the Indus Basin, Pakistan: sources, air–soil exchange fluxes and risk assessment. Sci Total Environ 497–498:113–122

Syed JH, Malik RN (2011) Occurrence and source identification of organochlorine pesticides in the surrounding surface soils of the Ittehad Chemical Industries Kalashah Kaku, Pakistan. Environ Earth Sci 62:1311–1321

Syed JH, Malik RN, Liu D, Xu Y, Wang Y, Li J, Zhang G, Jones KC (2013) Organochlorine pesticides in air and soil and estimated air-soil exchange in Punjab, Pakistan. Sci Total Environ 444:491–497

Tahir S, Anwar T (2012) Assessment of pesticide exposure in female population living in cotton growing areas of Punjab, Pakistan. Bull Environ Contam Toxicol 89:1138–1141

Tarcau D, Cucu-Man S, Boruvkova J, Klanova J, Covaci A (2013) Organochlorine pesticides in soil, moss and tree-bark from North-Eastern Romania. Sci Total Environ 456–457:317–324
Tariq MA (2002) Need to tap agriculture sector. Daily Dawn, Economic and Business Review
Tariq MI, Afzal S, Hussain I (2004a) Adsorption of pesticides by salorthids and camborthids of Punjab, Pakistan. Toxicol Environ Chem 86:247–264
Tariq MI, Afzal S, Hussain I (2004b) Pesticides in shallow groundwater of Bahawalnagar, Muzafargarh, D.G. Khan and Rajan Pur districts of Punjab, Pakistan. Environ Int 30:471–479
Tariq MI, Afzal S, Hussain I (2006) Degradation and persistence of cotton pesticides in sandy loam soils from Punjab, Pakistan. Environ Res 100:184–196
Tariq MI, Afzal S, Hussain I, Sultana N (2007) Pesticides exposure in Pakistan:a review. Environ Int 33:1107–1122
Tehseen WM, Hansen LG, Wood SG, Hanif M (1994) Assessment of chemical contaminants in water and sediment samples from Degh Nala in the province of Punjab. Pak Arch Environ Contam Toxicol 26:79–89
Tierney KB, Kennedy CJ, Gobas F, Gledhill M, Sekela M (2014) Organic contaminants and fish. Org Chem Toxicol Fish 33:1–52
Xiong T, Leveque T, Shahid M, Foucault Y, Mombo S, Dumat C (2014) Lead and cadmium phytoavailability and human bioaccessibility for vegetables exposed to soil or atmospheric pollution by process ultrafine particles. J Environ Qual 43:1593–1600
Yenisoy-Karakaş S (2006) Validation and uncertainty assessment of rapid extraction and clean-up methods for the determination of 16 organochlorine pesticide residues in vegetables. Anal Chim Acta 571:298–307
Zhou W, Zhu L (2007) Efficiency of surfactant-enhanced desorption for contaminated soils depending on the component characteristics of soil-surfactant–PAHs system. Environ Pollut 147:66–73

Iron Biofortification of Cereals Grown Under Calcareous Soils: Problems and Solutions

Pia Muhammad Adnan Ramzani, Muhammad Khalid, Muhammad Naveed, Ayesha Irum, Waqas-ud-Din Khan, and Salma Kausar

Contents

P.M.A. Ramzani (✉) • M. Khalid • M. Naveed • A. Irum
Institute of Soil and Environmental Sciences, University of Agriculture Faisalabad, Faisalabad, Pakistan
e-mail: piyya_muhammad@hotmail.com

W.-ud-D. Khan
Institute of Soil and Environmental Sciences, University of Agriculture Faisalabad, Faisalabad, Pakistan

Sustainable development study center, Government College University, Lahore-54000, Pakistan
e-mail: waqas1919@gmail.com

S. Kausar
Soil and Water Testing Laboratory, 63100 Bahawalpur, Pakistan

K.R. Hakeem et al. (eds.), *Soil Science: Agricultural and Environmental Prospectives*, DOI 10.1007/978-3-319-34451-5_10

Abstract Iron (Fe) deficiency is a prevalent nutritional deficiency throughout the world, affecting an estimated 3.7 billion people. Increasing Fe concentration in foodcrops is an important global challenge due to the high incidence of Fe deficiency in human population. Further, cereals grown on calcareous soil are low in Fe contents. High pH, high temperature, low organic matter and poorly managed soil with respect to fertility are the factors which cause low Fe availability to cereal crops in calcareous soil. Iron fertilization in calcareous soil is not effective due to their rapid conversion into unavailable forms and poor mobility of Fe in phloem. Iron-organic compounds in manure are effective in maintaining Fe availability to plants. Fe bioavailability inhibitors such as polyphenols and phytate inhibit iron absorption but it was concluded that their inhibitory effect on iron absorption can be limited by increasing iron content. Ferritin, an iron storage protein, can deposit thousands of iron atoms asnon-toxic form and ferritin iron is bioavailable to humans as ferrous sulphate. Improving both concentration and bioavailability of Fe in cereal grains is, therefore, an important challenge and a high-priority research area. Hence, there is a need for effective strategies to overcome Fe deficiency in cereals and to increase Fe bioavailabilty in cereals grains. Biofortification of food crops with Fe to combat iron deficiency problems in humans, is a cost-effective and sustainable agricultural strategy to alleviate malnutrition. We hypothesized that Fe nutrition management in calcareous soil can increase growth, yield and Fe bioavailability from cereals.

Keywords Biofortification • Bioavaliability • Fe deficiency • Calcareous soils

1 Introduction

The population of the world continues to increase with the passage of time. Thus, food demand is also increased. However, natural resources are limited (United Nations 2012). On the other hand malnutrition is becoming a serious threat to the people in the poor communities, especially in the developing countries (Sperotto et al. 2012). In developing countries, total of 805 million people are not leading a healthy life and suffering from hunger. State of Food Insecurity in the World, FAO (2015) states that about 13.5 % of the total population is not having enough food for their daily intake of calorie. Iron (Fe) deficiency causes 0.8 million deaths annually and is ranked third among the risk factors (WHO 2007). Iron malnutrition is leading to anemia in human's especially pregnant women and preschool children (WHO 2010). Iron deficiency anemia is the most severe type of iron deficiency (Lozoff and Georgieff 2006). It can result in a low resistance to infection, impaired psychomotor development, impaired cognitive function in children, poor academic performance, fatigue, fetal resorption, low productivity, and increased risk of maternal mortality (Bothwell and MacPhail 2004; Murray-Kolb and Beard 2009). The main cause of micronutrient malnutrition is consumption of Cereals food that is poor in microelements (Bouis et al. 2011). In developing countries majority of population depends on cereals based foods that are low in bioavailable Fe (Gibson et al. 2010). Thus low Fe contents from staple food result in Fe deficiency in humans, especially in developing countries (Sperotto et al. 2012).

Fe as a micronutrient required by all organisms and in plants it plays an important role in metabolic pathways including photosynthesis, respiration, chlorophyll formation and several redox reactions (Briat et al. 1995; Briat and Lobréaux 1997). Thus Fe deficient crops show interveinal chlorosis, stunting growth and reduced yield (Kabir et al. 2013). Fe plays an important role in functioning of cell's power house that impaired under Fe deficiency resultantly poor plant health (Bashir et al. 2013a).

Iron deficiency is a common problem for crops growing in calcareous soil (Laird et al. 2010). High pH and high HCO_3contents are two most important factors that limit Fe availability in calcareous soil (Bloom and Inskeep 1988; Alcantara et al. 2002). Soil pH, complexing ligands and redox potential, affects the availability of iron for crops (Beckwith et al. 1975). Fe exists mainly as oxidized Fe^{+3} in mineral soil under aerobic environment and slightly soluble if soil pH is more than 7 (Marschner 1995). Soils with pH above 5.5–6.5 cannot maintain inorganic Fe^{+3} oxides to the level of soluble Fe. Due to high pH and calcareousness, Pakistani soils promote the precipitation of Fe^{+3} oxides which are insoluble and not available to plants.

Phytic acid and polyphenols in food act as anti-nutrients and are considered to be the major inhibitors of iron bioavailability while ascorbic acid, inulin, garlic and onion have been reported to be the major enhancers of iron bioavailability (Fairweather-Tait et al. 2005; Scholz-Ahrens and Schrezenmeir 2007; Gautam et al. 2010). Fe bioavailability inhibitors such as polyphenols and phytate inhibit iron absorption but it was concluded that their inhibitory effect on iron absorption can be limited by increasing iron content (Carlson et al. 2012; Tako and Glahn 2011; Tako et al. 2013). Ferritin, an iron storage protein, can deposit thousands of iron atoms asnon-toxic form and ferritin iron is bioavailable to humans as ferrous sulphate (Lonnerdal 2007; Arosio et al. 2009). Improving both concentration and bioavailability of Fe in cereal grains is, therefore, an important challenge and a high-priority research area (Bouis and Welch 2010; Cakmak et al. 2010a). Hence, a variety of approaches are being used including supplementation, fortification, , food modification/diversification and biofortification to combat iron deficiency in humans (Frossard et al. 2000; Winichagoon 2002; Bouis et al. 2011)

A lot of problem has been reported with supplementation as Fe tablets cause's adverse side effects in humans (Winichagoon 2002; Mimura et al. 2008). Fortification of common foods with Fe, is also not effective because poor Fe bioavailability from the iron-fortified foods, while also changes in product taste may cause resistance by consumers to the fortified products (Frossard et al. 2000; Powell et al. 2013). Iron deficiency could also be mitigated by enhancing grain Fe concentration and bioavailability of Fe in grains, such as wheat and rice (Bouis and Welch 2010). In this regard, biofortification, which is a process of enhancing the bioavailable nutrient in the edible portion of crops is considered as the most suitable approach (Mayer et al. 2008; Bouis et al. 2011).

Approaches for Fe biofortification of crops such as genetic engineering, agronomic, transgenic and plant breeding have been developed (Cakmak et al. 2010a; Wei et al. 2012). It is reported by Bashir et al. (2013) that plant breeding has failed so far in developing Fe biofortified polished rice. On the other hand transgenic verities that sometime may provide more nutrients than genotype selection but many countries have strict regulations in commercialization of these transgenic verities (Saltzman et al. 2013). Biofortification of food crops with Fe through agricultural approach is a widely applied strategy (Cakmak et al. 2010a). Agronomic biofortification of food crops is considered as most sustainable approach.

In high pH calcareous soil agronomic biofortification will be ineffective unless we lower down the soil pH. Because the major issue in calcareous soil is quick transformation of more soluble Fe compounds to less soluble oxides and hydroxides. Hence, to reduce its rapid transformation and to increase its availability soil pH manipulation using some acidifying material could be a useful approach (Malakouti and Gheibi 1988). Different studies reported that microbial oxidation of elemental sulfur decreases soil pH and leads to mineral solubilization that increases mineral availability to crops (Kaplan and Orman 1998; Iqbal et al. 2012).

In calcareous soil, Fe biofortification with integrated use of Fe fertilizer and organic amendments in pH manipulated calcareous soil are new and have not been estimated. Therefore, present review focus on Fe biofortification in cereals with different strategies. There is a dire need to form new agriculture policies that not only meet the food requirements of the growing world but also fulfill nutritional requirements (Welch and Graham 2004). Cereals grown in calcareous soil are poor in essential micronutrients mineral contents i.e. Fe, Zn causing severe threat to population. Here in this chapter we discussed about the status of soil Fe, its significance in plant and human, strategies to combat Fe deficiency in soil, plant and ultimately in humans.

2 Status and Forms of Fe in Soil: An Overview

The predominant forms of Fe in soil are hematite, goethite, leidocrocite, ageetite, maghemite and ferrihydrite (Schwertmann and Taylor 1989). Each of these oxides contains Fe^{+3} form iron except magnetite that contains Fe (II) as well as Fe (III).

These exist predominately in clay-size fraction of soil but magnetite can exist in silt and sand size fraction of soil. In arid and semi-arid region goethite and hematite are the dominant minerals. While in slightly weathered or reduce soil magnetite in dominant form of Fe. Soils with the changing redox condition, ferrihydrite is likely to exist in most of the soils in small fraction but as a major component. All these mineral oxides found in well crystalline structure, but ferrihydrite shows short order of crystal structure with particle size of $\leq$10 nm. Particle size and surface reactivity of mineral oxides are the major factors that influence its solubility and availability. Schwertmann (1991) gave the stability order of different mineral oxides as follows: hematite = goethite > lepiddocrocite = magnetite > ferrihydrite. Iron is an important component of layer silicate in soil (20–50 g kg^{-1}) but not available for plant growth. In acid soil pH <4.5, some fraction of Fe is present as exchangeable ion. As primary or secondary minerals Fe exists as pyrite, amphibolite, pyroxenes and siderite. These are mostly present in reduce or relatively less weathered soils.

3 Iron Deficiency in Calcareous Soil

Soil pH is most important factor that determined nutrient availability. Most suitable pH for crops is around 6.0–6.5. Only few crops can tolerate high pH. Most of the nutrients are available at 6.5–7.5 pH. High pH soil make most micronutrients unavailable thus cause micronutrient deficiency in crops. Micronutrient deficiency is widespread in high pH calcareous soils. In calcareous soil HCO_3 and CO_3 make different complexes with Fe, P and Zn thus reduce availability of mineral nutrients. Calcium carbonate provides a reactive surface which acts as a sink for protons during acid/base reactions involving dissolved Fe species in the soil solution (Loeppert 1988; Abd El-Haleem 1996). Singh and Dahiya (1975) found that chemically available iron was decreased with increasing of $CaCO_3$ and time of incubation. They also reported that increasing levels of $CaCO_3$ to soil resulted in a decrease in some forms of iron (exchangeable, available); the decrease in exchangeable iron was probably from the release of Ca^{2+} from hydrolysis of $CaCO_3$. The decrease in the other forms of iron might be due to oxidation of soluble-native and added iron through direct reaction with $CaCO_3$. Singh and Dahiya (1975) found that excessive amounts of $CaCO_3$ decreased the availability of Fe to plants and it is often necessary to supply additional Fe to the soil.

Iron makes complexes with organic matter i.e. humic substances that are influenced by redox condition and pH. At low pH, Fe (III) complexes with organic matter are present (Goodman 1987) but at high pH values greater than 6, humic substances make less significant because of hydrolysis of Fe and precipitation of Fe oxides. In calcareous soil where pH is high Fe^{+3} is not retained by soil humates against hydrolysis. In pH range of 7.5–8.5 the solution Fe^{+3} concentrations remains are minimum (Lindsay 1979). In calcareous soil as values of redox potential and pH > 0.75 except under condition of flooding or saturation of soil pores for longer period of time therefore total Fe concentration is insufficient to meet the requirement of plants. Hence, plant and microbes evolved different mechanisms i.e. exudation of protons, reducing agents, organic acids and phytosiderophores to cope with this Fe limiting conditions.

Soil deficiency of nutrient leads to nutrient deficient plants (Cifuentes and Lindemann 1993). Iron is 4th most abundant element present on earth crust. But its availability is limited in most soils. Iron changes its oxidation state very quickly. Celik and Katkat (2007) studied that in high pH ferrous ion converts into insoluble ferric ion. High lime contents in calcareous soil also decrease Fe uptake. Redox conditions are also detrimental to Fe limitation. Mengel and Kirkby (1987) reported that Fe solubility and availability is pH dependent, it is also reported that Fe^{+3} solubility and activity decease 1000 time with every one unit increase in pH (Latimer 1952).

4 Strategies to Overcome Iron Deficiency in Soil

In cultivated calcareous soil, the potential of applied Fe depends upon the solubility of applied compound and the capacity of plant roots to assimilate Fe from these compounds (Garcia-Mina et al. 2003). Therefore different strategies are adopted to maintain available Fe status of soil.

4.1 Sulfur Effect in Lowering of Soil pH and Fe Solubilization

Decreasing soil pH is considered, as an effective way to deal with the stabilization of nutrients in calcareous and alkaline soil. As phosphorus, iron and zinc in calcareous and alkaline soil due to high pH and high concentration of calcium ions, become unavailable (Deluca et al. 1989; Tisdale et al. 1993; Kaplan and Orman 1998).

Common methods for dealing with these deficiencies, is the use of chemical fertilizers that in addition to the high cost and low efficiency, also have the risk of environmental pollution (Malakouti et al. 1991).

Different chemical fertilizers have been used in order to lower down the soil pH. Any fertilizer that contains ammonium or produces ammonium can reduce the pH i.e. ammonium sulfate, ammonium phosphate aluminum sulfate, calcium sulphate and potassium sulphate (Extension Education Center 423 Griffing Avenue, Suite 100 Riverhead, New York 11901-307). Along with these fertilizers elemental sulfur is considered as cost effective approach in lowering of soil pH. Because these chemical fertilizers have some constrains, i.e. aluminum sulfate is more expensive also the amount of aluminum sulfate needed to achieve the same decrease in pH is 6 times the amount of elemental sulfur required also aluminum sulfate is toxic to crops like blueberries. Gypsum and potassium sulfate can also be used to lower pH but it is documented that sulfate – sulfur containing fertilizers are not very effective in decreasing soil pH. Ferrous sulfate can also be used but it is cost effective and the total amount of ferrous sulfate is 8 times more than the amount of elemental sulfur needed. Thus, elemental sulfur is most common and cost effective acidify matter (Tisdale et al. 1993) that after oxidizing each mole producing two moles of hydrogen ions (H^+) in the soil and reducing soil pH that leads to dissolution nutrients in the root (Modaihsh et al. 1989; Kaplan and Orman 1998; Besharati and Salehrastin 1999). The capacity of oxidized sulfur and produce sulfuric acid reduce the soil pH. The oxidation of sulfur has a significant positive correlation with soil pH (Lawrence and Germida 1988; Kaplan and Orman 1998).

4.2 Water Soluble, Exchangeable and DTPA Extractable Fe

Extraction of soil with neutral (i.e. $CaCl_2$, $MgCl_2$ or KCl) or buffered (e.g. $NH_4C_2H_3O_2$) salt solution will result in displacement of both solution and exchangeable Fe. Acid buffered extraction may not be suitable for calcareous soil as it may results in considerable release of carbonates and oxides along with release of Fe from mineral structure. The extractability of Fe can be increased by using reducing agents (reductive dissolution). However these extractants are not that much in practice.

4.3 Use of Synthetic Fe Chelates

Use of synthetic chelates may help in combating Fe deficiency in plants (Pestana et al. 2003). Fe cheaters increase Fe availability in soil by increasing its concentration in soil solution, by increasing its diffusion and rapid replenishment of depleted zone and by providing ease to uptake by roots (Lindsay 1995). Chelates are very effective in improving Fe status of alkaline and calcareous soil but due to high stability constant root may impair and may take up poorly available Fe (Rodrıguez-Lucena et al. 2010). Another constrain in applying chelates in soil is applied chelates may leach down where frequent irrigation is applied. Most frequently used iron chelates are ethylenediaminetetraacetic acid (ETDA), and diethylinetriaminepentaacetic acid DTPA are characterized by low stability in soil while ethylenediamine-di(o-hydroxy-p-sulphoxyphenylacetic) acid (EDDSHA), ethylenediamine-di(o-hydroxy-p-methylphenylacetic) acid (EDDHMA), ethylenediamine-di(o-hydroxy-phenylacetic) acid (EDDHA) can be used as soil application as these are having high stability constant. But Fe chelates are expensive and out of reach of most of the farmers in developing countries (Akinrinde 2006).

4.4 *Vivianite*

Soil application of synthetic Fe-phosphate $Fe_3 (PO_4)_2.8H_2O$ has shown significant results in supplying of Fe over time. Fe (II)-phosphate is analogous to mineral vivianite. Vivianite is cheap compound that can be easily made by mixing ferrous sulphate and di-ammonium phosphate or mono-ammonium phosphate and vigorous shaking. Del Campillo et al. (1998) and Rosado et al. (2000) found vivianite as an effective compound to combat Fe deficiency in plants that are grown in calcareous soil.

5 Significance of Iron for Plants

Iron plays a vital role in variety of physiological and metabolic functions not only in plants but also in human body. Iron is a transitional metal and does change its oxidation state very rapidly thus it can be used as cofactor for many important processes like oxidative phosphorylation, electron transfer and DNA synthesis (Hass et al. 2005). Fe act as catalyst of chlorophyll formation, important component of cytochrome, involve in nitrogen fixation and component of ferridoxin. Ferridoxin (iron-sulfur protein)act as an electron transmitter in many basic metabolic processes (Marschner 1995). Iron also plays its role in some degradation processes i.e. reactions of peroxides. Being a part of heme protein, Fe plays the key role in hemoglobin formation, oxygen transport via hemoglobin. Fe is also important for binding of oxygen to RBCs, formulation of cytochrome and myoglobin, brain development and function, contraction and relaxation of muscles (Başar 2005).

6 Severity of Iron Deficiency in Crops

Iron is key component of chlorophyll ring structure. Any change in Fe availability leads to major alteration in overall plant metabolism. Iron deficiency cause yellowing of young leaves. Most common visible Fe deficiency symptom is leaf chlorosis and soil calcareousness favors this (Tagliavini and Rambola 2001). Plant Fe deficiency is common in many regions (Wiersma 2005). In severe conditions interveinal chlorosis cause serious damage to crops. Leaves yellowing result in poor photosynthetic activity. In Fe limitation crop growth will reduce. Plant life cycle become slows leading to reduce yield. Susceptibility to disease also increase in Fe limiting conditions (Rashid and Ryan 2004; Chatterjee et al. 2006).

Severe yield losses also occur in Fe deficient condition. It is estimated that soybean yield loss in USA is about 300.000 tons a year (Hansen et al. 2004). Peach production was reduced upto 20–30 % due to interveinal chlorosis (Başar 2003a). Total Fe content might be higher in chlorotic leaves (Marschner 1995) but DTPA extractable Fe contents are low that are important for completion of life cycle (Katkat et al. 1994; Başar 2000). Yields of iron deficient crops can, in principle, be increased through application of iron to soils. However, as the uptake of iron from soils is highly complex, improving crop yields through fertilization with iron has been shown to be difficult (Schulte and Kelling 2004). For example, application of iron to soils in the form of ferrosulfate ($FeSO_4$) has generally resulted in at most limited effects on crop yields (Frossard et al. 2000). Other forms in which iron might be added to soils (e.g. as chelates) are possibly more effective, but also expensive, and generally too costly for use on low value staple crops (Akinrinde 2006).

7 Strategies to Overcome Fe Deficiency in Plants

7.1 Soil pH Manipulation and Fe Bioavailability

Nutrients like Fe, Zn and P become deficient at high pH soil. Soil acidity favors the solubilization of mineral cations in soil with high calcium carbonate contents. Malakouti and Gheibi (1988) reported that consumption of sulfur in calcareous soil and with neutralizing lime increased solubility and availability of iron.

Wu et al. (2014) conducted an experiment to check that sulfur has effect on iron accumulation using different levels of sulfur. Results indicated that concentration of iron was increased in rice and then decreased by increasing sulfur concentration. This study suggested that sulfur application may improve Fe contents in rice when cultivated in low S contents soils, while Fe contents may decreased in rice with S inputs (fertilizers, atmospheric deposition) in high-sulfur soils. Similar studies were conducted by Heydarnezhad et al. (2012) and Kavamura et al. (2013) to investigate nutrient i.e. Fe, Zn and P concentration in calcareous soils. Results showed that sulfur application increased the concentration of Fe, Zn and P in soil. This study suggested that sulfur application not only increased concentration of Fe, Zn and P but also increased solubility of Fe, Zn and P by soil sulfur oxidizing microbes. The acidity produced during S oxidation increases the availability of nutrients such as P, Fe, Mg, Mn, Ca, and SO_4 in soils (Lindemann et al. 1991).

Experiments have demonstrated that S in soil affects Fe uptake in rice because S can regulate the formation of Fe plaque on the root surface of rice (Hu et al. 2007; Gao et al. 2010; Fan et al. 2010), influence Fe uptake by rice (Liu and Zhu 2005), and influence the formation of phytosiderophore, which is closely linked with Fe uptake by plants (Cao et al. 2002; Jin et al. 2005). Sulfur can increase the Fe transport in plants xylems (Hu and Xu 2002; Na and Salt 2011) and phloem, as well as accelerate the activation of deposited Fe in the apoplast (Holden et al. 1991; Toulon et al. 1992). Previous studies showed that S supply can increase the accumulation of Fe in rice seedlings (Min et al. 2007). Hassan and Olson (1966) suggested that applied sulfur directly increased the amount of Fe and Mn removed from the neutral and calcareous soils by production and consumption of sulphides.

7.2 Injection of Ion Salts

Application of Fe salts in liquid form ($FeSO_4$ and ammonium citrate) has been injected in to plant xylem vessels and has significant results in reducing iron induced chlorosis in fruits like pear, kiwifruit, peach, olives and apple (Wallace and Wallace 1986; Wallace 1991). Application of Fe as bullets into trunks by making holes is also an effective and long lasting (2–3 years) to cure Fe chlorosis (Wallace 1991). However, it may cause phytotoxicity when iron concentration and injection times are wrongly chosen.

7.3 Blood Meal

Use of blood meal is also considered as an effective approach to combat Fe chlorosis in trees (Taglivaini et al. 2000). Blood meal is a byproduct of slaughter house and an excellent source of Fe for plants. Effectiveness of blood meal was investigate by Kalbasi and Sharimatmadari (1993), they found the application of blood meal proved to be effective as Fe source. Blood meal contain Fe ranges from 20–30 mg kg^{-1}. In blood meal Fe is found as ferrous sulphate and in complex with heme group of hemoglobin.

7.4 Foliar Application of Fe

Under field condition acidic solution sprays (e.g. citric, ascorbic and H_2SO_4) are proved be effective to re-green Fe chlorotic leaves without applying exogenous iron (Aly and Soliman 1998; Taglivaini et al. 1995). There was a decrease in apoplastic pH and re-greening of Fe chlorotic leaves by applying citric and sulfuric acid (Kosegarten et al. 2004). But there are some constrains in adopting this techniques as studied by Taglivaini et al. (2000).

8 Organic Amendments and Nutrient Availability

Organic matter also supplies organic chemicals to the soil solution that can serve as chelates and increase metal availability to plants, providing metal chelates and increasing the solubility of nutrients in soil solution (Du Laing et al. 2009; McCauley et al. 2009).

8.1 Animal Manure

Animal manure has been used for many years alone and in combination with chemical fertilizers. It is observed that its effectiveness was enhanced when applied with mineral fertilizers than its separate application. Animal manure has ability to dissolves soil insoluble organic compounds. Incubation of Fe salts with organic manure can improve efficiency of Fe sources before application (Taglivaini et al. 2000).

Chemical composition of poultry manure revealed that it has high concentration of nitrogen compared to various other organic amendments (Bujoczek et al. 2000) and high nitrogen contents change the dynamics of Fe contents in wheat as activity of YS1 protein needed for Fe transport was promoted in root cell membrane (Murata et al. 2008; Curie et al. 2009).

With increasing trend of people towards poultry industry disposal of poultry waste is becoming an issue of great concern. Poultry manure is used as source of organic fertilizer as it contains many nutrients (Moore et al. 1995). It also contains many secondary elements, micronutrients or some heavy metals (Gupta and Gardner 2005). Fraction of plant available nutrients can be changed by applying manure as manure can change the soil biota and physical properties of the soil (Demir et al. 2010). It was reported by Zhou et al. (2013) that animal manure not only improves microstructure of soil but also improves soil aggregation (de Cesare Barbosa et al. 2015).

8.2 Compost

A sustainable approach to manage municipal waste is the use of compost made from municipal solid waste (Aggelides and Londra 2000; Soumare et al. 2003a). Physicochemical properties of soil and plant nutrient status can be improved by use of compost because mature compost contains plant nutrients. Application of compost improved readily available Fe, Zn and Cu contents soil (Soumare et al. 2007). Application of compost influences the nutrient dynamics, due to change in physicochemical condition of soil nutrient mobility and bioavailability; availability of few nutrients can increase or vice versa (Gardiner et al. 1995). Upon decomposition, release of mineral elements from organic complexes also increases availability of mineral nutrients (Dudley et al. 1986; Nyamangara 1998). However, complete knowledge of mineral behavior should be understood before application of compost in order to maintain a sustainable and reliable agro-ecosystem.

8.3 Biochar

Biochar affect mineral forms of Fe by acting as an electron shuttle in redox-mediated reactions (Kappler et al. 2014). Recently, Graber et al. (2014) noted how the redox catalytic activity associated to biochar solubilized Fe from a sandy soil, increasing the metal release with decreasing pH. One of the main mechanisms proposed to justify the benefits of biochar is its positive impact on the availability of soil nutrients (Xu et al. 2013). Direct effects of biochar on soil fertility have been mainly related to the presence of nutrients in mineral form on biochar surface (Kimetu et al. 2008). However, indirect effects of biochar on soil fertility are to change soil physico-chemical and biological properties (such as pH, redox conditions, porosity, water retention capacity, and biotic interactions at the rhizosphere), leading to nutrient mobilization (Lorenz and Lal 2014; Ngo et al. 2014; Jeffery et al. 2015). In addition, changes in soil properties induced by biochar may enhance root growth these changes can constitute an effective mechanism for enhancing nutrient mobilization and uptake in the rhizosphere via increasing the exploratory capacity of the root system and modifying nutrient solubility (Lehmann et al. 2011).

9 Iron for Human Health

It is recognized that micronutrient deficiency causes harmful impact on public health (Black et al. 2008; Stein 2010). Iron scarcity causes fatigue, poor work performance, reduce immunity, deficient oxygen supply to RBCs and death. In most part of the world, iron deficiency particularly affects preschool children and women (Benoist et al. 2008). Mortality rate and overall burden of disease has increased due to micronutrient deficiency. Common periods of high iron demand include pregnancy, period of blood loss during surgery or iron demand due to insufficient iron absorption (Trost et al. 2006).

Rice crop is usually very low in iron contents as compared to recommended dietary allowance. Daily dietary dose is about 0.06 g day^{-1} for adult women with a low bioavailability of iron 5 % and 0.02 g day^{-1} with a high bioavailable iron, in developing countries (WHO 2004). According to UNSSCN (2010) about 88 % of all pregnant women of 16 years and 63 % of 5 and 14 years children are considered to be anemic in South Asia.

10 Strategies to Combat Iron Deficiency in Humans

In order to improve Fe supply, recently three strategies namely food diversification, supplementation and food fortification are in practice. These three are aim to improve Fe supply and bioavailability in food (Bothwell et al. 2004; McDonagh et al. 2015). Diversification of food increase intake and provision of Fe rich diet that is bioavailable to humans. On the other hand supply of Fe in the form of medicine and pills is supplementation. While either addition of bioavailable Fe or reduction of inhibitory effect of different compounds to most frequently consumed dietary products is categorize as food fortification. In the long run, biofortification of plant based food is most recent strategy to improve bioavailable Fe contents. All these approaches need certain conditions to be fulfilled.

10.1 Food Diversification

Food diversification aims to increases Fe contents of frequently consumed daily diet. Diversifications can improve Fe bioavailability in variety of foods i.e. fruits, vegetables and meat. But this has some limitations too: meat and fish that are rich in haem iron are quite expensive while fruits and vegetables rich in vitamin C are seasonal and available for short period of time only. Thus, making it bit difficult to enhance their intake.

10.2 Supplementation

A therapeutic approach that is utilizes either to treat or for prevention in severe micro nutrient deficiency is supplementation (Imdad and Bhutta 2012). In certain countries significant results have been shown by supplementation programs organized with health department. Vitamin A supplementation for the cure of night blindness and newborn mortality has been shown remarkable success. Fe-foliate supplementation to pregnant women is also a feasible approach and has been shown a positive impact on anemia. But in developing countries target is hard to achieve on daily compliance, also lack of proper infrastructure, disruption of stocks, lack of proper contact with target population and official health care department make it difficult timely supply (Bothwell et al. 2004). Another limitation to supplementation approach is not addressing to the root cause of malnutrition. It is a short term solution to malnutrition and nutrient deficiencies. Supplemented food shows variety of physiological and absorption responses of nutrients compared to nutrients find in food i.e. zinc, Fe and folic acid (Bailey et al. 2015). Iron supplemented food may not be the solution for iron malnutrition. Different trails on Supplementation and screening for iron-deficiency anemia (IDA) in young children reported that no clear benefits of supplementation were observed (McDonagh et al. 2015). It is also reported that supplemental Fe in human diet with higher doses may cause serious health hazards i.e. gastric problems, gastric upset, vomiting and nausea, faintness, abdominal pain or constipation (Murray-Kolb and Beard 2009; Aggett et al. 2012). Oxidative stress that lead to damage of cellular components due to lack of supply of some antioxidants also result due to Fe introduction to the diet (Ibrahim et al. 1997).

10.3 Fortification

A more long term strategy to overcome Fe malnutrition is fortification of food items. Fortification is meant for large number of population while supplementation is exposed for a certain group of individuals. Fortification is more beneficial where micro nutrient deficiency is widespread. However it need more time for implementation than supplementation. Fortification program needs some industrial engagement and policy matter. Food vehicle and amount of fortificant added is equally critical. Many food products are also sensitive to color or flavor changes and oxidative damage of some nutrients (Hurrell 2002). Recently published data from Powell et al. (2013) has reported that dietary fortified iron intake is negatively associated with quality of life in patients, probably as a result of low bioavailability as well as an antagonistic mechanism with other metals. The toxic effect of high doses of iron is also known (Golub et al. 2009).

10.4 Biofortification

Biofortification with Fe in staples provides an economical tool to reduce Fe malnutrition (Jeong and Guerinot 2008; Nagesh et al. 2012). Enrichment of crops with micronutrient before harvest is known as biofortification. Biofortification enhance micronutrients thus important tool to overcome micronutrient malnutrition (Brinch-Pedersen et al. 2007).

Biofortification is considered as long term solution to combat Fe malnutrition (Zimmermann and Hurrell 2002). Biofortification is only one time investment of money. For a healthy body a dose of 2–6 ppm is enough to improve Fe level (Hass et al. 2005). Successful biofortification needs the acceptance from consumers, adaptation by farmers and cost effectiveness.

Biofortified staple foods may not deliver equally high levels of minerals and vitamins per day, but they can increase micronutrient intake for the resource-poor people who consume them daily, and therefore complement existing approaches (Bouis et al. 2011). Cereals are rich in anti-nutrients, plant genome along with their growth conditions determine the level of these inhibitors (Hunt 2003).

11 Approaches for Iron Biofortification

For Fe biofortification of crops genetic engineering, agronomic, transgenic and plant breeding approaches have been developed (Bouis et al. 2011; Sperotto et al. 2012)

11.1 Breeding and Genetics Approaches

Breeding and genetics approaches have been used for many years to obtain such genotypes that are rich in micronutrients. Aung et al. (2013) and Masuda et al. (2012, 2013) studied three combined approaches to biofortify rice grains. Results indicated combination of genes is involved in Fe homeostasis that can be used to enrich rice grain with Fe. A successful Fe biofortified rice and vitamin A-fortified named "Golden Rice" was introduced by Goto et al. (1999) and Ye et al. (2000). Conventional and modern plant breeding and biotechnological approaches suggested Fe contents in rice are a genotypic character that is significantly different for different genotype hence, new Fe enrich varieties can be screened or bred (White and Broadley 2005; Wen et al. 2005).

But these approaches are not always very successful either because of some environmental and genotypic interactions or there may be lack of target genome (Palmgren et al. 2008; Zhao 2010) Source and sink strategy were prioritized in order to biofortify rice grain with Fe and Zn (Wirth et al. 2009; Masuda et al. 2013). Traditional breeding efforts to biofortify polished rice have not proven that much effective as there are limited variations in Fe contents. Over 20,000 rice accessions from Latin America Asia and Caribbean were evaluated for Fe and Zn contents. It revealed that maximum concentration was only 8 mg kg^{-1} in polished grains (Graham 2003; Martínez et al. 2010). It is reported by Bashir et al. (2013a) that plant breeding has failed so far in developing Fe biofortified polished rice.

11.2 Transgenic Approaches

Goto et al. (1999) was the first to explore transgenic approaches in order to enrich Fe in endosperm over a decade ago. Since then, countless efforts have been made to improve grain Fe contents by Fe homeostasis gene expression that either increase the Fe uptake from soil ultimately accelerate Fe translocation from root, shoot to grains, or by improving efficacy of Fe storage protein (Kobayashi and Nishizawa 2012; Lee et al. 2012). Studies also suggested that stability of selected trait over number of plant generations nevertheless is still a challenging task, furthermore, to motivate the farmers, adoption and consumers, acceptance. Oliva et al. (2014) introduced an *indica* variety with phytoferritin over expresser events without selectable marker genes; however, the level of Fe was not sufficient to reach the target. Transgenic verities that sometime may provide more nutrients than genotype selection but many countries don't allow commercialization of these transgenic verities (Saltzman et al. 2013).

11.3 Soil and Crop Management

Rice grain Fe contents are regulated by soil and other environmental factors (Barikmoa et al. 2007; Zuo and Zhang 2011). Several sources of micronutrients can be used such as inorganic salts, natural organic polymers and synthetic cheaters. Foliar application of micronutrients is considered as very effective as it requires fewer amounts of fertilizers and quick response crop response than soil application (Mortvedt 2000).

Synthetic Fe chelates are also considered as an effective approach to biofortify crops with Fe and are used both in soil and foliar application. There initial cost may be prohibitive but these are proved cost effective for the high value crops (Fageria et al. 2002). Studies suggested that foliar application of Fe was effective in increasing Fe contents of wheat in arid climate (Habib 2009; Pahlavan-Rad and Pessarakli 2009), but foliar application remained ineffective in humid areas. Pahlavan-Rad and Pessarakli 2009) and Habib (2009) evaluated the effectiveness of complex micronutrients application as foliar sprays and suggested that complex micronutrient foliar application is superior to single application as wheat grain concentration of Fe and Zn were improved by complex micronutrient foliar application.

Application of organic amendments, such as farmyard manure, increases nutrients concentration, improve nutritional quality and enhance nutrient balance of crops (Graham et al. 2001). On decomposition of organic matter different organic acids i.e. oxalic, phenolic, citric and malic are released. These organic acids form complexes with Fe hence enhance its mobility and bioavailability (Lindsay 1995). Most recent approach is to enhance bioavailable Fe contents while reducing phytate contents and to increase total Fe content, but these are not that much practical at this time (Raboy et al. 2000; Hurrell et al. 2003).

Most Fe biofortification studies were conducted under favorable glasshouse conditions, with only limited studies performed under field conditions (Masuda et al. 2008, 2012).

12 Nutritional Factors Affecting Fe Bioavailability

Apart from high pH and alleviated lime contents there are some other factors that affect Fe bioavailability. Among these factors phytic acid and poly-phenolics are most important.

12.1 Phytic Acid (Phytate)

Phytate is stored form of seed phosphorous deposited during seed development (Doria et al. 2009). Phytic acid act as binding agent in intestinal tract of human as it makes strong bonding with Ca, Zn, Fe and other essential mineral elements during digestion (Garcia et al. 1999). Anti-nutrient phyate reduces bioavailability of important nutrients and cause micronutrient malnutrition (Welch 2002). The main challenge is to reduce phyate contents to assure maximum ferritin concentration. It is the only way we can enrich crops with micronutrients like Fe. Total Fe contents are of no meaning unless we decrease phytic acid concentration that limits its bioavailability.

12.2 Polyphenol

Like phytate, polyphenol is also considered as antinutrient that interacts with essential mineral contents of food and make them unavailable for absorption (Idris et al. 2006; Abd El Rahaman et al. 2007). Sharma and Kapoor (1997) studied that nutrient absorption by human body was significantly influenced by polyphenols and phytate present in pearl millet. Studies also suggested that polyphenols act as chelating agent that effect Fe bioavailability by forming insoluble complexes (Hurrell and Egli 2010).

Many cereals contain sufficient quantity of polyphenols i.e. maize. Liyana-Pathirana and Shahidi (2005) reported that food digestion enhances the antioxidant capacity of cereals and cereal based food. The solubility and functionality of polyphenols present in cereals increases in stomach and duodenum.in vitro digestion studies showed that the amount of antioxidants released by the array of cereals in the human gut may be higher than expected (Perez-Jimenez and Saura-calixto 2005). In plant cell wall, lignin is present that is known to have polyphenolic properties. About 30 % of plant biomass and 3–7 % of bran is made up of lignin. Lignin compounds were considered to be inert during digestion but, their polyphenolic structure gives them antioxidant properties (Fardet et al. 2008).

Del Pozo-Insfran et al. (2006) evaluated the varietal difference in antioxidant fraction. He demonstrated that Mexican purple maize showed a significantly higher antioxidant capacity than American purple and white varieties. However, these were attributed to the specific anthocyanins and/or the composition of polyphenols in the plants.

Thu, fermentation, malting, sprouting, soaking and cooking have long been documented by many researchers in order to lower down antinutrient concentration (Lewu et al. 2010; Osman 2011) but such information still needs more investigation.

12.3 *Ferritin*

Ferritin is a stable iron storage protein consisting of a 24- subunit shell around a 4500-atom iron core (Theil and Briat 2004). It is reported that Ferritin doesn't form complexes with other cations thus increase iron availability to humans. Ferritin is stable protein and doesn't denature in human elementary tract (Theil et al. 2001; Murray-Kolb et al. 2003).

In most seeds ferritin content ranges from 8 to 80 μg/g of seed. Several studies suggested that rice, wheat and corn have low bioavailable Fe contents (May et al. 1980) while nodule forming crops are rich in ferritin concentration (Ambe et al. 1987). Ferritin-Fe contents are bioavailable to human and source of iron for completion of human life cycle (Lonnerdal 2007). Iron stored in ferritin is completely bioavailable (Goto and Yoshihara 2001).

Ferritin is present in all crops but differing in concentration. Biofortification also aims to enhance ferritin concentration of crops. Ferritin binds to free radicals that are damaging to cells.

Ferritin also acts as temporary storage form of Fe that is available in Fe limiting conditions (Briat et al. 2010a). It is reported that ferritin is the key component in alleviating oxidative stress (Mata et al. 2001). The main function of ferritin in seeds is protection against free iron damage through Fenton oxidation (Briat et al. 2010b).

13 Models Used for Determination of Iron Bioavailability

The bioavailability and iron absorption from the daily diet are influenced by the type and quantity of iron present in food as well as by the presence of inhibitors and promoters of iron absorption in the diet and the individual's iron status (Duque et al. 2014). The urgency of addressing iron deficiency stems from its implication in a number of health conditions, some serious or even fatal. Rats model are most frequently used for testing of the effects of agents that are toxic or potentially hazardous to humans. This also refers to the toxicity of metals and a possible preventive and therapeutic effect (Brzóska et al. 2012; Al-Rejaie et al. 2013). It was observed by Zielińska-Dawidziak et al. (2012) that in iron deficient rats, decreased level of hemoglobin (Hb), hematocrit (HCT), mean corpuscular volume (MCV), mean corpuscular hemoglobin (MCH), mean corpuscular hemoglobin concentration (MCHC), serum and liver ferritin were increased to normal values or better after feeding ferritin isolate. Expression of soybean ferritin in rice resulted in Fe bioavailability similar to that of ferrous sulfate fortified rice when evaluated in a rat hemoglobin repletion model (Murray-Kolb et al. 2002) and human lactoferrin produced in rice had bioavailability similar to that of ferrous sulfate in young women (Lonnerdal et al. 2006). In mice iron biofortified rice feeding test and Caco-2 cell model confirmed that metal bioavailability in rats and humans increases with the increased level of metals in rice grains (Zheng et al. 2010; Lee et al. 2012). Recently it was shown that biofortification of rice with Zn significantly increases Zn uptake in Caco-2 cell model as well as in rat pups, and is suggested to have similar effect in human populations (Jou et al. 2012).

14 Conclusion

In the last 40 year agriculture research has focused only on produce but not on quality of produce that leads the farmers only to maximize profit by adopting latest production technology for fewer crops especially in 3rd world countries. This inclusion of cash crops and high yielding varieties in cropping scheme resulted in micronutrient malnutrition i.e. (Fe, Zn) and less food diversity. According to an estimate every 3rd person in the world is suffering from hidden hunger due to essential nutrients and vitamin deficiency resulting in poor health conditions.

Calcareous soils are one of the main reasons in increasing micronutrient (especially Fe) malnutrition problems. High pH and high HCO_3 contents are two most important factors that limit Fe availability in calcareous soil. Iron is not deficient in mineral soil but due to high pH and calcareousness it is merely available to plants because at high pH Fe^{+2} ions converted into less soluble Fe^{+3} oxides and hydroxides and disappear from the soil solution. Iron fertilization may not effective in calcareous soil due to their rapid conversion Mixing inorganic salts of micronutrients with different organic materials can enhance the efficacy of micronutrients. Soil acidity favors the solubilization of mineral cations in soil with high calcium carbonate contents. Thus, elemental sulfur is most common and cost effective acidifies matter. Microbial oxidation of sulfur leads to dissolution of nutrients by releasing H^+. To combat these limitations we developed a strategy in order to increase Fe solubility in soil solution and availability for plant uptake using some acidifying agent in calcareous soil. Phytate and polyphenols are major inhibitors of Fe present in cereals, that reduce Fe bioavailability that leads to anemia in both developed and developing countries; consequence is more in developing countries. Along with anemia Fe deficiency results in increased susceptibility to infection, mental and psychomotor retardation in children, impaired immune system. On the other hand ferritin is a stable iron storage protein present in cereals grains and doesn't form complexes with other cations thus increase iron availability to humans. Iron stored in ferritin is completely bioavailable because it doesn't denature in human elementary tract. Conventionally both, short term i.e. supplementation, fortification, food diversification and long term solutions i.e. breeding, genetic engineering and agronomic approaches have been purposed in order to combat Fe deficiency in crops and in humans. As compare to these approaches biofortified crops may not provide high level of nutrients as per day demand but it is a promising approach to provide adequate micronutrients throughout the life cycle. It is a cost effective approach that sustainably provide minerals and vitamins to target population.

References

Abd El Rahaman SM, El Maki HB, Idris WH, Hassan AB, Babiker EE, El Tinay AH (2007) Antinutritional factors content and hydrochloric acid extractability of minerals in pearl millet cultivars as affected by germination. Int J Food Sci Nutr 58:6–17

Abd El-Haleem AA (1996) Iron availability as affected by $CaCO_3$ Particle size and Fe:P ratio in Fe-P-containing compounds. Egypt J Appl Sci 11:308–327

Aggelides SM, Londra PA (2000) Effects of compost produced from town wastes and sewage sludge on the physical properties of a loamy and a clay soil. Bioresour Technol 71:253–259

Aggett PJ, Hathcock J, Jukes D, Richardson DP, Calder PC, Bischoff-Ferrari H, Nicklas T, Mu¨hlebach S, Kwon O, Lewis J, Lugard MJF, Prock P (2012) Nutrition issues in Codex: health claims, nutrient reference values and WTO agreements: a conference report. Eur J Nutr 51:1–7

Akinrinde EA (2006) Issues of optimum nutrient supply for sustainable crop production in tropical developing countries. Pak J Nutr 5:387–397

Alcantara E, Romera FJ, Canete M, Guardia MD (2002) Effect of bicarbonate andiron supply on Fe (III) reducing capacity of roots and leaf chlorosis of the susceptible peach rootstock nema guard. J Plant Nutr 23:1607–1617

Al-Rejaie SS, Aleisa AM, Sayed-Ahmed MM, Al-Shabanah OA, Abuohashish HM, Ahmed MM, Al-Hosaini KA, Hafez MM (2013) Protective effect of rutin on the antioxidant genes expression in hypercholestrolemic male Westar rat. BMC Complement Altern Med 13:136–144

Aly SSM, Soliman SM (1998) Impact of some organic acids on correcting iron chlorosis in two soybean geneotypes grown in calcareous soil. Nutr Cycl Agroecosyst 51:185–191

Ambe S, Ambe F, Nozuki T (1987) Mossbauer study of iron in soybean seeds. J Agric Food Chem 35:292–296

Arosio P, Ingrassia R, Cavadini P (2009) Ferritins: a family of molecules for iron storage, antioxidation and more. Biochim Biophys Acta 1790:589–599

Aung MS, Masuda H, Kobayashi T, Nakanishi H, Yamakawa T, Nishizawa NK (2013) Iron biofortification of Myanmar rice. Front Plant Sci 4:158. doi:10.3389/fpls.2013.00158

Bailey RL, West KP Jr, Black RE (2015) The epidemiology of global micronutrient deficiencies. Ann Nutr Metab 66:22–33

Barikmoa I, Ouattara F, Oshaug A (2007) Differences in micronutrients content found in cereals from various parts of Mali. J Food Comp Anal 20:681–687

Başar H (2000) Factors affecting iron chlorosis observed in peach trees in the Bursa region. Turk J Agric 24:237–245

Başar H (2003) Analytical methods for evaluating iron chlorosis in peach trees. Commun Soil Sci Plant Anal 34:327–341

Başar H (2005) Methods for estimating soil iron availability to chlorotic peach trees. Commun Soil Sci Plant Anal 36:1187–1198

Bashir K, Takahashi R, Nakanishi H, Nishizawa NK (2013) The road to micronutrient biofortification of rice: progressand prospects. Front Plant Sci 4:15. doi:10.3389/fpls.2013.00015

Beckwith RS, Tiiier KG, Suwadji E (1975) The effect of flooding on the availability of trace metals to tice in soils of differing organic matter status. Trac Elem Soil Plant Anim Syst 4:128–135

Benoist BD, Mclean E, Anderrson M, Rogers L (2008) Iodine deficiency in 2007: global progress since, 2003

Besharati H, Salehrastin N (1999) The effect of sulfur inoculated with Thiobacillus bacteria to increase P uptake. J Soil Water Sci 13:23–39

Black RE, Allen LH, Bhutta ZA, Caulfield LE, de Onis M, Ezzati M, Mathers C, Rivera J (2008) Maternal and child under nutrition: global and regional exposures and health consequences. Lancet 371:243–260

Bloom PR, Inskeep WP (1988) Factors affecting bicarbonate chemistry and iron chlorosis in soils. J Plant Nutr 9:215–228

Bothwell TH, MacPhail AP (2004) The potential role of NaFeEDTA as aniron fortificant. Food Chem Toxicol 74:421–434

Bothwell T, Cori H, Glahn R, Hertrampf E, Kratky Z, Miller D, Rodenstein M, Streekstra H, Teucher B, Turner E, Yeung CK, Zimmerman MB (2004) Enhancing the absorption of fortification iron – a sustain task force report. Int J Vitam Nutr Res 74:387–401

Bouis HE, Welch RM (2010) Biofortification: a sustainable agricultural strategy for reducing micronutrient malnutrition in the global south. Crop Sci 50:20–32

Bouis HE, Hotz C, McClafferty B, Meenakshi JV, Pfeiffer WH (2011) Biofortification: a new tool to reduce micronutrient malnutrition. Food Nutr Bull 32:31–40

Briat JF, Lobréaux S (1997) Iron transport and storage in plants. Trend Plant Sci 2:187–193

Briat JF, Fobis-Loisy I, Grignon N, Lobreaux S, Pascal N, Savino G, Thoiron S, Wiren NV, Wuytswinkel OV (1995) Cellular and molecular aspects of iron metabolism in plants. Biol Cell 84:69–81

Briat JF, Duc C, Ravet K, Gaymard F (2010a) Ferritins and iron storage in plants. Biochim Biophys Acta 1800:806–814

Briat JF, Ravet K, Arnaud N, Duc C, Boucherez J, Touraine B, Cellier F, Gaymard F (2010b) New insights into ferritin synthesis and function highlight a link between iron homeostasis and oxidative stress in plants. Ann Bot 105:811–822

Brinch-Pedersen H, Borg S, Tauris B, Holm PB (2007) Molecular genetic approaches to increasing mineral availability and vitamin content of cereals. J Cereal Sci 46:308–326

Brzóska MM, Gala˙zyn-Sidorczuk M, Dzwilewska I (2012) Ethanol consumption modifiesthe body turnover of cadmium: a study in a rat model of human exposure. J Appl Toxicol 33:784–798

Bujoczek G, Oleszkiewicz J, Sparling R, Cenkowski S (2000) High solid anaerobic digestion of chicken manure. J Agric Eng Res 76:51–60

Cakmak I, Pfeiffer WH, McClafferty B (2010) Biofortification of durum wheat with zinc and iron. Cereal Chem 87:10–20

Cao H, Han ZH, Xu XF, Zhang Y (2002) Iron nutrition in higher plants. Plant Physiol Commun 38:180–186

Carlson D, Nørgaard JV, Torun B, Cakmak I, Poulsen HD (2012) Bioavailability of trace elements in beans and zinc-biofortified wheat in pigs. Biol Trace Elem Res 150:147–153

Celik H, Katkat AV (2007) Some parameters in relation to iron nutrition status of peach orchards. J Biol Environ Sci 1:111–115

Chatterjee C, Gopal R, Dube BK (2006) Impact of iron stress on biomass, yield, metabolism and quality of potato (*Solanumtuberosum* L.). Scr Hortic 108:1–6

Cifuentes FR, Lindemann WC (1993) Organic matter stimulation of elemental sulfur oxidation in a calcareous soil. Soil Sci Soc Am J 57:727–731

Curie C, Cassin G, Couch D, Divol F, Higuchi K, Jean M, Misson J, Schikora A, Czernic P, Mari S (2009) Metal movement within the plant: contribution of nicotianamine and yellow stripe 1-liketransporters. Ann Bot Lond 103:1–11

de Cesare Barbosa GM, de Oliveira JF, Miyazawa M, Ruiz DB, Filho JT (2015) Aggregation and clay dispersion of an oxisol treated with swine and poultry manures. Soil Till Res 146:279–285

del Campillo M, Barron CV, Oglesias J, Dalmao R, Marce X (1998) Fertilization with iron (II) phosphate effectively prevents iron chlorosis in pear trees. 2–7 August 2008. In: Abstract XXV international horticulture congress ISHS, Bruxelles

Del Pozo-Insfran D, Percival SS, Talcott ST (2006) Acai (Euterpe oleracea Mart.) polyphenolics in their glycoside and aglycone forms induce apoptosis of HL-60 leukemia cells. J Agric Food Chem 54:1222–1229

Deluca TH, Skogley EO, Engle RE (1989) Band-applied elemental sulfur to enhance the phytoavailability of phosphorus in alkaline calcareous soils. Biol Fertile Soil 7:346–350

Demir K, Sahin O, Kadioglu YK, Pilbeam DJ, Gunes A (2010) Essential and non-essential element composition of tomato plants fertilized with poultry manure. Sci Hort 127:16–22

Doria E, Galleschi L, Calucci L, Pinzino C, Pilu R, Cassani E (2009) Phyticacid prevents oxidative stress in seeds: evidence from a maize (Zea mays L.) low phytic acid mutant. J Exp Bot 60:967–978

Du Laing G, Rinklebe J, Vandecasteele B, Meers E, Tack FM (2009) Trace metal behaviour in estuarine and riverine floodplain soils and sediments: a review. Sci Total Environ 407:3972–3985

Dudley LM, McNeazl BL, Baham JE (1986) Time-dependent changes in soluble organics, copper, nickel and zinc from sludge amended soils. J Environ Qual 15:188–192

Duque X, Martinez H, Vilchis-Gil J, Mendoza E, Flores-Hernández S, Morán S, Navarro F, Roque-Evangelista V, Serrano A, Mera RM (2014) Effect of supplementation with ferrous sulfate or iron bis-glycinate chelate on ferritin concentration in Mexican school children: a randomized controlled trial. Nutr J 13:71–80

Fageria NK, Aligar VCB, Clark RB (2002) Micronutrients in crop production. Adv Agric 77:85–268

Fairweather-Tait S, Lynch S, Hotz CH, Hurrell R, Abrahamse L, Beebe S (2005) The usefulness of in vitro models to predict the bioavailability of ironand zinc: a consensus statement from the harvest plus expert consultation. Int J Vitam Nutr Res 75:371–374

Fan JL, Hu ZY, Ziadi N, Xia X, Wu CYH (2010) Excessive sulfur supply reduces cadmium accumulation in brown rice (*Oryza sativa*L.). Environ Pollut 158:409–415

FAO (2015) The State of food insecurity in the world 2015. Meeting the 2015 international hunger targets: taking stock of uneven progress. FAO, Rome

Fardet A, Rock E, Remesy C (2008) Is the in vitro antioxidant potential of whole-grain cereals and cereal products well reflected in vivo? J Cereal Sci 48:258–276

Frossard E, Bucher M, Ma-chler F, Mozafar A, Hurrell R (2000) Potential for increasing the content and bioavailability of Fe, Zn, and Ca in plants for human nutrition. J Sci Food Agric 80:861–879

Gao MX, Hu ZY, Wang GD (2010) Effect of elemental sulfur supply on cadmium uptake into rice seedlings when cultivated in low and excess cadmium soils. Commun Soil Sci Plant Anal 41:990–1003

Garcia ER, Hernandez EG, Villanova BG (1999) Phytic acid content in milled cereal products and breads. Food Res Int 32:217–221

Garcia-Mina J, Cantera GR, Zamarreno A (2003) Interaction of different iron chelates with and alkaline and calcaresous soil: a complementaty methodology to evaluate the performance of iron compounds in the correction of iron chlorosis. J Plnat Nutr 26:1943–1954

Gardiner DT, Miller RW, Badamchian B, Azzari AS, Sisso DR (1995) Effects of repeated sewage sludge applications on plant accumulation of heavy metals. Agric Ecosyst Environ 55:1–6

Gautam S, Patel K, Srinivasan K (2010) Higher bioaccessibility of iron and zinc from food grains in the presence of garlic and onion. J Agric Food Chem 58:8426–8429

Gibson RS, Bailey KB, Gibbs M, Ferguson EL (2010) A review of phytate, iron, zinc, and calcium concentrationsin plant-based complementary foods used in low-income countries and implications for bioavailability. Food Nutr Bull 31:134–146

Golub MS, Hogrefe CE, Widaman KF, Capitanio JP (2009) Iron deficiency anemia and affective response in rhesus monkey infants. Dev Psychobiol 51:47–59

Goodman BA (1987) The characterization of Fe complexes with soil organic matter. pp. 677–687. In: Stucki JW, Goodman BA, Schwertmann U (eds) Iron in soil and clay minerals. Reidel, Dordrecht

Goto F, Yoshihara T (2001) Improvement of micronutrient contents by genetic engineering – development of high iron content crops. Plant Biotechnol 18:7–15

Goto F, Yoshihara T, Shigemoto N, Toki S, Takaiwa F (1999) Iron fortification of rice seed by the soybean ferritin gene. Nat Biotechnol 17:282–286

Graber ER, Tsechansky L, Lew B, Cohen E (2014) Reducing capacity of water extracts of biochars and their solubilization of soil Mn and Fe. Eur J Soil Sci 65:162–172

Graham RD (2003) Biofortification: a global challenge. Int Rice Res Notes 28:4–8

Graham RD, Welch RM, Bouis HE (2001) Addressing micronutrient malnutrition through enhancing the nutritional quality of staple foods: principles, perspectives and knowledge gaps. Adv Agron 70:77–142

Gupta G, Gardner W (2005) Use of clay mineral (montmorillonite) for reducing poultry litter leachate toxicity (EC50). J Hazard Mater 118:81–83

Habib M (2009) Effect of foliar application of Zn and Fe on wheat yield and quality. Afr J Biotechnol 8:6795–6798

Hansen NC, Jolley VD, Naeve SL, Goos RJ (2004) Iron deficiency of soybean in the north central US and associated soil properties. Soil Sci Plant Nutr 50:983–987

Hass JD, Beard JL, Murray-Kolb LE, del Mundo AM, Felix A, Gregorio GB (2005) Iron-biofortified rice improves the iron stores of non-anemic Filipino women. J Nutr 135:2823–2830

Hassan NA, Olson RA (1966) Influence of applied sulphur on availability of soil nutrients for corn nutrition. Soil Sci Am Proc 30:286–288

Heydarnezhad F, Shahinrokhsa P, Vahed HS, Besharati H (2012) Influence of elemental sulfur and sulfur oxidizing bacteria on some nutrient deficiency in calcareous soils. Int J Agric Crop Sci 4:735–739

Holden MJ, Luster DG, Chaney RL, Buckout TJ, Robinson C (1991) Fe^{3+}-chelate reducates activity of plasma membranes isolated from tomato (*Lycopersicon esculentum* Mill.) roots. Plant Physiol 197:537–544

Hu ZY, Xu CK (2002) Soil sulfur and environmental quality. In: Chen HM (ed) Behaviors of chemical sub-stances in soils and environmental quality. Science Press, Beijing, pp 283–307

Hu ZY, Zhu YG, Li M, Zhang LG, Cao ZH, Smith FA (2007) Sulfur (S)–induced enhancement of iron plaque formation in the rhizosphere reduces arsenic accumulation in rice (*Oryza sativa*L.) seedlings. Environ Pollut 147:387–393

Hunt J (2003) Bioavailability of iron, zinc, and other trace minerals from vegetarian diets. Am J Clin Nutr 78:633–639

Hurrell RF (2002) Fortification: overcoming technical and practical barriers. J Nutr 132:806–812

Hurrell R, Egli I (2010) Iron bioavailability and dietary reference values. Am J Clin Nutr 91:1461–1467

Hurrell RF, Reddy MB, Juillerat MA, Cook JD (2003) Degradation of phytic acid in cereal porridges improves iron absorption by human subjects. Am J Clin Nutr 77:1213–1219

Ibrahim W, Lee US, Yeh CC, Szabo J, Bruckner G, Chow CK (1997) Oxidative stress and antioxidant status in mouse liver: effects of dietary lipid, vitamin E and iron. J Nutr 127:1401–1406

Idris WH, Abd El Rahaman SM, El Maki HB, Babiker EE, El Tinay AH (2006) Effect of malt pretreatment on phytate and tannin level of two sorghum (Sorghum bicolor) cultivars. Int J Food Sci Technol 41:1229–1233

Imdad A, Bhutta ZA (2012) Routine iron/folate supplementation during pregnancy: effect on maternal anaemia and birth outcomes. Paediatr Perinat Epidemiol 26(s1):168–177

Iqbal M, Puschenreiter M, Oburger E, Santner J, Wenzel WW (2012) Sulfur-aided phytoextraction of Cd and Zn by Salix smithiana combined with in situ metal immobilization by gravel sludge and red mud. Environ Pollut 170:222–231

Jeffery S, Bezemer TM, Cornelissen G, Kuyper TW, Lehmann J, Mommer L, Sohi SP, van de Voorde TFJ, Wardle DA, Van Groenigen JW (2015) The way forward in biochar research: targeting trade-offs between the potential wins. Glob Chang Biol Bioenergy 7:1–13

Jeong J, Guerinot ML (2008) Biofortified and bioavailable: the gold standard for plant-based diets. Proc Natl Acad Sci U S A 105:1777–1778

Jin CW, Yu XH, Zhao SJ (2005) Latent function of microorganisms on plant iron acquisition. Plant Nutr Fertil Sci 11:688–695

Jou MY, Du X, Hotz C, Lönnerdal B (2012) Biofortification of rice with zinc: assessment of the relative bioavailability of zinc in a Caco-2 cell model and suckling Rat Pups. J Agric Food Chem 60:3650–3657

Kabir AH, Paltridge NG, Rossener U, Stangoulis JCR (2013) Mechanisms associated with Fe-deficiency tolerance and signalling in shoots of *Pisum sativum* L. Physiol Planta 147:381–395

Kalbasi M, Sharimatmadari H (1993) Blood powder, a source of iron for plants. J Plant Nutr 16:2213–2223

Kaplan M, Orman S (1998) Effect of elemental sulfur and sulfur containing waste in a calcareous soil in Turkey. J Plant Nutr 21:1655–1665

Kappler A, Wuestner ML, Ruecker A, Harter J, Halama M, Behrens S (2014) Biochar: as an electron shuttle between bacteria and Fe (III) minerals. Environ Sci Technol Lett 1:339–344

Katkat AV, Ozgumus A, Basar H, Altınel B (1994) Iron, manganese, zinc and copper nutrition of peach trees growing in the Bursa region. Turk J Agric For 18:447–456

Kavamura VN, Santos SN, Silva JL, Parma MM, Avila LA, Visconti A, Zucchi TD, Taketani RG, Andreote FD, Melo IS (2013) Screening of Brazilian cacti rhizobacteria for plant growth promotion under drought. Microbiol Res 168:183–191

Kimetu JM, Lehmann J, Ngoze SO, Mugendi DN, Kinyangi JM, Riha S, Verchot L, Recha JW, Pell AN (2008) Reversibility of soil productivity decline with organic matter of differing quality along a degradation gradient. Ecosystems 11:726–739

Kobayashi T, Nishizawa NK (2012) Ironuptake, translocation, and regulation in higher plants. Annu Rev Plant Biol 63:131–152

Kosegarten H, Hoffmann B, Rroco E, Grolig F, Glusenkamp K, Mengel K (2004) Apoplastie pH and Fe3+ reduction in young sunflower (*Helianthus annuus*) roots. Physiol Plant 122:95–106

Laird D, Fleming P, Wang B, Horton R, Karlen D (2010) Biochar impact on nutrient leaching from a Midwestern agricultural soil. Geoderma 158:436–442

Latimer WM (1952) The oxidation status of elements and their potentials in aqueous solutions, 2nd edn. Prentice, Hall Inc, New York

Lawrence JR, Germida JJ (1988) Relationship between microbial biomass and elemental sulfur oxidation in agricultural soils. Soil Sci Soc Am J 52:672–677

Lee S, Jeon JS, An G (2012) Iron homeostasis and fortification in rice. J Plant Biol 55:261–267

Lehmann J, Rillig MC, Thies J, Masiello CA, Hockaday WC, Crowley D (2011) Biochar effects on soil biota: a review. Soil Biol Biochem 43:1812–1836

Lewu MN, Adebola PO, Afolayan AJ (2010) Effect of cooking on the mineral contents and antinutritional factors in seven accessions of *Colocasia esculenta* (L.) Schott growing in South Africa. J Food Comp Anal 23:389–393

Lindemann WC, Aburtto JJ, Haffner WM, Bono AA (1991) Effect of sulfur source on sulfur oxidation. Soil Sci Soc Am J 55:85–90

Lindsay WL (1979) Chemical equilibrium in soil. Wileyinter-science, New York

Lindsay WL (1995) Chemical reactions in soils that affect iron availability to plants: a quantitative approach. In: Abadia J (ed) Iron nutrition in soil and plants. Kulwer Academic Publishers, Dordrecht, pp 7–14

Lindsay WL, Norvell WA (1978) Development of a DTPA soil test for zinc, iron, manganese and copper. Soil Sci Soc Am J 42:421–428

Liu WJ, Zhu YG (2005) Iron and Mn plaques on the surface of roots of wetland plants. Acta Ecol Sin 25:358–363

Liyana-Pathirana CM, Shahidi F (2005) Optimization of extraction of phenolic compounds from wheat using response surface methodology. Food Chem 93:47–56

Loeppert RH (1988) Chemistry of iron in calcareous systems. In: Stucki JW et al (eds) Iron in soils and clay minerals. D. Reidel, College Station, pp 689–713

Lonnerdal B (2007) The importance and bioavailability of phytoferritin- bound iron in cereals and legume foods. Int J Vitam Nutr Res 77:152–157

Lonnerdal B, Bryant A, Xiaofeng L, Theil L (2006) Iron absorption from soybean ferritin in nonanemic women. Am J Clin Nutr 83:103–107

Lorenz K, Lal R (2014) Biochar application to soil for climate change mitigation by soil organic carbon sequestration. J Plant Nutr Soil Sci 177:651–670

Lozoff B, Georgieff MK (2006) Iron deficiency and brain development. Semin Pediatr Neurol 13:158–165
Malakouti MJ, Gheibi N (1988) Determine the critical nutrients strategic and proper fertilizer recommendations in the country. Publication of Agricultural Education, Training and Equipping the Human Resources Department of Tat, The Ministry of Agriculture, Karaj
Malakouti MJ, Riazi Hamedani SAH (1991) Fertilizer and Fertility. University of Tehran press
Marschner H (1995) Mineral nutrition of higher plants, 2nd edn. Academic Press Inc, San Diego
Martínez C, Borrero J, Taboada R, Viana JL, Neves P, Narvaez L (2010) Rice cultivars with enhanced iron and zinc content to improve human nutrition. Paper Presenteda the 28th international rice research conference, Hanoi
Masuda H, Suzuki M, Morikawa KC, Kobayashi T, Nakanishi H, Takahashi M (2008) Increase in iron and zinc concentrations in rice grains via the introduction of barley genes involved in phytosiderophore synthesis. Rice 1:100–108
Masuda H, Ishimaru Y, Aung MS, Kobayashi T, Kakei Y, Takahashi M (2012) Iron biofortification in rice by the introduction of multiple genes involved in iron nutrition. Sci Rep 2:1–7
Masuda H, Kobayashi T, Ishimaru Y, Takahashi M, Aung MS, Nakanishi H, Mori S, Nishizawa NK (2013) Iron-biofortification in rice by the introduction of three barley genes participated in mugineic acid biosynthesis with soybean ferritin gene. Front Plant Sci 4:132:1–12
Mata CG, Lamattina L, Cassia RO (2001) Involvement of iron and ferritin in the potato- Phytophera infestans interaction. Eur J Plant Pathol 107:557–562
May L, Morris ER, Ellis R (1980) Chemical identity of iron in wheat by Mossbauer spectroscopy. J Agric Food Chem 28:1004–1006
Mayer JE, Pfeiffer WH, Beyer P (2008) Biofortified crops to alleviate micronutrient malnutrition. Curr Opin Plant Biol 11:166–170
McCauley A, Jones C, Jacobsen J (2009) Soil pH and organic matter. (Nutrient management modules 8, #4449-8). Montana State University Extension Service, Bozeman, Montana, pp 1–12
McDonagh MS, Blazina I, Dana T, Cantor A, Bougatsos C (2015) Screening and routine supplementation for iron deficiency anemia: a systematic review. Pediatrics 135:723–733
Mengel K, Kirkby EA (1987) Principles of plant nutrition, 4th edn. International Potash Institute, Bern
Mimura ÉCM, Breganó JW, Dichi JB, Gregório EP, DichiI (2008) Comparison of ferrous sulfate and ferrous glycinate chelate for the treatment of iron deficiency anemia in gastrectomized patients. Nutrition 24:663–668
Min L, Li-gan Z, Zheng-yi H (2007) Elemental sulphur oxidation in the rhizosphere of rice and its impact on iron, manganese, phosphorus, and sulphur uptake in rice. J Anhui Agric Univ 34:426–431
Modaihsh S, Al-mustafa WA, Metwally AE (1989) Effect of elemental sulfur on chemical changes and nutrient availability in calcareous soils. Plant Soil 116:95–101
Moore PA, Daniel TC, Sharpley AN, Wood CW (1995) Poultry manure management: environmentally sound options. J Soil Water Conserv 50:321–327
Mortvedt JJ (2000) Bioavailability of micronutrients. In: M. E. Sumer (ed) Handbook of soil science. Boca Raton, CRC press, pp 78–81
Murata Y, Ma JF, Yamaji N, Ueno D, Nomoto K, Iwashita T (2008) A specific transporter for iron (III)- phytosiderophore in barley roots. Plant J 46:563–572
Murray-Kolb LE, Beard JL (2009) Iron deficiency and child and maternal health. Am J Clin Nutr 89:946–950
Murray-Kolb L, Takaiwa F, Goto F (2002) Transgenic rice is a source of iron for iron-depleted rats. J Nutr 132:957–960
Murray-Kolb LE, Welch R, Theil EC, Beard JL (2003) Women with low iron stores absorb iron from soybeans. Am J Clin Nutr 77:180–184
Na GN, Salt DE (2011) The role of sulfur assimilation and sulfur-containing compounds in trace element homeostasis in plants. Environ Exp Bot 72:18–25

Nagesh RV, Usharani G, Dayakar RT (2012) Heterosis studies for grain iron and zinc content in rice (*Oryza sativa* L.). Ann Biol Res 3:179–184

Ngo PT, Rumpel C, Thu TD, Tureaux THD, Dang DK, Jouquet P (2014) Use of organic substrates for increasing soil organic matter quality and carbon sequestration of tropical degraded soil: a 3-year mesocosms experiment. Carbon Manage 5:155–168

Nyamangara J (1998) Use of sequential extraction to evaluate zinc and copper in a soil amended with sewage sludge and inorganic metal salts. Agric Ecosyst Environ 69:135–141

Oliva N, Chadha-Mohanty P, Poletti S, Abrigo E, Atienza G, Torrizo L (2014) Large scale production and evaluation of marker-free indica riceIR64 expressing phytoferritin genes. Mol Breed 33:23–37

Osman MA (2011) Effect of traditional fermentation process on the nutrient and antinutrient contents of pearl millet during preparation of Lohoh. J Saudi Soc Agric Sci 10:1–6

Pahlavan-Rad M, Pessarakli M (2009) Response of wheat plants to zinc, iron, and manganese applications and uptake and concentration of zinc, iron, and manganese in wheat grains. Commun Soil Sci Plant Anal 40:1322–1332

Palmgren MG, Clemens S, Williams LE, Kramer U, Borg S, Schjorriing JK, Sanders D (2008) Zinc biofortification of cereals: problems and solutions. Trends Plant Sci 13:464–473

Perez-Jimenez J, Saura-Calixto F (2005) Literature data may underestimate the actual antioxidant capacity of cereals. J Agric Food Chem 53:5036–5040

Pestana M, de Vsrennes A, Faria EA (2003) Diagnosis and correction of iron chlorosis in fruit tress: a review. Food Agric Environ 1:46–51

Powell JJ, Cook WB, Hutchinson C, Tolkien Z, Chatfield M, Pereira DI, Lomer MCE (2013) Dietary fortificant iron intake is negatively associated with quality of life in patients with mildly active inflammatory bowel disease. Nutr Metab 10:9–22

Raboy V, Gerbasi PF, Young KA, Stoneberg SD, Pickett SG, Bauman AT, Murthy PPN, Sheridan WF, Ertl DS (2000) Origin and seed phenotype of maize low phytic acid 1–1 and low phytic acid 2–1. Plant Physiol 124:355–368

Rashid A, Ryan J (2004) Micronutrient constraints to crop production in soils with Mediterranean-type characteristics: review. J Plant Nutr 27:959–975

Rodrıguez-Lucena P, Ropero E, Hernandez-Apaolaza L, Lucena JJ (2010) Iron supply to soybean plants through the foliar application of IDHA/Fe^{3+} effect of plant nutritional status and adjuvants. J Sci Food Agric 90:2633–2640

Rosado R, del Campillo MC, Torrent J (2000) Long term effectiveness of vivinaite in reducing iron chlorois in olive trees. Plant Soil 241:139–144

Saltzman A, Birol E, Bouis HE, Boy E, De Moura FF, Islam Y, Pfeiffer WH (2013) Biofortification: progress toward a more nourishing future. Glob Food Secur 2:9–17. doi:10.1016/j.gfs.2012.12.003

Scholz-Ahrens KE, Schrezenmeir J (2007) Inulin and oligofructose and mineral metabolism: the evidence from animal trials. J Nutr 137:2513–2523

Schulte EE, Kelling KA (2004) Understanding plant nutrients: soil and applied sulfur. University of Wisconsin-Extension, Cooperative Extension Publications, Lancaster

Schwertmann U (1991) Solubility and dissolution of iron-oxide. Plant Soil 130:1–25

Schwertmann U, Taylor RM (1989) Iron-oxides. In: Dixon JB, Weed SB (eds) Minerals in soil environment. SSSA, ASA, Madison, pp 379–438

Sharma A, Kapoor AC (1997) Effect of processing on the nutritional quality of pearl millet. J Food Sci Technol 34:50–53

Singh M, Dahiya SS (1975) Effect of calcium carbonate and iron on availability of iron in a light textured soil. J Ind Soc Soil Sci 23:247–252

Soumare M, Tack FMG, Verloo MG (2003) Ryegrass response to mineral fertilization and organic amendment with municipal waste compost in two tropical agricultural soils of Mali. J Plant Nutr 26:1169–1188

Soumare M, Tack FMG, Verloo MG (2007) Redistribution and availability of iron, manganese, zinc, and copper in four Malian agricultural soils amended with solid municipal waste compost. Commun Soil Sci Plant Anal 38:2567–2579
Sperotto RA, Ricachenevsky FK, Waldow VA, Fett JP (2012) Iron biofortification in rice: it's a long way to the top. Plant Sci 190:24–39
Stein AJ (2010) Global impacts of human mineral malnutrition. Plant Soil 335:133–154
Tagliavini M, Rambola AD (2001) Iron deficiency and chlorosis in orchard and vineyard ecosystems. Eur J Agron 15:71–92
Taglivaini M, Rombola AD, Marangoni B (1995) Response to iron-deficiency stress of pear and quince genotypes. J Plant Nutr 18:2465–2482
Taglivaini M, Zavalioni C, Romola AD, Quartieri M, Malahuti D, Mazzanti F, Millard P, and Marangoni B (2000) Mineral nutrients partitioning to fruits of deciduous tres. In: Proceedings XXV international horticultural congress ISHS, Bruxelles, Belgium, 2–7 August 1998 Acta Hortic 512:131–140
Tako E, Glahn RP (2011) White beans provide more bioavailable iron than redbeans: studies in poultry (Gallus gallus) and an in vitro digestion/Caco-2model. Int J Vitam Nutr Res 81:1–14
Tako E, Hoekenga OA, Kochian LV, Glahn RP (2013) High bioavailability iron maize (*Zea mays* L.) developed through molecular breeding provides more absorbable iron in vitro (Caco-2 model) and in vivo (Gallus gallus). Nutr J 12. doi:10.1186/1475-2891-12-3
Theil EC, Briat JF (2004) Plant ferritin and non-heme iron nutrition in humans. International Food Policy Research Institute and International Center for Tropical Agriculture, Washington, DC
Theil EC, Huber A, Poulos R, Weighards T (2001) Ferritin. Handbook of metalloproteins. Messerschmidt Wiley, Chichester, pp 771–781
Tisdale SL, Nelson WL, Beaton JD, Havlin JL (1993) Soil fertility and fertilizers, 5th edn. Mcmillon publishing co, New York
Toulon V, Sentenac H, Davidian JC, Moulineau C, Grignon C (1992) Role of apoplast acidification by the H^+pump: effect on the sensitivity to pH and CO_2 of iron reduction by roots of *Brassica napus* L. Planta 186:551–556
Trost LB, Bergfeld WF, Calogeras E (2006) The diagnosis and treatment of iron deficiency and its potential relationship to hair loss. J Am Acad Dermatol 54:824–844
United Nation System Standing Committee on Nutrition (UNSSCN) (2010) 6th Report on the World 20 Nutrition Situation, Geneva
United Nations (2012) World population prospects: the 2010 revision
Wallace A (1991) Rational approaches to control iron deficiency other than plant breeding and choice of resistant cultivars. Plant Soil 130:281–288
Wallace GA, Wallace A (1986) Correction of iron deficiency in tress by injection with ferric ammonium citrate solution. J Plant Nutr 9:981–986
Wei Y, Shohag M, Yang X, Zhang Y (2012) Effects of foliar iron application on iron concentration in polished rice grain and its bioavailability. J Agric Food Chem 60:11433–11439
Welch R (2002) The impact of mineral nutrients in food crops on global human health. Plant Soil 247:83–90
Welch RM, Graham RD (2004) Breeding for micronutrients in staple food crops from a human nutrition perspective. J Exp Bot 55:353–364
Wen JC, Zhang ZL, Jin SL, Tang L, Xu J, Tan XL (2005) Iron and zinc contents in Iaponica hybrid rice based on CMS-D1 system. Sci Agric Sin 38:l182–l1187
White PJ, Broadley MR (2005) Biofortifying crops with essential mineral elements. Trend Plant Sci 10:586–593
WHO (2004) Iron, vitamin and mineral requirement in human. 12 nutrition, 2nd edn. World Health Organization, Geneva, pp 246–272
WHO (2007) Micronutrient deficiencies: iron deficiency anaemia. WHO, Geneva
WHO (2010) Progress on the health-related Millennium Development Goals MDGs
Wiersma JV (2005) High rates of Fe-EDDHA and seed iron concentration suggest partial solutions to iron deficiency in soybean. Agron J 97:924–934

Winichagoon P (2002) Prevention and control of anemia: Thailand Experiences. J Nutr 132:862–866

Wirth J, Poletti S, Aeschlimann B, Yakandawala N, Drosse B, Osorio S (2009) Rice endosperm ironbiofortification by target edandsynergistic action of nicotianamine synthase and ferritin. Plant Biotechnol J 7:631–644

Wu CYH, Lu J, Hu ZY (2014) Influence of sulfur supply on the iron accumulation in rice plants. Commun Soil Sci Plant Anal 45:1149–1161

Xu G, Wei LL, Sun JN, Shao HB, Chang SX (2013) What is more important for enhancing nutrient bioavailability with biochar application into a sandy soil: direct or indirect mechanism? Ecol Eng 52:119–124

Ye X, Al-Babili S, Klöti A, Zhang J, Lucca P, Beyer P, Potrykus I (2000) Engineering the provitamin A (β-carotene) biosynthetic pathway into (carotenoid- free) rice endosperm. Science 87:303–305

Zhao G (2010) Phytoferritin and its implications for human health and nutrition. Biochim Biophys Acta 1800:815–823

Zheng L, Cheng Z, Ai C, Jiang X, Bei X, Zheng Y (2010) Nico-tianamine, a novel enhancer of rice iron bioavailability to humans. PLoS One 5:e10190. doi:10.1371/jour-nal.pone.0010190

Zhou H, Peng X, Perfect E, Xiao T, Peng G (2013) Effects of organic and inorganic fertilization on soil aggregation in an ultisol as characterized by synchrotron based X-ray micro-computed tomography. Geoderma 195–196:23–30

Zielińska-Dawidziak M, Hertig I, Piasecka-Kwiatkowska D, Staniek H, Nowak KW, Twardowski T (2012) Study on iron availability from prepared soybean sprouts using an iron-deficient rat model. Food Chem 135:2622–2627

Zimmermann MB, Hurrell R (2002) Improving iron, zinc and vitamin A nutrition through plant biotechnology. Curr Opin Biotechnol 13:142–145

Zuo Y, Zhang F (2011) Soil and crop management strategies to prevent iron deficiency in crops. Plant Soil 339:83–95

Boron Toxicity in Salt-Affected Soils and Effects on Plants

Tayyaba Naz, Javaid Akhtar, Muhammad Mazhar Iqbal, Muhammad Anwar ul Haq, and Muhammad Saqib

Contents

T. Naz (✉)
School of Veterinary and Life Sciences, Murdoch University, Perth, Australia

Institute of Soil and Environmental Sciences, University of Agriculture Faisalabad, Faisalabad, Pakistan
e-mail: tayyaba.uaf@gmail.com

J. Akhtar • M.A. ul Haq • M. Saqib
Institute of Soil and Environmental Sciences, University of Agriculture Faisalabad, Faisalabad, Pakistan

M.M. Iqbal
Institute of Soil and Environmental Sciences, University of Agriculture Faisalabad, Faisalabad, Pakistan

Department of Soil and Environmental Sciences, University College of Agriculture, University of Sargodha, Sargodha, Pakistan

Laboratory of Analytical Chemistry and Applied Ecochemistry, Department of Applied Analytical and Physical Chemistry, Ghent University, Ghent, Belgium

K.R. Hakeem et al. (eds.), *Soil Science: Agricultural and Environmental Prospectives*, DOI 10.1007/978-3-319-34451-5_11

Abstract Modern civilization with the rapid growth in population, large scale urbanization and industrialization around the globe, results in deterioration of soil and irrigation water quality. It is becoming necessary to understand the responses of crops towards these environmental issues. Among various abiotic stresses, salinity is considered an important limiting factor for worldwide wheat production. High levels of boron (B) and salinity are a serious constraint to crop production around the world. Cropping on saline and B toxic land is restricted by the low tolerance of agricultural crops to these abiotic factors. Frequently, B and salt occur together, however, it is unknown whether the interactions of B and salt increase or decrease the tolerance of a plant to both of these stresses. Low concentration of B is essential to plant growth and may limit the plant growth and development in excess quantity especially under saline conditions. In this chapter the individual and interactive effects of salinity and boron toxicity on physiological and biochemical process in plants have been reviewed.

Keywords Salt stress • Boron • Abiotic stress • Agriculture

1 Introduction

Salinity is a major soil related problem that affects agricultural production. (Yamaguchi and Blumwald 2005; Ahmad et al. 2013; Hakeem 2015). According to an estimate, about 7 % of land area is affected by salinity in the world, which amounts to more than 800 mha (FAO 2008). Salt affected area is increasing and a global study of land use indicates that 6 % had become saline over 45 years (Ghassemi et al. 1995). Pakistan is situated in arid and semi-arid region which leads to insufficient leaching of soluble salts rather promote salt accumulation in irrigated soils. Soil salinity occupies a prominent place amongst soil problems that threaten the sustainability of agriculture in Pakistan. Pakistan has about 22.94 mha land under cultivation and irrigated agriculture is practiced round the year on about 18.78 mha (GOP 2014). About 6.67 mha of salt-affected soils occur in Pakistan (Khan 1998) of which 56 % are saline-sodic and 84 % of such soils in Punjab are saline-sodic. In Punjab province about 75–80 % of the pumped ground water is unfit for irrigation owing to high residual sodium carbonate (RSC), EC, and/or sodium adsorption ratio (SAR) (Ghafoor et al. 2001). A range of dissolved salts are present in the soil solution of salt affected soils including NaCl, $CaSO_4$, Na_2SO_4, $MgCl_2$, Na_2SO_4, KCl, and Na_2CO_3 but NaCl is the most soluble and prevalent salt (Rengasamy 2002; Munns and Tester 2008).

The plant growth and development is deleteriously affected due to the presence of excess amount of salts in the soil. High salt concentration affects seed germination, seedling growth, vegetative growth, flowering and fruit set, resulting in reduced yield and poor quality of produce (McCue and Hanson 1990). Salinity reduces the osmotic potential of the soil solution and thus decreases the ability of plant to absorb water from soil, thereby exposing the plants to osmotic stress. In addition to osmotic effect, salinity may cause specific ion toxicities, such as those associated with sodium (Na^+), chloride (Cl^-) and boron (B). Salinity also results in nutritional imbalance in plants (Munns 2002; Yamaguchi and Blumwald 2005). Moreover, salinity stress is expressed as an oxidative stress and all these factors contribute to the harmful effects of salinity on plants (Hernandez et al. 1995; Gueta-Dahan et al. 1997). When plants exposed to salted environment, the biochemical changes occur, for instance, assembly of reactive oxygen species (ROS) such as superoxide, hydrogen peroxide and hydroxyl radicals (Dat et al. 2000). Plants evolve an operational defence scheme consisted of enzymatic antioxidants like superoxide dismutase (SOD) and catalase (CAT) and non-enzymatic antioxidants (e.g., carotenoids, ascorbate etc.) in order to escape the detrimental possessions of the ROS. It has been previously reported that the tolerance to salinity is generally linked with the most effective antioxidative system (Gossett et al. 1994; Gueta-Dahan et al. 1997). In plants, superoxide anions (O_2^-) is scavenged by SOD that rummages O_2^- and are further converted to hydrogen peroxides (Zhu et al. 2004). Moreover, CAT, the 2nd contour of defense, transforms the harmful hydrogen peroxides to water and oxygen. In plants, the activity of antioxidant enzymes is a test measure for the efficacy of different oxidative defense system (Geebelen et al. 2002; Ahmad et al. 2012).

It is reported that all the major physiological and biochemical processes such as photosynthesis, synthesis of protein and lipid metabolism are affected during the inception and expansion of salinity trauma within the plant (Hasanuzzaman et al. 2012; Rahdari et al. 2012). Salinity stress decreased the internal CO_2 partial pressure, exchange of gases via stomata (Iyengar and Reddy 1996) as well as stomatal conductance (Parida et al. 2004; Yan et al. 2012), reduced the photosynthetic rate, which results in rigorous decline in plant growth and productivity (Manikandanm and Desingh 2009). The productivity of plants is decreased by salt stress in the first phase due to osmotic stress and later because of the built up of toxic ions in leaves (Munns 2002). The plants have evolved various mechanisms to tolerate the presence of salts within the cells or to exclude from their cells. The synthesis of osmolytes or increased activity of antioxidative enzymes help plants to counter the toxic effects of salinity (Ashraf and Foolad 2007).

Boron is categorized among essential plant nutrients and is required for optimum plant growth and development. Boron plays several functions in plants. It is involved in the transport of sugars, synthesis and lignification of cell wall, cell wall structure, metabolism of carbohydrate and RNA, respiration, functions of membranes, synthesis of DNA and regulation of metal activated enzymes (Marschner 1995; Dordas and Brown 2005). The range between the concentration of B essential for plant growth and the concentration causing toxicity is narrow (Gupta et al. 1985). In the regions of the world where climate is arid and semi-arid, high B concentrations are usually found in saline soils (Tsadilas 1997) and irrigation water (Ben-Gal and Shani 2002). It is reported that B toxicity reduced photosynthetic efficiency, leaf expansion and fruit set which ultimately decreased crop productivity (Nable et al. 1997). Plants exposed to B toxicity have reduced root growth than at optimal B application (Gunes et al. 2006). Boron toxicity causes abnormalities in the process of cell division in the meristemtic regions of root (Liu et al. 2000). Boron toxicity causes oxidative stress in which the production of O_2^- and H_2O_2 is increased, resultantly CAT and SOD activity is enhanced. The oxidative stress caused by B toxicity increases O_2^- and H_2O_2 production and, subsequently, CAT and SOD activity (Wang et al. 2011). The excess B induces a significant reduction in photosynthetic rate and increase in interacellular CO_2 concentration (Sotiropoulos et al. (2002).

Although it has been known that excess B inhibits photosynthesis, yet the information regarding the mechanism is still scarce (Ardic et al. 2009; Chen et al. 2012). The CO_2 assimilation decreased in different species including kiwifruit (Sotiropoulos et al. 2002), citrus (Sheng et al. 2010), *Clementine mandarin* (Papadakis et al. 2004) and pear (Wang et al. 2011) in response to B toxicity. The various factors including decrease in photosynthetic enzymatic activities, oxidative load and diminished electron transport seem to be related to reduced CO_2 assimilation under excess B conditions (Han et al. 2009).

There are a few research reports that describe the effect of salinity and excess B in combination on crop growth (Yermiyahu et al. 2008). Some evidences indicate that increased salinity increases the toxic effects of B in different crops including chickpea, cucumber and tomato (Alpaslan and Gunes 2001). Other studies show that B toxicity is reduced by increasing salinity in wheat, chick pea and vegetables (El-Motaium et al. 1994; Ferreyra et al. 1997). In general, the plant growth is reported to be less affected by collective salinity and B toxicity than their individual stresses.

2 Salinity and Agriculture

Salinity is an important problem for both rainfed and irrigated agriculture. More than 6 % of the total land area and 20 % of the irrigated lands in the world are presently salt affected (Munns 2002). Soil or water salinity is among the important disorders and severely limit crop production particularly in arid and semi-arid regions (Shannon 1998). The presence of high concentration of inorganic salts in growing medium retard the growth in most of the crop plants depending on the nature of the salts present, the growth stages and the salt tolerance or avoidance mechanism of the plants. Several cations and anions are associated with salinity such as Na^+, Ca^{2+}, Mg^{2+}, Cl^-, SO_4^{2-} and HCO^-. However, Na^+ and Cl^- ions are the most important because of their toxicity to plants the deleterious effects on soil properties (Dudley 1994; Hasegawa et al. 2000). Even though the plant salinity tolerance differs greatly, mostly crop are intolerant to one third concentration of salts present in seawater (Flower 2004).

In semiarid and arid regions where rainfall is inadequate, temperature and evapotranspiration are high; the effects of salinity become more apparent. Poor soil and water management practices and lack of drainage are other contributing elements (Azevedo Neto et al. 2006). The secondary salinization in Pakistan started with the introduction of canal irrigation system after Indus Basin Treaty. An important cause of this secondary salinization is insufficient and uneven application of irrigation water. This problem is further aggravated by poor drainage followed by water logging, poor quality of ground water, lack of proper water management, and use of brackish water to supplement the water requirements of crops. These factors lead to salt accumulation, sodication and water-logging of soil. Salinity has overwhelming socio-economic effects on farming communities in Pakistan which leads to lower standards of living, health problems, migration and damage to communications and transport (Ijaz and Davidson 1998). To maximize the productivity in country, proper management of the presently salt-affected soils is required. One solution is to make soil environment more suitable for healthy growth of plants. This can be achieved through agriculture engineering approach in combination with chemical and biological reclamation of salt affected soils. The reclamation of these soils is not feasible on large scale due to lack of good quality water, low permeability of soil and high price of amendments (Akhtar et al. 2003). The mismanagement of soil and water resources, saline ground water and the increasing prices of energy-intensive inputs are other limitations. Introduction of salt tolerant species and/or genotypes in such areas can be an alternative approach. Qureshi and Barrett-Lennard (1998) recommended that improving wheat salt tolerance is very much economical for the farmers of developing countries including Pakistan.

3 Effects of Salinity on Plant Growth

There are two reasons for the inhibited plant growth in salt-affected soils: first, reduced ability of plants to take up water from the soil due to high concentration of salts in the soil solution. It is the osmotic effect of salinity which causes water deficit. Second, in transpiration stream, excessive amounts of salts enter which ultimately injure cells in leaves and growth is further reduced. It is due to the specific ion toxicity (Yamaguchi and Blumwald 2005; Munns 2005). This results into two phase plant growth response to salt stress. The 1st response is because of the effect of salts present in soil solution. These salts decrease leaf and root growth. Salinity in the soil reduces plant water uptake, leading to slower growth (Munns et al. 2006) along with a clear plant stunting (Wang and Lin 2000). The components of this effect on plant growth are the reduction in plant height, fresh and dry weights of roots, stem and leaves, lower yield and deterioration of the quality of the product (Chartzoulakis and Klapaki 2000). The 2nd phase is the result of salt present within the plant. The salts entering in the plant accumulates in older leaves: constant transportation of salts into leaf for long time ultimately results into greater Na^+ and Cl^- accumulation due to which leaves may die away. Salts quickly increase in the cytoplasm and reduce enzyme action. The salts might deposit in cell wall and cause cell dehydration (Munns 2005). In a number of plants Na^+ and Cl^- levels increase and Ca^{2+}, K^+ and Mg^{2+} levels decrease due to Increased NaCl concentration in soil (Khan and Ungar 2000; Bayuelo-Jimenez et al. 2003).

Salts present in excess amount in soil also interact with mineral nutrition which may result in nutrient imbalance and ultimately nutrient deficiencies occur. As a result of growth reduction and molecular damage caused by all of these factors, plant death may occur (McCue and Hanson 1990). Plant acquisition and utilization of necessary nutrients particularly K^+ and Ca^{2+} is also impaired under saline conditions causing changes in ratios of K^+/Na^+ and Ca^{2+}/Na^+, thus further affecting growth and productivity of plants (Zhu 2001). Under saline conditions, growth reduction has been observed in almost all plants, but the rate of growth reduction and the tolerance levels are different for different plant species (Hasegawa et al. 2000). Different processes in plants such as growth, water relations, energy and lipid metabolism, protein synthesis, photosynthesis and mineral uptake are influenced by salt stress.

The most prevalent ions in saline soils or water are Na^+ and Cl^-, and these account for most of the harmful effects related to specific ion toxicities (Levitt 1980). Salinity induced reduction in plant growth varies greatly with species and to a lesser extent with varieties (Ghoulam et al. 2002). The severity of salinity response is also affected by environmental interactions such as temperature, relative humidity, air pollution and radiation (Shannon et al. 1994). Salt accumulation in leaves lead to premature senescence, decreasing the availablity of assimilates to the growing regions and consequently reduces plant development (Munns et al. 1995). Reduction in growth of plant owing to salt stress is reported by many scientists. Meloni et al. (2001) described that high concentrations of NaCl resulted in reduced shoot and root biomass, leaf growth and increased root/shoot ratio in cotton. The salts induced formation of reactive oxygen species (ROS), alterations in nutrient uptake and turgor loss (Ashraf 2009), hormonal imbalance (Ashraf et al. 2010) and cytoplasmic enzymes inhibition (Pitann et al. 2009) lead to reduced cell division and elongation which impair plant growth and yield.

It has been reported that tolerance indices of wheat, rice and barley seedlings were inversely related to NaCl treatments. This happened due to greater inhibition of seed germination followed by decrease in root length and shoot growth in response to adverse effects of salinity (Mahmood et al. 2007). Similarly, Asmare (2013) reported that among two varieties of bean, cultivar Chercher showed higher seedling tolerance index and lower root and shoot phytotoxicity. This explains that different cultivars have genetic variability for tolerance and toxicity effects of salinity. There is large genetic variation in the cultivated gene pool of wheat for salt tolerance (Royo and Abio 2003), and identification of genotypes that have tolerance to salinity is relatively simple and practical approach for improving crop yield on saline soils.

4 Salinity and Oxidative Stress

The generation of ROS is important biochemical response of plants to environmental stresses (Dat et al. 2000; Mittler 2002; Ahmad et al. 2012). The ROS are produced during normal aerobic metabolism as a result of electrons leakage from electron transport chain in chloroplast and mitochondria and the reaction of these leaked electrons with O_2 (Thompson et al. 1987). These species include superoxide

(O_2^{-1}), hydroxyl radical (•OH) and hydrogen peroxide (H_2O_2). These oxygen species are highly reactive and can seriously disrupt normal metabolism when some protective mechanism is not present by damaging proteins, lipids and nucleic acid (Molassiotis et al. 2006).

The defense system in plants against these reactive species consists of both non-enzymatic antioxidants and enzymatic antioxidants. The non-enzymatic antioxidants include α-tocopherol, glutathione and ascorbate. Plants also have different antioxidant enzymes in addition to the non-enzymatic antioxidants; such as catalase (CAT), peroxidase and superoxide dismutase (SOD) which prevent the deleterious effects of ROS (Niknam et al. 2003; Agarwal and Pandey 2004). Superoxide is eliminated in plants with the help of SOD and as a result of its enzymatic reactions H_2O_2 and O_2 are formed. The hydrogen peroxide thus formed is destroyed by CAT and peroxidases (Chen and Asada 1989). The exposure of plants to different environmental stresses results in an increased production of ROS (Garratt et al. 2002). As a result the equilibrium between production and quenching of ROS is disturbed and oxidative damage occurs (Molassiotis et al. 2006).

Plants having high levels of antioxidants show greater resistance to the damage imposed by the ROS (Garratt et al. 2002). The activities of antioxidative enzymes in plants subjected to stress are commonly considered as an indicator of genotype tolerance to stress conditions (Gueta-Dahan et al. 1997; Hernandez et al. 2001). It has been established that stress tolerant genotypes showed higher activities of antioxidant enzyme (Sairam and Srivastava 2002). The increased antioxidant enzyme activities under saline conditions have been described in several studies. Moghadam et al. (2014) reported that CAT activity in wheat genotypes increased in response to salinity. The wheat genotypes showed variation in CAT activity but the tolerant genotypes had higher CAT activity. Sairam and Srivastava (2002) investigated the impact of salinity on the antioxidant responses of salt tolerant and sensitive wheat varieties and found an increase in CAT and SOD activities in both varieties in response to salt stress. The tolerant variety had higher activities of enzymes SOD and CAT than salt sensitive one. In another study, it was found that salinity significantly increased CAT activity in pot grown wheat (Barakat et al. 2013). Salt stress enhanced the contents of H_2O_2 and SOD activity in the leaves of rice plants (Lee et al. 2001).

5 Physiological Responses of Plants Exposed to Salinity

Salinity stress induces alterations in different metabolic and physiological processes, depending on exposure and intensity of salinity and ultimately inhibits crop production (James et al. 2011; Rahnama et al. 2010). Salinity affects plant physiological processes and results in various changes such as decreased efficiency of photosynthesis (Ashraf and Shahbaz 2003), increased ion toxicity and respiration rate, changes in mineral distribution and reduced membrane stability due to the displacement of K^+ and Ca^{2+} by Na^+ (Grattan and Grieve 1992). Photosynthesis is the most important process affected by salinity (Hayat et al. 2010). Decrease in photosynthesis in response to salt stress is attributed to stomatal closure leading to reduced intercellular CO_2 assimilation and to non-stomatal factors such as reduction in leaf area and green pigment. Excess salts affect photosynthetic enzymes, ionic contents and chlorophyll (Misra et al. 1997). The salt stress reduced the chlorophyll contents in *Brassica juncea* (Hayat et al. 2011), mustard (Ahmad et al. 2012), wheat (Ghogdi et al. 2012) and basil genotypes (Heidari 2012). Wani et al. (2013) reported significant decrease in SPAD value in the *Brassica Juncea* in response to soil salinity. Either the synthesis of chlorophyll molecules is inhibited or the degradation of existing molecules is accelerated by salt stress (Iyengar and Reddy 1996).

Plants subjected to soil salinity exhibited reduced photosynthesis, caused by a substantial reduction in transpiration rate, internal CO_2 concentration and stomatal conductance (Wani et al. 2013). According to Saleem et al. (2011), under salinity stress, photosynthetic CO_2 assimilation is decreased due to closure of stomata caused by salt-induced abcisic acid accumulation. Therefore gaseous exchange through stomata, internal CO_2 partial pressure (Iyengar and Reddy 1996) and stomatal closure (Yan et al. 2012) are important factors that severely affect photosynthesis during salinity. The inadequate photosynthesis is considered as an important factor limiting plant productivity under saline conditions (Manikandam and Desingh 2009).

Salinity decreases CO_2 assimilation into carbohydrate through reduced efficacy of photosynthetic enzymes (Reddy et al. 1992), leaf area (Munns et al. 2000) and stomatal conductance (Agastian et al. 2000; Parida et al. 2003). Salt stress also affects the repair of Photosystem-II by suppressing protein synthesis at the transcriptional level (Allakhverdiev et al. 2002). Salinity also significantly decreases the total chlorophyll content and the degree of decrease depends on salt tolerance of plant species and salt concentrations (Ashraf and McNeilly 1988). The reduction in the rate of leaf expansion is an instant plant response to salinity. Consequently, the total leaf area of the plant is reduced. The common reduction in leaf expansion is due to the decreased cell turgor pressure. In the salt-sensitive genotypes, inefficient salt exclusion from transpiration stream causes accumulation of salts in leaves, resulting in death of old leaves and new leaves becoming injured and succulent (Munns and James 2003). Consequently, the number of green and healthy leaves will ultimately decrease.

6 Physiological and Biochemical Mechanisms of Plants for Salinity Tolerance

Genetic variations in salt tolerance exist, and the degree of tolerance differs with plant species and varieties within a species. Plants develop various physiological and biochemical modifications to survive in salt-affected soils. Principle mechanisms include (1) ion homeostasis and compartmentalization, (2) ion transport and uptake, (3) biosynthesis of osmo-protectants and compatible solutes, (4) activation of antioxidant enzyme and synthesis of antioxidant compounds, (5) synthesis of polyamines, (6) generation of nitric oxide (NO), and (7) hormone modulation (Hakeem et al. 2012; Gupta and Huang 2014).

Maintaining ion homeostasis by ion uptake and compartmentalization is essential process for normal plant growth during salt stress (Hasegawa 2013). Irrespective of their nature, both glycophytes and halophytes cannot tolerate high salt concentration in their cytoplasm. Hence, the excess salt is either transported to the vacuole or sequestered in older tissues which eventually are sacrificed, thereby protecting the plant from salinity stress (Zhu 2003). Two main strategies adapted by plants to deal with saline environment are the sequestration of Na^+ into the vacuole and exclusion of Na^+ at plasma membrane. During osmotic adjustment, plant cells compartmentalize the absorbed ions in vacuoles and also synthesize compatible organic solutes in the cytoplasm for maintaining osmotic equilibrium among these two compartments (Serrano and Gaxiola 1994; Hasegawa et al. 2000). Antioxidant metabolism plays crucial role in detoxification of ROS being generated due to salinity. Salinity tolerance is positively correlated with the activity of antioxidant enzymes, such as SOD, CAT etc. and with the accumulation of non-enzymatic antioxidant compounds (Gupta et al. 2005).

Accumulations of carbohydrates (e.g., glucose, fructose, fructans, and trehalose) and starch also occur in plant cells in response to salt stress (Parida et al. 2004). The major role played by these carbohydrates in stress mitigation involves osmoprotection, carbon storage and scavenging of ROS. It was observed that salt stress increases the level of reducing sugars (sucrose and fructans) within the cell in a number of plants belonging to different species (Kerepesi and Galiba 2000). Besides being a carbohydrate reserve, trehalose accumulation protects organisms against several physical and chemical stresses including salinity stress. They play an osmoprotective role in physiological responses (Ahmad et al. 2013). Sucrose content was found to increase in tomato (*Solanum lycopersicum*) under salinity due to increased activity of sucrose phosphate synthase (Gao et al. 1998). During salinity stress, sugar content has been reported to both increase and decrease in various rice cultivars (Alamgir and Ali 1999). It has been observed that starch content decreased in rice roots in response to salinity while it remained fairly unchanged in the shoot. In *Bruguiera parviflora* leaves, increased reducing and non-reducing sugar contents were noted under salinity stress (Parida et al. 2004). Thus, improved understanding of these processes and mechanisms will be helpful to improve salt tolerance of crops.

7 Boron: Forms, Behavior and Sources

Boron exists as an intermediate between metals and non-metals and is adherent to the semi-conductor group of the elements in periodic table. The B atom has three outer valence electrons, and has distinctive and complex chemistry than other elements (Greenwood and Earnshaw 1984). It is a metalloid element (i.e., atomic number = 5, atomic weight = 10.81 g mol^{-1}, oxidation state +3), member of Group IIIA in periodic table. Boron is randomly occur in earth's crust (Nable et al. 1997). Tourmaline and volcanic emanations are the prime B sources B in the furthermost soils (Chesworth 1991). Tourmaline is unaffected to chemical weathering and therefore, masses in the clastic portion of rocks. Boron occur as borosilicates i.e., resistant to chemical breakdown, not easily phyto-available, among igneous, sedimentary and metamorphic types of rocks. In the pedosphere, mobilization of immobile forms of rock B, occurs by weathering processes and reactions. The $B(OH)_3$ (boric acid) is the most dominant species and due to its mobility easily available to vegetation. Moreover, the other fates includes $B(OH)_3$ can be retained by organic matter or adsorbed on fine fractions of mineral (Nable et al. 1997). At neutral pH, boric acid is the foremost compound and it exists as colorless, odorless, white granules or translucent crystals or powder at ambient temperature (O'Neil et al. 2004).

The abundance of B is enormously less in environment: about 9–10 times compared to hydrogen (H) and about 6–10 that of carbon (C), nitrogen (N) or oxygen (O). However, B is commonly distributed both in the crust of Earth such as from 5 mg kg^{-1} in basalts to 100 mg kg^{-1} in shales (Shorrocks 1997) and in the ocean (~4.5 mg L^{-1}) (Lemarchand et al. 2000). In the environment, B is mainly derived from the weathering of minerals (Kot 2009). Furthermore, geothermal steams processes considerably contribute to natural existence of B. In sea water, high B concentrations are also observed with an average of 4.5 mg L^{-1} as boric acid, that denotes the main cause of B pollution in coastal areas because of seawater intrusions into fresh water aquifers. Furthermore, boric acid is often found in volcanic spring waters and in the material released by an erupting volcano. Globally, elemental release of B via volcanic, geothermal processes and weathering are estimated to be approximately 360,000 metric tons annually (Moore 1991). Sources of B resulting from human activities are less important than natural processes. Anthropogenic B sources includes power generation using coal, oil, agriculture, borate mining/processing, glass product manufacture, waste and fuelwood burning, leaching of treated wood/paper, industrial or household usage of borates/perborates and sewage or sludge dumping (HSDB 2003). Other sources of excess B in soil include municipal and other waste water effluents used for the purpose of irrigation (Tsadilas 1997). Addition of B in the soils also occurs from the application of fertilizers and mining operations (Nable et al. 1997).

Boric acid as well as borate minerals are extensively utilized in industrial process like leather production, glass and porcelain manufacturing operations, photographic chemicals, carpets and fertilizers. The usage of sodium perborate as an oxidation bleaching agent in industrial and domestic cleaning products is however, the main application of B. During production and end point usage of the detergents, the release of sodium perborate into the environment has resulted in the accretion of B in waste effluent and subsequently in natural aquatic systems and groundwater (Vengosh et al. 1994).

Boron occurs as an uncharged molecule, i.e., boric acid at physiological pH. The presence of a negative or positive charge for other nutrients restricts the permeability of membrane and permits a degree of control over their influx or efflux via membrane transporters (Reid 2010). The boric acid can pass directly across the phospholipid bilayers (Dordas et al. 2000). It has recently been revealed that fluxes of boric acid can be further accelerated by movement via aquaglyceroporins (Fitzpatrick and Reid 2009). When soils contain high B, this more pronounced membrane permeability and lack of control over B entry into the plant generates problems. Nevertheless, like most of elements, B becomes toxic to plant growth at high levels. The solubility of B is high and it can be leached down in high rainfall areas. However, it can persist in agricultural soils when rainfall is low. In these areas, partial leaching moves B down the soil profile where roots then encountered B toxicity in their search for water. Irrigation water can increase B toxicity in the regions where soil B is only moderate, if it contains high concentrations of B. In most cases, amelioration of B-toxic soils is impractical, thus solutions based on improved B tolerance by crop plants have been intensively investigated in the last two decades by research scientists. In crop genotypes, B tolerance has been recognized and can be conveyed to cultivars by conventional breeding or molecular means, however, substantial enhancement in the crop yields on B-toxic soils have still to be realized (Reid 2010).

8 Boron Toxicity in Soils and Plants

Boron affects cellular and metabolic functions in plants (El-Hamdaoui et al. 2003). It is required in small amount for optimum plant growth (Gupta et al. 1985). High B concentrations are usually found in semiarid and arid regions where salt affected soils and saline irrigation water are prevalent. The soils of various countries such as USA, Australia, Iraq, Peru, Pakistan and India have been reported to have excessive B (Nable 1992). High levels of B in irrigation water or soil are responsible for the suppression of plant growth (Alpaslan and Gunes 2001). Hot water soluble B concentration of 5 mg kg^{-1} or more in soil is considered toxic for plants (Nable et al. 1997).

Boron present in high concentration in soil is an important concern because of its narrow range between deficient and toxic concentrations (Marschner 1995). Secondly, there are several reports that B is found in toxic concentrations in salt-affected soils in several parts of the world (Bastias et al. 2004). Thirdly, B has higher affinity to soil compared to the other salts. Therefore, more water is required to decrease soil B to nontoxic levels than to decrease salinity (Ayers and Westcot 1985).

In many crops, critical values for B toxicity have been previously recognized (Gupta 1993). The species in which B is accumulated in leaves, about 40 to 100 mg B kg^{-1} dry weight is found in their leaves. However, it can reach to 250 mg kg^{-1} dry weight and may exceed 700 to 1000 mg kg^{-1} when B is present in toxic concentration in soil. The critical values in wheat and barley are reported to range from 10 to 130 mg B kg^{-1} dry weight (Riley et al. 1994). Therefore, a broad range of critical values for B toxicity are found, occasionally even for the same species. Within leaf blades the sharp B gradient with higher accumulation in tips and margins is the main cause of this problem (Oertli 1994). For example, leaves accumulate enormously different amounts of B, depending upon the rates of transpiration, though mostly concentrated in the tips and margins, with the majority of the leaf blades having similar concentrations. The overall concentrations of B in leaves vary under these diverse transpiration situations, though the consequence on growth can be similar (Nable 1988). Concentration of leaf B of tolerant and susceptible species has been found to vary as much as ten times (Gupta 1993).

9 Boron Toxicity Symptoms in Plants

In plants, excessive B concentration produces many abnormalities in the main growth stream (Reid et al. 2004) and this was due to toxic B levels. According to Loomis and Durst (1992), the existence of B toxicity symptoms in plants may be due to soluble B concentration. Boron toxicity symptoms depicted the B build up at the last part of transpiration stream and fictitious B distribution in the majority of species.

The leaf burn and the development of chlorotic and necrotic areas is the typical observable symptom of B toxicity in a variety of plant species. The symptoms are frequently found at the tips and margins of older leaves (Eaton 1944). These symptoms reveal the distribution pattern of B in plant species, that B accumulate at the end of the transpiration stream (Nable et al. 1997). It has been reported that the apical parts of leaves contained toxic B concentrations in wheat (Wimmer et al. 2003). Some studies showed that the necrotic or chlorotic leaf segments have extremely high B contents then adjoining leaf tissues and barley exhibited distinctive sketches for various cultivars (Oertli and Roth 1969). Unlike to the common observation, leaf burn does not provide an indication of toxicity of B in all species. In some plant species (e.g., *Prunus, Malus, Pyrus*) B is phloem mobility and tends to accumulate in the developing sinks. In these species the B toxicity symptoms include fruit disorders, stem die back and bark necrosis which occurs due to the death of cambial tissues (Brown and Hu 1996). In orange plants, B toxicity symptoms were observed only in older leaves in the form of chlorosis starting from leaf tips. The chlorotic area then extended toward the base from the tip and towards the mid rib from the margins of the leaf. Both salinity and drought have been reported to worsen B toxicity symptoms in wheat. Increase in B concentration in wheat and variety of other species is associated with salinity (Grieve and Poss 2000). Observable B toxicity symptoms have not been found in roots. Probably toxic B concentration do not arise in root tissues as the concentrations of B in roots remain relatively less than leaves B concentration (Nable 1988).

10 Boron Toxicity and Its Effects on Plant Growth and Physiology

At physiological pH, B exists as uncharged boric acid molecules. The neutrality of boric acid enables it to easily pass through the membranes. When soils contain toxic levels of B, this free entry of B into the plant results in serious problem as B at high concentrations become toxic to plants.

Boron toxicity affects several metabolic processes in plants. As a result of B toxicity, the supply of photosynthate is reduced to the developing areas of plants. In mature tissues necrotic areas are formed and leaf area is reduced because of reduced cell expansion in meristematic tissues (Nable et al. 1997). Root growth is also inhibited due to B toxicity (Reid et al. 2004). Boron is taken up by plants as boric acid which travels with the transpiration flow. Within plants, B is immobile and accumulates in leaves. Boron concentration is found to be high in mature leaves, particularly at leaf tips and margins. Boron toxicity symptoms include chlorosis and necrosis in older leaves, usually at the tips and margins where B accumulates (Nable et al. 1997). Older leaves become burned at the edges with yellow tips. The severity of toxicity symptoms depend on the tolerance of plant and the toxicity level of B (Nadav 1999).

It has been described that B toxicity leads to abnormal cell division in root meristem (Liu et al. 2000). The root growth of plants subjected to B toxicity was reduced in comparison with plants grown at optimum B levels (Gunes et al. 2006). It was found that excess B supply resulted in inhibition of tomato root growth and tomato genotypes responded variably to high B (Cervilla et al. 2007). In barley cultivars, the root and leaf fresh and dry weights reduced in response to B toxicity (Karabal et al. 2003).

11 Activity of Antioxidant Enzymes in Response to Boron Toxicity

Boron toxicity has been reported to trigger the activity of antioxidant enzymes (Han et al. 2009). The increased activities of SOD and CAT were found in the roots of barley cultivars (Karabal et al. 2003). The SOD detoxify Superoxide radicals (O^{2-}) by converting it to H_2O_2. The H_2O_2 thus formed is also toxic must be eradicated in subsequent reactions (Mittler 2002; Azevedo-Neto et al. 2006). Generally, SOD activity was found to increase during B toxicity (Sotiropoulos et al. 2006). An increase in SOD activity has been previously reported in crops e.g., in tomato (Kaya et al. 2009) and chickpea (Ardic et al. 2009) under B toxicity and in citrus leaves in response to both B deficiency and excess (Han et al. 2009). However increased SOD activity was found only in one tomato cultivar in response to toxic B levels in growth medium while the other cultivar did not show the similar increase (Cervilla et al. 2007). According to Ardic et al. (2009) there was no increase in the constitutive isoenzymes and increased SOD acivity was due to the induction of new isoforms of the enzyme.

The toxicity of B caused oxidative stress and thus increased O^{2-} production and subsequently, SOD activity. Therefore, this enzyme plays an important role to catalyze radicals scavenging. In order to detoxify H_2O_2 generated by SOD, the other enzymes are also imperative. In plants, H_2O_2 is detoxified by a number of enzymes, but ascorbate peroxidase (APX) and CAT are the most important (Noctor and Foyer 1998). The activity of CAT and APX can be promoted by certain concentrations of H_2O_2 (Bowler et al. 1992). For example, in pear, activity of these two enzymes increased in response to H_2O_2 production at 100 and 300 μM B in comparison with control plants at 10 μM (Wang et al. 2011).

The increased CAT activity has been detected in several plant species in response to B excess (Gunes et al. 2006; Soylemezoglu et al. 2009). On the contrary, CAT activity decreased in citrus leaves due to B toxicity (Keles et al. 2004; Han et al. 2009). An enhancement of CAT was also recorded in tomato (Cervilla et al. 2007), tobacco (Garcia et al. 2001) and hot pepper (Lee 2006). Increased CAT activity and reduced peroxidase activity was noted in sunflower due to excess B (Dube et al. 2000). Landi et al. (2013) reported significant increase in CAT and SOD in two cultivars of sweet basil in response to high concentration of B in nutrient solution. Comparable results were also stated by Wang et al. (2011) in pea. However, no previous report is available on activities of antioxidant enzymes in Pakistani wheat varieties exposed to B toxicity.

12 Boron Toxicity and Plants Photosynthetic Features

It has been well-known from the previous studies that excess of B in the growth medium inhibited the photosynthetic rate and reduced CO_2 assimilation in different plant species such as kiwifruit (Sotiropoulos et al. 2002), pear (Wang et al. 2011) and citrus (Sheng et al. 2010). Different factors in combination are related to the reduced CO_2 assimilation under B toxicity including decreased photosynthetic and enzymatic activities, oxidative load and impaired transport of electrons (Han et al. 2009).

Boron toxicity is reported to decrease transpiration rate and ultimately reduce yield (Ben-Gal and Shani 2002). The gas exchange rate, stomatal conductance, photosynthesis and chlorophyll contents also decreased due to B toxicity (Papadakis et al. 2004). The photosynthetic rate is reduced and intracellular CO_2 is increased in kiwifruit in response to B toxicity (Sotiropoulos et al. 2002). The effects of developing B toxicity in young squash plants included reduced chlorophyll contents, leaf area, growth and CO_2 fixation (Lovatt and Bates 1984). The inhibition of ureide metabolism in the leaves of nodulated soybean (Lukaszewski et al. 1992) and metabolic disturbance caused by complexing of ribonucleotides may also occur in response to B toxicity (Loomis and Durst 1992).

According to Pereira et al. (2000) the structural damage of thylakoids caused by B toxicity was responsible for reduction in photosynthesis. As a result rate of electron transport and CO_2 photo-assimilation was influenced. A noteworthy reduction in Fv/Fm ratio (maximum chlorophyll fluorescence quantum yield) was found in many species in response to B toxicity (Guidi et al. 2011). The reduced Fv/Fm ratio is an indication of photo-inhibition of leaves. In such conditions molecular oxygen can serve as an alternative acceptor for unused light and electrons (Velez-Ramirez et al. 2011). As a result ROS are produced. It also explains the observed chloroplast damage and reduction in chlorophyll contents of plants exposed to B toxicity. (Han et al. 2009; Chen et al. 2012).

Thus, modifications in the mesophyll cells induced by B cause reduction in the transport of electron light utilization. The activity of enzymes of CO_2 assimilation pathway is also reported to be decreased. This reduced enzyme activity along with reduced ATP and NADPH utilization also contribute to inhibit transport of electrons (Han et al. 2009). Therefore, the reduced rate of electron transport regulates oxidative stress that results in ROS generation in chloroplast. The cell death is induced because of the ROS induced oxidation of organic molecules as lipids and chlorophyll. As a result these events, visible symptoms of injury appear typically at the leaf tips and margin where B is stored to greater extent (Guidi et al. 2011).

13 Occurrence of Boron Toxicity in Saline Environment

Boron toxicity is considered as significant limiting factor for the production of crops in many areas of the world. Furthermore, B toxicity often coincides with salinity especially in arid and semiarid environments (Grieve and Poss 2000). Plants can encounter concurrent stress caused by salinity and B toxicity when grown in B rich saline soil or irrigated with water having high salt and B concentrations. These conditions usually occur in arid and semiarid regions with low rainfall (Nable et al. 1997). The affinity of B to the soil is higher than common salts, resulting in higher requirement of water for the reclamation of B enrich soil than that required for salinity reclamation (Ayers and Westcot 1985). Irrigation with saline ground water having high concentration of B is considered the main reason for higher concentrations of B in soil. Several scientists reported high B concentrations when irrigating with saline groundwater throughout the world (Manchanda and Sharma 1991). Another source of combined salinity and B toxicity is recycled waste water used for irrigation (Tsadilas 1997). Boron toxicity has become a serious concern in the recent years, because of increasing utilization of desalinated water containing very high B concentration (Parks and Edwards 2005). During leaching process, B is removed more slowly as compared to other ions (Yermiyahu et al. 2008). Therefore some reclaimed soils may have unnecessary high B levels (Ben-Gal and Shani 2002). Crops vary in the sensitivity to excess boron and salinity and those that are tolerant to salinity may also be tolerant to boron (Grieve et al. 2012). A thorough understanding about the responses of plants to high soil B present in combination with salinity is needed to manage for better crop performance.

14 Salinity and Boron Toxicity: Consequences

14.1 Growth, Yield and Ionic Composition

There are several research reports on the crop behavior against the individual stresses of B toxicity or salinity. The co-occurrence of salinity and B in agricultural and natural systems has resulted in research on the interaction of collective stresses of B toxicity and salt stress. The toxic effects of B on plant growth may be further increased or decreased by salt stress (Yermiyahu et al. 2008). There are some reports that increasing salinity can increase toxic effects of B in crop plants (Alpaslan and Gunes 2001; Wimmer et al. 2005). On the contrary, increased salinity reduced toxicity of B in some crops and vegetables (Holloway and Alston 1992; Ferreyra et al. 1997). According to Yermiyahu et al. (2008), there may be three types of interaction between two factors: antagonistic, additive or synergistic. An antagonistic effect designates that the consequence of the two factors, applied simultaneously, is less than the summation of the effects of both factors, when applied individually. An additive impact describes that the result of the two factors, applied concurrently, is equal to the sum of the effects of both factors when applied independently and a synergistic effect designates that the consequence of the two factors, applied together, is more than sum of the effects of both factors while applied individually. However, relationships between B toxicity and salinity are relatively complex. Mostly, an antagonistic interaction has been reported when both stresses are present together (Bastias et al. 2004; Yermiyahu et al. 2008). However, there is no harmony about mutual associations between B toxicity and salt stress (Grieve et al. 2010).

The responses of different crop plants to the concurrent behaviuor of salinity and excess of B has been reported previously, for instance in: cucumber (Alpaslan and Gunes 2001), carrot (Eraslan et al. 2007a), melon (Edelstein et al. 2005), tomato (Guidi et al. 2011), bell pepper (Yermiyahu et al. 2008), spinach (Eraslan et al. 2008) and lettuce (Eraslan et al. 2007b). In general, it was described that combined salinity and B toxicity causes less damage on crop growth compared to what would be predicted if the impacts of the individual factors were additive. It has also been suggested that boric acid might affect the activity of defined membrane components (Martinez-Ballesta et al. 2008a, b). In fact, significant B transport via the plasma membrane aquaporins occurs at high external B levels (Dordas and Brown 2001).

The production of biomass, yield components and grain yield of wheat was decreased significantly due to combined effect of B toxicity salinity (Grieve and Poss 2000). Wimmer et al. (2003) reported higher reduction of root or shoot growth of wheat exposed to combined B toxicity and salt stress than their odd effects. There was higher reduction of shoot growth in response to salinity and root growth in response to high B as compared to control. In another study, it was reported that Broccoli shoot biomass and head yield were significantly decreased by both salinity and B. A salinity-B interactive effects was found significant, wherever increasing B was comparatively less damaging in saline environment (Smith et al. 2010a, b. Masood et al. (2012a, b) reported an additive effect of combine salinity and high B regarding the reduction in shoot growth of wheat in solution culture experiments. Smith et al. (2013) reported that shoot dry weights of broccoli were affected adversely by increasing B and pH and significant salinity boron interaction was observed for shoot dry weight. It was reported by Yemiyahu et al. (2008) that increased NaCl or B, each in the growth medium resulted in gradual decrease on peppers growth and harvest yield. They reported an antagonistic impact of excess B and salinity on growth and yield of peppers. The effect of combined toxicities of B and salinity was reported to be antagonistic in tomato. The mechanism of this interaction is not clear yet, however the possible mechanisms may be the role of B in preventing nutrient imbalances under saline conditions or functions of aquaporins (Bastias et al. 2010). Alpaslan and Genes (2001) found that the salt tolerant tomato was affected less than salt sensitive cucumber plant while investigating the effect of salinity and B and concluded that response of plant varies according to the relative salt tolerance. They also described that salt tolerant plants may have higher resistance to B toxicity as the salt exclusion mechanisms also reduce uptake of B.

The reduction in leaf B with the increased NaCl concentration in the irrigation water have been reported widely. As an elucidation it has been suggested that decreased rates of transpiration might reduce the transported and accumulation of B in leaves (Yermiyahu et al. 2008). Smith et al. (2010b) found that at low level of B, the increasing salinity increased shoot B concentration in broccoli. However, at higher level of B, the increased salinity decreased shoot B concentration. Masood et al. (2012a) also reported increase in apoplastic and symplastic soluble B concentrations with NaCl salinity at adequate B supply; while reduction in soluble B at high B level in the nutrient solution. Salinity reduced B uptake by plants but the impact was mainly evident at high substrate B concentrations, an observation also found in wheat by Wimmer and Goldbach (2012). Yermiyahu et al. (2008) determined that decreased B concentration in leaves in the presence of salinity was due to reduced transpiration rates in plants. Masood et al. (2012b) also reported increased Na^+ and reduced K^+ in wheat leaves exposed to salinity in nutrient solution. Boron had no effect on leaf Na^+ concentration in combined salinity and high B treatment. A significant increase in Na^+ concentration in wheat shoot tissue was observed by Grieve and Poss (2000) exposed to increasing salinity in irrigation water. Substrate B also increased shoot Na^+ concentration and shoot K^+ was significantly decreased with salinity and further depressed by increasing substrate B. It has been reported that under salinity, the activity of definite membrane components can be affected by B maintaining the functions of defined aquaporin isoforms, as likely components of the salinity tolerance mechanism (Martínez-Ballesta et al. 2008b). Furthermore, it has been suggested that salt-tolerant plants may have more resistance to toxic B ratess because their salt-exclusion mechanisms also contribute to decrease the uptake of B (Alpaslan and Gunes 2001). However, more studies are required to investigate the impacts and mechanisms of salinity and high B tolerance among wheat varieties, for the reduced uptake of B in B-toxic saline soils.

14.2 Photosynthetic Functions, Antioxidant Enzymes and Carbohydrate Contents

Salinity and high B in the growth medium affect photosynthesis in plants (Guidi et al. 2011). Masood et al. (2012a) reported a significant decrease in transpiration rate in wheat exposed to combined salinity and B toxicity. The several functions of membrane are altered due to salt and B stress. The stomatal resistance is reported to be increased and the synthesis of photosynthetic pigments is disturbed, resultantly photosynthesis is also affected (Sairam et al. 2005). The stomatal conductivity was increased while chlorophyll contents were decreased under combined stresses of salinity and B in carrots plants (Eraslan et al. 2007b). It has been reported that the formation of necrotic areas on tomato leaves increased as the B concentration increased in growth medium under saline conditions. The leaf margins developed more necrosis while the central areas of leaves remained green. The photosynthesis was decreased in the affected areas of the leaves (Guidi et al. 2011).

It is reported that plants react to the oxidative impairment caused by a enormous number of abiotic and biotic stresses (Mittler 2002) by improving antioxidant systems consisting of antioxidant molecules like ascorbate, and antioxidant enzymes such as CAT and SOD. The activity of antioxidant enzymes in plants is not only affected by individual salinity or B toxicity but also by their combined presence. When carrot plants were subjected to salinity and excess B simultaneously, the activity of antioxidant enzymes APX and CAT was increased (Eraslan et al. 2007b). Peroxidation of lipid and the actions of CAT and SOD were increased in grapevine in response to combined stresses of high B and salts. Increased H_2O_2 indicate membrane damage caused by oxidative stress and as a result membrane permeability increases under excess B and salinity (Soylemezoglu et al. 2009). Lee (2006) described that the antioxidant enzyme action was greatly increased by salinity, B or both in hot pepper plants. Lopez-Gomez et al. (2007) also reported increased activity of SOD in the presence of combined B toxicity and salinity in loquat plants indicating increased capability to scavenge superoxide radicals. Eraslan et al. (2007) found an increased concentrations of SOD and CAT in lettuce plants, when exposed to combine B toxicity and salinity. Gunes et al. 2006 also reported similar results under sodic-B toxic conditions. It is well-established that stress tolerant genotypes showed higher antioxidant enzyme activities (Sairam and Srivastava 2002). The tolerance of plants might be closely associated to enhanced capacity of the antioxidative systems (CAT and SOD etc.) to scavenge ROS and thus reduce lipid peroxidation in stress conditions (Ardic et al. 2009).

Water soluble carbohydrates (WSC) in stressed plants may have a significant role in osmotic adjustments (Fayez and Bazaid 2014). When photosynthesis is reduced, pre-anthesis WSC can contribute significantly to grain yield (Dreccer et al. 2009). Ruuska et al. (2006) also reported that the WSC gathers in the stems of wheat during growth, can be an important contributor to grain filling, particularly under conditions when assimilates is restricted. Two novel evolving roles have been projected for fructan component of WSC. First, fructans may add to cellular ROS homeostasis by direct ROS scavenging mechanisms. Secondly, small fructans may act as phloem-mobile signaling compounds in stress (Van den Ende 2013). Such signaling and antioxidant mechanisms might take part to stress tolerance and disease prevention (Van den Ende 2013). Kerepesi et al. (1998) investigated the role of WSC in genotypic variances in response to drought and salinity stresses in wheat seedlings. They found higher accumulation of soluble carbohydrate in tolerant genotypes than the sensitive ones. The stem fructan content decreased in sensitive genotypes but increased in tolerant one under NaCl treatment. Amini and Ehsanpour (2005) reported an increase in total carbohydrates in stem and leaf of cv. Shirazy tomato with increasing salt concentration in the growth medium, but contents of carbohydrate decreased in cv. Isfahani. The carbohydrate increased in roots when explants form two cultivars were subjected to the higher rates of salts. Fayez and Bazaid (2014) also found that salt and water deficit stresses results in a considerable increase in soluble carbohydrate level of barley leaves. With increasing salt doses, the rate of increase in soluble carbohydrate content was increased, confirming the character of soluble carbohydrate in the osmotic regulation. Picchioni and Miyamoto (1991) reported that concentration root starch and total leaf sugars such as sucrose, glucose or fructose etc. increased, root glucose concentration reduced and root carbohydrates remained unaffected in *Pistacia vera* seedlings upon B addition to the growth medium. Leaf carbohydrate supply limitations and altered contents of root carbohydrate status might be consequences of toxic B in *Pistacia vera* leaves.

Choi et al. (2007) reported that sugar transport and utilization metabolism was severely affected with higher levels of B. The results indicated that enhancement of sugar contents in the root tips of SloopVic (B tolerant) occurred with the application of excess B. This accorded with an enhancement in the root elongation rate and with a 2.7 times increase in sucrose level within mature leaf tissue. A gradual decrease in contents of reducing sugar was measured in the leaf tissue and root tips of Clipper (B intolerant) at higher B rates. The variations in total WSC during salinity and high B stress was attributed to variation in the major soluble carbohydrates such as sucrose, hexose and fructan components.

In brief, due to the unique role of WSC during stress conditions, more detailed studies are still needed, to further investigate the composition of WSC in wheat varieties in collective stresses of salinity and B toxicity. The carbohydrate dynamics, related enzymatic activities and gene expression levels should be followed. Moreover, the successful breeding for high WSC may be possible for increased and safe production in wheat.

15 Conclusion and Future Prospects

Overall, it can be concluded from the above review that salinity and excess B affects plant growth responses, physiological functions, biochemical processes and ultimately yields in B toxic saline agricultural lands. Their continued increase can be a risk for crop production, food safety and sustainable agriculture worldwide. The variation exists in plant species and genotypes for tolerance to salinity, high B and their combined toxicities. These variations can be exploited to produce higher yields from salt-affected B toxic soils. The plant species capable of maintain lower concentration of toxic ions in their root and leave, higher photosynthetic rate, higher scavenging of ROS through enhanced activity of antioxidant enzymes and higher leaf and root WSC have better survival and yields on salt-affected B toxic soil. Such tolerant varieties can be used to produce higher yields in combined salinity and high B stresses and can have significant contribution to food supply for increasing population.

Amelioration of B-toxic soils is not an easy task under most circumstances and remediation of salt-affected soils is also not possible due to limitations of availability of good quality irrigation water, high cost of amendments and low soil permeability. Therefore, saline agriculture i.e., the genetic differences among plants species/genotypes for tolerance to salinity and B toxicity or their combination of both, can be used as the most feasible approach to provide genetic material for future wheat breeding programs. It is also a very viable alternative to soil amelioration in order to get economical crops yields from such problematic soils. Thus, the changeover from soil intrusion to plant adaptation in order to solve an obdurate crop nutrition constraint signifies an innovative paradigm in agricultural sciences.

References

Agarwal S, Pandey V (2004) Antioxidant enzyme responses to NaCl stress in cassia angustifolia. Biol Plant 48:555–560

Agastian P, Kingsley SJ, Vivekanandan M (2000) Effect of salinity on photosynthesis and biochemical characteristics in mulberry genotypes. Photosynthetica 38:287–290

Ahmad P, Hakeem KR, Kumar A, Ashraf M, Akram NA (2012) Salt-induced changes in photosynthetic activity and oxidative defense system of three cultivars of mustard (Brassica juncea L). Afr J Biotechnol 11:2694–2703

Ahmad M, Zahir ZA, Nazli F, Akram F, Muhammad A, Khalid M (2013) Effectiveness of halotolerant, auxin producing pseudomonas and rhizobium strains to improve osmotic stress tolerance in mung bean (*Vigna radiata* L.). Braz J Microbiol 44:1341–1348

Akhtar J, Haq T, Shahzad A, Haq MA, Ibrahim M, Ashraf N (2003) Classification of different wheat genotypes in salt tolerance categories on the basis of biomass production. Int J Agric Biol 5:322–325

Alamgir AN, Ali MY (1999) Effect of salinity on leaf pigments, sugar and protein concentrations and chloroplast ATPAase activity of rice (*Oryza sativa* L.). Bangladesh J Bot 28:145–149

Allakhverdiev SI, Nishiyama Y, Miyairi S (2002) Salt stress inhibits the repair of photodamaged photosystem II by suppressing the transcription and translation of psbA genes in synechocystis. Plant Physiol 130:1443–1453

Alpaslan M, Gunes A (2001) Interactive effects of boron and salinity stress on the growth, membrane permeability and mineral composition of tomato and cucumber plants. Plant Soil 236:123–128

Amini F, Ehsanpour AA (2005) Soluble proteins, proline, carbohydrates and Na^+/K^+ changes in two tomato (*Lycopersicon esculentum* Mill.) cultivars under in vitro salt stress. Am J Biochem Biotechnol 1:204–208

Ardic M, Sekmen AH, Tokur S, Ozdemir F, Turkan I (2009) Antioxidant response of chickpea plants subjected to boron toxicity. Plant Biol 11:328–338

Ashraf M (2009) Biotechnological approach of improving plant salt tolerance using antioxidants as markers. Biotech Adv 27:84–93

Ashraf M, Foolad MR (2007) Roles of glycine betaine and proline in improving plant abiotic stress tolerance. Environ Exp Bot 59:206–216

Ashraf M, McNeilly TM (1988) The potential for exploiting variation in salinity tolerance in pearl millet (*Pennisetum americanum* (L.) Leeke). Plant Breed 108:234–240

Ashraf M, Shahbaz M (2003) Assessment of genotypic variation in salt tolerance of early CIMMYT hexaploid wheat germplasm using photosynthetic capacity and water relations as selection criteria. Photosynthetica 41:273–280

Ashraf M, Akram NA, Arteca RN, Foolad MR (2010) The physiological, biochemical and molecular roles of brassinosteroids and salicylic acid in plant processes and salttolerance. Crit Rev Plant Sci 29:162–190

Asmare HA (2013) Impact of salinity on tolerance, vigor, and seedling relative water content of haricot bean (*Phaseolus vulgaris* L.) cultivars. J Plant Sci 1:22–27

Ayers RS, Westcot DW (1985) Water quality for agriculture. FAO irrigation drainage paper 29(1). FAO UN Rome 174 pp

Azevedo-Neto AD, Prisco JT, Eneas-Filho J, Braga-de-abreu-Eneas CE, Gomes-Filho E (2006) Effect of salt stress on antioxidative enzymes and lipid peroxidation in leaves and roots of salt-tolerant and salt-sensitive maize genotypes. Environ Exp Bot 56:87–94

Barakat N, Laudadio V, Cazzato E, Tufarelli V (2013) Antioxidant potential and oxidative stress markers in wheat (*Triticum aestivum*) treated with phytohormones under salt-stress condition. Int J Agric Biol 15:843–849

Bastias E, Fernandez-Garcia N, Carvajal M (2004) Aquaporin functionality in roots of *Zea mays* in relation to the interactive effects of boron and salinity. Plant Biol 6:415–421

Bastias E, Alcaraz-Lopez C, Bonilla I, Martinez-Ballesta MC, Bolanos L, Carvajal M (2010) Interactions between salinity and boron toxicity in tomato plants involve apoplastic calcium. J Plant Physiol 167:54–60

Bayuelo-Jimenez JS, Debouck DG, Lynch JP (2003) Growth, gas exchange, water relations and ion composition of Phaseous species grown under saline conditions. Field Crop Res 80:207–222

Ben-Gal A, Shani U (2002) Yield, transpiration and growth of tomatoes under combined excess boron and salinity stress. Plant Soil 247:211–221

Bowler C, Von MM, Lnze D (1992) Superoxide dismutase and stress tolerance. Annu Rev Plant Mol Biol 43:83–116

Brown PH, Hu H (1996) Phloem mobility of boron is species dependent: evidence for phloem mobility is sorbitol-rich species. Ann Bot 77:497–505

Cervilla LM, Blasco B, Rios JJ, Romero L, Ruiz JM (2007) Oxidative stress and antioxidants in tomato (*Solanum lycopersicum*) plants subjected to boron toxicity. Ann Bot 100:747–756

Chartzoulakis K, Klapaki G (2000) Response of two greenhouse pepper hybrids to NaCl salinity during different growth stages. Sci Hortic 86:247–260

Chen GX, Asada K (1989) Ascorbate peroxidase in tea leaves: occurrence of two isozymes and the differences in their enzymatic and molecular properties. Plant Cell Physiol 30:987–998

Chen LS, Han S, Qi YP, Yang LT (2012) Boron stresses and tolerance in citrus. Afr J Biotechnol 11:5961–5969

Chesworth W (1991) Geochemistry of micronutrients. In: Micronutrients in agriculture, 2nd edn. Soil Science Society of America, Madison, pp 1–30

Choi EY, Kolesil P, Mcneill A, Collins H, Zhang QS, Huynh BL, Graham R, Stangoulis J (2007) The mechanism of boron tolerance for maintenance of root growth in barley (*Hordeum vulgare* L.). Plant Cell Environ 30:984–993

Dat J, Vandenabeele S, Vranova E, Van-Montagu M, InzeD D, Van-Breusegem F (2000) Dual action of the active oxygen species during plant stress responses. Cell Mol Life Sci 57:779–795

Dordas C, Brown PH (2001) Evidence for channel mediated transport of boric acid in squash (Cucurbita pepo). Plant Soil 235:95–103

Dordas C, Brown PH (2005) Boron deficiency affects cell viability, phenolics leakage and oxidative burst in rose cell cultures. Plant Soil 268:293–301

Dordas C, Chrispeels MJ, Brown PH (2000) Permeability and channel-mediated transport of boric acid across membrane vesicles isolated from squash roots. Plant Physiol 124:1349–1362

Dreccer MF, van Herwaarden AF, Chapman SC (2009) Grain number and grain weight in wheat lines contrasting for stem water soluble carbohydrate concentration. Field Crop Res 112:43–54

Dube B, Sinha P, Chatterjee C (2000) Boron stress affects metabolism and seed quality of sunflower. Trop Agric 77:89–92

Dudley LM (1994) Salinity in the soil environment. In: Pessarakli M (ed) Handbook of plant and crop stress. Mercel Dekker, New York, pp 13–30

Eaton FM (1944) Deficiency, toxicity and accumulation of boron in plants. J Agric Res 69:237–277

Edelstein M, Ben-Hur M, Cohen R, Burger Y, Ravina I (2005) Boron and salinity effects on grafted and non-grafted melon plants. Plant Soil 269:273–284

El-Hamdaoui A, Redondo-Nieto M, Rivilla R, Bonilla I, Bolanos L (2003) Effects of boron and calcium nutrition on the establishment of the Rhizobium leguminosarum-pea (*Pisum sativum*) symbiosis and nodule development under salt stress. Plant Cell Environ 26:1003–1011

El-Motaium R, Hu H, Brown PH (1994) The relative tolerance of six prunus rootstocks to boron and salinity. J Am Soc Hortic Sci 119:1169–1175

Eraslan F, Inal A, Savasturk O, Gunes A (2007a) Changes in antioxidative system and membrane damage of lettuce in response to salinity and boron toxicity. Sci Hortic 114:5–10

Eraslan F, Inal A, Gunes A, Alpaslan M (2007b) Impact of exogenous salicylic acid on the growth, antioxidant activity and physiology of carrot plants subjected to combined salinity and boron toxicity. Sci Hortic 113:120–128

Eraslan F, Inal A, Gunes A, Alpaslan M (2007c) Boron toxicity alters nitrate reductase activity, proline accumulation, membrane permeability and mineral constituents of tomato and pepper plants. Sci Hortic 30:981–994

Eraslan F, Inal A, David JP, Gunes A (2008) Interactive effects of salicylic acid and silicon on oxidative damage and antioxidant activity in spinach (*Spinacia oleracea* L. cv. Matador) grown under boron toxicity and salinity. Plant Growth Regul 55:207–219

FAO (2008) FAO land and plant nutrition management service. Available at http://www.fao.org/ag/agl/agll/spush

Fayez KA, Bazaid SA (2014) Improving drought and salinity tolerance in barley by application of salicylic acid and potassium nitrate. J Saudi Soc Agric Sci 13:45–55

Ferreyra RE, Alijaro AU, Ruiz RS, Rojas LP, Oster JD (1997) Behavior of 42 crop species grown in saline soils with high boron concentrations. Agric Water Manag 32:111–124

Fitzpatrik KL, Reid RJ (2009) The involvement of aquaglyceroporins in transport of boron in barley roots. Plant Cell Environ 32:1357–1365

Flowers TJ (2004) Improving crop salt tolerance. J Exp Bot 55:307–319

Gao S, Luo TC, Zhang BR, Zhang HF, Han YW, Zhao ZD, Hu YK (1998) Chemical composition of the continental crust as revealed by studies in East China. Geochim Cosmochim Acta 62:1959–1975

Garcia POC, Rivero RM, Lopez-Lefebre LR, Sanchez E, Ruiz JM, Romero L (2001) Response of oxidative metabolism to the application of carbendazim plus boron in tobacco. Aus J Plant Physiol 28:801–806

Garratt AM, Schmidt L, Mackintosh A, Fitzpatrick R (2002) Quality of life measurement: bibliographic study of patient assessed health outcome measures. Br Med J 324:1417–1421

Geebelen W, Vangronsveld J, Adriano DC, Carleer R, Clijsters H (2002) Amendment-induced immobilization of lead in lead-spiked soil: evidence from phytotoxicity studies. Water Air Soil Pollut 140:261–277

Ghafoor A, Nadeem SM, Hassan A, Sadiq M (2001) Reclamation response of two different textured saline sodic soils to EC: SAR ratios. Pak J Soil Sci 19:84–91

Ghassemi F, Jakerman AJ, Nix HA (1995) Salinization of land water resources: human causes, extent, management and case studies. CAB International, Wallingford

Ghogdi EA, Izadi-Darbandi A, Borzouei A (2012) Effects of salinity on some physiological traits in wheat (*Triticum aestivum* L.) cultivars. Indian J Sci Technol 5:1901–1906

Ghoulam C, Foursy A, Fares K (2002) Effects of salt stress on growth, inorganic ions and proline accumulation in relation to osmotic adjustment in five sugar beet cultivars. Environ Exp Bot 47:39–50

Gossett DR, Millhollon EP, Lucas MC (1994) Antioxidants responses to NaCl stress in salt-tolerant and salt-sensitive cultivars of cotton. Crop Sci 34:706–714

Grattan SR, Grieve CM (1992) Mineral element acquisition and growth response of plants grown in saline environments. Agric Ecosyst Environ 38:275–300

Greenwood NN, Earnshaw A (1984) Chemistry of the elements. Pergamon press, Oxford 1542 seiten, 7 anhänge Preis

Grieve CM, Poss JA (2000) Wheat response to interactive effects of boron and salinity. J Plant Nutr 23:1217–1226

Grieve CM, Poss JA, Grattan SR, Suarez DL, Smith T (2010) The combined effects of salinity and excess boron on mineral ion relations in broccoli. Sci Hortic 125:179–187

Grieve CM, Grattan SR, Maas EV (2012) Plant salt tolerance. In: Wallender WW, Tanji KK (eds) Agricultural salinity assessment and management, 2nd edn. ASCE, Reston, pp 405–459

Gueta-Dahan Y, Yaniv Z, Zilinskas BA, Ben-Hayyim G (1997) Salt and oxidative stress: similar and specific responses and their relation to salt tolerance in citrus. Planta 203:460–469

Guidi L, Degl'Innocenti E, Carmassi G, Massaand D, Pardossi A (2011) Effects of boron on leaf chlorophyll fluorescence of greenhouse tomato grown with salinewater. Environ Exp Bot 73:57–63

Gunes A, Soylemezoglu G, Inal A, Bagci EG, Coban S, Sahin O (2006) Antioxidant and stomatal responses of grapevine (*Vitis vinifera* L.) to boron toxicity. Sci Hortic 110:279–284

Gupta UC (1993) Deficiency, sufficiency, and toxicity levels of boron in crops. In: Gupta UC (ed) Boron and its role in crop production. CRC Press, Boca Raton, pp 137–145

Gupta B, Huang B (2014) Mechanism of salinity tolerance in plants: physiological, biochemical and molecular characterization. Int. J Genom Article ID 701596, 18 pages Available online http://dx.doi.org/10.1155/2014/701596

Gupta UC, Jame YW, Campbell CA, Leyshon AJ, Nicholaichuk W (1985) Boron toxicity and deficiency: a review. Can J Soil Sci 65:381–409

Hakeem KR (2015) Crop production and global environmental issues. Springer International Publishing AG, Gewerbestrasse 11, 6330 Cham, Switzerland, 598 pp

Hakeem KR, Khan F, Chandna R, Siddiqui TO, Iqbal M (2012) Genotypic variability among soybean genotypes under NaCl stress and proteome analysis of salt tolerant genotype. Appl Biochem Biotechnol 168:2309–2329

Han S, Tang N, Jiang HX, Yang LT, Lee Y, Chen LS (2009) CO2 assimilation, photosystem II photochemistry, carbohydrate metabolism and antioxidant system of citrus leaves in response to boron stress. Plant Sci 176:143–153

Hasanuzzaman M, Hossain MA, Teixeira da Silva JA, Fujita M (2012) Plant responses and tolerance to abiotic oxidative stress: antioxidant defenses is a key factors. In: Bandi V, Shanker AK, Shanker C, Mandapaka M (eds) Crop stress and its management: perspectives and strategies pp. Springer, Berlin, pp 261–316

Hasegawa PM (2013) Sodium (Na^+) homeostasis and salt tolerance of plants. Environ Exp Bot 92:19–31

Hasegawa PM, Bressan RA, Zhu JK, Bohnert HJ (2000) Plant cellular and molecular responses to high salinity. Annu Rev Plant Physiol Plant Mol Biol 51:463–499

Hayat R, Ali S, Amara U, Khalid R, Ahmed I (2010) Soil beneficial bacteria and theirrole in plant growth promotion: a review. Ann Microbiol 60:579–598

Hayat S, Mir BA, Wani AS, Hasan SA, Irfan M, Ahmad A (2011) Screening of salt tolerant genotypes of Brassica juncea based on photosynthetic attributes. J Plant Int 6:53–60

Heidari M (2012) Effects of salinity stress on growth, chlorophyll content and osmotic components of two basil (*Ocimum basilicum* L.) genotypes. Afr J Biotechnol 11:379–384

Hernadez JA, Olmos E, Corpas FJ, Sevilla F, del-Rio LA (1995) Salt-induced oxidative stress in chloroplasts of pea plants. Plant Sci 105:151–167

Hernandez JA, Ferrer MA, Jimenez A, Ros-Barcelo A, Sevilla F (2001) Antioxidant systems and O_2- /H_2O_2 production in the apoplast of *Pisum sativum* L. leaves: its relation with NaCl-induced necrotic lesions in minor veins. Plant Physiol 127:817–831

Holloway RE, Alston M (1992) The effects of salt and boron on growth of wheat. Aus J Agric Res 43:987–1001

HSDB (2003) Information from the hazardous substances data bank. Xylene. Hazardous Substances Data Bank. Toxicology Data Network System. National Library of Medicine. Bethesda, MD. Available at http://www.plantstress.com/Articles/index.asp

Ijaz K, Davidson AP (1998) Baseline socio-economic survey, Joint Satiana Project. IWASDI, Lahore, p 77

Iyengar ER, Reddy MP (1996) Photosynthesis in highly salt-tolerant plants. In: Pessaraki M (ed) Handbook of photosynthesis. Marcel Dekker, Baten Rose, pp 897–909

James RA, Blake C, Byrt CS, Munns R (2011) Major genes for Na^+ exclusion, Nax1 and Nax2 (wheat HKT1;4 and HKT1;5), decrease Na^+ accumulation in bread wheat leaves under saline and waterlogged conditions. J Exp Bot 62:2939–2947

Karabal E, Yucel M, Oktem HA (2003) Antioxidant responses of tolerant and sensitive barleycultivars to boron toxicity. Plant Sci 164:925–933

Kaya C, Tuna AL, Dikilitas M, Ashraf M, Koskeroglu S (2009) Supplementary phosphorus can alleviate boron toxicity in tomato. Sci Hortic 121:284–288

Keles Y, Oncel I, Yenice N (2004) Relationship between boron content and antioxidant compounds in citrus leaves taken from fields with different water source. Plant Soil 265:345–353

Kerepesi I, Galiba G (2000) Osmotic and salt stress-induced alteration in soluble carbohydrate content in wheat seedlings. Crop Sci 40:482–487

Kerepesi I, Galiba G, Banyai E (1998) Osmotic and salt stresses induced differential alteration in water-soluble carbohydrate content in oat seedlings. J Agric Food Chem 46:5347–5354

Khan G (1998) Soil salinity/sodicity status in Pakistan. Soil survey of Pakistan, Lahore, 59p

Khan MA, Ungar IA (2000) Alleviation of salinity-enforced dormancy in *Atriplex griffithii* Moq. var. stocksii Boiss. Seed Sci Technol 28:29–37

Kot FS (2009) Boron sources, speciation and its potential impact on health. Rev Environ Sci Bio Technol 8:3–28

Landi M, Pardossi A, Remorini D, Guidia L (2013) Antioxidant and photosynthetic response of a purple-leaved and a green-leaved cultivar of sweet basil (*Ocimum basilicum*) to boron excess. Environ Exp Bot 85:64–75

Lee HW (2006) The relationship between boron content and crack properties in FCAW deposited metal. Weld J 85:131–136

Lee DH, Kim YS, Lee CB (2001) The induction response of antioxidant enzymes by salt stress in the rice (*Oryza sativa* L.). J Plant Physiol 158:737–745

Lemarchand D, Gaillardet J, Lewin E, Allegre CJ (2000) The influence of rivers on marine boron isotopes and implications for reconstructing past ocean pH. Nature 408:951–954

Levitt J (1980) Responses of plants to environmental stresses, vol 1. Acad Press, New York, p 496

Liu D, Jiang W, Zhang L, Li L (2000) Effects of boron ions on root growth and cell division of broad bean (*Vicia faba* L.). Isr J Plant Sci 48:47–51

Loomis WD, Durst RW (1992) Chemistry and biology of boron. Bio Factors 3:229–239

Lopez-Gomez E, San Juan MA, Diaz-Vivancos P, Mataix Beneyto J, Garcia-Legaz MF, Hernandez JA (2007) Effect of rootstocks grafting and boron on the antioxidant systems and salinity tolerance of loquat plants (*Eriobotrya japonica* Lindl). Environ Exp Bot 60:151–158

Lovatt CJ, Bates LM (1984) Early effects of excess boron on photosynthesis and growth of Curubita pepo. J Exp Bot 5:297–305

Lukaszewski KM, Blevins DG, Randall DD (1992) Asparagine and boric acid cause allantoate accumulation in soybean leaves by inhibiting manganese-dependent allantoate amidohydrolase. Plant Physiol 99:1670–1676

Mahmood IA, Ali A, Shahzad A, Salim M, Jamil M, Akhtar J (2007) Yield and quality of Brassica cultivars as affected by soil salinity. Pak J Sci Ind Res 50:133–137

Manchanda HR, Sharma SK (1991) Boron tolerance in wheat in relation to soilsalinity. J Agric Sci Camb 116:17–21

Manikandanm K, Desingh R (2009) Effect of salt stress on growth, carbohydrate and proline content of two finger millet varieties. Recent Res Sci Technol 1:48–51

Marschner H (1995) Mineral nutrition of higher plants, 2nd edn. Academic, London

Martinez-Ballesta MDC, Bastias E, Zhu C, Schaffner AR, GonzAlez-Moro B, Gonzalez-Murua C, Carvajal M (2008a) Boric acid and salinity effects on maize roots. Response of aquaporins ZmPIP1 and ZmPIP2, and plasma membrane H^+-ATPase, in relation to water and nutrient uptake. Physiol Plant 132:479–490

Martinez-Ballesta MC, Cabanero FJ, Olmos E, Periago P, Maurel C, Carvajal M (2008b) Two different effects of calcium on aquaporin in salinity stressed pepper plants. Planta 228:15–25

Masood S, Saleh L, Witzel K, Plieth C, Muhling KH (2012a) Determination of oxidative stress in wheat leaves as influenced by boron toxicity and NaCl stress. Plant Physiol Biochem 56:56–61

Masood S, Wimmer MA, Witzel K, Zorb C, Muhling KH (2012b) Interactive effects of high boron and NaCl stresses on subcellular localization of chloride and boron in wheat leaves. J Agron Crop Sci 198:227–235

McCue KF, Hanson AD (1990) Drought and salt tolerance: towards understanding and application. Trends Biotechnol 8:358–362

Meloni DA, Oliva MA, Ruiz HA, Martinez CA (2001) Contribution of proline and inorganic solutes to osmotic adjustment in cotton under salt stress. J Plant Nutr 24:599–612

Misra AN, Sahu M, Misra P, Singh I, Meera N, Das N, Kar M, Sahu P (1997) Sodium chloride induced changes in leaf growth, and pigment and protein contents in two rice cultivars. Biol Plant 39:257–262

Mittler R (2002) Oxidative stress, antioxidants and stress tolerance. Trends Plant Sci 7:405–410

Moghadam SF, Ebrahimi M, Najafabadi MF, Ramshini H, Noori SAS, Mansouri S (2014) The effect of salinity on antioxidant enzyme activity in bread wheat genotypes. The first international conference on new ideas in agriculture. Islamic Azad university khorasgan branch. 26–27 January, 2014. Isfahan, Iran

Molassiotis A, Sotiropoulos T, Tanou G, Diamantidis G, Therios L (2006) Boron induced oxidative damage and antioxidant and nucleolytic responses in shoot tips culture of the apple rootstock EM9 (*Malus domestica* Borkh). Environ Exp Bot 56:54–62

Moore JW (1991) Inorganic contaminants of surface water: research and mining priorities, 1st edn. New York, Springer

Munns R (2002) Comparative physiology of salt and water stress. Plant Cell Environ 25:239–250

Munns R (2005) Salinity stress and its impact. In: Blum A (ed) Plant stress. Available at http://www.plantstress.com/Articles/index.asp.

Munns R, James RA (2003) Screening methods for salinity tolerance: a case study with tetraploid wheat. Plant Soil 253:201–218

Munns R, Tester M (2008) Mechanisms of salinity tolerance. Annu Rev Plant Bio 59:651–681

Munns R, Schachtman DP, Condon AG (1995) The significance of a two-phase growth response to salinity in wheat and barley. Aus J Plant Physiol 22:561–569

Munns R, Guo J, Passioura JB, Cramer GR (2000) Leaf water status controls day-time but not daily rates of leaf expansion in salt-treated barley. Aus J Plant Physiol 27:949–957

Munns R, James RA, Lauchli A (2006) Approaches to increasing the salt tolerance of wheat and other cereals. J Exp Bot 57:1025–1043

Nable RO (1988) Resistance to boron toxicity amongst several barley and wheat cultivars: a preliminary examination of the resistance mechanism. Plant Soil 112:45–57

Nable RO (1992) Mechanism of tolerance to boron toxicity in plants. In: Mann CE, Rerkasem B (eds) Boron deficiency in wheat. CIMMYT Mexico, Mexico, pp 98–109

Nable RO, Banuelos GS, Paull JG (1997) Boron toxicity. Plant Soil 193:181–198

Nadav N (1999) Boron removal from seawater reverse osmosis permeate utilizing selective ion exchange resin. Desalination 124:131

Niknam V, Razavi N, Ebrahimzadeh H, Sharifizadeh B (2003) Effect of NaCl on biomass, protein and proline contents, and antioxidant enzymes in seedlings and calli of twoTrigonella species. Biol Plant 50:591–596

Noctor G, Foyer C (1998) Ascorbate and glutathione: keeping active oxygen under control. Annu Rev Plant Physiol Plant Mol Biol 49:249–279

O'Neill MA, Ishii T, Albersheim P, Darvill AG (2004) Rhamnogalacturonan II: structure and function of a borate cross-linked cell wall pectic polysaccharide. Annu Rev Plant Biol 55:109–139

Oertli JJ (1994) Non-homogeneity of boron distribution in plants and consequences for foliar diagnosis. Commun Soil Sci Plant Anal 25:1133–1147

Oertli JJ, Roth JA (1969) Boron nutrition of sugar beet, cotton and soybean. Agron J 61:191–195

Papadakis IE, Dimassi N, Bosabalidis AM, Therios IN, Patakas A, Giannakoula A (2004) Boron toxicity in 'Clementine' mandarin plants grafted on two rootstocks. Plant Sci 166:539–547

Parida AK, Das AB, Mittra B (2003) Effects of NaCl stress on the structure, pigmentcomplex composition, and photosynthetic activity of mangrove Bruguiera parviflorachloroplasts. Photosynthetica 41:191–200

Parida AK, Das AB, Mittra B (2004) Effects of salt on growth, ion accumulation photosynthesis and leaf anatomy of the mangrove, *Bruguiera parviflora*. Trees Struct Funct 18:167–174

Parks JL, Edwards M (2005) Boron in the environment. Crit Rev Environ Sci Biotechnol 35:81–114

Pereira WE, de-Siqueira DL, Martinez CA, Puiatti M (2000) Gas exchange and chlorophyll fluorescence in four citrus rootstocks under aluminium stress. J Plant Physiol 157:513–520

Picchioni GA, Miyamoto S (1991) Boron uptake and effects on growth and carbohydrate partitioning of pistachio seedlings. Soc Hortic Sci 116:706–711

Pitann B, Schubert S, Muhling KH (2009) Decline in leaf growth under salt stress is due to an inhibition of H^+ pumping activity and increase in apoplastic pH of maize leaves. J Plant Nutr Soil Sci 172:535–543

Qureshi RH, Barrett–Lennard EG (1998) Salt and waterlogging effects on plants. In: Saline Agriculture for Irrigated Lands in Pakistan. A H book, ACIAR, Canbera, Australia, pp 37–49

Rahdari P, Tavakoli S, Hosseini SM (2012) Studying of salinity stress effect on germination, proline, sugar, protein, lipid and chlorophyll content in Purslane (*Portulaca oleraceae* L) leaves. Stress Physiol Bio J 8:182–193

Rahnama A, Poustini K, Munns R, James A (2010) Stomatal conductance as a screen for osmotic stress tolerance in durum wheat growing in saline soil. Funct Plant Biol 37:255–263

Reddy MP, Sanish S, Iyengar ERR (1992) Photosynthetic studies and compartmentation of ions in different tissues of Salicornia brachiata Roxb. Under saline conditions. Photosynthetica 26:173–179

Reid R (2010) Can we really increase yields by making crop plants tolerant to boron toxicity? Plant Sci 178:9–11

Reid RJ, Hayes JE, Post A, Stangoulis JCR, Graham RD (2004) A critical analysis of the causes of boron toxicity in plants. Plant Cell Environ 25:1405–1414

Rengasamy P (2002) Transient salinity and subsoil constraints to dryland farming in Australian sodic soils: an overview. Aus J Exp Agric 42:351–361

Riley MM, Robson AD, Deller GA, Gartrell JW (1994) Critical toxic concentrations of boron are variable in barley. J Plant Nutr 17:1701–1719

Royo A, Abio D (2003) Salt tolerance in durum wheat cultivars. Span J Agric Res 1:27–35

Ruuska SA, Rebetzke GJ, van Herwaarden AF, Richards RA, Fettell NA, Tabe L, Jenkins CLD (2006) Genotypic variation in water-soluble carbohydrate accumulation in wheat. Funct Plant Biol 33:799–809

Sairam RK, Srivastava GC (2002) Changes in antioxidant activity in sub-cellular fractions of tolerant and susceptible wheat genotypes in response to long term salt stress. Plant Sci 162:897–904

Sairam RK, Srivastava GC, Agarwal S, Meena RC (2005) Differences in antioxidant activity in response to salinity stress in tolerant and susceptible wheat genotypes. Biol Plant 49:85–91

Saleem M, Ashraf M, Akram NA (2011) Salt (NaCl) induced modulation in some key physio-biochemical attributes in okra (*Abelmoschus esculentus* L). J Agron Crop Sci 197:202–213

Serrano R, Gaxiola R (1994) Microbial models and salt stress tolerance in plants. Crit Rev Plant Sci 13:121–138

Shannon MC (1998) Adaptation of plants to salinity. Adv Agron J 60:75–119

Shannon MC, Grieve CM, Francois LE (1994) Whole-plant response to salinity. In: Wilkinson RE (ed) Plant-environment interactions. Dekker, New York, pp 199–244

Sheng O, Zhou GF, Wei QJ, Peng SA, Deng XX (2010) Effects of excess boron on growth, gas exchange, and boron status of four orange scion–rootstock combinations. J Plant Nutr Soil Sci 173:469–476

Shorrocks VM (1997) The occurrence and correction of boron deficiency. Plant Soil 193:121–148

Smith TE, Grattan SR, Grieve CM, Poss JA, Suarez DL (2010a) Salinity's influence on boron toxicity in broccoli: I. Impacts on yield, biomass distribution and water use. Agric Water Manag 97:777–782

Smith TE, Grattan SR, Grieve CM, Poss JA, Suarez DL (2010b) Salinity's influence on boron toxicity in broccoli: II. Impacts on boron uptake, uptake mechanisms and tissue ion relations. Agric Water Manag 97:783–791

Smith T, Grattan S, Grieve C, Poss J, Lauchli A, Suarez D (2013) pH dependent salinity-boron interactions impact yield, biomass, evapotranspiration and boron uptake in broccoli (*Brassica oleracea* L.). Plant Soil 370:541–554

Sotiropoulos TE, Therios NI, Dimassi NK, Bosbalidis A, Kofilids G (2002) Nutritional status, growth, CO_2 assimilation and leaf anatomical responses in two kiwi fruit species under boron toxicity. J Plant Nutr 25:1249–1261

Sotiropoulos TE, Molassiotis A, Almaliotis D, Mouhtaridou G, Dimassi K, Diamantidis G (2006) Growth, nutritional status, chlorophyll content, and antioxidant responses of the apple rootstock MM 111 shoots cultured under high boron concentrations in vitro. J Plant Nutr 29:575–583

Soylemezoglu G, Demir K, Inal A, Gunes A (2009) Effect of silicon on antioxidant and stomatal response of two grapevine (*Vitis vinifera* L.) rootstocks grown in boron toxic, saline and boron toxic-saline soil. Sci Hortic 123:240–246

Thompson JE, Legge RL, Barber RF (1987) The role of free radicals in senescence and wounding. New Phytol 105:317–344

Tsadilas CD (1997) Soil contamination with boron due to irrigation with treated municipal waste water boron in soils and plants. In: Bell RW, Rerkasem B (eds) Boron in soils and plants, volume 76 of the series developments in plant and soil sciences. Kluwer Academic Publishers, Dordrecht, pp 265–270

Van den Ende W (2013) Multifunctional fructans and raffinose family oligosaccharides. Front Plant Sci 4:247

Velez-Ramirez AI, Van-Ieperen W, Vreugdenhil D, Millenaar FF (2011) Plants under continuous light. Trends Plant Sci 16:310–318

Vengosh A, Heumann KG, Juraske S, Kasher R (1994) Boron isotope application for tracing sources of contamination in groundwater. Environ Sci Technol 28:1968–1974

Wang WQ, Lin P (2000) Study on membrane lipid peroxidation of the leaves of kandelia candel seedlings to longterm and short-term salinity (in Chinese). Acta Oceanol Sin 22:49–54

Wang JZ, Tao ST, Qi KJ, Wu HQ, Zhang SL (2011) Changes in photosynthetic properties and antioxidative system of pear leaves to boron toxicity. Afr J Biotechnol 10:19693–19700

Wani AS, Ahmad A, Hayat S, Fariduddin Q (2013) Salt-induced modulation in growth, photosynthesis and antioxidant system in two varieties of *Brassica juncea*. Saudi J Biol Sci 20:183–193

Wimmer MA, Goldbach HE (2012) Boron-and-salt interactions in wheat are affected by boron supply. J Plant Nutr Soil Sci 175:171–179

Wimmer MA, Muhling KH, Lauchli A, Brown PH, Goldbach EH (2003) The interaction between salinity and boron toxicity affects the subcellular distribution of ions and proteins in wheat leaves. Plant Cell Environ 26:1267–1274

Wimmer MA, Bassil ES, Brown PH, Läuchli A (2005) Boron response in wheat is genotype-dependent and related to boron uptake, translocation, allocation, plant phenological development and growth rate. Funct Plant Biol 32:507–515

Yamaguchi T, Blumwald E (2005) Developing salt-tolerant crop plants: challenges and opportunities. Trends Plant Sci 10:615–620

Yan K, Chen P, Shao H, Zhao S, Zhang L, Xu G, Sun J (2012) Responses of photosynthesis and photosystem II to higher temperature and salt stress in Sorghum. J Agron Crop Sci 198:218–226

Yermiyahu U, Ben-Gal A, Keren R, Reid RJ (2008) Combined effect of salinity and excess boron on plant growth and yield. Plant Soil 304:73–87

Zhu JK (2001) Plant salt tolerance. Trends Plant Sci 6:66–71

Zhu JK (2003) Regulation of ion homeostasis under salt stress. Curr Opin Plant Biol 6:441–445

Zhu Z, Wei G, Li J, Qian Q, Yu J (2004) Silicon alleviates salt stress and increases antioxidant enzymes activity in leaves of salt-stressed cucumber (*Cucumis sativus* L.). Plant Sci 167:527–533

Silicon: A Beneficial Nutrient Under Salt Stress, Its Uptake Mechanism and Mode of Action

Waqas-ud-Din Khan, Tariq Aziz, Muhammad Aamer Maqsood, M. Sabir, Hamaad Raza Ahmad, Pia Muhammad Adnan Ramzani, and M. Naseem

Contents

Abstract Food security is a serious issue in this era of rapidly growing population. Food security is threatened due to low crop yields around the world due to different biotic and abiotic stresses. The arable lands are decreasing due to different soil degradation process and thus prospect for increasing crop yields through extending areas under cultivation are not very bright. Hence, to achieve the food security on a sustainable basis, it is necessary to utilize degraded soils productively. Among the degraded soils, Salt-affected soils share the major fraction and their presence is prevalent in all continents. The adoption of different management techniques for productive use of salt-affected soils is thus pre-requisite for enhancing crop yields on such soils. These management strategies include management of irrigation water,

W.-ud.-D. Khan (✉)
Institute of Soil and Environmental Sciences, University of Agriculture Faisalabad, Faisalabad, Pakistan

Sustainable development study center, Government College University, Lahore-54000, Pakistan
e-mail: waqas1919@gmail.com

T. Aziz • M.A. Maqsood • M. Sabir • H.R. Ahmad • P.M.A. Ramzani • M. Naseem
Institute of Soil and Environmental Sciences, University of Agriculture Faisalabad, Faisalabad 38040, Pakistan

K.R. Hakeem et al. (eds.), *Soil Science: Agricultural and Environmental Prospectives*, DOI 10.1007/978-3-319-34451-5_12

reclamation techniques, raising beds, organic matter and salt tolerant crops like kallar grass. In addition to these, better fed plants have good potential to withstand with salinity and other stresses. Exogenous application of nutrients can alleviate the detrimental effects of salts. Silicon (Si) as a beneficial nutrient is known to improve plant growth particularly under abiotic stresses. It is helpful for plants in many ways as it improves plant water status in the context of relative water content and transpiration rate. The role of Si to promote the plant growth under salt-affected soils is reviewed in detail in this chapter.

Keywords Silicon • Salt stress • Degraded soils • Sustainable agriculture

1 Introduction

Food security is a serious issue in developing countries because of ever-increasing population. According to recent estimates, more than 2 billion people lack food security occasionally due to poverty and/or natural calamities. The problem can be tackled by; (a) efficient utilization of the natural resources (b) proper application of physical inputs (c) increasing production of major food crops through either increasing area under crop production or improving crop yields per acre. To increase area under crop production is not feasible rather not possible. The area available for agriculture is decreasing day by day due to: (a) urbanization of arable land (b) poor soil management practices with intensive use of cultivation (Gruhn et al. 2000; Cakmak 2002) (c) degradation of the existing arable land due to various abiotic factors including salinity and drought. Rengasamy (2006) reported that out of 13 billion hectares of total land, one billion is salt affected, including 30 % of all irrigated land. Hence, for sustaining food security, a high priority should be given to safe use of these salt affected soils. Salinity, not only, induces ion toxicity and physiological drought, but also reduces water use efficiency and photosynthesis due to interveinal chlorosis which ultimately decrease crop yields. Salinity poses the osmotic stress to the plants so excess of soluble salts present around the root zone creating 'physiological drought' . When these salts are taken up by plant they cause ion toxicity particularly due to Na^+. Salt stress also causes oxidative stress in plants which leads to the production of reactive oxygen species. Reactive oxygen species (ROS) are highly reactive species produced inside the plant cell by more than one ways like; free oxygen radical (O_2•, OH•, HO_2•, RO•, alkoxy radicals, superoxide radicals; hydroxyl radical; perhydroxy radical) and non-radical (molecular) forms (H_2O_2, $^1O_2$, hydrogen peroxide and singlet oxygen). There are different mechanisms adapted by plants to detoxify ROS and maintain their healthy growth in saline environment. Production of antioxidant enzymes like superoxide dismutase (SOD) which dismutated super oxide radical O_2• which is reduced to H_2O_2 and O_2 (Prashanth 2008). Then this H_2O_2 is converted into H_2O by catalases (CAT), ascorbate peroxidase (APX) and glutathione peroxidase (GPX). The major reason of reduced growth of crops in salt stressed condition is specific ion toxicity (certain ions like Na and Cl uptake at elevated level) (Chinnusamy et al. 2005; Tahir et al. 2011, 2012; Ahmad et al. 2012). Therefore, to enhance Na^+ exclusion and K^+ absorption there by maintaining optimum K^+/Na^+ratio, is crucial for salt tolerance in higher plants (Zhu 2001; Liu et al. 2015).

As crop yield and its sustainability is a pre-requisite for food security and self-sustainability. It is therefore pre-requisite that degraded soils particularly salt-affected soils should be brought under cultivation. It is highly recommended to adopt strategies aiming at increased crop production on salt affected lands (Irrigation water, raising beds, organic matter and salt tolerant crops like kallar grass). All of these strategies have some advantages and disadvantages. Judicious use of mineral nutrition is a recommended shotgun strategy as it strengthens the cereals to cope against salt stress. Exogenous application of K^+ in wheat (Akram et al. 2010) ameliorated the detrimental effects of salts. Silicon (Si)as a beneficial nutrient is known to improve plant growth particularly under abiotic stresses. It is helpful for plants in many ways as it improves plant water status in context of relative water content and transpiration rate (Romero-Aranda et al. 2006), ameliorates the harmful effects of salinity on chlorophyll content and plant biomass (Tuna et al. 2008) in both leaves and roots, it lowers significantly the Na^+ concentrations (Kafi and Rahimi 2011; Tahir et al. 2011). The beneficial effects may vary within the cereal species. Cereals that accumulate maximum Si in their shoots usually performed better than the others (Ma et al. 2001). Rice is a typical example, which accumulates up to 10 % Si on a dry weight basis in the shoot. Higher contents of Si in rice have been revealed to be essential for healthy growth, stable yield and production. For this reason, Si has been acknowledged as an "agronomically essential element" in Japan and silicate fertilizers have been applied to paddy soils (Ma et al. 2011). The other typical beneficial effects of Si are usually expressed more clearly when plants are subjected to various abiotic and biotic stresses. In this chapter, role of silicon under abiotic stresses particularly salt stress is discussed in detail.

2 Silicon as a Beneficial Nutrient

Silicon is the most prevalent nutrient after oxygen in earth crust (Epstein 2000) and can make up the plant body as much as 0.1–10 %. Its availability in plant body remains satisfactory in normal conditions and no exogenous application of Si is required by plant to complete its life cycle (Epstein and Bloom 2005). But in stress conditions, one cannot deny the fact that Si is quassi essential for plant. Silicic acid is the available form of Si in the soil solution (concentration upto 0.1–0.6 mM) and Si absorption takes place in monosilicic acid form by the plant roots from soil solution via transpiration stream. The polymerization of Si in the form of phtoliths ($SiO_2.nH_2O$) bodies takes place, when accumulation of silicic acid reaches upto a critical level of 100 ppm, that comprise the bulk of a plant's Si content (Exley 1998). In that context, significant amount of Si is present in the tissues of all plants growing in the soil medium (Ma et al. 2011). There are two types of Si deposited layers formed within cell wall of leaves and stem (1) silica-cuticle double layers and (2) silica-cellulose double layer (Raven 2001). Never the less, plants accumulate it in higher amounts and it can contribute upto 0.1 to 10 % of the dry matter of plants. This wide variation in Si concentration in plant tissues is attributed mainly to differences in the characteristics of Si uptake and transport (Liang et al. 2005).

Si-enriched plants are quite different from Si-deficient plants in their structure, mechanical strength, chemical composition, enzymatic activities, yield and yield contributing factors, metal toxicity, pest and disease resistance, drought and salt tolerance etc. (Epstein 2000). Adverse effects of salinity can be minimized by the application of Si; as it plays a multitude of roles in crop performance and plant existence. Na^+ uptake is reduced inside the plant when Si is present in soil solution (Tahir et al. 2012). Different silicate sources were compared with sulfate sources by Abou-Baker et al. (2011) to determine the effectiveness of salt for crop. They pointed out that silicon solutions significantly increased all measured parameters as compared with sulphate solutions, although potassium silicate was the best. Potassium silicate gave the highest K% values in plant tissue in contrast to $MgSO_4$ solution which gave the highest values of N% and P%. Silicon always remain an under rated nutrient and its role in plant growth and physiology never get acknowledgement until the beginning of the twentieth century. There are many reasons that most plant physiologists overlooked the beneficial effects of Si on plant body: First, Si remains an un-reactive element in soil plant system and secondly, it is present in the nature in quite abundantly and also present as a major inorganic constituent of plants, therefore, no visible symptoms of either Si toxicity or deficiency were appeared on plants (Richmond and Sussman 2003). There was a large amount of Si build up by certain crops especially, from Poaceae family (Mitani et al. 2005) so healthy and better growth ensured by its application to these crops. Usually higher amount of Si is accumulated and deposit in the tissues of graminaceous plants relative to other species (Matichenkov and Kosobrukhov 2004).

3 Mechanism of Silicon Uptake in Cereals

Silicon present in soil solution can provide great benefits to the tissues of cereal plants. Silicic acid [$Si(OH)_4$] is the soluble form of Si which is taken up by the plant roots through a systematic process (Ma and Takhashai 2002). In recent years, it has been reported in rice and maize that some genes are involved in Si uptake. Transporters are the specific genes inside the plants which express themselves with the uptake of specific nutrient. Different transporters are involved in Si uptake, its translocation, and distribution inside the cereal plant body. In 2006, a breakthrough in cereals occurs when first Si transporter (OsLsi1) was identified in rice (Ma et al. 2006). Since then, different types of Si transporters have been identified in rice (OsLsi1, OsLsi6) (Ma et al. 2006, 2007), wheat (TaLsi1), barley (HvLsi1, HvLsi6) [Chiba et al. 2009] and maize (ZmLsi1, ZmLsi6) (Mitani et al. 2009).

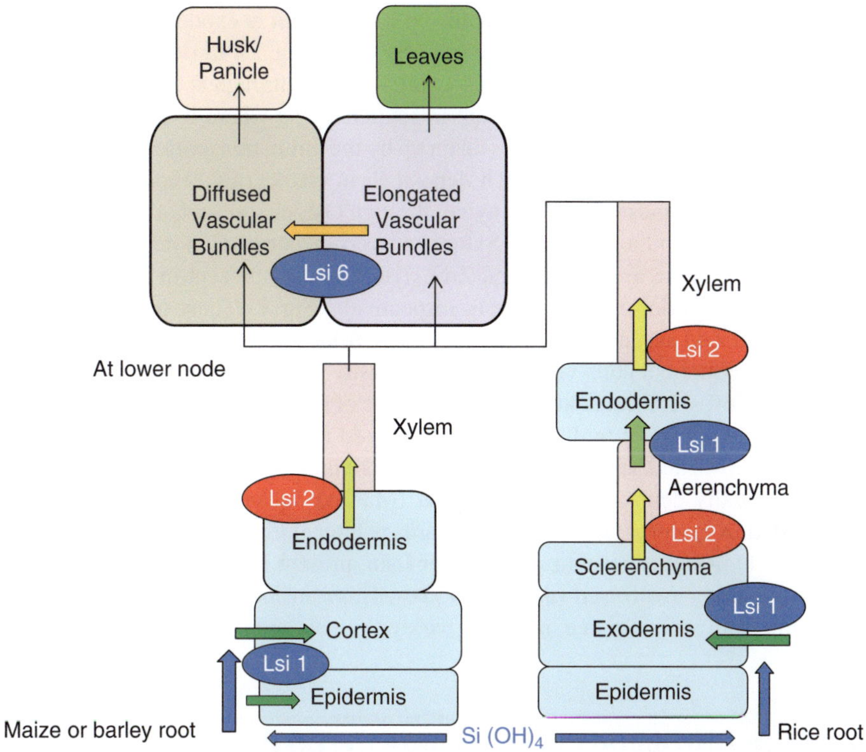

Fig. 1 Schematic diagram of Si transport, its uptake from soil solution, xylem loading and unloading and distribution inside cereal plants. LSi 1 is a low Si influx transporter, present in exodermis and aerenchyma cells of rice roots; while in cortex of maize or barley root. LSi 2 is a low Si efflux transporter present in endodermis of rice, maize or barley root. Lsi 6 is a low Si influx transporter localized at the the xylem in the leaf sheaths and leaf blades is involved in xylem unloading, transporting Si out of the xylem in rice, maize and barley (Modified from Ma and Yamaji 2015)

Lsi1 and Lsi6 are categorized as low Si influx transporters (Fig. 1) and associated to aquaporin protein (subfamily: NIP). There is a passive transport of water and different solutes (ammonia, glycerol, boric acid and silicic acid) occur along the concentration gradient through these transporters. Lsi1 is mainly restricted and shows its expression in the roots of different cereal crops i.e rice, wheat maize and barley while its homolog, Lsi 6 identified in rice (OsLsi6) and barley (HvLsi6) express itself both in root and shoot. Lsi2 is mainly Si efflux transporter involved in Si transport from the plant cell to external solution inside plant body (Fig. 1). They are first identified in rice (Ma et al. 2007) and then there homologs were identified in maize (ZmLsi2) and barley (HvLsi2) [Mitani et al. 2009]. Though, Lsi2 has been localized on plasma membrane with Lsi1, but it has no resemblance with Lsi1. Infect, Lsi2 shows resemblance with arsenite efflux transporter ArsB, which has been reported in archaea and bacteria (Ma et al. 2007). Lsi2 is involved in active transport of Si across the membrane against the concentration gradient.

In cereals, Si uptake has been initiated by Lsi1 present in exodermal cells of rice roots; while in cortical cells of maize or barley root (Fig. 1). In rice, Si is taken up by the OsLsi1 present at the distal ends exodermis and through apoplastic pathway, it reaches to the proximal side of Lsi2 in sclerenchyma where it is transported to aerenchyma cells (Fig. 1). Here Si is taken up by the influx transporter LSi1 (present at distal ends of endodermis) which deposit Si in endodermis. From endodermal cells, Si is further translocated into xylem through OsLsi2 (present at proximal ends of endodermis). Similar process of Si uptake has been occurred in maize and barley roots (Fig. 1). There are HvLsi1 and ZmLsi1transporters present in epidermal and cortical cells of barley and maize roots, respectively (Fig. 1). These two transporters (HvLsi1 and ZmLsi1) are involved in Si uptake from soil solution. This Si is transported to endodermal cells via symplastic pathway and then imported to xylem by HvLsi2 or ZmLsi2, located at the proximal ends of endodermis (Fig. 1).

After xylem loading by Lsi2, Si is translocated to the shoots via transpirational bypass flow (Ma and Yamaji 2015). There is more than 90 % Si taken up by the rice roots eventually deposited into the shoots (Ma and Takhashai 2002). Then Si is transported from lower node of shoot towards palnt leaf via elongated vascular bundles (Fig. 1). Another Si influx transporter Lsi6, present in leaf sheath involved in xylem unloading to diffused vascular bundles. This translocation of Si resulted in Si exclusion from xylem in rice, maize or barley (Yamaji, et al. 2008).

4 Silica Distribution in the Mature Cereal Plant

There was continuous deposition of silica in the plant top organs which resulted in the increase in total silica content of cereals in all parts of the shoot with increasing age (Jones and Handreck 1969). The significance of tissue and organ Si location was shown by the consistent increases of total plant silica, starting in the roots of cereals through the leaf sheaths to the leaf blades. There were 2.07, 12.3 and 13.4 % SiO_2 observed in root, leaf sheath and blade of rice on a dry matter basis, respectively (Yoshido et al. 1962). The highest silica levels generally occur in the inflorescence bracts (Jones and Handreck 1969). There was 90 % of the total silicon constituted as a solid silica gel in rice and the rest as soluble or colloidal forms. Silica gel was present as an extracellular layer under the cuticle and compartmentalized between the cell walls and cell lumens (Sangster et al. 2001). This cuticle-silica layer was the heaviest silica deposition site in the rice leaf and inflorescence husk (Yoshido et al. 1962). The basic silica distribution pattern was unaffected by varying the monosilicic acid content of the soil solution over the range of 7 to 67 ppm SiO_2 (Jones and Handreck 1969). There was variation of silica content occurring in wheat and barley leaves, where the younger leaves had more silica than do the mature (Sangster et al. 2001).

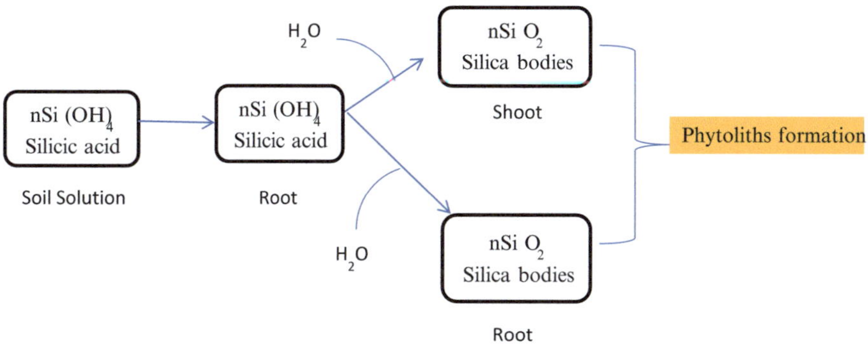

Fig. 2 Formation of phytoliths beneath the cell wall of root and shoot of cereals. Different molecules of Silicic acid when combined with each other, they form long chain of silica cemented with each other by removal of water

5 Silicon Mediated Mechanisms Improving Salinity Tolerance

5.1 Formation of Phytoliths

The deposition of silica can take place anywhere in the plant body as phytoliths or discrete silica bodies present in different shapes when they occupy the intercellur spaces (Fig. 2). Phytoliths size and shape determined on the basis of deposition location in the cell and size and shape of cell (Cooke and Leishman 2011). In cereals, they adopted different shapes like dumb-bell shaped silica cells in maize (Cheng and Kim 1989) and a leaf shaped short cell in rice (Cooke and Leishman 2011). It is assumed that phytoliths might have the ability to bind Na^+ with their surface so roots can enhance the uptake of K^+ from soil solution (Tahir et al. 2011). Silica deposition as phytoliths in rice and barley shoots (Yeo et al. 1999) improving the water flow through transpiration stream and reduces the translocation of Na gives mechanical strength to the stem (Epstein and Bloom 2003). When these silica bodies present beneath the cell wall of rice and barley leaves, they not only maintain its turgor pressure but also linked with the protection of photosynthetic apparatus under salinity stress (Yeo et al. 1999).

5.2 Growth and Morphology

Silicon-salinity interaction has been investigated by many scientists in different plant species like maize (Khan et al. 2015b), barley (Liang et al. 2003), wheat (Tahir et al. 2012; 2011), rice (Yeo et al. 1999; Gong et al. 2006; Shi et al. 2013), tomato, canola (Hashemi et al. 2010), cucumber (Zhu et al. 2004), and sugarcane (Ashraf et al. 2010).

It decreases the chloride transport to the shoot by minimizing the transpirational bypass flow in rice roots (Shi et al. 2013). Among the yield components, it is involved in enhancing the ripened grains percentage in barley and rice plants under water stress (Ma and Takahashi 2002). Silicon inclusion to a salt stressed rice plant increased the shoot growth but no effect on root was studied by Gong et al. (2006).

Silicon enhanced suberization and lignification in roots of rice (Fleck et al. 2011) so, radical oxygen loss in the stress condition could be minimized and plant survived under unfavorable environment. Shoot dry and fresh weight and plant height was also enhanced significantly by Si treatment. Similar findings were observed by different scientists in other crops like Si additions resulted in enhancement of dry root and shoot weight, leaf number and chlorophyll content in lettuce (Milne et al. 2012), fennel (Rahimi et al. 2012), alfalfa (Wang et al. 2011), tomato (Romero-Aranda et al. 2006) and grapevine (Soylemezoglu et al. 2009) under salinity stress. Addition of NaCl to the root growing medium decreased the root fresh and dry weights and shoot length of maize cultivars (Parveen and Ashraf 2010). Exogenously applied Si significantly enhances these parameters under saline regimes. Dry and fresh fennel plant weight, 1000 grain weight and grain yield had no significant influence by Si application under salinity stress (Rahimi et al. 2012).

Savvas et al. (2009) also reported that the salinity-associated suppression was alleviated by the inclusion of 1 mM of Si in the salinized nutrient solution, so both yield and growth suppression was because of reduced net photosynthesis rate at elevated salinity level. Silicon significantly improves wheat dry biomass when added to the salt treatment especially at the higher salt levels (100 mM NaCl) where reduction in total plant dry weight in the NaCl treatment were 39 and 54 % for salt-tolerant (Izmir-85) and sensitive (Gediz-75) cultivars, respectively (Tuna et al. 2008). Silicon addition mitigated the negative effect of Na^+ on different growing parts of the tomato and enhanced its biomass yield (Al-Aghabary et al. 2004).

5.3 *Physiological and Biochemical*

a) Osmolyte concentration

Salt stress significantly enhances H_2O_2, free proline level and malondialdehyde concentration in different crops like maize but different scientists revealed that Si has a potential to mitigate the toxic effects of salinity on plant cellular level like root ion accumulation and proline content (Kafi and Rahimi 2011; Moussa 2006). They revealed that application of Si improved root dry weight, root area, and leaf and root K content in the presence of salinity, but decreased leaf and root sodium (Na) content and leaf proline content in maize. Additionally, it decreased electrolyte leakage as Si made salt dilution by improving the water storage within plant tissues, which allows a higher growth rate that, in turn, mitigating salt toxicity effects (Xu et al. 2015; Romero-Aranda et al. 2006). Silicon also significantly alleviated the NaCl adverse effects by enhancing bioactive gibberellins (GA_1 and GA_4) contents but

Jasmonic acid (JA) contents sharply declined when the plants were supplemented with Si which were increased under salinity stress (Hamayun et al. 2010). Wang and Galletta (1998) investigated that ratios of fatty acid unsaturation ((18: 2 + 18: 3)/18: 1) enhances by foliar application of silicate in phospholipids and glycolipids; as a result higher amounts of membrane lipids are calculated in strawberry.

Tuna et al. (2008) revealed that plasma membrane permeability and membrane lipid peroxidation decreases with Si application so it maintains the functions and membrane integrity of salt stressed barley, thus improving the plant growth and mitigating salt toxicity. When environmental stress was exerted on rice plants, Si improved the lipids stability in cell membranes confirming that Si prohibited the functional deterioration of cell membranes (Agarie et al. 1998).

b) Photosynthetic activity

Adding Ca-silicate in salt stressed plants maintains transpiration, membrane permeability, stomatal conductance, chlorophyll content, net photosynthesis, intercellular CO_2 and reduced Na in leaves with increased Na uptake by improving growth, balanced nutrition, physiological parameters and increased nutrient uptake (Murillo-Amador et al. 2007). Different photosynthetic parameters like stomatal conductance, net CO_2 assimilation rate, leaf internal CO_2 concentration and transpiration rate of maize cultivars were checked by Parveen and Ashraf (2010) in which they concluded that exogenously applied Si improved all those parameters both under non-saline and saline regimes. Similarly, Si improved photosynthetic activity by enhancing RuBisCO activity and the ultrastructure of leaf cells in barley (Liang et al. 1998) and reduced electrolyte leakage in the leaves enhancing the plant growth at high salinity.

Salts accumulation inside the plant body lead to the water shortage for the normal functioning of plant cells which causes physiological drought and ultimately plant cell death takes place (Mateos-Narenjo et al. 2013). Major cereal crops like wheat, rice and maize are pretty responsive to the applied Si under stress conditions. Maize growth was reported by Kaya et al. (2006) under water stress conditions by the addition of Si as a mineral nutrient. Silicon was found to increase the total dry matter, relative water content, and chlorophyll content.

c) Antioxidants

Activity of antioxidant enzyme could be increase by exogenous Si application which simultaneously reduced the lipid peroxidation in roots of salt stressed barley (Liang et al. 2003). Gong et al. (2005) reported in wheat that supplementation of Si under water stress conditions increased some antioxidant enzymes activities: SOD, CAT and GR which ultimately lead to the amelioration of oxidative damage caused by reactive oxygen species. Salinity stress significantly reduced the activity of SOD, CAT and APX in maize plants by enhancing the level of H_2O_2 and MDA but Si addition enhanced the SOD, CAT and APX enzymes activity (Moussa 2006). When Si was applied to saline growth medium, it not only enhanced the chlorophyll content, photosynthetic activity and ribulose bis-phosphate carboxylase (RUBP) activity in organelles of leaf cell but also diminished the salt-induced H_2O_2 production

(Gunes et al. 2007). More-over, Si supplementation was enhancing activities of SOD, APX, GPX and GR enzymes in salt stressed plant leaves by alleviating salt toxicity and helped plants to withstand oxidative damage under salt stress in wheat (Saqib et al. 2008) and maize (Moussa 2006; Khan et al. 2015b).

There are different mechanisms adapted by cereals to reduce oxidative damage resulting from salt stress, through the biosynthesis of a cascade of antioxidants. Among them, phenolic compounds such as phenolic acids, flavonoids and proanthocyanidins play an important role in scavenging free radicals. Salinity stress reduces root and shoot total phenolics (Ashraf et al. 2010) in wheat cultivars. Silicon application enhances the total phenolic compounds in root and shoot of maize cultivars under stress conditions thereby increasing the tolerance against salinity by scavenging free radicals.

d) Na, K homeostasis

Similarly, K^+/Na^+ with reduced Na^+ and enhanced K^+ uptake and increase in wheat shoot growth was observed with Si addition under salt stress by Ali et al. (2009). The possible mechanism behind reduced Na^+ uptake is that Si deposited beneath the cell wall of roots in the form of long chains called phytoliths where these phytoliths bind Na and restrict its translocation to the upper parts of plant. Potassium concentration in salt stressed wheat genotypes Auqab 2000 and SARC-5 shows significant improvement when Si was applied which also reduced Na concentration by enhancing K/Na uptake (Tahir et al. 2006). Silicon also worked as a plant Na^+ detoxification by increasing cell-wall Na^+ binding in both salt-resistant wheat genotype SARC-1 and salt sensitive 7-Cerros (Saqib et al. 2008). Many nutrients showed synergistic effects with each other like zinc and phosphorus in wheat (Khan et al. 2015a) as they facilitate each other to increase their concentration within plant body. Similarly, Si showed synergistic effect with K by increasing its concentration within plant cells like maize (Khan et al. 2015b) and wheat (Tahir et al. 2012).

6 Future Prospects/Missing Links

Cereals are very sensitive to grow in saline conditions. As salinity hampers their growth and minimize yield potential, so the application and accumulation of Si has increased the ability of cereal crops to maximize their growth and yield in the world where the human population is increasing and their land use activity are likely to lead increased salinization. Silicon accumulation inside plant body is almost as much as other macronutrients (Ma et al. 2001), and it is categorized as 'quasi-essential' element in both abiotic and biotic stress conditions. But exact mechanisms for its uptake in salt stress condition are still debatable.

There is a lot of literature available on the N, P, K solubilizing bacteria (Duponnois et al. 2006; Chuang et al. 2007) but very few studies has been reported on Si solubilizing bacteria (Sheng et al. 2008). As Si concentration is 0.1–0.6 mM in soil solution, there might be Si solubilizing bacteria involved in increasing Si concentration in soil solution which must be investigated under salt stress condition.

Similarly, there is a variety of soil microorganisms, particularly Arbuscular mycorrhizal fungi (AMF) can help plants to survive under salt stress conditions. Cereal crops showed significant improvements in physiological mechanisms when they are inoculated with AMF under stress conditions; like in maize, there is significant increase in efficiency of photosystem II, stomatal conductance, enzymatic activities (SOD, CAT) and decrease in transpiration, hydrogen peroxide and the oxidative damage to lipids (Estrada et al. 2013). All these attributes are also related to the Si concentration inside plant body. Silicon enhances the fungal growth in different nutrient solutions (Wainwright et al. 1997), so there might be fungi involved in Si translocation during its symbiotic relationship with plants, which must be explained.

Foliar application of salts is a shotgun approach to combat abiotic stress. Foliar Si application was already used to combat heavy metal toxicity like cadmium in pots (Liu et al. 2009), but no such study was yet reported on soil and foliar Si application to reduce salt toxicity in cereals under field conditions. Korndörfer et al. (2004a, b) reported that the Si deposition under the leaf epidermis is directly related to the foliar Si application on the plants. It not only increases crop yield but also provides a physical mechanism of defense which minimizes transpiration losses, reduces lodging and enhances photosynthetic activity.

Different Si transporters have been identified in cereals like ZmLsi1 and ZmLsi6 in maize (Mitani et al. 2009) and OsLsi1, OsLsi2, and OsLsi6 from rice (Ma et al. 2006, 2007; Yamaji et al. 2008). Silicon accumulation has been attributed to the ability of the roots to take up Si (Takahashi et al. 1990), but still there is no such literature available which describes expression of the genes under salt stress condition. They must be tested under stress condition to check whether they are the main players during stress condition or some other mechanisms are being activated.

7 Conclusion

Salinity stress generally imposes severe impacts both on human and plants by degradation of land and poor crop growth. Silicon proves to be essential in such cases under salt stress condition. It is present in higher amounts inside the plant body and prevents the crop to be transpired and lowers the activity of reactive oxygen species. It helps crops to overcome stress condition in their critical growth stage and improves many physiological and biochemical mechanisms of plants. It also increases the crop biomass and yield so prove to be necessary for the cereal crop to accomplish healthy reproductive stage under salt stress. Thus, salt stress can be minimized and salt affected area must be utilized by growing cereal crops with foliar or soil Si application as an amendment. To attain this target, more field trials are required from researchers.

References

Abou-Baker NH, Abd-Eladl M, Abbas MM (2011) Use of silicate and different cultivation practices in alleviating salt stress effect on bean plants. Aust J Basic Appl Sci 5:769–781

Agarie S, Uchida H, Agata W, Kubota F, Kaufman PB (1998) Effects of silicon on transpiration and leaf conductance in rice plants (*Oryza sativa* L.). Plant Product Sci 1:89–95

Ahmad P, Hakeem KR, Kumar A, Ashraf M, Akram NA (2012) Salt induced changes in photosynthetic activity and oxidative defense system of three cultivars of mustard (*Brassica juncea* L.). Afr J Biotechnol 11:2694–2703

Akram M, Ashraf MY, Ahmad R, Waraich EA, Iqbal J, Mohsan M (2010) Screening for salt tolerance in maize (*Zea mays* L.) hybrids at an early seedling stage. Pak J Bot 42:141–154

Al-Aghabary K, Zhu Z, Shi Q (2004) Influence of silicon supply on chlorophyll content, chlorophyll fluorescence and anti-oxidative enzyme activities in tomato plants under salt stress. J Plant Nutr 27:2101–2115

Ali A, Basra SMA, Ahmad R, Wahid A (2009) Optimizing silicon application to improve salinity tolerance in wheat. Soil Environ 28:136–144

Ashraf MA, Ashraf M, Ali Q (2010) Response of two genetically diverse wheat cultivars to salt stress at different growth stages: leaf lipid peroxidation and phenolic contents. Pak J Bot 42:559–565

Cakmak I (2002) Plant nutrition research: priorities to meet human needs for food in sustainable ways. Plant Soil 247:3–24

Cheng PC, Kim HG (1989) The use of x-ray contact microradiography in the study of silica deposition in the leaf blade of maize. Am J Bot (Suppl) 76:29

Chiba Y et al (2009) HvLsi1 is a silicon influx transporter in barley. Plant J 57:810–818

Chinnusamy V, Jagendorf A, Zhu JK (2005) Understanding and improving salt tolerance in plants. Crop Sci 45:437–448

Chuang CC, Kuo YL, Chao CC, Chao WL (2007) Solubilization of inorganic phosphates and plant growth promotion by Aspergillus niger. Biol Fertil Soils 43:575–584

Cooke J, Leishman MR (2011) Is plant ecology more siliceous than we realise? Trends Plant Sci 16:61–68

Duponnois R, Kisa M, Plenchette C (2006) Phosphate solubilizing potential of the nematofungus Arthrobotrys oligospora. J Plant Nutr Soil Sci 169:280–282

Epstein E, Bloom AJ (2005) Mineral nutrition of plants: principles and perspectives, 2nd edn. Sinauer, Sunderland

Epstein E (2000) The discovery of the essential elements. In: Kung S-D, Yang S-F (eds) Discoveries in plant biology, vol 3. World Scientific Publishing, Singapore, in press

Epstein E, Bloom AJ (2003) Mineral nutrition of plants: principles and perspectives, 2nd edn. Wiley, New York

Estrada B, Aroca R, Barea JM, Ruiz-Lozano JM (2013) Native arbuscular mycorrhizal fungi isolated from a saline habitat improved maize antioxidant systems and plant tolerance to salinity. Plant Sci 201:42–51

Exley C (1998) Silicon in life: a bioinorganic solution to bioorganic essentiality. J Inorg Biochem 69:139–144

Fleck AT, Nye T, Repenning C, Stahl F, Zahn M, Schenk MK (2011) Silicon enhances suberization and lignification in roots of rice (*Oryza sativa*). J Exp Bot 62:2001–2011

Gong H, Zhu X, Chen K, Wang S, Zhang C (2005) Silicon alleviates oxidative damage of wheat plants in pots under drought. Plant Sci 169:313–321

Gong HJ, Randall DP, Flowers TJ (2006) Silicon deposition in the root reduces sodium uptake in rice (*Oryza sativa* L.) seedlings by reducing bypass flow. Plant Cell Environ 29:1970–1979

Gruhn P, Goletti F, Yudelman M (2000) Integrated nutrient management, soil fertility and sustainable agriculture: current issues and future challenges. Food, Agriculture, and the Environment Discussions Paper 32, International Food Policy Research Institute, Washington

Gunes A, Inal A, Bagci EG (2007) Silicon mediated changes of some physiological and enzymatic parameters symptomatic for oxidative stress in spinach and tomato grown in sodic-B toxic soil. Plant Soil 290:103–114

Hamayun M, Sohn E, Khan SA, Shinwari ZK, Khan AL, Lee I (2010) Silicon alleviates the adverse effects of salinity and drought stress on growth and endogenous plant growth hormones of soybean (*Glycine max* L.). Pak J Bot 42:1713–1722

Hashemi A, Sadeghipour HR, Abdolzadeh A (2010) Beneficial effects of silicon nutrition in alleviating salinity stress in hydroponically grown canola, *Brassica napus* L., plants. Soil Sci Plant Nutr 56:244–253

Jones LHP, Handreck K (1969) Silica in soils, plants and animals. Adv Agron 19:107–149

Kafi M, Rahimi Z (2011) Effect of salinity and silicon on root characteristics, growth, water status, proline content and ion accumulation of purslane (*Portulaca oleracea* L.). Soil Sci Plant Nutr 57:341–347

Kaya C, Tuna L, Higgs D (2006) Effect of silicon on plant growth and mineral nutrition of maize grown under water-stress conditions. J Plant Nutr 29:1469–1480

Khan WD, Faheem M, Khan MY, Hussain S, Maqsood MA, Aziz T (2015a) Zinc requirement for optimum grain yield and zinc biofortification depends on phosphorus application to wheat cultivars. Rom Agric Res 32: DII 2067–5720 RAR 2015–124

Khan WD, Aziz T, Warraich EA, Khalid M (2015b) Silicon application improves germination and vegetative growth in maize grown under salt stress. Pak J Agric Sci 52(4): 937–944

Korndorfer AP, Cherry R, Nagata R (2004a) Effect of calcium silicate on feeding and development of tropical sod webworms (Lepidoptera: Pyralidae). Fla Entomol 87:393–395

Korndorfer GH, Pereira HS, Nolla A (2004b) Silicon analysis in soil, plant and fertilizers. GPSi/ICIAG/UFU, Brazil

Liang YC, Chen Q, Liu Q, Zhang WH, Ding RX (2003) Exogenous silicon (Si) increases antioxidant enzyme activity and reduces lipid peroxidation in roots of salt-stressed barley (*Hordeum vulgare* L.). J Plant Physiol 160:1157–1164

Liang YC, Zhang WQ, Chen J, Ding R (2005) Effect of silicon on H^+-ATPase and H^+- PPase activity, fatty acid composition and fluidity of tonoplast vesicles from roots of salt stressed barley (*Hordeum vulgare* L.). Environ Exp Bot 53:29–37

Liang YC, Zhang YC, Yin SX, Yang MC (1998) Feasibility of using trichloroacetic acid soluble silicon as a diagnostic index of silicon deficiency in rice. Pedosphere 8:21–26

Liu C, Li F, Luo C, Liu X, Wang S, Liu T, Li X (2009) Foliar application of two silica sols reduced cadmium accumulation in rice grains. J Hazard Mat 161:1466–1472

Liu P, Lina Y, Shiwen W, Meijuan Z, Xiping D, Suiqi Z, Kiyoshi T (2015) Enhanced root hydraulic conductance by aquaporin regulation accounts for silicon alleviated salt-induced osmotic stress in *Sorghum bicolor*L. Environ Exp Bot 111:42–51

Ma JF, Yamaji N (2015) A cooperative system of silicon transport in plants. Trends Plant Sci 20:435–442

Ma JF, Miyake Y, Takahashi E (2001) Silicon as a beneficial element for crop plants. In: Datnoff LE, Snyder GH, Korndorfer GH (eds) Silicon in agriculture. Elsevier Science, Amsterdam, pp 17–39

Ma JF, Tamai K, Yamaji N, Mitani N, Konishi S, Katsuhara M, Ishiguro M, Murata Y, Yano M (2006) A silicon transporter in rice. Nature 440:688–691

Ma JF, Takahashi E (2002) Soil, fertilizer, and plant silicon research in Japan. Elsevier Science, Amsterdam

Ma JF, Yamaji N, Mitani-Ueno N (2011) Transport of silicon from roots to panicles in plants. Proc Jpn Acad Ser B Phys Biol Sci 87:377–385

Ma JF, Yamaji N, Mitani N, Tamai K, Konishi S, Fujiwara T, Katsuhara M, Yano M (2007) An efflux transporter of silicon in rice. Nature 448:209–212

Mateos-Naranjo E, Andrades-Moreno L, Davy AJ (2013) Silicon alleviates deleterious effects of high salinity on the halophytic grass *Spartina densiflora*. Plant Physiol Biochem 63:115–121

Matichenkov VV, Kosobrukhov AA (2004) Silicon effect on the plant resistance to salt toxicity: 13th international soil conservation organization conference. Conserving soil and water for society. Brisbane, Australia

Milne CJ, Charles PL, Patrick V, Jeanine LM, Rautenbach F (2012) Salinity induced changes in oxidative stress and antioxidant status as affected by applications of silicon in lettuce (*Lactuca sativa*). Int J Agric Biol 14:763–768

Mitani N, Ma JF, Iwashita T (2005) Identification of the silicon form in xylem sap of rice (*Oryza sativa* L.). Plant Cell Physiol 46:279–283

Mitani N, Yamaji N, Ma JF (2009) Identification of maize silicon influx transporters. Plant Cell Physiol 50:5–12

Moussa HR (2006) Influence of exogenous application of silicon on physiological response of salt-stressed maize (*Zea mays* L.). Int J Agric Biol 2:293–297

Murillo-Amador B, Yamada S, Yamaguchi T, Rueda-vPuente E, Avila-Serrano N, Garcia-Hernandez JL, Lopez-Aguilar R, Troyo-Dieguez E, Nieto-Garibay A (2007) Influence of calcium silicate on growth, physiological parameters and mineral nutrition in two legume species under salt stress. J Agron Crop Sci 193:413–421

Parveen N, Ashraf M (2010) Role of silicon in mitigating the adverse effects of salt stress on growth and photosynthetic attributes of two maize (*Zea Mays* L.) cultivars grown hydroponically. Pak J Bot 42:1675–1684

Prashanth SR, Sadhasivam V, Parida A (2008) Over expression of cytosolic copper/zinc superoxide dismutase from a mangrove plant Avicennia marina in indica Rice var Pusa Basmati-1 confers abiotic stress tolerance. Transgenic Res 17:281–291

Rahimi R, Mohammakhani A, Roohi V, Armand N (2012) Effects of salt stress and silicon nutrition on cholorophyll content, yield and yield components in fennel (*Foeniculum vulgar* Mill.). Int J Agric Crop Sci 4:1591–1595

Raven JA (2001) Silicon transport at the cell and tissue level. In: Datnoff LE, Snyder GH, Korndörfer GH (eds) Silicon in agriculture, vol 8, Studies in plant science. Elsevier, Amsterdam

Rengasamy P (2006) World salinization with emphasis on Australia. J Exp Bot 57:1017–1023

Richmond KE, Sussman M (2003) Got silicon? The non-essential benefical plant nutrient. Curr Opi Plant Biol 6:268–272

Romero-Aranda MR, Jurado O, Cuartero J (2006) Silicon alleviates the deleterious salt effect on tomato plant growth by improving plant water status. J Plant Physiol 163:847–855

Sangster AG, Hodson MJ, Tubb HJ (2001) Silicon deposition in higher plants. In: Datnoff LE, Snyder GH, Korndörfer GH (eds) Silicon in agriculture, vol 8, Studies plant sci. Elsevier, Amsterdam, pp 85–113

Saqib M, Zorb C, Schubert S (2008) Silicon-mediated improvement in the salt resistance of wheat (*Triticum aestivum*) results from increased sodium exclusion and resistance to oxidative stress. Funct Plant Biol 35:633–639

Savvas D, Giotis D, Chatzieustratiou E, Bakea M, Patakioutas G (2009) Silicon supply in soilless cultivations of zucchini alleviates stress induced by salinity and powdery mildew infections. Environ Exp Bot 65:11–17

Sheng XF, Zhao F, He LY, Qiu G, Chen L (2008) Isolation and characterization of silicate mineral-solubilizing Bacillus globisporus Q12 from the surfaces of weathered feldspar. Can J Microbiol 54:1064–1068

Shi Y, Wang Y, Flowers TJ, Gong H (2013) Silicon decreases chloride transport in rice (*Oryza sativa* L.) in saline conditions. J Plant Physiol 170:847–853

Soylemezoglu G, Demir K, Inal A, Gunes A (2009) Effect of silicon on antioxidant and stomatal response of two grapevine (*Vitis vinifera* L.) rootstocks grown in boron toxic, saline and boron toxic-saline soil. Sci Hortic 123:240–246

Tahir MA, Rahmatullah Aziz T, Ashraf M, Kanwal S, Muhammad A (2006) Beneficial effects of silicon in wheat under salinity stress-pot culture. Pak J Bot 38:1715–1722

Tahir MA, Aziz T, Farooq M, Sarwar G (2012) Silicon-induced changes in growth, ionic composition, water relations, chlorophyll contents and membrane permeability in two salt-stressed wheat genotypes. Arch Agron Soil Sci 58:247–256

Tahir MA, Aziz T, Rahmatullah (2011) Silicon induced growth and yield enhancement in two wheat genotypes differing in salinity tolerance. Comm Soil Sci Plant Anal 42:395–407

Takahashi E, Ma JF, Miyake Y (1990) The possibility of silicon as an essential element for higher plants. Comm Agric Food Chem 2:99–122

Tuna AL, Kayab C, Higgs D, Murillo-Amador B, Aydemir S, Girgin AR (2008) Silicon improves salinity tolerance in wheat plants. Environ Exp Bot 62:10–16

Wainwright M, Al-Wajeeh K, GraystonS J (1997) Effect of silicic acid and other silicon compounds on fungal growth in oligotrophic and nutrient-rich media. Mycol Res 101:933–938

Wang SY, Gallettaa GJ (1998) Foliar application of potassium silicate induces metabolic changes in strawberry plants. J Plant Nutr 21:157–167

Wang X, Wei Z, Liu D, Zhao G (2011) Effects of NaCl and silicon on activities of antioxidative enzymes in roots, shoots and leaves of alfalfa. Afr J Biotechnol 10:545–549

Xu CX, Ma YP, Liu YL (2015) Effects of silicon (Si) on growth, quality and ionic homeostasis of aloe under salt stress. S Afr J Bot 98:26–36

Yamaji N, Mitatni N, Ma JF (2008) A transporter regulating silicon distribution in rice shoots. Plant Cell 20:1381–1389

Yeo AR, Flowers SA, Rao G, Welfare K, Senanayake N, Flowers TJ (1999) Silicon reduces sodium uptake in rice (*Oryza sativa* L.) in saline conditions and this is accounted for by a reduction in the transpirational bypass flow. Plant Cell Environ 22:559–565

Yoshido S, Ohnishi Y, Kitagishi K (1962) Chemical forms, mobility and deposition of silicon in rice plant. Soil Sci Pl ant Nutr Tokyo 8:15–20

Zhu JK (2001) Plant salt tolerance. Trends Plant Sci 6:66–71

Zhu Z, Wei G, Li J, Qian Q, Yu J (2004) Silicon alleviates salt stress and increases antioxidant enzymes activity in leaves of salt-stressed cucumber (*Cucumis sativus* L.). Plant Sci 167:527–533

Soil Microflora – An Extensive Research

Sameen Ruqia Imadi, Mustafeez Mujtaba Babar, Humna Hasan, and Alvina Gul

Contents

Abstract Soil is the most complicated biomaterial present on earth. It is composed of a variety of substances and provides a habitat to various organisms. Different chemical reactions take place in soil that ensures the sustainability of life. Microorganisms including bacteria, fungi, actinomyces and algae are widely distributed in soil. These natural micro flora have several advantages. They contribute to the growth and development of plants, decomposition of organic materials, nutrient cycling, soil nitrification, sustenance of pedological system and production of bioactive compounds. Soil fungi develop mutualistic associations with plants and increase their surface area for absorption. Rhizosphere of soil, the area in which

S.R. Imadi • A. Gul (✉)
Atta-ur-Rahman School of Applied Biosciences, National University of Sciences and Technology, Islamabad, Pakistan
e-mail: alvina_gul@yahoo.com

M.M. Babar
Shifa College of Pharmaceutical Sciences, Shifa Tameer-e-Millat University, Islamabad, Pakistan

H. Hasan
Quaid-I-Azam University, Islamabad, Pakistan

K.R. Hakeem et al. (eds.), *Soil Science: Agricultural and Environmental Prospectives*, DOI 10.1007/978-3-319-34451-5_13

micro flora is present, is rich not only in diverse micro flora but also plant roots and nutrients. Soil pollution and anthropogenic activities used for higher yield of agricultural crops negatively affect the soil micro flora. Pesticides kill micro flora and reduce soil biodiversity. The focus of this chapter is on the advantages of natural flora of soil and various factors causing their degradation. The chapter also sheds light on the changes in micro floral communities due to changes in environment. Towards the end, the future perspectives in which soil micro flora can be used for further benefit of mankind have also been discussed.

Keywords Soil microflora • Symbiosis • Microbe-environment interaction • Microflora applications • Agriculture improvement

1 Introduction

Soil is considered as one of the most complicated biomaterial on earth. It not only provides home to a variety of microorganisms but also serves as a natural laboratory where different kinds of experiments can be performed. In general, soil is formed by the weathering of parent rock material. Thereafter, microbes adapt this soil as their habitat, hence, being characterized as 'soil micro flora'. The upper 16–17.6 cm plow layer is considered to be the most active in terms of the population density of these micro flora (Suneetha and Khan 2010). Of the different layers of soil, rhizosphere is the region that is directly under the influence of various biological and physiological factors including plant roots and microbes (Chaudhry et al. 2012; Hakeem et al. 2016).

Soil micro flora have been observed to play a fundamental role in the organic matter mineralization in the soil. Plant nutrition and growth in agro systems is also largely dependent on soil micro flora. These microbial processes take place in the rhizosphere (Bérrard et al. 2015). Among the various soil microorganisms, the most prevalent fungi worldwide include species of genus *Aspergillus, Fusarium, Verticillum, Penicillium*, and *Trichoderma*. Largest fungal colonies residing in the soil are that of *Aspergillus niger* (Rajik et al. 2011). A number of experiments have revealed that upon the application of the microbial fertilizers, the number of soil bacteria and actinomycetes is significantly increased. This, ultimately, results in an increase in the crop yield (Wang et al. 2012).

A number of factors contribute to the microbial diversity in the rhizosphere of various biogeographical regions. It has been observed that the soil microbial communities are quite similar among the different cultivated soils, for instance the microbes in the cultivated deciduous forests and mowed grasslands are generally the same. Though there are major differences in soil properties and vegetation conditions in various cultivated lands yet the soils share similar microbial population. There are, however, a few reports in which the conversion of deciduous forest to tilled croplands resulted in a change in the soil properties leading to an alteration in the soil micro flora as well. It can, hence, be assumed that history of land use is a major determinant of soil micro flora along with the vegetation and soil properties

(Jangid et al. 2011). Moreover, other biological factors, like the species of the earthworm, have also been found to contribute significantly to the nature and amount of bacteria and actinomyces present in soil (Lin et al. 2016).

Quantities of soil bacteria and actinomyces have been observed to significantly decrease by conversion of plantation from evergreen broad leaved forests to Chinese fir plantation (Xia et al. 2012). Similarly, in a recent study, the effect of fertilizers was observed on soil micro flora of watermelon fields. It was seen that the treatment with both organic and inorganic fertilizers resulted in a decrease in the concentration of normal fungi levels in soil. On the other hand, the amount of bacteria and actinomyces was significantly increased as a result of fertilizer treatment (Zhu et al. 2013). Moreover, it has been observed, through paraffin baiting technique based experiments that several species of *Nocardia* genus are present in the soil that also contributes to a diverse microbial community in many cultivated regions (Habibnia et al. 2015).

Effects of application of different organic manures were observed on soil micro flora of mulberry in another recent study. On the treatment of roots of mulberry plants with yard manure biofertilizers, it was observed that root tips had communities of *Azotobacter chrococcum, Gluconacobacter diazotrophicus, Bacillus sonorensis, Bacillus pumilis, Bacillus coagulans* and *Pseudomonas putida.* The root tips of plants treated with *Azospirillium* biofertilizer were enriched in *Azosprillium brasilense, Bacillus coaculans, Bacillus subtilis, Azotobater chrococcum* and *Azotobacter vinelandii.* Root tips of phosphobacteria-fertilized mulberry contained *Streptomyces thermonitrificans, Pseudomonas putida, Brevibacillus borslelansis* and *Bacillus stearothermophillus.* Vermicompost biofertilized root tips had *Bacillus sonorensis, Bacillus megaterium, Brevibacillus borslelansis, Bacillus subtilis, Bacillus stearothermophillus, Glucanobacter diazotrophicus, Azotobacter vinelandi, Pseudomonas putida* and *Azotobacter chrococcum* (Mary et al. 2015; Hakeem et al. 2016). Similarly, *Listeria monocytogenes* is a food borne pathogen and is the causative agent of listeriosis in humans. It has been observed that *L. monocytogenes* naturally occupies a place in soil as micro flora. This pathogen has been isolated from a variety of garden vegetables including carrots, tomato, lettuce, spinach and radish (Locatellia et al. 2013).

2 Effects of Environment on Soil Micro Flora

The variation in the soil microbial community can partly be explained by the alteration in soil chemistry and plants growing in the region (Mitchell et al. 2012). In addition, the environment is also observed to affect the micro floral conditions of the soil. It has been reported that as the moisture content in the environment is decreased, the decomposing activity mediated by the microbes is also diminished. Soil micro flora, hence, undergo stress due to reduction in environmental moisture. Therefore, it can be safely concluded that changes in rainfall pattern result in a change in the carbon and nutrient cycling, owing to the role of soil micro flora in these cycles

(Manzoni et al. 2012). Moreover, the micro flora have an influence on the levels of zinc, copper and total sulphur in soil which ultimately affects the soil pH as well (Sullivan et al. 2013).

Tree species' transition also has a large impact on the biomass, in general, and soil microflora, in particular (Huang et al. 2013). Application of integrated agricultural management has shown that the amounts of beneficial bacteria, like *Bacillus, Streptomyces, Paenibacillus* and *Arthrobacter*, are significantly increased in response to these transitions. Contrarily, the colonization of *Ralstonia* is considerably decreased (Wu et al. 2014). Park grass experiment, which has been ongoing in United Kingdom since 1856, helps in measuring the long term response of soil microflora to variations in soil pH and other environmental parameters. The most prevalent microfloral species in these experiments are *Clostridium, Rhodoplanes, Bacteriodes, Paenibacillus, Bradyrhizobium, Rumniococcus* and *Mycobaterium.* Based upon these experiments, it has been reported that the soil micro floral diversity is directly dependent upon the plant species richness (Zhalnina et al. 2015).

With respect to the temperature, the warmer the environmental conditions, the greater is the concentration of *Actinobacteria* in soil. Increase in the concentration of ectomycorrhizal fungi is also observed with an increased atmospheric temperature (Deslippe et al. 2012). However, an increase in temperature is negatively related to the concentration of *Gimmatimonadaceae* and *Proteobacteria* in soil. Seasonal and inter-annual variability in chemical resources, temperature and precipitation results in alteration of micro floral fungal community in soil (Burke 2015). Normal field dose of pesticides possesses minor effects on soil micro flora whereas a 10-fold increased dose of pesticides has shown to be detrimental on the microfloral populations (Sofo et al. 2012).

Among the other biological factors, different effects of transgenic plants have been observed on the soil microflora. The natural flora is differentially affected by plants that have been modified to express traits considered beneficial to food industry, phytoparasite resistance and phytopathogen resistance (Turrini et al. 2015). Similarly, adverse effects of pesticides on soil microflora have been observed in a number of studies. Complete degradation of some species of soil micro flora are seen as a result of pesticide application (Singh et al. 2015). On applying cadmium and pyrene on soil, it has been observed that the metabolic quotient, basal respiration and microbial biomass were influenced. The microflora are highly sensitive towards heavy metals including zinc and lead. It has been observed that actinomyces are most sensitive to these metals followed by fungi and bacteria (Jin et al. 2015). These studies helped in concluding that different physicochemical characteristics of soil change the physiological response of soil micro flora (Lu et al. 2013).

In another study, the effect of intercropping of sugarcane and soybean cultivation on soil micro flora was tested. It was observed that in comparison to the monocultures, microbial concentrations in polycultures were relatively quite high. In lands where only sugarcane was cultivated, after microbial enrichment, the bacterial, fungal and actinomyces population was increased by 42.62 %, 14.5 % and 78.5 %

respectively. However, bacteria, fungi and actinomyces recorded a growth of 188 %, 183 % and 73 % in sugarcane-soybean intercropped systems. Therefore, it can be established that the growth of sugarcane with soybean is an excellent agricultural management practice (Li et al. 2013).

Furthermore, the application of photosynthetic bacterial inoculum in agricultural lands tends to increase both the quantity and diversity of soil micro flora. The number of fungi, bacteria and actinomyces was significantly high in soil treated with photosynthetic bacteria in comparison to the untreated soil (Fang et al. 2014). Recently, the effect of grafted tomato on the quantity of soil microflora was studied. The study showed that the rhizospheric microbial population including soil bacteria, cellulose decomposing bacteria, actinomyces, aerobic abiogenous azotobacter, ammonifying bacteria and nitrifying bacteria was significantly increased. However, the fungal population was decreased (Na et al. 2014). Treatment of coal mine disturbed soil with arbuscular ecto-mycorrhizal fungi has also been explored. The study has led to establish the symbiotic relationship of these fungi with plants. Moreover, an increase in the number of other soil bacteria, actinomyces and fungi has also been reported (Jin et al. 2014). Based upon these studies, it can be safely concluded that a number of environmental factors, both biotic and abiotic, help in the establishment of the complex, yet highly beneficial, relationship between various plants and microbes.

3 Microflora of Soil

A number of experiments have been carried out to ascertain the relative population of various microbes in soil. In one such study, the soils of tomato greenhouse were collected to determine the microflora in it. It was observed that the bacteria were the highest in number whereas fungi had the lowest number. Similarly, soil experiments were conducted over a long term to measure the changes in soil micro flora as a function of time. A significant decrease was seen in these microfloral communities over a period of 10–20 years. Similarly, on studying the effect of the soil thickness, microflora was largely present in upper 20 cm thick layer. Concentrations of actinomyces and fungi were observed to decrease with increasing depth of soil (Tang 2012). When quantity of soil micro flora was determined in different types of tobacco soil, it was seen that quantities of bacteria, actinomyces and fungi as well as nitrogen fixing microbes was higher in yellow brown soil as compared to purple soil (Xu et al. 2013). Majority of micro flora found in soil rhizosphere is bacteria but significant number of fungi are also present (Bhatt et al. 2015; Hakeem et al. 2016). The major types of microbes and various factors affecting their relative abundance have been discussed in the subsequent sub-sections.

3.1 *Actinomyces*

Actinomyces, especially *Streptomyces*, are largely present in soil as natural micro flora. It has been observed that the ultrasonic treatment of soil samples resulted in an increase in the quantity as well as the type of actinomyces. Quantity of actinomyces in soil increased by 280 % whereas that of *Streptomyces* increased by around 375 %, as a result of 50 s of treatment with ultrasonic. In comparison, the ultrasonic treatment for 20 s resulted in an increase in actinomyces concentration by merely 12.07 %. However, increasing the treatment duration to 100 s also had a negative effect on the *Actinomyces* species where a population growth of only 40 % was noted (Mao et al. 2013).

Root exudates of Chinese onion cultivar were applied to study their effects on cucumber seedling growth. It was reported that there was a significant increase in the concentration of actinomyces in the treated soil (Yang et al. 2013). Application of biological organic fertilizer, also, causes an increase in the number of actinomyces and the structure of soil microbial community (Yuan et al. 2011). Actinomyces also stimulate the soil functionality by enhancing the concentration of different soil enzymes as they are a great source of keratinases, lipases, xylanases and pectinases (Suneetha and Khan 2010).

3.2 *Bacteria*

Bacteria present in soil play an essential role in maintaining a number of biogeochemical cycles. They occupy small spaces present between the pores in soil. The impact of bacteria and their function is dependent on their spatial arrangement (Archana et al. 2015). Much similar to the actinomyces, the application of root exudates of Chinese onion cultivar enhances the concentrations of bacteria in the soil. The different bacterial species increased as a result of this treatment include various strains of genus *Anaerolineaceae, Actinobacteria* and *Proteobacteria* (Yang et al. 2013). Soils treated with organic fertilizers are observed to possess higher quantities of beta lactam resistant bacteria in comparison to those treated with inorganic fertilizers. Similarly, the number of *Pseudomonas* species is also increased in organic fertilizer treated soil. Manure treatment also increases the concentration of *Psychrobacter pulmonis* and *Janthinobacterium* species. This helps in hypothesizing that the application of organic fertilizers can lead to an increase in the concentration of antibiotic resistant strains of bacteria (Udikovic-Kolic et al. 2014). However, further experimentation in this regard is needed. The soil is also enriched in bacterial species including *Burkholderia terrestris, Burkholderia humi, Burkholderia udeis, Burkholderia choica* and *Burkholderia telluris* (Vandamme et al. 2013).

3.3 Fungi

The most common form of fungi found in soil are the arbuscular mycorrhizal fungi. These fungi are known to form mutualistic ecto-mycorrhizal associations with plants aiding their growth and development. Fungi have a prominent role in the ecosystem as they serve as mutualists, decomposers and, oftentimes, as pathogens (Taylor et al. 2014). The soil population of fungi and fusarium is significantly increased in the rhizosphere after treatment with root exudates of Chinese onion cultivar in cucumber fields (Yang et al. 2013). Similarly, on analyzing the soil samples from bat hibernacula in United States, it was reported that *Geomyces* species of fungi was the most significant among all the fungal isolates from the respective soil. The fungi account for almost 33 % of all fungal species found in soil (Lorch et al. 2012). Nearly 90 % of terrestrial plants are in symbiotic relationship with soil fungi (Miransari 2011). These fungi play an important role in the degradation and decomposition of organic matter and formation of soil infrastructure. Moreover, it has been reported that when these symbionts are inhibited, other beneficial soil microbes serve the same function to maintain the food web (Helfrich et al. 2015).

Arbuscular mycorrhizal fungi are well known soil organisms. They develop mutualistic relationship with plants by promoting the phosphorus uptake by the plant roots (Sharma and Buyer 2015). In a recent study, different types of fertilization treatment were provided to soil cultivated with sage to determine the distribution of arbuscular mycorrhizal fungi (AMF) in soil ecosystem. It was observed that the AMF colonization of sage was higher in case of plants treated with manure fertilization as compared to those treated with mineral fertilization. Moreover, *Olpidium* species and dark septate endophytes have also been isolated from the manure-treated fields (Zubek et al. 2012).

One of the most common genus of soil fungi is *Aspergillus.* It has been found that even after the treatment of soil with 600 ppm of cypermethrin, a highly toxic pesticide, at least 10 fungal species of *Aspergillus* tend to survive in the treated soil. *A. flavus* and *A. candidus* are the major isolates among these varieties. At 1000 ppm of cypermethrin, *A. flavus* is the single most dominant species. However, *A.* terries, *A. niger. A. awamori* have also been found to inhabit these soils (Sethi et al. 2015). Similarly, twelve different fungal species were isolated from crop fields of Mysore district in India. These included *A. niger, C. lunata, A. terreus, F. solani, A. fumigatus, F. oxysporum, A. flavus, T. harzianum, R. stolanifer, T. viride, P. chrysogenum* and *P. fumiculosum.* In general, *Aspergillum species* and *Penicillium* species are most dominant fungal species in agricultural soil. The primary reason for their high population density is their defense and survival mechanisms. *Aspergillus* produces toxins and *Penicillium* produces penicillin antibiotics which cause a retardation in the growth of other microbes. Moreover, they are also involved in the biodegradation of compounds and enhancement of soil nutrients (Sharma and Raju 2013).

3.4 Algae

A diversity of cyanobacteria have been isolated from rice fields. In a recent study carried out in Andhra Pradesh, India, at least 38 different species of the algae were isolated from rice paddies. They included both heterocystous and non-heterocystous. Heterocystous are specialized cells that are produced under nitrogen-stressed conditions. In rice fields, cyanobacteria are the most abundant nitrogen fixing organisms (Danaboyina and Sivakumar 2013). Unicellular, filamentous and colonial, all types of cyanobacteria have been found in these fields. However, a number of physiological parameters affect the nature and concentration of cyanobacteria in soil. These include temperature, moisture content and soil pH.

4 Advantages of Microflora

The microflora of soil is rich in a variety of organisms. These can be exploited to attain various benefits including production of enzymes, industrial products, secondary metabolites and biopharmaceuticals.

4.1 Living Soil

Living soil is a term coined for the soil that contains diverse microflora. Microorganisms especially filamentous fungi are one of the most important constituents of a well living soil. These fungi are prominent sources for of food as well as biosynthesis of a variety of products including antibiotics and other medicinal agents (Sethi et al. 2015). The soil microorganisms contribute to the modification of biological, physical and chemical properties of the soil. The health of soil is, hence, maintained by this micro flora. The characteristic earthy smell of soil is also a gift of bacterial diversity present in it (Bhatt et al. 2015).

4.2 Growth and Development of Plants-Agricultural Sustainability

The sustainability of the environment and agriculture depends upon the soil micro flora as well. Microflora have been observed to enhance the productivity of agricultural system in a natural and organic way. Soil microbial communities, hence, have multiple positive impacts on growth and development of plants (Singh et al. 2011). Mycorrhizal fungi are one of the groups of organisms living in rhizosphere. These fungi form associations with plants and contribute to their growth by symbiosis. It is due to mycelium of these fungi that the absorption of nutrients and water from soil is enhanced. With this plant-fungal association, soil micro flora plays a significant role and bind to these fungi to complement their development. The interaction which is formed between ectomycorrhizal fungi and other microbes of the soil flora leads to the development of a sustainable microbial complex ecosystem which helps in proper growth and development of plants. This, ultimately, enhances yield and quality of crops (Duponnois et al. 2012).

Fungi improve the availability of soil nutrients for plants. It also helps in the degradation of cellulose to release carbohydrates including pectin, hemicelluloses and starch (Bhatt et al. 2015). Soil micro flora have a direct relation with soil fertility (Marathe et al. 2012). Soil bacteria are the only known life forms which perform nitrogen fixation during nitrogen cycle (Bhatt et al. 2015). They are, hence, excellent candidates to increase zinc bioavailability to plants. Improvement of microbial activity in soil yields free zinc which can then be easily absorbed by plants (Imran et al. 2014). The soil microbes are involved in a number of biochemical transformations and mineralization activities taking place in soil (Yadav 2014).

4.3 Decomposition of Organic Matter

Soil microorganisms are known to possess a primary role in environment because they degrade plant and animal residues. Hence, global cycling of nutrients is possible due to diverse micro flora of soil (Sonia and Saksham 2013). Soil natural flora like *Pseudomonas* and *Trichoderma* are, also, involved in the control of a large variety of harmful plant pathogens including *Rhizoctonia, Pythium, Fusarium* and *Phytophthora* (Jagtap 2012).

4.4 Bioremediation of Soil

Soil micro flora have been observed to convert pesticidal toxic compounds to non-toxic forms. Bacteria present in soil, for instance *Bacillus stearothermophilus*, are involved in detoxification of largely used pesticides including chlorpyrifos (Savitha and Raman 2012). Pristine and oil contaminated soils naturally have micro flora which utilize hydrocarbons and possess heavy metal resistance. Bacterial species including *Chromobacterium orangum, Bacillus subtilis, Enterobacter aerogenes, Corynebacterium pseudotuberculosis, Nocardia paraffinea, Nocardia coralline, Micromonospora chalcea, Streptomyces flavovirens, Alcaligenes faecelis* and *Brevibacterium linens* are of prime importance in this respect. These species degrade aromatic hydrocarbons and yield nutrients that can be utilized by plants. Indigenous soil micro flora are quite diverse. Due to these facts, the development of genetically modified micro floral organisms for bioremediation of soil is not needed (Ali et al. 2012). Addition of organic fertilizers in soil enhances the amount of soil micro flora.

The process, hence, causes a decrease in the soil pollution. Bioremediation of polluted soil can, therefore, be performed using microbe-mediated organic fertilization process (Osaigbovo et al. 2013).

5 Deteriorating Effects on Soil Micro Flora by Anthropogenic Activities

Increased herbicide concentrations decrease the microbial concentrations in soil. However, with the degradation of applied herbicides, microbial populations are greatly enhanced. After 60 days of application of herbicides, soil micro flora gain their pre-treatment concentrations (Bera and Ghosh 2013a, b). Frequent long term application of pesticides on agricultural lands might cause severe detrimental effects on soil micro flora. Organophosphorus pesticides are one of the most widely used pesticides. They have long term degrading effects on soil microbes. Application of pesticides, like atrazine, greatly reduces the concentration of cellulose producing bacteria. Similar adverse effects have been noted with the use of other pesticides including paraquat and trifluralin (Sethi et al. 2015).

Though nearly all kinds of herbicides possess detrimental effects on soil microbes yet their persistent application has been associated with the development of resistance in these microbial species. This, ultimately, results in the recovery of microbial count in the soil (Bera and Ghosh 2013a, b). Chemical methods of controlling insects, fungi and herbs are gaining fame because they are highly efficient. They increase the crop yield by protecting them from pests. Repetitive, high-concentration treatment and the resulting accumulation of these chemicals in soil, however, poses significant harmful effects on micro flora even leading to their extinction in certain cases (Das et al. 2014). Among the biological threats, the soil micro flora are also degraded by continuous cultivation of *Bt* cotton in a particular area. *Bt* toxin which leaches in to the rhizosphere kills the microorganisms in soil (Marathe et al. 2012). Similarly, bacterial population in soil cultivated with soybean is inhibited by pesticide alachlor. Excessive use of chemical fertilizers and compounds might contribute to generating a biological imbalance. This makes the soil prone to degradation. Rate of biodegradation of soil is directly related to the diversity and density of microbial community in it (Yadav 2014).

6 Conclusion and Future Prospects

A variety of microbes are found in soil. It can be rightly said that microbial communities are the most significant component of soil. The microfloral diversity in soil is more diverse than any other environmental habitat (Bhatt et al. 2015). Richness of soil micro flora is dependent on cropping pattern, soil moisture and organic matter present in soil (Marathe et al. 2012). Integrated crop management techniques use fungicides, insecticides and herbicides in a manner that once they deteriorate soil micro flora, the concentration of micro flora rise again with the passage of time. These approaches can be used for the maintenance of the living soil (Jagtap 2012).

Modern day agriculture is not possible without exploiting the soil micro flora. Hence, this microbial community needs to be protected. Therefore, biological methods need to substitute the currently prevalent chemical techniques. This would help in prevention of soil degradation and preservation of microbial communities residing in the soil. Moreover, composition of soil microfloral communities also needs to be monitored. Further research on soil microflora is required to get maximum benefits from them. They can then be exploited to attain pharmaceutical drugs and super foods. They can also be employed for the biofertilization process. Therefore, the scientific expertise of botany, microbiology, molecular biology, molecular genetics, biotechnology, pharmaceutics and systems biology need to be exploited for attaining the maximum benefits from the plant-microbe interactions.

References

Ali N, Dashti N, Al-Mailem D, Eliyas M, Radwan S (2012) Indigenous soil bacteria with the combined potential for hydrocarbon consumption and heavy metal resistance. Environ Sci Poll Res 19(3):812–820

Archana J, Thilo E, Ruth F, Philippe B, Wilfred O (2015) Quantification of spatial distribution and spread of bacteria in soil at microscale. EGU General Assembly 2015. Vienna, Austria. Id: 15420

Bera S, Ghosh R (2013a) Soil micro flora and weed management as influenced by atrazine 50 % WP in sugarcane. Univ J Agric Res 1(2):41–47

Bera S, Ghosh RK (2013b) Soil physico-chemical properties and micro flora as influenced by bispyribac sodium 10 % SC in transplanted kharif rice. Rice Sci 20(4):298–302

Bérrard A, Ruy S, Coronel A, Toussaint B, Czarnes S, Legendre L, Doussan C (2015) Rhizosphere: a leverage for tolerance to water deficits of soil micro flora?. EGU General Assembly 2015. In Vienna, Austria. Id: 9481

Bhatt M, Patel S, Prajapti P, Jasrai YT (2015) Isolation and identification of soil micro flora of national parks of Gujarat, India. Int J Curr Microbiol Appl Sci 4(3):421–429

Burke DJ (2015) Effects of annual and inter-annual environmental variability on soil fungi associated with an old-growth, temperate hardwood forest. FEMS Microbiol Ecol. Fiv053

Chaudhry V, Dang HQ, Tran NQ, Mishra A, Chauhan PS, Gill SS, Nautiyal CS, Tuteja N (2012) Impact of salinity-tolerant MCM6 transgenic tobacco on soil enzymatic activities and the functional diversity of rhizosphere microbial communities. Res Microbiol 163(8):511–517

Danaboyin a KSB, Sivakumar K (2013) Distribution of heterocystous and non heterocystous soil micro flora and related physico- chemical parameters of the paddy fields of Guntur District, Andhra Pradesh. J Multidiscip Sci Res 1(3):01–08

Das R, Ghosh RK, Bera S, Poddar R (2014) Bioefficacy studies of chlorimuron ethyl 25 % WP in transplanted rice and its effects on soil micro flora in inceptisol of West Bengal. J Crop Weed 10(2):350–354

Deslippe JR, Hartmann M, Simard SW, Mohn WW (2012) Long-term warming alters the composition of Arctic soil microbial communities. FEMS Microbiol Ecol 82:303–315

Duponnois R, Baudoin E, Thioulouse J, Hafidi M, Galiana A, Lebrun M, Prin Y (2012) The impact of mycorrhizosphere bacterial communities on soil biofunctioning in tropical and

mediterranean forest ecosystems. In: Bacteria in agrobiology: plant probiotics, Springer Berlin Heidelberg. pp 79–95
Fang Y, JunLing T, PanPan Y, Hong F, GuangSheng H, XuDong C, YuSheng L, ZhiYuan T, GuiXiang P (2014) Effects of inoculant of photosynthetic bacteria on tomato quality, soil fertility and soil microbial characteristics. J South Chin Agric Univ 35(1):49–54
Habibnia S, Nasab MR, Heidarieh P, Bafghi MF, Pourmand MR, Eshraghi SS (2015) Phenotypic characterization of Nocardia spp. isolated from Iran soil micro flora. Int J Environ Health Eng 4(1):20–24
Hakeem KR, Akhtar MS, Abdullah SNA (2016) Plant, soil and microbes – vol 1, implications in crop science. Springer International Publishing AG, Gewerbestrasse 11, 6330 Cham, Switzerland. 366 pp
Helfrich M, Ludwig B, Thoms C, Gleixner G, Flessa H (2015) The role of soil fungi and bacteria in plant litter decomposition and macroaggregate formation determined using phospholipid fatty acids. Appl Soil Ecol 96:261–264
Huang Z, Wan X, He Z, Yu Z, Wang M, Hu Z, Yang Y (2013) Soil Biol Biochem 62:68–75
Imran M, Arshad M, Khalid A, Kanwal S, Crowley DE (2014) Perspectives of rhizosphere micro flora for improving Zn bioavailability and acquisition by higher plants. Int J Agric Biol 16:653–662
Jagtap GP (2012) Effect of agrochemicals on micro flora in soybean rhizospheric soil. Sci J Microbiol 1(2):55–62
Jangid K, Williams MA, Franzluebbers AJ, Schmidt TM, Coleman DC, Whitman WB (2011) Land-use history has a stronger impact on soil microbial community composition than aboveground vegetation and soil properties. Soil Biol Biochem 43(10):2184–2193
Jin W, YinLi B, YanXu Z, TianCai H, Lang Q, ShuLin C (2014) Effects of Arbuscular mycorrhiza on soil microorganisms and enzyme activities in disturbed coal mine areas. J South Agric 45(8):1417–1423
Jin Z, Li Z, Li Q, Hu Q, Yang R, Tang H, Li M, Huang B, Zhang J, Li G (2015) Canonical correspondence analysis of soil heavy metal pollution, micro flora and enzyme activities in the Pb–Zn mine tailing dam collapse area of Sidi village, SW China. Environ Earth Sci 73(1):267–274
Li X, Mu Y, Cheng Y, Liu X, Nian H (2013) Effects of intercropping sugarcane and soybean on growth, rhizosphere soil microbes, nitrogen and phosphorus availability. Acta Physiol Plant 35(4):1113–1119
Lin Z, Zhen Z, Wu Z, Yang J, Zhong L, Hu H, Luo C, Bai J, Li Y, Zhang D (2016) The impact on the soil microbial community and enzyme activity of two earthworm species during the bioremediation of pentachlorophenol-contaminated soils. J Hazard Mater 301:35–45
Locatellia A, Spor A, Jolivet C, Piveteau P, Hartmann A (2013) Biotic and abiotic soil properties influence survival of Listeria monocytogenes in soil. PLoS One 8(10):e75969
Lorch JM, Lindner DL, Gargas A, Muller LK, Minnis AM, Blehert DS (2012) A culture-based survey of fungi in soil from bat hibernacula in the eastern United States and its implications for detection of Geomyces destructans, the causal agent of bat white-nose syndrome. Mycologia 12:207
Lu M, Xu K, Chen J (2013) Effect of pyrene and cadmium on microbial activity and community structure in soil. Chemosphere 91(4):491–497
Manzoni S, Schimel JP, Porporato A (2012) Responses of soil microbial communities to water stress: results from a meta-analysis. Ecology 93(4)
Mao N, Zhang Z, Zhang X, Zhang F, Zhao H, Xi Q (2013) On the effects of ultrasonic treatment of soil samples on actinomyces and bacteria isolation. J Longdong Univ 3:012
Marathe SA, Sayyed RZ, Somani VJ (2012) Impact of Bt Cotton cultivation on bacterial soil micro flora in North Maharshtra. Int J Biotechnol Biosci 2(4):266–271
Mary LCL, Sujatha R, Chozhaa AJ, Navas PMA (2015) Influence of organic manures (Biofertilizers) on soil microbial population in the rhizosphere of mulberry (Morus Indica L.). Int J Appl Sci Biotechnol 3(1):61–66

Miransari M (2011) Interactions between arbuscular mycorrhizal fungi and soil bacteria. Appl Microbiol Biotechnol 89(4):917–930

Mitchell RJ, Hester AJ, Campbell CD, Chapman SJ, Cameron CM, Hewison RL, Potts JM (2012) Explaining the variation in the soil microbial community: do vegetation composition and soil chemistry explain the same or different parts of the microbial variation? Plant Soil 351(1):355–362

Na L, WeiMin Z, Bo L, ShouBo T, LongYing Z, Hui Z, Li Y (2014) Effects of grafting on the soil micro-ecological environment of tomato rhizosphere. Acta Agric Shanghai 30(5):6–10

Osaigbovo AU, Law-Ogbomo KE, Agele SO (2013) Effects of spent engine oil polluted soil and organic amendment on soil chemical properties, micro-flora on growth and herbage of Telfairia Occidentalis (hook f). Bayero J Pure Appl Sci 6(1)

Rajik M, Pathak SP, Biswas SK, Naresh P (2011) Effect of organic amendment on soil micro flora and soil borne diseases of potato. Ind Phytopathol 64(3)

Savitha K, Raman DNS (2012) Isolation, identification, resistance profile and growth kinetics of chlorpyrifos resistant bacteria from agricultural soil of Bangalore. Res Biotechnol 3(2)

Sethi BK, Pradhan S, Behera N, Sahoo SL (2015) Effect of cypermethrin, a pyrethroid insecticide on dynamics of soil micro flora. J Appl Biol Biotechnol 3(5):19–25

Sharma MP, Buyer JS (2015) Comparison of biochemical and microscopic methods for quantification of arbuscular mycorrhizal fungi in soil and roots. Appl Soil Ecol 95:86–89

Sharma R, Raju NS (2013) Frequency and percentage occurrence of soil mycoflora in different crop fields at HD Kote of Mysore district. Int J Environ Sci 3(5):1569–1576

Singh JS, Pandey VC, Singh DP (2011) Efficient soil microorganisms: a new dimension for sustainable agriculture and environmental development. Agric Ecosyst Environ 140(3–4):339–353

Singh S, Gupta R, Sharma S (2015) Effects of chemical and biological pesticides on plant growth parameters and rhizospheric bacterial community structure in Vigna radiate. J Hazard Mater 291:102–110

Sofo A, Scopa A, Dumontet S, Mazzatura A, Pasquale V (2012) Toxic effects of four sulphonylureas herbicides on soil microbial biomass. J Environ Sci Health Part B Pest Food Contam Agric Wastes 47(7):653–659

Sonia S, Saksham G (2013) Impact of pesticides and biopesticides on soil microbial biomass carbon. Univ J Environ Res Technol 3(2):326–330

Sullivan TS, McBride MB, Thies JE (2013) Soil bacterial and archaeal community composition reflects high spatial heterogeneity of pH, bioavailable Zn, and Cu in a metalliferous peat soil. Soil Biol Biochem 66:102–109

Suneetha V, Khan ZA (2010) Actinomycetes: sources for soil enzymes. Soil Enzymol 22:259–269

Tang C (2012) Study on population structure of microorganism and their dynamics in continuous cropping tomato soil. J Anhui Agric Sci 36:052

Taylor DL, Hollingsworth TN, McFarland JW, Lennon NJ, Nusbaum C, Ruess RW (2014) A first comprehensive census of fungi in soil reveals both hyperdiversity and fine-scale niche partitioning. Ecol Monogr 84(1)

Turrini A, Sbrana C, Giovannetti M (2015) Belowground environmental effects of transgenic crops: a soil microbial perspective. Res Microbiol 166(3):121–131

Udikovic-Kolic N, Wichmann F, Broderick NA, Handelsman J (2014) Bloom of resident antibiotic-resistant bacteria in soil following manure fertilization. PNAS USA 111(42):15202–15207

Vandamme P, De Brandt E, Houf K, Salles JF, van Elsas JD, Spilker T, LiPuma JJ (2013) Burkholderia humi sp. nov., Burkholderia choica sp. nov., Burkholderia telluris sp. nov., Burkholderia terrestris sp. nov. and Burkholderia udeis sp. nov.: Burkholderia glathei-like bacteria from soil and rhizosphere soil. Int J Syst Evol Microbiol 63:4707–4718

Wang J, Pang H, Li Y, Zhao Y, Zhang J (2012) Effects of microbial fertilizer on soil micro flora and sunflower yield in saline soil. J Agro Environ Sci 11

Wu K, Yuan S, Wang L, Shi J, Zhao J, Shen B, Shen Q (2014) Effects of bio-organic fertilizer plus soil amendment on the control of tobacco bacterial wilt and composition of soil bacterial communities. Biol Fertil Soils 50(6):961–971

Xia ZC, Kong CH, Wang P, Chen LC, Wang SL (2012) Characteristics of soil microbial community structure in Cunninghamia lanceolata plantation. Eur PMC 23(8):2135–2140

Xu R, Li J, Kong W, Li C, Guo J, Hu G, Chen L, Li X (2013) Effect of soil types on quantity of soil microbe in Hubei Tobacco-growing Region. Hubei Agric Sci 1:019

Yadav AM (2014) Comparative study of mycoflora of paddy field soil in bhandra district. Int J Life Sci A2:59–61

Yang Y, Liu SW, Pan K, Wu FZ (2013) Effects of Chinese onion's root exudates on cucumber seedlings growth and rhizosphere soil microorganisms. Eur PMC 24(4):1109–1117

Yuan Y, Li M, Hu W, Zhang J, Zhao L, Li H (2011) Effect of biological organic fertilizer on tomato bacterial wilt and soil microorganism. J Agro Environ Sci 30(7):1344–1350

Zhalnina K, Dias R, de Quadros PD, Davis-Richardson A, Camargo FAO, Clark IM, McGrath SP, Hirsch PR, Triplett EW (2015) Soil pH determines microbial diversity and composition in the park grass experiment. Microbiol Ecol 69(2):395–406

Zhu P, Zhang X, Ren Z (2013) Effects of different fertilizer application on soil micro flora of continuous cropping watermelon. Northern Hortic 11

Zubek S, Stefanowicz AM, Blaszkowski J, Niklińska M, Seidler-Łożykowska S (2012) Arbuscular mycorrhizal fungi and soil microbial communities under contrasting fertilization of three medicinal plants. Appl Soil Ecol 59:106–115

Arbuscular Mycorrhizal Fungi Boon for Plant Nutrition and Soil Health

Mehraj ud din Khanday, Rouf Ahmad Bhat, Shamsul Haq, Moonisa Aslam Dervash, Asma Absar Bhatti, Mehru Nissa, and Mohd Ramzan Mir

Contents

M. ud din Khanday
Division of Soil Science, Sher-e-Kashmir University of Agricultural Sciences and Technology-Shalimar, Srinagar, Kashmir 1900025, India

R.A. Bhat (✉) • S. Haq • M.A. Dervash • A.A. Bhatti
Division of Environmental Sciences, Sher-e-Kashmir University of Agricultural Sciences and Technology-Shalimar, Srinagar, Kashmir 1900025, India
e-mail: rufi.bhat@gamil.com

M. Nissa
Department of Environmental Sciences, University of Kashmir, Srinagar 190006, India

M.R. Mir
Department of Botany, Aligarh Muslim University, Aligarh 202002, Uttar Pradesh, India

K.R. Hakeem et al. (eds.), *Soil Science: Agricultural and Environmental Prospectives*, DOI 10.1007/978-3-319-34451-5_14

Abstract Arbuscular mycorrhiza-forming fungi make necessary feeding upon the host plant cells without killing the plant and derive their food from host-plant. Arbuscular mycorrhiza generates the mass on the root cells and forms some peculiar structures (arbuscules) within the cell. These cell structures are actually the sites where the give-and-take of nutrients between the fungi and plant takes place. Extracellular networks which spreads in all possible directions into the surrounding soil and reach to locations where nutrient concentration is very low and make available food, minerals and other nutrients exclusively PO_4^{2-} and NO_3^- for the plant. The host plant provides the symbiotic partner food for life activities. Arbuscular mycorrhiza-forming fungi provides sustainable soil health by improving soil structure, defence mechanism against plant root pathogens, provide availability of essential plant nutrients (Zn, Cu, Fe, Ca, Mg and NPK), tillage, bio-control of pests and play efficient role in phytoremediation.

Keywords Arbuscular mycorrhizal fungi • Bio-control • Phosphorus uptake • Nutrition and soil health

1 Introduction

The word 'Mycorrhiza' literally *means* fungus-root in Greek language and is characterized as the nutrient transferring association existing between a group of soil fungi and host plants. The host plant provides essential and other fundamental life supporting materials to the fungi. The fungi in-return makes available essential minerals and life supporting nutrients to the host plant. Mycorrhizal interactions have significant role in plant productivity by way of extending host plant roots and controlling diseases around the root environment (Hakeem et al. 2016). There are various forms of mycorrhizal interactions; among the all, ectomycorrhizal and endomycorrhizal arbuscular- vesicular (AV) associations are the two economically and ecologically important. Ectomycorrhizal fungi infection in host plant constantly leads to changes in their roots. The roots developed by the fungi are structurally and functionally different from unaffected roots. The modified roots by fungi are different from unmodified roots with respect colour, structure and number of branches. The roots developed by fungi have more and more branches and are more elongated than rest of the roots. In endomycorrhizal relationship between fungi and host plant, the fungi enter the root cells and creates cluster of thinly divided root structures

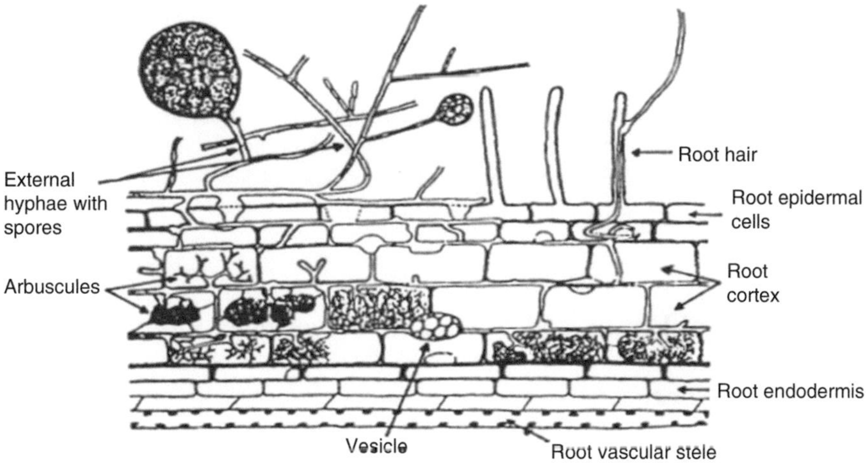

Fig. 1 A root colonized by AM fungi showing the diagnostic features of the fungi

(arbuscules) in root cells (cortex) as shown in Fig. 1. Simultaneously they generate other membrane-bound structures called vesicles, having different shapes, diameter and colour within or exterior to the root cortex. Arbuscule cell structures are actually the sites where the give-and-take of nutrients between the fungi and plant takes place. The main function of vesicles in developed root structure is to provide material storage facility and when fully developed they can act as multiplication organ. Vesicles, arbuscules and spores are the indicative characters of the arbuscular-vesicular mycorrhizas. Arbuscular mycorrhiza fungi forms diagnostic structures in host plant that helps the host plant to absorb more and more nutrients from the soil.

1.1 Host Characteristics

Arbuscular mycorrhiza associations take place in an open range of tropical and temperate plant species, while as ectomycorrhizas occur mainly in forest plant classes and partial in humid class of tree species.

1.2 Taxonomy of Arbuscular Mycorrhiza Fungi

Arbuscular mycorrhiza fungi show the strange distinctiveness in morphology and physiology. Spores of arbuscular mycorrhiza fungi are normally created in soil and their diameter are in range of 40–400 μm and are quite bigger than other type of fungi. Mass formation on plant roots is vital factor for production of arbuscular mycorrhiza fungi. This is the basic main reason that arbuscular mycorrhiza fungi are known as necessitate symbiotic fungi. The relationship between host plant and arbuscular mycorrhiza fungi in general is give-and-take of nutrient relationship. An Arbuscular mycorrhiza fungus belongs to the member of Zygomycota of a new Phylum Glomeromycota. Classification has been given on the basis of rRNA gene sequence that differentiates the genera under this phyllum. Under this classification large number of different genera has been raised with comparatively small records of isolates. The records of morphological characteristics of few genera are shown in the Fig. 2.

2 Structural Features of Arbucular Mycorrhiza Fungi

2.1 Arbuscules

Arbuscules are formed by continual dichotomous branching and reductions in hyphal width. They start to produce more than 1 day following root incursion and are considered the major junction where of nutrient give and take place between the fungus and host plant. Structural features of arbuscular are as under:

2.2 Vesicles

Vesicles are rounded intercellular structures which act as storage organs of phosphorus, oil droplets and also function as propagules.

2.3 Spores

Spores generally form bulky walls and have more than one layer, swollen structure with more than one subtending hyphae which form in the soil or in roots. They function as propagules. Spores of VAM fungi are called chlamydospores (Fig. 3).

2.4 External Hyphae

They are thick and runner type. There are different nutrient supplying hyphae as well as thin nutrient absorptive hyphae and responsible for nutrient acquisition and spore formation (Fig. 4).

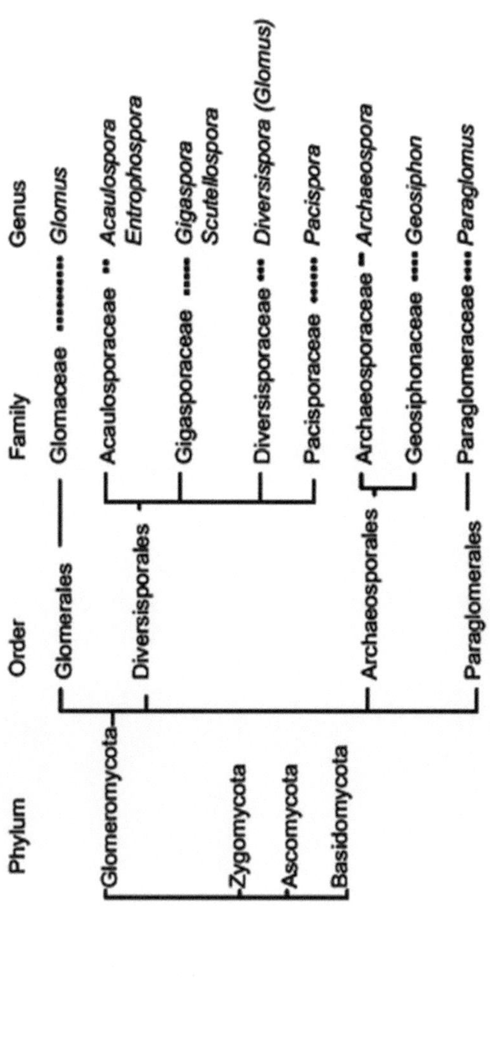

Fig. 2 Arbucular mycorrhiza classification

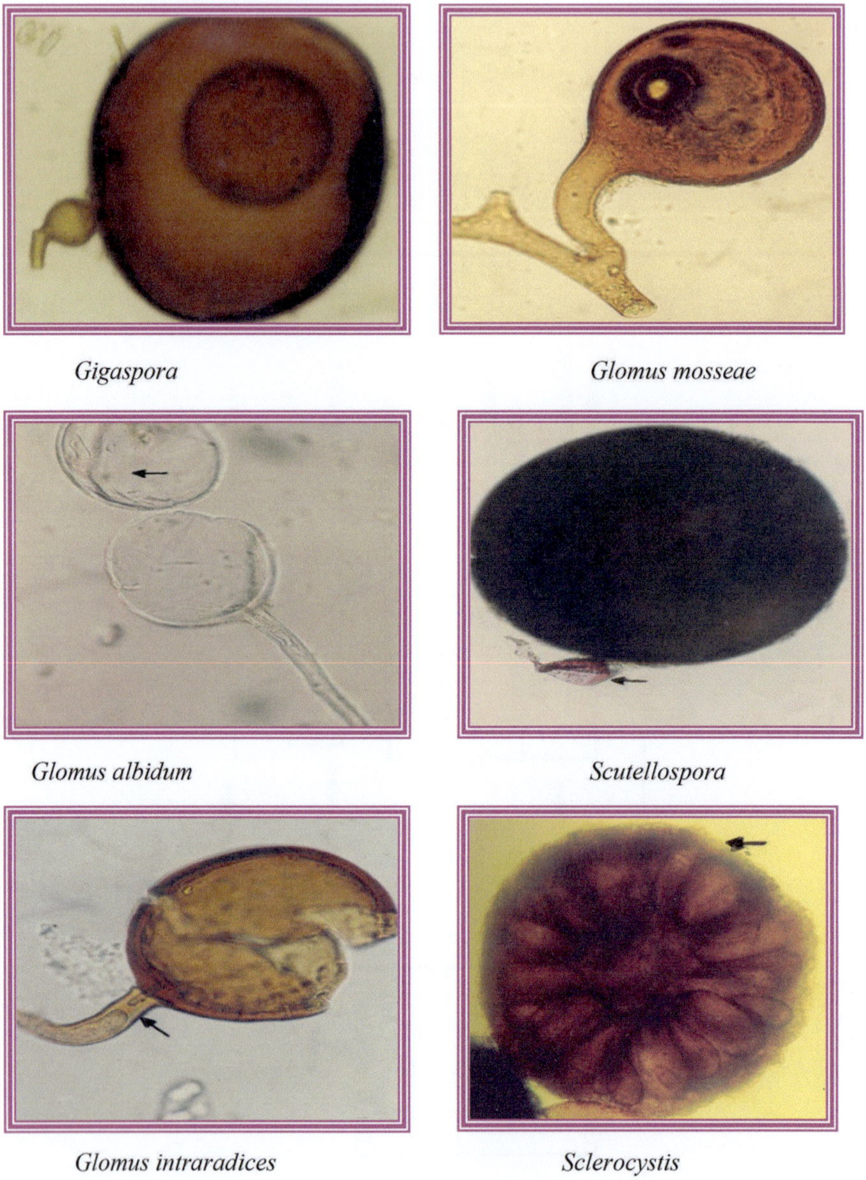

Fig. 3 Some arbucular mycorrhiza fungi species producing spores

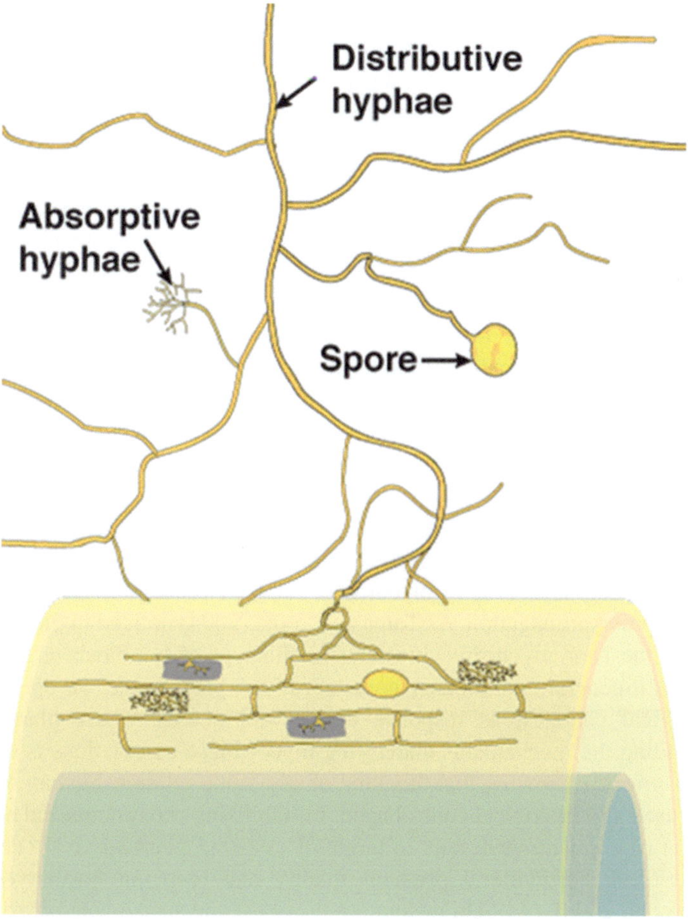

Fig. 4 External hyphae

3 Role of Arbuscular Mycorrhiza to Maintain Plant and Soil Nutrient Balance

The world is on the edge of a new agriculture that involves the marriage of plant biology with agroecology under the umbrella of biotechnology and germ plasm improvement (Vance et al. 2003). The developing countries face tremendous problems persistent food shortage. So, food security is a basic issue at local, regional and global levels in developing countries. The challenge to face against inadequate food supply to every individual in developing countries has been reached at peak. The urgency is to neutralize the problems associated with food inadequacy. To maintain or increase agricultural productivity and management of soil resources, agroecosystems should be practicing a sustainable agriculture, both economically and environmental protection view point, that increases the food productivity to feed every individual in developing countries. Therefore, we need to know the better awareness of the system and factors that govern the bioavailability of soil nutrients to plants, including the root-soil interactions, understanding of microorganisms in the rhizosphere (Martinez and Johnson 2010). As Samuil (2007) mentioned, organic farming promotes sustainable production systems, diversified and balanced crop, to prevent pollution and the environment damages. Moreover, the importance of the mycorrrhizal arbuscular fungi in organic farming and farmers' potential to increase the benefits of arbuscular mycorrhizae (AM) associations in such systems represented interesting subjects as it was synthesized by Gosling et al. (2006). Research accumulated over the years on the chemistry, microbiology and soil pedoenzimology allowed to shape their own ideas about soil fertility and possibility to assess its level (Ghinea et al. 2007). Amalgamation of usual propagation techniques, molecular biology, genetic engineering and natural deviation will improve the ability of plants to nutrients uptake and fixation, and improve roots system performance. In this context, equally valuable relations between different organisms play an important role in the ecology of natural ecosystemsand also in terms of sustainable agriculture (Hause and Schaarschmidt 2009). Many horticultural plants form AM. However, other types of mycorrhiza are important in specific situations: ecto-mycorrhiza, for reforestation programs; ericoid mycorrhiza for fruit production, orchid's mycorrhiza, to improve propagation. As Toma and Toma (2004) noticed in the first bibliographic synthesis on mycorrhiza in Romania, this type of symbiosis large extended in nature is a general evolutionary approach in the plant world, an option to root morphological changes, as response to the soil quality where plants grown. Understanding the mechanisms underlying mycorrhizae associations is one of the biologists priority, knowing that this kind of symbiotic relationship can be considered and used as a biological control agent in stabilizing ecosystems and ecological restoration of contaminated soils in terms of assisting plant survival (Popa et al. 2008), assuring horticultural products quality and reducing postharvest losses (Paliyath et al. 2008), old-field restoration, low cost and sustainable phytoremediation techniques, or anthropogenic soils rehabilitation (Cardinale et al. 2010) etc. Initiative of eco-friendly agriculture practices in emergent countries shall play a crucial role to achieving food safety measures. Therefore, need is to enhance savings in agriculture, widen admittance to food, get better governance to comprehensive agricultural employment and add to productivity, while conserving the natural resources foundation (Diouf 2011). Taking into consideration that one of the main goals of sustainable horticulture is to benefit farmers and consumers; arbuscular mycorrhizae act as biofertilizer and bioprotectants. Rhizosphere characteristic feature related to arbuscular mycorrhizae symbiosis promoting arbuscular mycorrhizae benefits related to plants performance (in terms of physiological importance for the producer, plants productivity, plants resistance to stress factors, satisfy consumer needs by high crop quality etc.); the role of arbuscular mycorrhizae in terms of soil health is not only a support and nutritive media supply for cultivated plants, but also as soil attributes in the agroecosystems (structure, compaction, diversity of microorganisms, rehabilitation etc.)

4 Rhizosphere: Concept and Molecular Signaling in the Context of Promoting Mycorrhizal Symbiosis

Plants growth and development is under the control of internal signals that depend on assuring adequate amounts of mineral elements from soil, to roots. Rhizosphere is the quantity of soil under the pressure of roots and roots itself. This area is by far a more complex environment from the physical, chemical and biological view point, than the aboveground area, around the aerial part. Rhizosphere is under the control of root and microflora (Aragno 2005). Living together is a common situation of the biological world, particularly in the case of plants and root environmental conditions. The underground plant systems and their environment need the same care as above ground systems, as is the case of arbuscular mycorrhiza which is crucial in sustainable management practices. Extremes changes of the chemical composition within minutes and spaces of millimeters, and the density and diversity of soil organisms play a major role. Anthropogenic action on it, namely the introduction of fertilizers and pesticides, including adding organic amendments should be also considered in terms of its influence on microbial activity, on the agricultural soils biological indicators, considering that the soil biodiversity determine economic benefits for society, as part of a broader strategy of conservation and use of the agro biodiversity. Roots are plants organs which very easily adapt their growth to the most profitable areas, thanks to a chemical communication between plant roots and between these and other rhizosphere organisms (Hause and Schaarschmidt 2009). Products of the living roots into soil (rhizodeposits) are not only a source of carbon substrate for microorganisms growth, but also signal molecules that promote chemotaxis of microorganisms in the soil, to rhizosphere (Dennis et al. 2010). Rhizodeposition is a carbon and energy flow which gives to the rhizosphere the characteristics of a rich habitat, as regard as nutrients supply to the heterotrophic micro-organisms. As Jones et al. (2004), rhizodeposition is under the influence of many factors. For instance, amino acids concentration is a result of flows between roots, soil and plants, with physiological and ecological effects on plant-soil-microorganisms interactions. During the most energetic stages of root colonization there were registered high levels of amino acids and of some fatty acids, corresponding to the plastids metabolism activation. Also, there was noticed an accumulation palmitvaccenic and vaccenid acids that can be used as markers for fungal root colonization. At the final stages of colonization it was observed a stimulation of apocarotenoids accumulation (cyclohexenone and mycorradicin), without registering a change of the typical cell wall bound specific phenols. Thus, understanding the molecular signaling processes and functions that are regulated in root environment shall have central responsibility in promoting coexistence of plant-beneficial microorganisms to overcome existing limits and setting new strategies to ensure inoculants, with applications in biotechnology with a view to promote agroecosystem sustainability.

4.1 General Aspects as Regard as Arbuscular Mycorrhizae Symbiosis

Mycorrhizae are classic examples of mutualistic symbiosis, primarily characterized by carbon gain by the fungus from the plant, also a reciprocal nutrient transfer from fungus to plant, next to other effects relating to improving water relation and pathogens tolerance. The fungi cannot manufacture their own food, depends wholly and soley on the plants. The fact that the plants sugar production in photosynthesis is used by the fungi. The fungal web reaches beneath the plant roots and brings water and nutrients to the plants that are otherwise unavailable to the plant. The fungi are predominantly adept at mobilizing phosphorus (P) to the plant. As a result, the plants are more energetic and productive, with growth and yield often vastly improved. Arbuscular mycorrhizae symbiosis between roots of 80 % of all terrestrial plants and genera of phylum Glomeromycota (Wang et al. 2008) are diffuse and often nonspecific. When mycelium fungus colonizes roots and link together two or more plants, sometimes belonging to several species, common mycorrhizal network is formed (CMN), a major component of terrestrial ecosystems, with important effects on plant community, on invasive plants trajectory (Pringle et al. 2009), as well as on the mycorrhizal community structure and functionality in invaded habitats. Depending on their morphology and the involved plants and fungus species, it can be distinguish several types of mycorrhizae associations. The first mycorrhizae classification was made by Frank (1887) and subsequently there have been various classifications.

4.2 Plants Compatibility with Mycorrhizal Fungus

It is a widespread and an old experience. Fossil samples suggest that this kind of symbiosis have been with very long ago, in first terrestrial plants (Remy et al. 1994). Plants capability to create arbuscular mycorrhiza could be the direct mechanisms that were preserved in the new plant species that occurred during evolution, involving selective recognition processes and certain specific plants to differentiate between valuable microorganisms and dangerous ones.

4.3 Symbiotic Relationships

It involves the development of specific adaptations for both associates and organization of their development, with specific signals (Hause and Schaarschmidt 2009). Plants generate an early indication, to initiate interaction, and fungus produces at least three signals: one diffusible, the second locally, allowing the plant to detect the appresorium position and last one, an autonomously cell signal, in the colonized cells which induces gene expression. Flavonoids have been proposed as the active molecule, but they are not totally crucial in plant and fungi relations. First defined credit phenomenon by the host fungus is the hyphae ramification induction, followed by appresorium formation on the root surface. Epidermal and hypodermic cells are in touch with first contamination structures and do not express significant cytological changes or typical defense responses (Gianinazzi-Pearson et al. 1996). The fungus enters the root, both intercellular and intracellular, determines extensive cytological changes, as well as in terms of pathogenesis related proteins (PR proteins) (Bonfante et al. 1996; Gianinazzi-Pearson et al. 1996). Plastids appear to be key cell organites to establishing symbiotic interface in the case of arbuscular mycorrhiza (Fester et al. 2007).

4.4 *Arbuscular Mycorrhizae Benefits in the Context of Agroecosystems Sustainability*

Mycorrhizae have a particularly importance for plants and ecosystem. From the biological view point, mycorrhizal fungus affects sustainable agriculture in two ways: plant production and soil quality. The beneficial effects of mycorrhizae fungus on plants performances and soil physical conditions are vital for sustainable management of agricultural ecosystems (Celik et al. 2004). Arbuscular mycorrhizae have multiple functions in the successful utilization of soil resources and because of their protective role against many soil pathogens. They are suitable tools for survival of many species of plants in different ecosystems, including many species of cultivated plants. Besides these function it can be remember: assure nutrients cycle and prevent the loss of ecosystem; assure carbon transport from the plants roots to other soil organisms involved in the processes of soil nutrient cycle soil hyphae may intervene in the nutrients cycle, by taking substances from saprophytic fungi; mycorrhizal fungus fructifications and root with mycorrhiza represent food sources and habitat for invertebrates; influence microbial populations in soil and exudates from the micorhizosphere and hyphosphere areas VA contribute to soil structure, aggregation of soil particles, respectively (Kahiluoto et al. 2011).

5 Inoculation of Arbuscular Mycorrhizae Fungi to the Roots

Production of plantlets play a valuable role on their post-transplanting performance: growth of a better-quality root structure and system; increased photosynthetic efficiency; enhanced nutrient uptake; alleviate environmental stress; protect from harmful soil borne pathogens.

It will cause improved overall growth and higher rate of survival in plants. Inoculated plants can have higher values of dry mass of aerial part, root dry mass as well as higher amounts of P and Zn in the aerial organs. Also, inoculated seedlings will have an earlier flowering time, compared to those no inoculated. Differentiated behavior in relation to the used fungus species must to be further studied in order to sustain inoculum production protocol and to implement this kind of technology (Ortas et al. 2011). An Arbuscular mycorrhizal fungus determines a crucial role in balancing the fertility in soils in case of non-agricultural environment than in usual agriculture conditions. As the results obtained by Bainard et al. (2011) demonstrated, in urban conditions (with a weaker presence and diversity of mycorrhizal fungi, compared to rural or natural ecosystems) inoculation would increase levels of colonization and growth of trees. As fungi play several key sociological roles, their establishment may be essential for the integrity and sustainability of restoration projects. Arbuscular mycorrhiza a key factors to soil quality characteristic feature, heavy metal control and agroecosystems sustainability. Plant production is directly associated to soil health which is based on its physicochemical and biological properties. Bethlenfalvay and Barea (1994) have highlighted michorrhizae role in sustainable agriculture. They transfer mineral elements in plants from soil and food (Carbon) into the soil from plants, thus having a dual role, both on plant and soil microflora composition. Experiments performed using *G. mosseae* fungus do not significantly alter the seed production (8 %), but aggregation of soil particles has improved by 400 %, in a soil rich in organic matter and phosphorus. In a soil containing little organic matter and phosphorus, seed production increased significantly (57 %), but were only small changes in the aggregation of soil particles (50 %). So, plant carbon allocation (seed production) and soil (formation of water stable aggregates) was influenced by VAM.

6 Arbuscular Mycorrhizal Fungi for Sustainable Soil Health

Soil health is a backbone to carry plant and animal life, balancing environmental excellence with special stress on soil accumulating surplus carbon, minerals and maintains water purity. Structure of soil is frequently considered as the unit of strength of aggregates and a key factor which moderates physicochemical and biological attributes leading the soil dynamics (Bronick and Lal 2005). Arbuscular mycorrhizal fungi are commonly scattered and connected to large groups of higher and lower plant species. Arbuscular mycorrhizal fungi forms boundary between roots of plants and soil environment, and are susceptible to alter in plant and soil conditions. Among the fungi, arbuscular mycorrhizal fungi are crucial for maintaining soil health, plant yield, fertility and functioning of terrestrial ecosystems and can influence variety patterns in plants and ecosystems worldwide. Interactions of arbuscular mycorrhizal fungi improve soil moisture for plants construct porous configuration in soil that allows infiltration of water, air, prevents soil erosion and improve resistance to root pathogens. They have remarkable capability in the reinstatement of waste land and enhance fertility in soils. Borowicz (2001) showed that plants generally grow better when they are mycorrhizal and this is especially true when plants are challenged by pathogens. The work of arbuscular mycorrhizal fungi with plant roots helps the plant to obtain essential micronutrients, food materials and water for life activities. Numerous studies indicate that the mycorrhizal symbiosis is most important to plants when soil nutrients are limiting (Marschner and Dell 1994; Jonhson et al. 2010). Plants exchange carbon (C) for fungal phosphorus (P) and nitrogen (N) (Smith et al. 2009).

6.1 Arbuscular Mycorrhizae Fungi Defense Mechanism Against Plant Root Pathogens

The nutrient uptake by plants by way of extending roots to their normal growth with help of arbuscular mycorrhizae fungi is a primary positive relationship between plants and fungi. The researchers have proved that these organisms can also play a vital role in checking diseases by way of damaging the crop pests. Table 1 shows soil-borne diseases and their eradication with arbuscular mycorrhizae fungi. The noticeable resistance of a plant to a pest or disease may be simply the result of improved nutrition (Karagiannidis et al. 2002). Colonisation of a root cell by AMF helps the plants to remove the pathogen from the cell, with result plants root is free from the pathogen infection that destroys the root systems in plants.

6.2 Plant Nutrition

Efficient assimilation of phosphorus is primary benefit that arbuscular mycorrhizae fungi put together available to plant. An Arbuscular mycorrhizae fungus significantly enhances the assimilation of P; the symbiotic relationship also has positive impact on overall development of the plant and productivity. An Arbuscular mycorrhizae fungus association not only enhances the phosphorus uptake but increases the

Table 1 Soil-borne fungal diseases controlled by Arbuscular mycorrhizae fungi (Gosling et al. 2006)

Pathogen	Disease	Crop
Sclerotium cepivorum	White rot	Onions (*Allium cepa*)
Fusarium oxysporum	Fusarium root rot	Asparagus (*Asparagus officinalis*) French bean (*Phaseolus vulgaris*)
Verticillium dahliae	Verticillium wilt	Tomatoes (*Lycopersicon esculentum*) Aubergines (*Solanum melongena*)
Helicobasidium mompa	Violet root rot	Asparagus
Rhizoctonia solani	Root and stem rots	Mung bean (*Vigna radiate*)
Aphanomyces euteiches	Root rot	Pea (*Pisium sativum*)

rate of absorption of other nutrients *viz.*, Zinc (Zn) copper (Cu), iron(Fe), nitrogen (N), potassium (K), calcium (Ca) and magnesium (Mg) by the host plant. An arbuscular mycorrhizae fungus significantly increases availability of plant nitrogen from dead organic sources.

6.3 Soil Tillage

There are various factors that enhance the nutrient uptake, plant health and root aeration. Soil tillage is another factor that determines soil quality. AMF play a pivotal role on soil tillage. The secondary impact of increasing mass of arbuscular mycorrhizae fungus promotes nutrient absorption and plant development growth. Bond and Grundy (2001), reported that tillage forms an important part of weed control strategies in organic systems.

7 Soil Management

The development of plant diversity and ecosystems is significantly enhanced by arbuscular mycorrhizal fungi. The main importance of mycorrhizal fungi is that they make available N and P to the primary producers in ecosystems. Mycorrhizal fungi are beneficial for the ecosystem working and sustainability. Poor soils that are deficient of nutrients can be well managed and improved to produce large quantity of plant products by way of additional management of mycorrhizae.

8 Arbuscular Mycorrhizal Fungi Potential for Phytoremediation

Soil contains large quantity of chemical substances and toxic metals are among them. Most plants have tendency to store or accumulate toxic and other metals in their body tissues. The hyperaccumulation of toxic metals in plants is dangerous to plants life activities. Heavy metals absorbed by roots are translocated to different parts by plants, leading to impaired metabolism and slow development, reduces enzymatic functions, destroying normal protein structures and replacing crucial elements resulting in deficiency symptoms.

Arbuscular mycorrhizal fungi provide striking coordination to precede plant-based environmental clean-up. AM associations are essential functioning for plant roots and are widely recognized as enhancing plant growth on severely concerned sites contaminated with heavy metals. They play an important role in metal tolerance and accumulation. Bhalerao (2013) reported that an isolation of the indigenous and presumably stress-adapted AM fungi can be a potential biotechnological tool for inoculation of plants for successful restoration of degraded ecosystems.

9 Conclusion

Arbuscular mycorrhizae (AM) symbiotic associations where both partners benefit from the reciprocal nutrient exchange. The main importance of mycorrhizal fungi is that they make available N and P to the primary producers in ecosystems. Mycorrhizal fungi are beneficial for the ecosystem working and sustainability. Arbuscular mycorrhiza-forming fungi provides sustainable soil health by improving soil structure, defence mechanism against plant root pathogens, provide availability of essential plant nutrients (Zn, Cu, Fe, Ca, Mg and NPK), tillage, biocontrol of pests and play efficient role in phytoremediation. So, emphasis should be given for efficient utilisation of Arbuscular mycorrhizae fungi in plant nutrition and bio-control of pests for better crop yield.

References

Aragno M (2005) The rhizosphere: a hot spot of bacterial density? In: Satyanarayana T, Johri BN (eds) Microbial diversiy: current perspectives and potential applications. International Publishing House Pvt. Ltd, New Delhi, pp 261–284

Bainard LD, Kliromonas JN, Gordon AM (2011) The mycorrhizal status and colonization of 26 tree species growing in urban and rural environments. Mycorrhiza 21:91–96

Bethlenfalvay GJ, Barea JM (1994) Mycorrhizae in sustainable agriculture. 1. Effects on seed yield and soil aggregation. Am J Altern Agric 9:157–161

Bhalerao SA (2013) Arbuscular mycorrhizal fungi: a potential biotechnological tool for phytoremediation of heavy metal contaminated soils. Int J Sci Nat 4:1–15

Bond W, Grundy AC (2001) Non-chemical weed management in organic farming systems. Weed Res 41:383–405

Bonfante P, Bergero R, Uribe X, Romera C, Rigau J, Pulgdomenech P (1996) Transcrptional activation of a maize α-tubulin gene in mycorrhizal maize and transgenic tobacco plants. Plant J 9:737–743

Borowicz V (2001) Do arbuscular mycorrhizal fungi alter plant-pathogen relations? Ecology 82:3057–3068

Bronick CJ, Lal R (2005) Soil structure and management: a review. Geoderma 124:3–22

Cardinale M, Brusetti L, Lanza A, Orlando S, Daffonchio D, Puglia AM, Quatrini P (2010) Rehabilitation of mediteranean anthropogenic soils using symbiotic wild legume shrubs: plant establishment and impact on the soil bacterial community structure. Appl Soil Ecol 46:1–8

Celik I, Ortas I, Kilic S (2004) Effect of compost mycorrhizal, manure and fertilizer on some physical properties of a chromoxerert soil. Soil Till Res 78:59–67

Dennis PG, Miller AM, Hirsch PR (2010) Are root exudates more important than other sources of rhizodeposits in structuring rhizosphere bacterial coomunities ? FEMS Microbiol Ecol 72:313–327

Diouf J (2011) Biotechologies for agricultural development. Food Agriculture Organization of the United Nations, Rome

Fester T, Lohsea S, Halfmanna K (2007) Chromoplast development in arbuscular mycorrhizal roots. Phytochemistry 68:92–100
Frank B (1887) Ueber neue Mycorhiza-Formen. Ber deut Bot Ges 5:395–409
Ghinea L, Ştefanic G, Popescu A, Oprea G (2007) Cercetări în domeniul chimiei şi biologiei solului. Analele I.N.C.D.A. Fundulea. LXXV. Volum Jubiliar
Gianinazzi-Pearson V, Dumas-Gaudot E, Gollotte A, Tahiri-Alaoul A, Gianinazzi S (1996) Cellular and molecular defence related root ressponses to invasion by aruscular mycorrhizal fungi. New Phytol 133:45–47
Gosling P, Hodge A, Goodlass G, Bending GD (2006) Arbuscular mycorrhizal fungi and organic farming. Agric Ecosyst Environ 113:17–35
Hakeem KR, Akhtar MS, Abdullah SNA (2016) Plant, soil and microbes – vol 1, Implications in crop science. Springer International Publishing AG, Gewerbestrasse 11, 6330 Cham, Switzerland, 366 pp
Hause B, Schaarschmidt S (2009) The role of jasmonates in mutualistic symbioses between plants and soil-born microorganisms. Phytochemistry 70:1589–1599
Jones DL, Hodge A, Kuzyakov Y (2004) Plant and mycorrhizal regulation of rhizodeposition. New Phytol 163:459–480
Jonhson NC, Graham JH, Smith FA (2010) Functioning of mycorrhizal associations along the mutualism-parasitism continuum. New Phytol 135:575–585
Kahiluoto H, Ketoja E, Vestberg M (2011) Plant-available P supply is not the main factor determining the benefit from arbuscular mycorrhiza to crop P nutrition and growth in contrasting cropping systems. doi:10.1007/s11104-011-0884-x
Karagiannidis N, Bletsos F, Stavropoulos N (2002) Effect of Verticillium wilt (*Verticillium dahliae* Kleb.) and mycorrhiza (*Glomus mosseae*) on root colonization, growth and nutrient uptake in tomato and eggplant seedlings. Sci Hortic 94:145–156
Marschner H, Dell B (1994) Nutrient uptake in mycorrhizal symbiosis. Plant Soil 159:89–102
Martinez TN, Johnson NC (2010) Agricultural management influences propagule densities and functioning of arbuscular mycorrhizas in low and high input agroecosystems in arid environment. Appl Soil Ecol 46:300–306
Ortas I, Sari N, Akpinar C, Yetisir A (2011) Screening mycorrhiza species for growth, P and Zn uptake in pepper seedling grown under greenhouse conditions. Sci Hortic 128:92–98
Paliyath G, Murr DP, Handa AK, Lurie S (2008) Postharvest biology and technology of fruit, vegetables and flowers. Wiley-Blackwell. Phylogeny and evolution. Myco Res 105:1413–1421
Popa D, Hănescu V, Grebenişan I, Câmpeanu C, Constantin C, Apostol T (2008) Aplicaţii ale micorizelor arbusculare în stabilizarea haldelor de zgură şi cenuşă. Universitaria, Craiova, 120 pp
Pringle A, Bever JD, Gardes M, Parrent JL, Rilling MC, Klironomos JN (2009) Mycorrhizal symbioses and plant invasion. Annu Rev Ecol Syst 40:699–715
Remy W, Taylor TN, Hass H, Kerp H (1994) Four hundred million year old vesicular arbuscular mycorrhizae. Proc Natl Acad Sci U S A 91:11841–11843
Samuil C (2007) Tehnologii de agricultură ecologică. U.S.A.M.V, Iaşi
Smith FA, Grace EJ, Smith SE (2009) More than a carbon economy: nutrient trade and ecological sustainability in facultative arbuscular mycorrhizal symbioses. New Phytol 182:347–358
Toma C, Toma I (2004) Simbioza micoriziană. Lucrare prezentată la Zilele Academice Ieşene, 25
Vance CP, Uhde-Stone C, Allan DL (2003) Phosphorous acquisition and use: critical adaptations by plants for securing a nonrenewable resource. New Phytol 157:423–447
Wang YY, Vestberg M, Walker C, Hurme T, Zhang X, Lindstrom K (2008) Diversity and infectivity of arbuscular mycorrhizal fungi in agricultural soils of the Sichuan Province of mainland China. Mycorrhiza 18:59–68

Azotobacter chroococcum – A Potential Biofertilizer in Agriculture: An Overview

Sartaj A. Wani, Subhash Chand, Muneeb A. Wani, M. Ramzan, and Khalid Rehman Hakeem

Contents

Abstract Research on *Azotobacter chroococcum spp.* in crop production has manifested its significance in plant nutrition and its contribution to soil fertility. The possibility of using *Azotobacter chroococcum* in research experiments as microbial inoculant through production of growth substances and their effects on the plant has markedly enhanced crop production in agriculture. Being free living N_2-fixer diazotroph, *Azotobacteria* genus synthesizes auxins, cytokinins, and GA like substances and these growth materials are the primary substances regulating the enhanced growth. It stimulates rhizospheric microbes, protects the plants from phyto-pathogens,

S.A. Wani (✉) • S. Chand
Division of Soil Science, Sher-e-Kashmir University of Agricultural Sciences and Technology-Shalimar, Srinagar, Kashmir 190001, India
e-mail: sartajmess@gmail.com

M.A. Wani
Division of FLA, Sher-e-Kashmir University of Agricultural Sciences and Technology-Shalimar, Srinagar, Kashmir 190001, India

M. Ramzan
Department of Botany, Aligarh Muslim University, Aligarh, Uttar Pradesh, India

K.R. Hakeem (✉)
Faculty of Forestry, Universiti Putra Malaysia, 43400 Serdang, Selangor, Malaysia
e-mail: kur.hakeem@gmail.com

K.R. Hakeem et al. (eds.), *Soil Science: Agricultural and Environmental Prospectives*, DOI 10.1007/978-3-319-34451-5_15

improves nutrient uptake and ultimately boost up biological nitrogen fixation. These hormonal substances, which originate from the rhizosphere or root surface, affect the growth of the closely associated higher plants. In order to guarantee the high effectiveness of inoculants and microbiological fertilizers it is necessary to find the compatible partners, i.e. a particular plant genotype and a particular *Azotobacter* strain that will form a good association.

Keywords *Azotobacter chroococcum* • Nitrogen fixation • Microbial inoculant • Soil fertility

1 Introduction

Biofertilizers also called as bio-inoculants, the organic preparations containing microorganisms are beneficial to agricultural production in terms of nutrient supply particularly with respect to N and P. When applied as seed treatment or seedling root dip or as soil application, they multiply rapidly and develop a thick population in rhizosphere. Biofertilizers can fix atmospheric N through the process of biological nitrogen fixation (BNF), solubilize plant nutrients like phosphates and stimulate plant growth through synthesis of growth promoting substances. They have C: N ratio of 20:1 indicating the capacity of the biofertilizer to release nutrients. Being eco-friendly, non hazardous and non-toxic products, biofertilizers are nowadays gaining the importance in agriculture (Sharma et al. 2007; Hakeem et al. 2016). They are cheaper and low capital intensive.

Biofertilizers benefiting the crop production include *Azotobacter*, *Azospirillum*, *Blue green algae*, *Azolla, P-solubilizing microorganisms*, *mycorrhizae* and *sinorhizobium* (Selvakumar et al. 2009). *Azotobacter chroococcum* and *Azotobacter agilis* were first of all studied by Beijerinck (1901). The first species of the genus *Azotobacter*, named *Azotobacter chroococcum* family *Azotobacteriaceae*, was isolated from the soil in Holland in 1901. In subsequent years several other types of *Azotobacter* group have been found in the soil and rhizosphere such as *Azotobacter vinelandii*, Lipman (1903); *Azotobacter beijernckii*, Lipman (1904); *Azotobacter nigricans*, Krassilnikov (1949); *Azotobacter paspali*, Dobereiner (1966), *Azotobacter armenicus*, Thompson and Skerman (1981); *Azotobacter salinestris*, Page and Shivprasad (1991).

2 Taxonomy, Morphology and Distribution of *Azotobacter*

The genus *Azotobacter* includes 6 species, with *A. chroococcum* most commonly inhabiting many soils all over the world (Mahato et al. 2009). Among the saprophytes along with nodular bacteria, genus *Azotobacter* was considered to be the most extensively studied (Horner et al. 1942). Aerobic bacteria belonging to the genus *Azotobacter* represent a diverse group of free-living diazotrophic (with the ability to use N_2 as the sole nitrogen source) microorganisms commonly inhibiting the soil. The taxonomic classification of *Azotobacter* is shown below.

Domain	:	Bacteria
Kingdom	:	Bacteria
Phylum	:	Proteobacteria
Class	:	Gammproteobacteria
Order	:	Pseudomonas
Family	:	Pseudomonadaceae/Azotobaceriaceae
Genus	:	Azotobacter

Azotobacter represents the main group of heterotrophic, non-symbiotic free living nitrogen-fixing bacteria principally inhabiting the neutral or alkaline soils. These bacteria are Gram negative and vary in shape. They are generally large ovoid pleomorphic cells of 1.5–2.0 um or more in diameter ranging from rods to coccoid cells. The cells can be dispersed or form irregular clusters or occasionally chains of varying lengths in microscopic preparations. In fresh cultures, the cells are mobile due to the numerous flagella present on their body surface but later the cells lose their mobility, become almost spherical and produce a thick layer of mucus, forming the cell capsule. The shape of the cell is affected by the amino acid glycine which is present in the nutrient medium peptone. Their distribution of existence is diverse and occurs either singly, in paired or irregular clumps and sometime in chains of varying length. Fig. 1 shows different stained *Azotobacter* species cells. *Azotobacter* possesses some unique features among the biofertilizers. They possess more than one type of nitrogenase enzymes (Joerger and Bishop 1988). They do not produce endospores but form cysts, oval or spherical bacteria that form thick-walled cysts (means of asexual reproduction under favorable condition) (Salhia 2013). The formation of cysts is induced by changes in the concentration of nutrients in the medium and addition of some organic substances such as ethanol, n-butanol, or β-hydroxybutyrate. The formation of cysts is also induced by chemical factors and is accompanied by metabolic shifts, changes in catabolism, respiration and biosynthesis of macromolecules (Sadoff 1975). The cysts of *Azotobacter* are spherical and consist of the so-called 'central body'a reduced copy of vegetative cells with several vacuoles and the 'two-layer shell'. The inner part of the shell has a fibrous structure and is called intine and outer part has a hexagonal crystalline structure called as exine (Page and Sadoff 1975). The central body can be isolated in a viable state by some chelation agents (Parker and Socolofsky 1968). The main constituents of the outer shell are alkyl resorcinol composed of long aliphatic chains and aromatic rings.

The population of *Azotobacter* is generally low in the rhizosphere of the crop plants in uncultivated soils. Jensen's N-free medium is frequently used for the mass multiplication of *Azotobacter. Azotobacter* grows well at an optimum temperature range between 20 and 30 °C and grows best in neutral to alkaline soil (pH of 6.5–

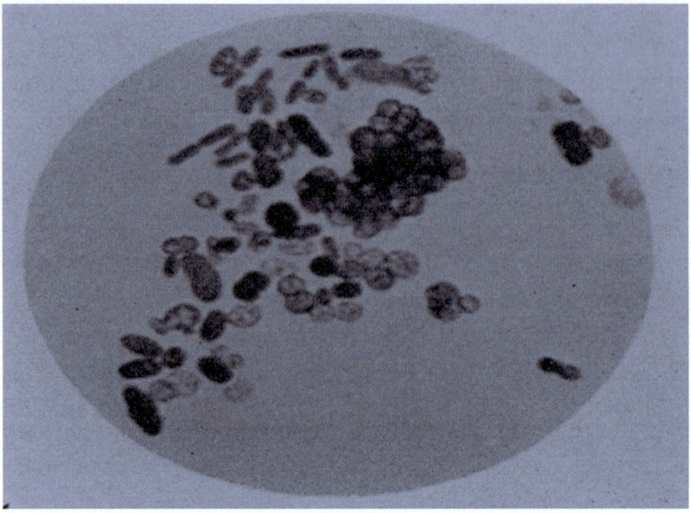

Fig. 1 *Azotobacter* species cells, stained with Heidenhain's iron hematoxylin, ×1000

7.5), but does not thrive when the pH is below 6 and hence not present in acidic soil. This organism has been reported to occur in the rhizosphere of a number of crop plants such as rice, maize, sugarcane, bajra, vegetables and plantation crops (Arun 2007) hence called rhizobacteria and or occurs endophytically (Hecht-Buchholz 1998). They work better in the root region of crop non-symbiotically when sufficient organic matter is present. They are reported to occur also in parenchymatous cells of root cortex and leaf sheath. *Azotobacter* is generally used in any non-legume crop (Singh and Dutta 2006). They can exhibit a variety of characteristics responsible for influencing the overall plant growth (Tippannavar and Reddy 1989). The *Azotobatcteria* also categorised as Plant growth Promoting Rhizobacteria (PGPB) are considered to promote plant growth directly or indirectly. These rhizobacteria derive their food and energy from the organic matter present in the soil and root exudates and fix atmospheric N (Maryenko 1964) depending on the amount of carbohydrates utilized by them. These non-specific associative nitrogen-fixing rhizobacteria are important for ecology and play a great role in soil fertility in agriculture.

3 Mode of Action of *Azotobacter* on Plant Growth

Despite the considerable amount of experimental data available concerning *Azotobacter* stimulation of overall plant development, however the exact mode of action by which *Azotobacter* enhances plant growth is not yet fully understood

(Wani et al. 2013). Three possible mechanisms have been proposed to explain the action: N_2 fixation; delivering combined nitrogen to the plant; the production of phytohormone – like substances that change the plant growth and morphology and bacterial nitrate reduction, thereby increasing nitrogen accumulation in inoculated plants.

3.1 Nitrogen Fixation

Nitrogen fixation is considered as one of the most important biological processes and interesting microbial activity on the surface of earth after photosynthesis as it makes the recycling of nitrogen and gives a fundamental contribution to nitrogen homeostasis in the biosphere. Biological nitrogen fixation plays an important role in maintaining soil fertility (Vance and Graham 1995). *Azotobacteria* is used for studying nitrogen fixation and inoculation of plants due to its rapid growth and high level of nitrogen fixation. They are extremely tolerant to oxygen while fixing nitrogen and this is due to respiration protection of nitrogenase (Robson and Postgate 1980; Hakeem et al. 2016). They have respiratory protection, uptake of hydrogenases and switch on-off mechanisms for protection of nitrogenase enzyme from oxygen (Chhonkar et al. 2009). *Azotobacter chroococcum* is shown to have uptake hydrogenase which metabolises hydrogen (H_2) evolved during nitrogen fixation (Partridge et al. 1980). *Azotobacter* is capable of converting nitrogen to ammonia, which in turn is taken up by the plants (Kamil, et al 2008). Iswaran and Sen (1960a) reported that the presence of optimum levels of calcium nutrient is essential for better growth of *Azotobacter* and its nitrogen fixation. However, the efficiency of *Azotobacter* was found to decrease with increased N level as reported by Soleimanzadeh (2013). *Azotobacter* spp. are non-symbiotic heterotrophic bacteria and capable of fixing about 20 kg N/ha/per year (Kizilkaya 2009) and it may be used in crop production as a substitute for a portion of mineral nitrogen fertilizers (Hajnal et al. 2004). According to Soliman et al. (1995) inoculation with *Azotobacter* replaced up to 50 % of urea-N for wheat grown in a greenhouse trial under aseptic conditions. The isolated culture of *Azotobacter* can fix about 10 mg nitrogen g^{-1} of carbon source under *in-vitro* conditions. The schematic representation of nitrogen fixation involved in nitrogen cycle in the biosphere by diazotrophs (*Azotobacteria*) is shown in Fig. 2.

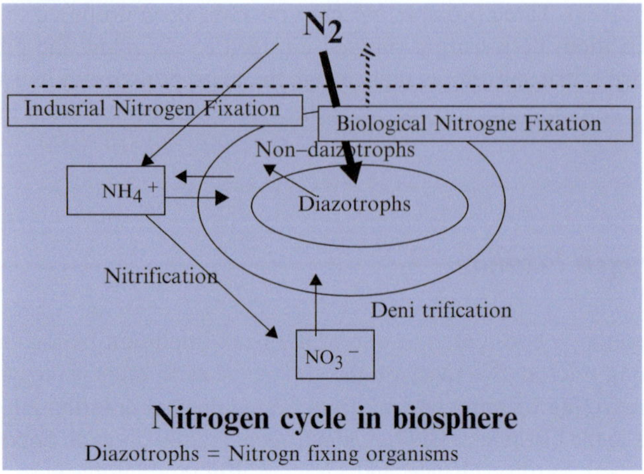

Fig. 2 Nitrogen fixation by diazotrophs (*Azotobacter spp.*)

3.2 *Growth Promoting and Other Substances Produced by Azotobacter*

Although most specifically noted for their nitrogen fixing ability, *Azotobacter* spp. have also been noted for their ability to produce different growth hormones (IAA and other auxins, such as gibberllins and cytokinins) (Barea and Brown 1974), vitamins, antibacterial and antifungal compounds and siderophores (Pandey and Kumar 1989b) which directly or indirectly effect the plant growth and microbial activity. Growth substances or plant hormones are natural substances that are produced by microorganisms and plants alike. They have stimulatory or inhibitory effects on certain physiological and biochemical processes in plants and microorganisms. These hormonal substances which originate from the rhizosphere or root surface affect the growth of the closely associated higher plants. Brakel and Hilger (1965) showed that *Azotobacter* produced indol-3-acetic acid (IAA) when tryptophan was added to the medium while as Hennequin and Blachere (1966) found only small amounts of IAA in old cultures of *Azotobacter* to which no tryptophan was added. Bacteria of the genus *Azotobacter* synthesize auxins, cytokinins, and GA-like substances and these growth materials are the primary substance controlling the enhanced growth of tomato (Azcorn and Barea 1975). *Azotobacter* spp. can also produce antifungal compounds to fight against many plant pathogens (Jen-Hshuan 2006). Many strains of *Azotobacter* have been reported to produces pigments which are involved in the metabolism of microbial organisms. For example, *A. chroococcum* forms a dark-brown water-soluble pigment melanin. This process occurs at high levels of metabolism during the fixation of nitrogen and is thought to protect the nitrogenase system from oxygen (Shivprasad and Page 1989). Other *Azotobacter* species produce pigments from yellow-green to purple colours (Jensen 1954)

Table 1 Effect of inoculation on dehydrogenase activity (μg TPF/10 g of soil)

Variant	Maize hybrids			
	ZP 555 su	620 k	NS 609b	NS 6030
Control	30.25	945.00	877.50	942.00
100 ml *A.chroococcum*	724.50	1244.00	1018.50	943.00
75 ml *A.chroococcum*	1000.75	985.00	914.75	950.00
50 ml *A.chroococcum*	962.75	1063.00	949.00	1055.00

including a green pigment which fluoresces with a yellow-green light and a pigment with blue-white fluorescence. *Azotobacter* acts as indicator hence enhances the microbial activity in soils. Jafari et al. (2012) reported that a more realistic indicator of total microbiological activity of soil is dehydrogenase activity which points to the intensity of oxidation reduction, i.e. the intensity of metabolic activity of microorganisms. In this research, dehydrogenase activity increased in all the variants where *Azotobacter* was applied (Table 1). Mutant *A. vinelandi* is more suitable for the biosynthesis of alginate in view of its latent utilization as a food stabilizer has better qualitative properties (Chen et al. 1985). Alginate the polymers are linear polysaccharides, which are composed of variable amounts of (1–4)-a-D-mannuronic acid and its epimer, a-L-guluronic acid. Several strains of *Azotobacter* are capable of producing amino acids when grown in culture media amended with different carbon and nitrogen sources (Lopez et al. 2005). Substances like amino acid produced by these rhizobacteria are involved in many processes that explain plant-grown promotion. Biochemical analysis of chlorophyll, nitrogen, phosphorous, potassium and protein content was higher in *Azotobacter* inoculated plants as compared to non-inoculated control plants (Naseri et al. 2013).

3.3 Response of Crops to Growth Promoting Substances

Large number of field trials and various experiments carried throughout India and whole world have convincingly established the importance of *Azotobacter* as microbial inoculant. Various crops like wheat, barley, maize, sugar beet, carrot, cabbage, potato were inoculated with *Azotobacterin* during its preparation. In India many crops like wheat, rice, onion, brinjal, tomato and cabbage were tested during experiment to study the effects of *Azotobacter*. Eklund (1970) demonstrated that the presence of *A. chroococcum* in the rhizosphere of tomato and cucumber is correlated with increased germination and growth of seedlings. Elgala et al. (1995) concluded that with microbial inoculation rock phosphate could be used as cheap source of P in alkaline soils and that combined inoculation could reduce the rate of fertilizer required to maintain high productivity. A study conducted by Govedarica et al. (1993) on the production of growth substances by nine *A. chroococcum* strains isolated from a chernozem soil has showed that these strains have the ability to produce auxins, gibberelins, and phenols and in association with the tomato plant, increase plant length, mass and nitrogen content. Puertas and Gonzales (1999) reported that dry weight of tomato plants inoculated with *A. chroococcum* and grown in phosphate-deficient soil was significantly greater than that of non inoculated plants. Phytohormones (auxin, cytokinin, gibberellin) can stimulate root development. *A. chroococcum* produces an antibiotic which inhibits the growth of several pathogenic fungi in rhizosphere thereby seedling mortality (Subba Rao 2001). Incidence of some diseases of mustard and rapeseeds could be reduced by inoculating with *Azotobacter* (Singh and Dutta 2006) Vijayan et al. (2007) observed that foliar application of *A. chroococcum* to mulberry grown under saline soil conditions showed significant level of improvement in biochemical and morphological parameters of leaf. Under greenhouse conditions inoculation of *A. chroococcum* recorded a significant N and P uptake in both seed and stover in brown sarson (*Brasssica campestris* L.) over the control (Wani 2012). Dual inoculation of *Azotobacter and Azospirillum* showed synergistic effects by improving growth prompting hormones, controlling pathogenesis and growth reducing agents due to producing fungicide antibiotics and compounds (antagonistic effect) and also air molecular N fixing and also producing growth prompting hormones such as oxine, cytokenine and gibberellins and solving mineral compound (Naseri et al. 2013).

4 Interaction of *Azotobacter* with Other Microorganisms

4.1 Interaction with Rhizobium

A synergistic relation of *Azotobacter* with *Rhizobium* interaction as co-inoculants have been observed in a majority of studies conducted under conditions like laboratory, greenhouse or field crops. Combined inoculation of *Azotobacter* and Rhizobium spp. has observed a positive response from crops. By significantly increasing nodulation *Azotobacter* spp. greatly influence Rhizobium activity. Increasing N_2 content within roots and shoots of respiring/metabolizing plant cells improves conditions within the rhizosphere and enhances synergistic interactions between the host and *Azotobacter sp*. in an open field conditions. Associative effect of *A. chroococcum* on Bradyrhizobium strains (BM 42 and BM 43) specific to moong bean (*Vigna radiata*) was also observed (Yadav and Vashishat 1991). The effect was more pronounced when *A. chroococcum* was co-inoculated with both the strains of *Bradyrhizobium*.

4.2 Interaction with *Azospirillum*

The beneficial effects of *Azotobacter* and *Azospirillum* interaction on plants are mainly attributed to improvements in root development, an increase in the rate of water and mineral uptake by roots, the displacement of fungi and plant pathogenic bacteria and to a lesser extent, biological nitrogen fixation (Okon and Itzigsohn 1995). Associative effect of *Azospirillum lipoferum* and *Azotobacter chroococcum* with Rhizobium spp. improved the growth of chick pea grown on both loamy sand and sandy soils (El-Mokadem et al. 1989). Both *Azotobacter* and *Azospirillum* have been shown to improve growth yields in various soil mineral compositions. This suggests that a mutualistic relationship exists between *Azotobacter* and *Azospirillum* where both interact with the Rhizobium to improve *Cicer arietinum* (chick pea) yields (Parmar and Dadarwal 1997). However, maximum values were obtained with *Azospirillum* application. Similarly, positive reports on application of *Azotobacter* and *Azospirillum* on the yield of mustard (*Brassica juncea*) are available (Tilak and Sharma 2007). Yasari et al. (2009) reported that inoculation of seed with *Azotobacter chroococcum*, *Azospirillium brasilense* and *Azospirillium lipoferum* recorded 1000 seed weight of 4.10 g, pods plant $^{-1}$ of 125.10 and seed yield of 1668 kg ha^{-1} in rapeseed (*Brassica napus. L*) at maturity in a field experiment conducted at Gharakheil Agricultural Research Station in Mazandaran province (Iran) during *rabi* season.

Some of thc studies have shown that a relationship exists between chemotactic behaviour and Azotobacter's influence on plant growth such as cotton (*Gossypium hirsutum* L.) and wheat (*Triticum aestivum* L.) (Kumar et al. 2007). In the areas of soil where plant root exudates or secretions such as sugars, glucose, amino acids and organic acids have been deposited, bacteria mobilize towards these exudates through chemotactic attraction. Increased yields and enhanced growth using *A. chroococcum* indicate a positive response attributed to nitrogen fixation, phosphorus mobilization, bacterial production and the release of phytohormones (Kumar et al. 2007).

5 Possibility of Using *Azotobacter* in Crop Production

Azotobacter has beneficial effects on crop growth and yield through biosynthesis of biologically active substances, stimulation of rhizospheric microbes, producing phyopathogenic inhibitors (Lenart 2012). *Azotobacter* makes availability of certain nutrients like carbon, nitrogen, phosphorus and sulphur through accelerating the mineralization of organic residues in soil and avoid uptake of heavy metals (Levai et al. 2008). *Azotobacter* can be an important alternative of chemical fertilizer because it provides nitrogen in the form of ammonia, nitrate and amino acids without situation of over dosage, which might be one of the possible alternatives of

Table 2 Effect Of *Azotobacter* On Crop Yield

S. No	Crop	Increase in yield over yield obtained with chemical fertilizers (%)
1	Wheat	8–10
2	Rice	5
3	Sorghum	15–20
4	Maize	15–20
5	Potato	13
6	Carrot	16
7	Cauliflower	40
8	Tomato	2–24
9	Cotton	7.27
10	Sugarcane	9–24

(Bhattacherjee And Dey 2014)

inorganic nitrogen source (eg. Urea). *Azotobacter* as nitrogen biofertilizer increases the growth and yield of various crops under field conditions (Table 2).

5.1 Effects of Azotobacter on Growth and Yield of Crops

There is increment in dry matter accumulation in Azotobacter inoculated plants; it stimulates development of foliage, roots, branching, flowering and fruiting which is triggered by fixed nitrogen and plant growth regulator like substance produced. It also increases plant tolerance to lack of water under adverse condition (Zena and Peru 1986). The rate of increase in the leaf area determines the photosynthetic capacity of plant, which leads to better assimilation of produce and towards yield. Using *Azotobacter spp*. potato yield has been increased by 33.3 % and 38.3 % (Zena and Peru 1986). Triplett (1996) concluded that the development of the diazotrophic endophytic association in maize appears to be the most likely route to success in the development of a corn plant which does not require nitrogen fertilization for optimum growth and yield. Yield increased ranges from 2 to 45% in vegetables, 9 to 24% in sugarcane, 0 to 31% in maize, sorghum, mustard etc., on *Azotobacter* inoculation (Pandey and Kumar 1989a, b). Tandon (1991) estimated the fertilizer equivalent of important biofertilizers. According to the estimate, fertilizer equivalent of 19–22 kg ha^{-1} for rhizobium 20 kg N ha^{-1} for *Azotobacter* and *Azospirillium*, 20–30 kg N ha^{-1} for blue green algae (BGA) and 3 to 4 kg N ha^{-1} of Azolla. Results of pot experiments a under greenhouse conditions with onion showed that application of *G. fasciculatum* + *A. chroococcum* + 50 % of the recommended P rate resulted in the greatest root length, plant height, bulb girth, bulb fresh weight, root colonization and P uptake (Mandhare et al. 1998). Laxminarayan (2001) reported that seed inoculation with *Azotobacter* produced higher grain and stover yield compared to

Table 3 Effect of inoculation on the grain yield of maize (t/ha)

Variant	Maize hybrids			
	ZP555 su	620 k	NS 609b	NS 6030
Control	12.27	4.27	8.88	10.59
100 ml *A.chroococcum*	13.32	4.97	8.39	10.90
75 ml *A.chroococcum*	13.24	4.89	8.87	10.75
50 ml *A.chroococcum*	13.31	4.30	8.92	10.96

uninoculated treatments. Singh and Dutta (2006) reported a significant in seed yield (7.86q ha^{-1}) in rapeseed and mustard (var. yella) due to inoculation with Azotobacter. Sharma (2002) reported the effect of biofertilizers and nitrogen on growth and yield of cabbage cv. Pride of India. Biofertilizer application significantly increased the leaf number, weight of non-wrapper leaves per plant, head length and width, gross and net weight of head per plant and yield per hectare over no biofertilizer application. Azotobacter in balanced nutrient condition results in 3.5 % increment in LAI at rosette stage of canola crop and additional application of Azotobacter shot up the yield by 21.17 % over the control (chemical fertilizers) (Yasari and Patwardhan 2007). According to Das and Saha (2007) combined inoculation of Azotobacter, Azospirillium along with diazotrophs increased grain and straw yield of rice by 4.5 and 8.5 kg ha^{-1}, respectively. The dual inoculation of *A. chroococcum* and *P. indica* had beneficiary response on shoot length, root length, fresh shoot and root weight, dry shoot and root weight, and panicle number that affect growth of rice plant (Kamil et al. 2008). Similar result put forwarded by Sandeep et al. (2011) which revealed that there is better growth response of Azotobacter inoculated plants as compared to non-inoculated control plants. Jafari et al. (2011) reported that the use of azotobacter had a positive effect on the grain yield of maize. In the variants where *Azotobacter* was applied, the grain yield increased in three maize hybrids (Table 3). In ZP 555 su, the yield increased by 1000 kg/ha, in NS 6030 by 280 kg/ha and in 620 k by 450 kg/ha. In NS 609b hybrid, the inoculation did not have any effect. An investigation was conducted under field conditions Milosevic et al. (2012), on a chernozem soil to study the effect of wheat seed inoculation (the cultivars Renesansa and Zlatka) with *A. chroococcum*, strain 86 (2–5 × CFU 108 ml^{-1}) reported that inoculation increased the energy of germination by 1 to 9 % and seed viability by 2 to 8 %. The largest increase in 1000 seed weight was obtained in the case of the cultivar Renesansa (16 %). *A. chroococcum* inoculation increased the seed yield of both cultivars and highest yield increase (74 %) was registered in the case of the cultivar Zlatka. According to Salhia (2013) azotobacter inoculants have a significant promoting effect on growth parameters like root, shoot length and dry mass of bamboo and maize seedlings in vitro and in pot experiments. Under green house conditions plant height, leaf number/plant, number of primary and secondary branches/plant, fresh and dry weight of whole plant, number of siliqua/plant, seeds/siliqua of brown sarson increased significantly with *Azotobacter* inoculation than no inoculation with seed and stover yield of 10.107 g pot^{-1} and 22.400 g pot^{-1} respectively

(Wani 2012). Naseri et al. (2013) while studying the effect of *A. chroococcum* and *Azospirillum brasilense* on grain yield, yield components of maize (S.C.704) as a second cropping in western Iran indicated that the dual inoculation with *Azotobacter and Azospirillum* on plant height, number of grain per row, 1000-grain weight, grain yield, biological yield and protein content was significant. Estiyar et al. (2014) reported that, number of branches, pod per plant and 1000 grain weight also increased with *Azotobacter* application.

6 Conclusion

Azotobacter spp. of bacteria, regarded as Plant Growth Promoting Rhizobacteria (PGPR) synthesize growth substances that greatly enhance plant growth and development and inhibit phytopathogenic growth by secreting inhibitors. There is a great significance of *A. chroococcum* in plant nutrition and its contribution to soil fertility. It is thus an important component of integrated nutrient management system due to its significant role in soil fertility. More research is necessary in future to explore the potentiality of *Azotobacter* in soil fertility using modern technology of soil genomics etc. The challenge to the research community will be to develop systems to optimize beneficial plant-endophyte bacterial relationships (Sturz et al. 2000) for long term effective role. In order to guarantee the high effectiveness of inoculants and microbiological fertilizers it is necessary to find compatible partners, i.e. a particular plant genotype and a particular *Azotobacteria* strain that will form a good association especially adapted to local edaphic and climatic conditions.

References

Arun KS (2007) Bio-fertilizers for sustainable agriculture, 6th edn. Agribios Publishers, Jodhpur, pp 76–77

Azcorn R, Barea JM (1975) Synthesis of auxins, gibberellins and cytokinins by *Azotobacter vinelandi* and *Azotobacter beijerinckii* related to effects produced on tomato plants. Plant Soil 43:609–619

Barea JM, Brown ME (1974) Effect on plant growth produced by *A paspali* related to the synthesis of plant growth regulating substance. J Appl Bacteriol 37:583–593

Beijerinck MW (1901) Über ologonitrophile mikroben. Zentralbl. Bakteriol. Parasitenkd. Infektionskr. II Abt. 9: 561–582

Bhardwaj D, Ansari MW, Sahoo RK, Tuteja N (2014) Biofertilizers function as key player in sustainable agriculture by improving soil fertility, plant tolerance and crop productivity. Microb Cell Fact 13:66

Bhattacherjee R, Dey U (2014) A way towards organic farming; A review. Afr J Microbiol Res 8(24):2332–2342

Brakel J, Hilger F (1965) Etude qualitative et quantitative de la synthese de substances de nature auxinique par *Azotobacter chroococcum in vitro*. Bull Inst Agron Stns Rech Gembloux 33:469–487

Chen WP, Chen JY, Chang SC, Su CL (1985) Bacterial alginate produced by a mutant of Azotobacter vinelandii. Appl Environ Microbiol 49:543–546

Chhonkar PK, Pareek RK, Rao DLN, ADiya TK (2009) Soil biology and biochemistry. Fundamentals of soil science, 2nd Edn. Indian Society of Soil Science, Pusa, New Delhi, pp 535–565

Das AC, Saha D (2007) Effect of diazotrophs on mineralization of organic nitrogen in the rhizosphere soils of rice (*Oryza sativa* L.). J Crop Weed 3:69–74

Döberainer J (1966) *Azotobacter paspali* sp. nov., uma bacteria fixadora de nitrogenio na rhizosfera de Paspalum. Pesquisa Agropecuaria Brasileira 1:357–365

Eklund E (1970) Secondary effects of some *Pseudomonads* in the rhizosphere of peat grown cucumber plant. In: Pharis RP, Reid DM (eds) Hormonal regulation of development, vol 3. Springer, New York, p 613

Elgala AM, Ishac YZ, Abdel Monem M, ElGhandour IAI (1995) Effect of single and combined inoculation with Azotobacter and VA micorrhizal fungi on growth and mineral nutrient contents of soil component interactions. In: Huang PM Berthilin J. Bollaj JM, McGill WB, Page AL (eds) Metals other inorganics and microbial activities (vol II). CRC Press, London

El-Mokadem MT, Helemish FA, Abou-Bakr ZYM, Sheteaws A (1989) Associative effect of Azotobacter lipoferum and Azotobacter chroococcum with Rhizobium sp. on mineral composition and growth of chick pea (*Cicer arietinum* L) on sandy soils. Zentralblatt fur Mikrobiologie 144: 255–265

Estiyar HK, Khoei FR, Behrouzyar EK (2014) The effect of nitrogen biofertilizer on yield and yield components of white bean (*Phaseolus vulgaris* cv. Dorsa). Int J Biosci 4(11):217–222

Govedarica M, Miliv V, Gvozdenovi DJ (1993) Efficiency of the association between *Azotobacter chroococcum* and some tomato varieties. Soil Plant 42:113–120

Hajnal T, Jarak M, Milosevic N (2004) Bacterization of maize: yield response of maize to inoculation. Book of abstracts of the 10th international symposium on microbial ecology: isme-10, cancun, mexico: 207

Hakeem KR, Akhtar MS, Abdullah SNA (2016) Plant, soil and microbes – vol 1, Implications in Crop Science Springer International Publishing AG, Gewerbestrasse 11, 6330 Cham, Switzerland. 366 pp

Hecht CB (1998) The apoplast-habitat of endophytic dinitrogen-fixing bacteria and their significance for the nitrogen nutrition of non leguminous plants. J Plant Nutr Soil Sci 161:509–520

Hennequin JR, Blachere H (1966) Recherches sur la synthese de phytohormones et de composes phenoliques par *Azotobacter* et des bacteries de la rhizosphere. Ann Inst Pasteur 3:89–102

Horner CK, Burk D, Allison FE, Sherman MS (1942) Nitrogen fixation by *Azotobacter* as influenced by molybdenum and vanadium. J Agric Res 65:173–193

Iswaran V, Sen A (1960a) Inactivation of Azotobacter by heat. Curr Sci 27(9):341–342

Jafari TH, Latkovic D, Duric S, Mrkovacki N, Najdenovska O (2012) Research. J Agric Sci 44(2)

Jen-Hshuan Chen (2006) The combined use of chemical and organic fertilizers and/or biofertilizer for crop growth and soil fertility. International workshop on sustained management of the soil-rhizosphere system for efficient crop production and fertilizer use 16(20):1–10

Jensen HL (1954) The Azotobacteriaceae. Bacteriol Rev 18(4):195–214

Joerger RD, Bishop PE (1988) Bacterial alternative nitrogen fixing systems. Crit Rev, Microbiol 16: 1–14

Kamil P, Yami KD, Singh A (2008) Plant growth promotional effect of *Azotobacter chroococcum*, *Piriformospora indica* and vermicompost on rice plant. Nepal J Sci Technol 9:85–90

Kizilkaya R (2009) Nitrogen fixation capacity of *Azotobacter* spp. Strains isolated from soils in different ecosystems and relationship between them and the microbiological properties of soils. J Environ Biol 30(1):73–82

Krassilnikov NA (1949) Opred Zlita baktzrij i aktinomyc etov. Mo skva

Kumar R, Bhatia R, Kukreja K, Behl RK, Dudeja SS, Narula N (2007) Establishment of Azotobacter on plant roots: chemotactic response, development and analysis of root exudates of cotton (*Gossypium hirsutum* L.) and wheat (*Triticum aestivum* L.). J Basic Microbiol 47:436–439

Laxminarayana K (2001) Effect of *Azotobacter* and *Azospirillum* on yield performance of maize in hilly regions of Mizoram. Indian J Hill Farm 14(2):134–137

Lenart A (2012) Occurrence characteristics and genetic diversity of *Azotobacter chroococcum* in various soils of Southern Poland. Pol J Environ Stud 21(2):415–424

Levai L, Szilvia V, Nora B, Eva G (2008) Can wood ash and biofertilizer play a role in organic agriculture? Agronomski Glasnic 3:263–271

Lipman JG (1903) Report on the New Jersey Agricultural Experiment Station 24: 217–285

Lipman JG (1904) Report on the New Jersey Agricultural Experiment Station 25: 237–289

Lopez GJ, Pozo RB, Lopez S, Toledo M, Salmeron V (2005) Liberation of amino acids by heterotrophic nitrogen fixing bacteria. Amino Acid 28(4):363–367

Mahato P, Anoop B, Chauhan JS (2009) Effect of *Azotobacter* and Nitrogen onseed germination and early seedling growth in tomato. Researcher 1(4):62–66

Mandhare VK, Patil PL, Gadekar DA (1998) Phosphorus uptake of onion as influenced by *Glomus fasciculatum*, *Azotobacter* and phosphorus levels. Agric Sci Digest 18:228–230

Maryenko VG (1964) Zavisimost lurozaja kukuruzyl balansa azota Aztobacter chroococcum V usbvijah monobakterialznoj kultury (Dependence of maize klyeid and nitrogen balance kon Azotobacter chroococcum in the jkcodnitions of monobacterial cultivatiions). Dokt ISHA 99:399–406

Milosevic N, Tintor B, Protic BC, Cvijanovig R (2012) Effect of inoculation with Azotobacter chroococcum on wheat yield and seed quality. Rom Biotechnol Lett 17(3):7352–7357

Nagananda GS, Das A, Bhattachrya S, Kalpana T (2010) In vitro studies on the effects of biofertilizers (*Azotobacter* and *Rhizobium*) on seed germination and development of *Trigonella foenum-graecum* L. using a novel glass marble containing liquid medium. Int J Bot 6(4):394–403

Naseri R, Moghadam A, Darabi F, Hatami A, Tahmasebei GR (2013) The Effect of deficit irrigation and *Azotobacter Chroococcum* and *Azospirillum brasilense* on grain yield, yield components of maize (S.C.704) as a second cropping in western. Iran Bull Environ Pharmacol Life Sci 2(10):104–112

Okon Y, Itzigsohn R (1995) The development of Azospirillum as a commercial inoculant for improving crop yields. Biotechnol Adv 13:415–424

Page WJ, Sadoff HL (1975) Relationship between calcium and uronic acids in the encystment of *Azotobacter vinelandiil*. J Bacteriol 122(1):145–151

Page WJ, Shivprasad S (1991) *Azotobacter salinestris* sp. nov., a sodium-dependent, microaerophili, and aeroadaptive nitrogen-fixing bacterium. Int J Syst Bacteriol 41: 369–376. Ann Microbiol 51: 145–158

Pandey A, Kumar SJ (1989a) Soil beneficial bacterial and their role in plant growth promotion. Sci Indian Res 48:134–144

Pandey A, Kumar S (1989b) Potential of *Azotobacters* and *Azospirilla* as biofertilizers for upland agriculture: a review. J Sci Ind Res 48:134–144

Parker LT, Socolofsky MD (1968) Central body of the Azotobacter cyst. J Bacteriol 91(1):297–303

Parmar N, Dadarwal KR (1997) Rhizobacteria from rhizosphere and rhizoplane of chick pea (*Cicer arietinum L.*). Indian J Microbiol 37:205–210

Patridge CDP, Walker CC, Yates MG, Postage JR (1980) The relationship between hydrogenase and nitrogenase in *Azotobacter chrocooccum* effect of nitrogen sources on hydrogenase activity. J Gen Microbiol 119:313–319

Puertas A, Gonzales LM (1999) Aislamiento de cepas nativas de *Azotobacter chroococcum* en la provincia Granmay evaluacion de su actividad estimuladora en plantulas de tomate. Cell Mol Life Sci 20:5–7

Robson RL, Postgate JR (1980) Oxygen and hydrgen biological nitrogen fixation. Ann Rev Microbio 34:183–207

Sadoff HL (1975) Encystment and germination in Azotobacter vinelandii. Microbiol Rev 39(4):516–539

Salhia B (2013) The effect of *Azotobacter chrococcum*as nitrogen biofertilizer on the growth and yield of *Cucumis sativus.* The Islamic University Gaza, Deanery of Higher Education Faculty of Science, Master of Biological Sciences, Botany

Sandeep C, Rashmi SN, Sharmila V, Surekha R, Tejaswini R (2011) Growth response of *Amaranthus gangeticus* to *Azotobacter chroococcum* isolated from different agroclimatic zones of Karnataka. J Phytology 3(7):29–34

Sariiv MR, Sariiv Z, Govedarica M (1988) Efficiency of strain combination of different genera of nitrogen fixing bacteria on sunflower genotypes. In: 12th international sunflower conference, Novi Sad, pp 187–191

Selvakumar G, Lenin M, Thamizhiniyan P, Ravimycin T (2009) Response of biofertilizers on the growth and yield of blackgram (*vigna mungo L.*). Recent Res Sci Technol 1(4):169.17

Sharma R. (2002) Ph.D. Thesis, AAU, Jorhat.

Sharma K, Dak G, Agrawal A, Bhatnagar M, Sharma R (2007) Effect of phosphate solubilizing bacteria on the germination of *cicer arietinum* seeds and seedling growth. J Herbal Med Toxicol 1(1):61–63

Shivprasad S, Page WJ (1989) Catechol formation and melanization by Na^+ dependent Azotobacter chroococcum: a protective mechanism for aeroadaptation? Appl Environ Microbiol 55(7):1811–1817

Singh MS, Dutta S (2006) Mustard and rapeseed response to Azotobacter – a review. Agric Rev 27(3):232–234

Soleimanzadeh H, Gooshchi F (2013) Effects of *Azotobacter* and nitrogen chemical fertilizer on yield and yield components of wheat (*Triticum aestivum* L.). World Appl Sci J 21(8):1176–1180

Solimam S, Seeda MA, Ally SSM, Gadalla AM (1995) Nitrogen fixation by wheat plants as affected by nitrogen fertilizer levels and non-symbiotic bacteria. Egypt J Soil Sci 35:401–413

Sturz AV, Christie BR, Nowak J (2000) Bacterial endophytes: potential role in developing sustainable systems of crop production. Crit Rev Plant Sci 19:1–30

Subba Rao NS (2001) An appraisal of biofertilizers in India. In: Kannaiyan S (eds) Biotechnology of biofertilizers. Maximising the use of biological nitrogen fixation in agriculture. Narosa Publising House, New Delhi, p 375

Tandon HLS (1991) Role of sulphur in plant nutrition. Fert News 36(69):79

Thompson JP, Skerman VBD (1981) Validation list no. 6. Int J Syst Bacteriol 31:215–218

Tilak K, Sharma KC (2007) Does *Azotobacter* help in increasing the yield. Indian Farm Digest 9:25–28

Tippanavar CM, Reddy TK (1989) Aztobacter in root stem and leaf tissuesof cells of *Triticum aestivum* and *Triticum durum* L. Curr Sci 58:1342–1343

Triplett GT, Dabney SM, Siefker JH (1996) Tillage systems for cotton on silty upland soils. Agron J 88:507–512

Vance CP, Graham PH (1995) Nitrogen fixation in agriculture: applications and perspectives. In: Tikhonovich IA, Provorov NA, Romanov VI, Newton WE (eds) Nitrogen fixation: fundamentals and applications, current plant science and biotechnology in agriculture 27: 77–86

Vijayan K, Chakraborti SP, Ghosh PD (2007) Foliar application of *Azatobacter chroococcum* increases leaf yield under saline conditions in mulberry (Morus spp.). Sci Hortic 113:307–311

Wani SA (2012) Effect of balanced NPKS, biofertilizer (*Azotobacter*) and vermicompost on the yield and quality of brown sarson (*Brassica rapa* L.), M. Sc thesis, Sher-e-Kashmir University of Agriculture Sciences and Technology, Kashmir, Srinagar

Wani SA, Chand S, Ali T (2013) Potential use of Azotobacter chroococcum in crop production: an overview. Curr Agric Res J 1:35–38

Yadav AS, Vashishat RK (1991) Associative effect of Bradyrhizobium and Azotobacter inoculation on nodulation, nitrogen fixation and yield of Moong bean (*Vigna radiata* L. Wilczek). Indian J Microbiol 31:297–300

Yasari E, Patwardhan AM (2007) Effect of (*Azotobacter* and *Azosprillium*) inoculants and chemical fertilizers on growth and productivity of canola (*Brassica napus* L.). Asian J Plant Sci 6(1):77–82

Yasari ES, Esmaeili MA, Azadgolch MS, Alasthi MR (2009) Enhancement of growth and nutrient uptake of rapeseed (*Brassica napus* L.) by applying mineral nutrients and biofertilizers. Pak J Biol Sci 112:127–133

Zena GG, Peru C (1986) Effect of different rates of *Azotobacter* and frequency of application of Agrispon on yield and quality in the growing of onion (*Allium cepa* L.) in Cajamarca. National University of Cajamarca Faculty of Agriculture Sciences and Forestry

Sources and Composition of Waste Water: Threats to Plants and Soil Health

Hamaad Raza Ahmad, Tariq Aziz, Muhammad Zia-ur-Rehman, Muhammad Sabir, and Hinnan Khalid

Contents

H.R. Ahmad (✉) • T. Aziz • M. Zia-ur-Rehman • M. Sabir • H. Khalid
Institute of Soil and Environmental Sciences, University of Agriculture Faisalabad, Faisalabad 38040, Pakistan
e-mail: khamaad@gmail.com

K.R. Hakeem et al. (eds.), *Soil Science: Agricultural and Environmental Prospectives*, DOI 10.1007/978-3-319-34451-5_16

Abstract Industrialization has caused huge changes in the global budget of critical chemicals at the earth's surface. Waste water is being added to the soil with or without treatment, causing accumulation of metals, salts, pathogens, toxins etc. in the soil. Accumulation of these substances in soil ultimately affects crop growth and human health. Metals, salts, pathogens are added into soils through various means like pesticides, fertilizer, waste water and municipal waste either remain in the soil by forming insoluble complexes with soil constituents or taken up by plants and/or may pass into drainage water. The waste water contains toxic material likely to affect plants and human health. For Agricultural irrigation may result sail salinity, sodicity and heavy metal accumulation. The untreated waste water coming from various sources contains nutrients and excess of these nutrients resulting eutrophication. However, the waste water might also bring benefits for agricultural crops as it contains organic matter and essential nutrients. The common processes which have been used to remove non-biodegradable pollutants from waste water are sedimentation, flocculation, membrane filtration, photo catalysis and use of different sorbent materials. The present chapter will provide general understanding about the different sources of waste water, its composition and impacts on soil and plant health.

Keywords Soil health • Waste water • Salinity stress • Industrial waste • Plant health

1 Introduction

Urbanization results in the introduction of pollutants into the environment as well as greater demands on fresh water resources. Agriculture is the dominant water user sector. On world level water tables are falling rapidly due to invariable rainfall. Waste water generated during various industrial and municipal activities carry suspended solids, salts, nutrients, bacteria, and oxygen-demanding material etc. According to estimation 90 % untreated wastewater generated in developing countries is discharged directly into rivers, lakes or the oceans. Such discharges are part of the reason why de-oxygenated dead zones are growing rapidly in the seas and oceans. Currently an estimated 245 000 km^2 of marine ecosystems are affected with impacts on fisheries, livelihoods and the food chain (Corcoran et al. 2010). Waste water mainly consists of heavy metals, soluble salts, radioactive isotopes, heat, macro nutrients, micro nutrients, bacteria and viruses. Wastewater discharged from industries when released into a water body changes chemical composition of water body and thus have toxic effects on flora and fana (Killi et al. 2014). For example oil discharging from refineries when enters into water, depletes the dissolved oxygen in water body rapidly and thus have drastic effect on aquatic life (Castege et al.

2014). Similarly hot waters released from the industries are harmful to vegetation and microbial population of the water (Bougherira et al. 2014). Sewage and domestic waste waters from houses may cause introduction of pathogens like protozoa, worms-eggs and bacteria into water (Arora and Kazmi 2015). Use of contaminated waste water causes jaundice, typhoid, dysentery, cholera, tuberculosis etc. (Curriero et al. 2001). In addition to above-mentioned issues related to wastewater, heavy metal contamination is one of the major threats to plants, animals and human beings. Although some metals like iron, zinc, manganese, nickel and copper are consider to be essential for plant health (Vamerali et al. 2010) however concentration of these metals when exceeds from threshold levels they become toxic to plants (Muhammad et al. 2011). In addition to essential metals, others like lead, mercury, cadmium and arsenic are not required for plant growth and these are toxic at very low concentrations (Martinez-Martinez et al. 2013).

It is evident from literature that distribution of heavy metals in water, soil and air lead to their accumulation in crops, which may affect food quality and safety (Martinez and Blasco 2012; Hakeem et al. 2014; Obiora et al. 2016). Therefore heavy metal toxicity receives much attention in present scenario due to its negative impacts on soil, plant, animals as well as environment (Pruvot et al. 2006; Wu et al. 2011). Now it is well documented that the concentration of heavy metals in the agricultural soils is increasing (Pandey and Pandey 2009; Zhang and Pu 2011). The major cause of this increasing concentration is the use of poor quality water due to the non availability of good quality irrigation water (Pruvot et al. 2006; Ahmad et al. 2011).

Waste water generated from agricultural, industrial or municipal activities are responsible for polluting soil and its environment. Extensive use of insecticides, pesticides and fertilizers gradually making soil and water resources unproductive (Hakeem 2015). Irrigated agriculture comprises of 13 % of the world's total arable land but the value of crop production from irrigated land is 34 % of the world's total. This potential is more pronounced in semi-arid and arid areas. The scarcity of water supplies to meet the needs of population growth and rapid development in agriculture as well as industry have given cause for concern in formulating national development plans in these countries towards the use of unconventional water resources in particular the sewage effluent. Waste water can be used as a source of irrigation due to the shortage of irrigation water but it requires management practices. Therefore, the present review takes in to account with an idea to review the impact of waste water on the soil and plant health.

2 Sources and Composition of Waste Water

Naturally all the human activities result in water pollution. Industrialization has caused a huge increase in water pollution. There are generally four types of waste water on basis of their production/source (McKinney et al. 2013) *viz.* Industrial wastes, domestic wastes, hospital waste and agriculture waste.

2.1 Industrial Wastes

Every industry has variable contribution into water contamination. Major contributors are manufacturing, power-generating, mining and construction and food processing industries (McKinney et al. 2013). In Pakistan, there are 6634 registered industries out of which 1228 are responsible for the severe pollution (Sial et al. 2006) while 400,000 factories are causing water pollution in the United States (Environmental Protection Agency). All these industries are discharging inorganic and organic pollutant. They are responsible for the serious water pollution in Pakistan (Nasrullah et al. 2006). Approximately two million tons of waste materials become the part of world's water on daily basis. Developing countries are facing more miserable situation where more than 90 % of sewage waste and almost 70 % of industrial raw wastes are becoming the part of water bodies. Industries are also releasing the waste water which have the contaminants that are harmful for all the living organisms e.g. nitrates, nitrites, different cations and anions like K^+, Ag^{2+}, Na^+, Mg^{2+}, Ca^{2+}, Cl^-, CO^{3-}, HCO^{3-}, Cl^- and hazardous metals such as arsenic, lead, iron, mercury, chromium, nickel, cadmium, copper, nickel, zinc and cobalt (Wang and Yang 2015; Alvarez et al. 2016).

2.1.1 Manufacturing Industries

Manufacturing industries like textile printing, dying, leather tanning, paint, plastics, pharmaceutical, and paper and pulp industry (Raja and Venkatesan 2010; Hakeem 2015) are adding waste material in the fresh water bodies. Beside these industries the most damaging industries which have the major role in the water pollution are chemical industry, oil refining and steel industry. These industries contaminate the water by adding toxic chemicals, organic pollutants and heavy metals (McKinney et al. 2013). In developing countries the disposal problem of manufacturing industrial wastes become very difficult because of greater production wastes per unit area and decrease in proportion of land available for its disposal. Mostly industries disposed their raw effluents that contain heavy metals, soluble salts, organic matter and pathogens in nearby unlined drains that affect the groundwater quality of that area. The deposition of heavy metals in the soils water may be naturally or due to anthropogenic activities including dying of clothes industry, paint production industry, soap manufacturing, electroplating and many more (Rattan et al. 2005; Balkhair and Ashraf 2016; Li et al. 2001). Mostly these industries discharge their effluent in to the soil that reaches to the ground water (Ahmad 2011; Suciu et al. 2008) and most of this effluent have concentration of certain elements above the critical limits that is harmful for soil and plants (Xia 1999).

Major industries along with their waste composition are:

2.1.2 Petroleum Manufacturing

It is one of major water polluting industry by adding its waste material in the water bodies. The oil materials contain the components that are toxic or even fatal for aquatic life as well as for the human beings and animals. Dispersed oil, aromatic hydrocarbons and alkyl phenols (AP), heavy metals, and naturally occurring radioactive material (NORM) are of basic environmental hazard (Neff et al. 2011) which are present in the petroleum products and are contaminating the water bodies. Moreover the other toxic compounds which are either the components of petroleum waste or the byproducts of petroleum industry are monocyclic aromatic hydrocarbons (BTEX: benzene, toluene, ethyl benzene, xylenes), polycyclic aromatic hydrocarbons (PAH), and related heterocyclic aromatic compounds (AMAP 2010; Neff et al. 2011).

Petroleum industries present at the bank of water bodies are frequently adding their waste directly into rivers, ponds, streams and seas without any treatment. As oil is the major demand, in present era every country is trying to be self-sufficient in this regard and thus launching more and more petroleum industries. The byproducts and waste materials of petroleum industries are not only contaminating the fresh water but other problems like oil leakage and spilling have also been recorded in this regard. During 2012 almost 122 small accidents were recorded with a total oil discharge of 16 m^3 only in Norwegian State and they become manifolds when talking about globally. Large spills of chemicals have been stable at 100–150 incidents every year on the Norwegian Continental Shelf (NCS) over the past decade (Norwegian Oil and Gas 2013). Data collected from field specific discharge reports for 2012 tells that the average BTEX (benzene, toluene, ethyl benzene, xylenes) concentrations in produced water (PW) on Norwegian Continental Shelf (NCS) establishments show variation from 2 to 58 mg L^{-1} from its total high concentration. Similarly barium and iron are also used in the preparation of different petroleum products and their amounts that were recorded in 2012 was 0.0017–1100 mg L^{-1} and 0.8–75 mg L^{-1} respectively (Azetsu-Scott et al. 2007; Lee et al. 2005).

2.1.3 Stainless Steel Manufacturing

Iron (Fe) and chromium (Cr) are primary components which are used in the preparation of stainless steel (Murthy et al. 2011) and its application and demand are increasing day by day (ISSF 2011). Three types of wastes are produced which are of chromium (Cr) coating nature and pollute the water bodies by excess amount of chromium (Cr) which is toxic heavy metal for all aquatic as well as land life. Slag is produced when smelting process occurs, bag filter dust (BFD) during the cleaning of the off-gas coming from semi-closed/open furnaces, and venture sludge during the scrubbing of the off-gas from closed furnaces (Beukes et al. 2010). No doubt slag is by volume the biggest waste produced in Fe, Cr production, BFD is the most important waste material from environmental point of view, as it have small amounts of Cr (VI) (Beukes et al. 2010). Venturi sludge is produced in same amount than BFD, but has very less amount of Cr (VI) (Beukes et al. 2010; Gericke 1995).

2.1.4 Paint Industry

Paint Industry is considered to be one of the biggest industries of the whole world. As paint has large area of application like automobile bodies, household furniture, walls of offices, houses and other commodities. Thus as the demand for paint is increasing, its waste materials are also increasing with the same proportion. Most common waste materials generated during paint manufacturing are toxic heavy metals such as chromium and lead and also the organic solvents like toluene and methyl ethyl ketone. Moreover, the waste material produced during the industrial paint application includes equipment cleaning wastes, volatile organic carbon (VOC) emissions during paint application, curing and drying, obsolete and unwanted paint, aqueous waste and spent solvents from surface preparations, scrubber water and paint sludge. All these components pollute the water directly and indirectly.

2.1.5 Power Generation Industries

Power generating industries cause the water pollution by raising the temperature and causing the radioactive pollution. Radioactivity that release from nuclear power plants can pollute water by discharging radioactive waste into ground water and soils (McKinney et al. 2013). Clays and clay stones are sometimes thought to be the best dumping house for the radioactive material in some countries including Belgium, France and Switzerland (Tsang et al. 2012). One of the drawbacks of this process is that the temperature of that site goes high and high due to the exothermic processes of radioactive materials. This elevated temperature not only affects the properties of clay and clay stone but also has severely bad impact on the ground water quality (Yu et al. 2013). For the power generation the use of nuclear material is increasing very rapidly and thus the radioactive waste material is also increasing with the same ratio. This increased radioactive waste making the water bodies more polluted. World Nuclear Association (WNA) reports show that 435 nuclear power reactors are operating in 31 countries, with 60 new units under construction in 13 countries. Color as well as quality of water changes due the presence of these radioactive waste material released by power generating industries (WNA 2013).

2.1.6 Mining and Construction Industries

Mining is ranked as the fifth largest industry in the world. 70 countries including 31 from Africa have come in the international mining corporations from start of 1990 (Europa 2003). For ease mining and construction wastes can be categorized separately as both have the different composition of their wastes.

Construction wastes: Construction and Demolition (C and D) waste materials contain the waste materials produced during the construction of new buildings, renovation, and demolition of buildings and structures. Major construction waste materials include extra wood, cement, bricks, paints, iron, steel, polyesters, heavy metals released during processing of different construction commodities and various chemicals used for the development of different infrastructures. Water is the key component of all the construction processes and all the solid wastes also get mixed in it. When this polluted water get mix with other water bodies deterioration of water quality occur. Several researchers and scientists had tried to evaluate the amount of the waste material and contaminated water produced during the construction and mining. It was determined that the landfills of the US contain one third Construction and Demolition (C and D) waste materials that reache to the ground water and contamination of ground water occur (Chun-Li et al. 1994; Kibert 2000). Similarly other countries also have large concentration of construction and building wastes like Austria, Denmark, Germany and Netherlands are producing about 300, over 500, about 2600 and about 900 kg/cap construction and demolition waste, respectively (Brodersen et al. 2002). All these landfills severely affect the ground water quality and get entry into in the food chain by becoming the part of drinking water.

Mining wastes: Mining industry plays important role in the economy of every country. Mining can be defined as a process in which breakdown of the rocks (either by machinery or labor) occur for the extraction of the coal and other useful minerals. When these minerals and coals are extracted in the raw form, they add hazardous chemicals into the water bodies which are either the part of these minerals or used for the extraction. Mining can also increase the concentration of toxic elements by releasing the contaminants into water and impose serious health issues. Mining activities produce metal waste in huge amount and also release sulphides from the rocks which are toxic for the fresh water. Mining industries are also causing water pollution by adding sediment and acids. There are four basic effects of mining waste on water quality i) acid mine drainage ii) heavy metal contamination and leaching iii) processing chemicals pollution and iv) erosion and sedimentation.

During the process of mining several volatile elements are also produced as waste materials which when go high in the atmosphere not only cause the air pollution but also water pollution when mix up with the water vapors present in the atmosphere. When these water vapors come to earth in the form of rain called the acid rain, pose serious effects on the all living creatures. This acid rain when fall on the rivers, ponds, streams and seas, add pollutants into water bodies and problem of water pollution occur.

2.1.7 Food Industry

Food Industry produce a huge quantity of waste both in form of solid and liquid that are released during the processing, preparation and consumption of food. Generally food industry produces the following types of wastes i) Organic wastes like carbohydrates, proteins and lipids ii) Suspended solid wastes based on their sources iii) high biological oxygen demand (BOD) or chemical oxygen demand (COD) waste water that is the indication of severe water pollution (Litchfield 1987).

Besides it, food industries also generate a huge volume of waste water that contains the toxic material and this water is released commonly into the surface water without any treatment which causes severe contamination. Poor quality water also affects the food products manufactured by food industry. Major food industries include fruit and vegetables industry, dairy industry, meat and fish industry, alcoholic and non-alcoholic beverage industry, oil industry, and packaged foods industries. The threat of water pollution due to food technology industry is that the chemicals and waste products released during the preparation and packaging of food products are increasing day by day and are becoming the part of water resources (McKinney et al. 2013). Every year, millions of tons of halal food wastes are discarded in the environment and it cause the land pollution as well as water pollution. The world municipal solid waste (MSW) production in 1997 was approximately 0.5 billion tons and its increasing rate is 2–3 % for the developing countries (Abud Manaf et al. 2009). The waste material derived from Halal bio finery includes vegetable market solid waste, agricultural materials and animal manures such as cow dung, chicken and goat manure and sewage sludge (Suthar 2009; Roca-Pérez et al. 2009; Suthar 2010), Municipal solid waste, dairy waste, cheese whey and dairy wastewater, pulp and paper wastewater and olive mill wastewater, palm oil mill effluent (Tabatabaei et al. 2010; Sulaiman et al. 2009a, b. Beside these packaging waste is released in huge quantity from food industry on daily basis. Latest data shows that in 2008 Europe generated 80 million tons of packaging waste (Eurostat, the statistical office of the EU while Italy generates 12 million tons of packaging waste and 2/3 of this waste comes from the food industry.

2.1.8 Dairy Wastes

Dairy industry produces large volumes of waste effluents such as dilutions of whole milk, separated milk, whey and sanitizers used in the clean-in-place (CIP) system. During the process of packaging drippings are released and become the part of food wastes. Moreover incidental leaks as well as CIP i-e clean in place system also generate waste water which finds their way into the sewer system and a serious cause of water pollution. During the processing of yogurt milk is heated, starter and sodium stearate are also added to enhance consistency followed by incubation. For the processing of white cheese skimmed and dry milk are mixed then the process of heating occur with the addition of salts and rennet. Mixture will be kept for separation. From the beginning the whey generated during the separation is disposed to the sewerage without treatment and thus is the cause of water pollution. The composition of whey is given in Table 1.

2.1.9 Fruits and Vegetables processing Industry

Wastes from food and vegetables processing unit are released in huge volume. This waste includes peel, pulp, molasses and a lot of other compounds which with the waste water become the part of water bodies and cause the contamination of these water bodies. Moreover fruit and vegetable processing also generate a large quantity

Table 1 Composition of whey

Substances	Amount
Lactose	5.00 %
Salt	7.50 %
Total carbohydrate	5.00 %
Protein	0.85 %
Oil	0.30 %
Ash	0.30 %
Riboflavin	0.10 mg/100 mL
Niacin	0.10 mg/100 mL
Fe	0.12 mg/100 mL
Na	14.5 mg/100 mL
K	30 mg/100 mL
Ca	13.5 mg/100 mL
Water	92 %
pH	5.50

of protein waste. Vinegar, citric acid and acetic acid are also produced during the fermentation of fruits and vegetables. Apricot is included in the major agricultural products and about. 60,000–80,000 tons dried apricots are produced every year 180–200 tons dried apricot wastes are generated in the food factories each year in Turkey. During the fermentation process of dried apricot, citric acid is generated as a waste substrate of dried apricot. Similarly brine is used in the preparation of pickles and olives and it is a critical environment concerning waste and causing the serious problem as there is almost 20 % and 7 % salt in the waste material of olive and cucumber pickles, respectively. Wastes coming from potato and wheat starch processing units are also the threat to the water quality.

2.1.10 Citrus By-Product Wastes

Processing of citrus into juice results release of residual waste material including peels, membranes, seeds, and other compounds. Wastes of citrus juice production are a source of water pollution as it contains dried pulp and molasses, fibber-pectin, cold-pressed oils, essences, limonene, juice pulps and pulp wash, ethanol, seed oil, pectin, limonoids and flavonoids. Fiber-pectins are produced from lime peels as a waste product and have significant amount of fiber contents. The major flavonoids present in citrus species are hesperidin, narirutin, naringin and eriocitrin. Peel and other solid residues of lemon waste are basically composed of hesperidin and eriocitrin, while the latter was predominant in liquid residues. Additionally citrus-juice processing is the one of the major food industry and it generates large amount residual wastes including peel and rag, pulp wash, seeds and citrus molasses. Citrus by-products and wastes also generate a large quantity of coloring and preservative waste material along with their complex polysaccharide content (Sreenath et al. 1995).

Table 2 Composition of black water produced during oil extraction

Substances	Amounts
Sugar	0.98 %
Protein	0.77 %
Oil	0.4 %
Dry matter	6.2 %
Ash	1.4 %
Fe	0.3 mg/100 mL
Ca	7.5 mg/100 mL
K	112 mg/100 mL
Na	39.5 mg/100 mL
pH	4.5

(source: Denaro et. al 2010)

2.1.11 Oil Wastes

The waste material generated from olive oil extraction includes the vegetation water which is also called black water or vegetable water, and the olive husk containing skins and stones. Varying with the processing conditions 50–110 kg of waste water generates from 100 kg of olives. Major factories of olive oil production and processing have no treatment systems and thus dispose of this dark brownish wastewater into the surface water bodies which pose hazardous environmental issues with high BOD (90–100 g/L). Huge efforts were made to know the composition of this Black Water and results show the following composition Table 2.

As these wastes are producing water pollution, a lot of researchers are trying to utilize these wastes for the production of useful products like single cell protein from black water. Moreover production of organic fertilizer by sludge has also been studied (Anaç et al. 1993).

2.1.12 Meat Industry's Waste

A large portion of the waste materials of animal by-products are generated from slaughterhouses. This waste material consists of heads, feet, offal, bone, hair, blood and feathers. This waste material is very harmful as it contains metals, chemical solvents, cleaning and disinfection compounds which are very harmful for living organisms. The inedible waste material is almost 40 % of the total weight of the killed animals. Almost half of the chicken meat production wastes have not been utilized and especially the blood in large quantity is discharged to outside that finally goes to the water bodies. In these water bodies they add the toxic and hazardous pollutants which are the reason of enhanced water pollution. Moreover there are also a lot of random units of meat that generate the waste materials in the form of meat meal, tallow and degreased bone.

Table 3 Composition of food packaging waste in 2005

Materials	million (tons)
Food scraps	29.20
Containers	76.70
Paper and paper board	39.00
Plastics	13.70
Glass	10.90
Food cans	2.10
Beer and soft drink bottles	7.10
Bags and sacks	1.60
Milk cartons	0.40

(source: EPA 2006)

2.1.13 Food Packaging Wastes

The major use of food packaging material is to safe the food items from environmental hazards and to provide the healthy and nutritious food to everyone (Coles 2003). But now this packaging material is the cause of great environmental concern as water pollution is rising due to this waste. It is said that food packaging wastes are the two third of the total waste material of packaging material (Hunt et al. 1990). This material consists of the following composition (Table 3).

2.2 Domestic Wastes

Domestic wastes include the waste material coming from the houses on daily, weekly and monthly basis. Waste materials coming from household are becoming the part of water resources such as rivers, ponds, lakes, streams and seas that cause severe water pollution. Domestic waste can be categorized on the basis of their nature.

2.2.1 Inorganic Waste

Different nutrients especially nitrate and phosphates that are the major components of washing powder and detergents are being released in fresh water. It cause harm not only to humans but also to animals by causing the severe water pollution. In addition to this excessive use of fertilizer and pesticides in the lawns and small home orchards are also the cause of water pollution (McKinney et al. 2013). Extra chemicals that may be the part of synthetic food products are also contributing in the water pollution enhancement. It also includes the batteries and cells used in houses for daily activities that are usually made of lead and cadmium. Lead and cadmium are heavy metals and thus cause serious threat for the living entities when come in contact with them by contaminated water.

2.2.2 Organic Waste

Plastic bags, papers, polythene bags, rotten fruits and vegetables along with non-degradable by products used in houses are major causes of water pollution that are organic in nature. Today most of people dump their waste products into the seas, streams, rivers, ponds and lakes as they thought it's the final home of waste material (Groundwater Quality 2003). But they don't realize the fact that it's the serious harm for not only their health but also for all living creature. Most commonly the waste material released by household are of organic nature. Large quantities of organic nature waste material was found (82 out of 95) in polluted water when a study was conducted in 30 different states during 1999 and 2000. A study was conducted in Libya on the composition of municipal waste and results show that organic matter represents 59 % of waste, followed by paper–cardboard 12 %, plastic 8 %, miscellaneous 8 %, metals 7 %, glass 4 %, and wood 2 % (Hamad et al. 2014). When this waste material will be added to water sources will become the cause of water pollution.

2.2.3 Liquid Waste

It includes mostly the dirty water coming from houses by different activities of daily routine. It also includes the liquid material which releases when waste material is dumped near or in the water bodies. Toxic liquid which is the major component of different cosmetic products is also included in this respect. When this toxic waste material leach down and become the part of fresh water it deteriorates the quality of water severely (Hudak 2012b).

2.2.3.1 Microbial Contamination

The untreated municipal waste contains biodegradable organic matter which is the house of pathogenic microorganisms. These microorganisms cause fatic water borne diseases like dysentery, hepatitis, typhoid, yellow fever and cholera. High rate of kid's mortalities especially the kids which are under 5 years age is due to these ever increasing waterborne diseases. The main sources of these pathogens are untreated municipal water discharged directly into unlined drains which affect the ground water quality. Water quality mainly assessed by the presence of coliform bacteria. These bacteria mostly attack the intestine of living organisms. The well documented among these are *Escherichia coli* (or *E. coli*). The most common assumption is where the *E. Coli* bacteria are present it also accompany other pathogens in water (Paruch and Maehlum 2012).

2.3 Pesticides and Insecticides

Use of pesticides, insecticides and herbicides impose severe damage to living organisms when they become the part of soil and water. Many of these chemicals produce dioxin. Most pesticides are chlorinated hydrocarbon which are persistent and non-biodegradable. Pesticide and insecticides when applied to get rid of noxious insects and pests, also kills the beneficial microorganisms and leads to groundwater pollution (Hakeem 2015). Water contamination also occurs due to inadequate storage and distribution of agro-chemicals. Chemicals like Aldrin/Aieldrin, DDT, DDD, DDE, hexachlorobenzene, chlordane, benzo pyrene do not readily biodegrade in the environment and tended to persist in the environment and accumulate in water bodies. The bioaccumulation of Dioxins and other chlorinated hydrocarbons (hydrocarbon molecules that contain chlorine atoms) have been shown to accumulate in high concentration in the fats of salmon, fish-eating birds and humans that cause health problems similar to those resulting from toxic metal compounds like methyl mercury.

2.4 Hospital Waste

Hospital waste can be defined as a waste material that has discarded and is of no further use in hospital. Different types of hospital waste like general waste, pathological waste, sharps, chemical waste, pharmaceutical waste and radioactive waste is produced in different amounts in hospitals.

2.4.1 General Waste

Mostly contain domestic or house hold like waste. It doesn't cause any harm to human beings, e.g. kitchen waste, packaging material, paper, wrappers and plastics.

2.4.2 Pathological Waste

It includes tissues, organs, body parts, human foetuses, blood and body fluids. It is dangerous waste and can cause severe harm to not only humans but also have bad effect on other living creature. These pathogens can cause the mild diseases like gastroenteritis to severe and sometime fatal diarrhea, dysentery, hepatitis or typhoid fever (WHO 1996).

2.4.3 Infectious Waste

It includes all the waste material of hospitals which have the organisms causing diseases. It is also dangerous for living organisms. It includes culture and stocks of infectious agents from laboratories, waste from surgery, waste originating from infectious patients. Diseases related with ingestion or penetration of human skin by infective organisms that need a snail, fish or other organisms living in polluted water can be schistosomiasis, clonorchiasis and paragonimiasis (Obasohan et al. 2010). Almost 15–25 % (by weight) of hospital waste is thought to be infectious waste (Shinee et al. 2008).

2.5 Pharmaceutical Wastes

Pharmaceutical wastes comprise of chemicals, plastic wastes, medicines and glass vials that have been released from hospital wards, have been spilled, are expired, or contaminated. It is released in large quantity and thus is the major cause of water pollution. According to Verlicchi et al. (2013) reported that a hospital effluents includes ciprofloxacin ranges 0.0015 and 0.026 mg m^{-3} while the effluent from waste treatment plant it ranges 0.0003 and 0.0011 mg m^{-3}. This is the huge amount for contamination of water bodies and thus will cause the serious hazards to the animals as well as to humans. A big environmental challenge is the presence of pharamactical drugs released from hospital waste in the aquatic environment. The concentration of drugs in surface and groundwater ranges from the ng L^{-1} to the g L^{-1} level, these compounds have been recognized as emerging contaminants. The main role of Pharmaceuticals is to affect biochemical functions in livestock and humans. The release of pharmaceutical in surface and groundwater is a matter of sever concern particularly in countries where groundwater use as a source for drinking water (Table 4).

Table 4 Presence of drugs in waste water/river water/ground water

Source	Location	Drug	Reference
Municipal waste water	Serbia	Ibuprofen (20.1 μg L^{-1})	Petrovic et al. (2014)
Allier River	France	Acetaminophen (200 ng·L^{-1})	Jeanton et al. (2014)
Drinking water	Netherlands	Metformin (26 ng·L^{-1})	Houtman et al. (2014)
Drinking water	Germany	Diclofenac (4.70 μg L^{-1})	Heberer (2002)
Surface water	Italy	Diclofenac (10 μg L^{-1})	Brambilla and Testa (2014)
Groundwater	California	acetaminophen (n 1.89 μg/L)	Fram and Belitz (2011)

2.5.1 Chemical Wastes

This comprised of solid, liquid and gaseous chemicals generating as result of cleaning, housekeeping, and disinfecting product.

2.5.2 Radioactive Wastes

It includes solid, liquid, and gaseous waste that is contaminated with radionuclides generated from in-vitro analysis of body tissues and fluid, in-vivo body organ imaging and tumor localization and therapeutic procedures.

2.5.3 Other Biological Wastes

It includes all the living organisms especially microorganisms present in the hospital waste. It is commonly present in the excreta of patients. It includes different types of bacteria, virus, fungi and protozoa. At least 600 species of fungi are linked with human beings in one or more way but still less than 50 are clearly identified in different studies (Khan et al. 2009). This fungus can contaminate the water when released by sick patients either in the form of excreta or blood effluents. Besides fungi a lot of antimicrobial residues contaminate the water released by hospitals. These microbial residues are the part of bandages of wounds, diseased body parts, attachments of certain used needles and syringes and also incorporated in the outdated or contaminated medicines. Harris et al. (2012) reported that the range of antimicrobial residue release through wastewater treatment plant effluent was 0.02–1.05 mg m^{-3} for the six basic antimicrobial groups used in Europe. Antimicrobial resistant bacteria are very harmful for the humans. It not only has the indirect effect but also effect directly when anyone come in contact with the contaminated water.

3 Nutrients

Nutrients present in waste water particularly nitrogen (N) and phosphorus (P) coming from different sources act an indicator of water quality (Badruzzaman et al. 2012). Nutrients play a vital role for the growth of all living organisms. High concentration of these nutrients in water bodies has great impact on the water quality (Bricker et al. 2007; Freeman et al. 2009). Surface runoff of nitrogen (N) and phosphorous (P) from agricultural fields lead to eutrophication which ultimately results in production of phytoplankton, promote the growth of epiphytic and benthic algae, in marine environment while production of toxins from algal species that ultimately changes the taste and odor of surface water. Research indicates that oxygen-depleted zones result from excessive inputs of nitrates and phosphates from runoff fertilizers and animal wastes deposition. A study conducted by Anderson et al. (2006), reported

Table 5 Nutrient composition in waste water

Source/medium	Location	Nutrient composition	References
Waste water	JERSEY	54%N, 98 % P	Stapleton et al. (2000)
Strom water	Florida	N 0.26 ppm, NO_x-N 0.14 ppm	Grave et al. (2004)
Pit latrines	Kampala, Uganda	2–20 % of total N, < 1 % of total P	Nyenje et al. (2013)
Municipal water	Zenne basin France	1047 kg TP day^{-1}, 8105 kg TN day^{-1}	Garnier et al. (2013)
River water	Morroco	5 mg L^{-1} TN, 0.20 mg L^{-1} TP, NO_3 2.3 mg L^{-1}, Cr 0.8 mg L^{-1},	Perrin et al. (2014)
Wastewater	Taichung, Taiwan	TN 56.7 g L^{-1}, P 2.1 g L^{-1}; K 32.8gL^{-1}; Mn 31.8 mg L^{-1},; Cu 9.22 mg L^{-1}, Zn 36.28 mg $^{-1}$	Singh et al. (2014)
Eastern Aegean Sea water	Turkey	NH_4 0.10–25.6, NO_2 0.01–1.5, NO_3 0.19–7.0, PO_4 0.17–6.8, TPO_4 0.32–9.6 and Si 0.30–13.8 l μ M	Onen et al. (2012)

that about 9.1×10^3 g N/year/system can be added from a normal septic system and a fraction of that loading may end up in the shallow groundwater, and consequently to the surface water bodies where interconnectivity exists. The nutrient pollution creates the dead zones. The largest dead zone was found in Gulf of Mexico covering an area of 5840 $mile^2$ in 2013. During summer season, it occurs as a result of nutrient pollution from the Mississippi River Basin (US-EPA 2014). According to Otis (2007), the environmental factors responsible for nitrogen removal in the subsurface are the soil texture, structure, mineralogy, soil drainage and infiltration rate. The following table (Table 5) indicate the nutrient composition of various types of water.

4 Waste Water Impact on Soil and Plant Health

Waste water discharging from various industries containing higher level of heavy metals, soluble salts and pathogens. Application of such waste water resulting soil salinity, sodicity and heavy metal accumulation (Hakeem 2015). When water contains heavy metals applied to agricultural crops it affects plant health. Waste water containing high Cd concentration when applied results chlorosis and reduced plant growth. As for as Cu is concerned though it is essential for plants but its higher level in irrigation water can effect plant growth particularly when grown on calcareous soils as plant absorb high concentration of Cu in the presence of Calcium (Misra and Mnai 1992). Exposure of plants to excess Cu generates oxidative stress and ROS (Stadtman and Oliver 1991). Oxidative stress causes disturbance of metabolic pathways and damage to macromolecules. Elevated level of Zn causes chlorosis in the younger leaves, which can extend to older leaves after prolonged exposure to high soil Zn levels (Ebbs and Kochian 1997). Another typical effect of Zn toxicity is the appearance of a purplish-red color in leaves, which is ascribed to phosphorus (P) deficiency (Lee et al. 1996). Many studies reported toxicity of Cr in plants. At higher concentration Cr causes inhibition of plant growth, chlorosis in young leaves, nutrient imbalance, wilting of tops, and root injury (Sharma et al. 2003; Scoccianti et al. 2006). In terrestrial plants retardation of chlorophyll synthesis has also been reported (Vajpayee et al. 2000). According to Skeffington et al. (1976) barley seedlings growth retarded 40 % when grown on 100 μM Cr growth medium. Toxic effects of Cr on plant growth and development include alterations in the germination process as well as in the growth of roots, stems and leaves. Hence, exposure to high level of Cr affected total dry matter production and yield of plants. Lead exerts adverse effect on morphology, growth and photosynthetic processes of plants. High level of Pb also causes inhibition of enzyme activities, water imbalance, alterations in membrane permeability and disturbs mineral nutrition (Sharma and Dubey 2005). In Rice excess of Ni in soil causes various physiological alterations and diverse toxicity symptoms such as chlorosis and necrosis in different plant species (Rahman et al. 2005), including rice.

5 Conclusions and Future Aspects

Urbanization and industrial development resulting decreasing availability of fresh water resources. So farming community will have to depend on the safe use the waste water available freely round the clock. For its safe use it is desirable that irrigation quality of waste waters must be assessed not only for EC, SAR and RSC but also for metal ions before their soil application to conserve the produce quality and preserve the agro-environment.

References

Ahmad HR, Aziz MA, Ghafoor A, Rehman MZ, Sabir M, Saifullah (2011) Wheat assimilation of nickel and zinc added in irrigation water as affected by organic matter. J Plant Nutr 34:1–7

Álvarez AM, Carral P, Hernández Z, Almendros G (2016) Hydrocarbon pollution from domestic oil recycling industries in peri-urban soils. Lipid molecular assemblages. J Envir on Chem Eng 4:695–703

AMAP (2010) Assessment 2007: oil and gas activities in the arctic effects and potential effects, vol II. Arctic Monitoring and Assessment Programme (AMAP), Oslo

Anac D, Hakarer H, Irget M (1993) Ege Üniv Ziraat Fak. Dergisi 30:3

Anderson BS, Phillips BM, Hunt JW, Worcester K, Adams M, Kapellas N, Tjeerdema RS (2006) Evidence of pesticide impacts in the Santa Maria river watershed, California, USA. Environ Toxicol Chem 25(4):1160–1170

Arora S, Kazmi AA (2015) The effect of seasonal temperature on pathogen removal efficacy of vermi filter for wastewater treatment. Water Res 74:88–99

Azetsu-Scott K, Yeats P, Wohlgeschaffen G, Dalziel J, Niven S, Lee K (2007) Precipitation of heavy metals in produced water: influence on contaminant transport and toxicity. Mar Environ Res 63:146–167

Badruzzaman M, Pinzon J, Oppenheimer J, Jacangelo JG (2012) Sources of nutrients impacting surface waters in Florida: a review. J Environ Manag 109:80–92

Balkhair KS, Ashraf MA (2016) Field accumulation risks of heavy metals in soil and vegetable crop irrigated with sewage water Saudi. J Biol Sci 23:S32–S44

Beukes JP, Dawson NF, Van Zyl PG et al (2010) Theoretical and practical aspects of Cr(VI) in the South African ferrochrome industry. J South Afr Inst Min Metall 110:743–750

Bougherira N, Hani A, Djabri L, Toumi F, Chaffai H, Haied N, Nechem D, Sedrati N (2014) Impact of the urban and industrial waste water on surface and groundwater, in the region of Annaba, (Algeria). Energy Procedia 50:692–701

Brambilla G, Testa C (2014) Food safety/food security aspects related to the environmental release of pharmaceuticals. Chemosphere 115:81–87

Bricker S, Longstaff BW, Dennison A, Jones K, Boicourt C, Woerner J (2007) Effects of nutrient enrichment in the nation's estuaries: a decade of change, vol 26, NOAA coastal ocean program decision analysis series. National Centers for Coastal Ocean Science, Silver Spring

Brodersen J, Juul J, Jacobsen H, Tsotsos D (2002). Review of selected waste streams (vol. 69). Technical Report. European Environment Agency. Copenhagen

Castege I, Milon E, Pautrizel F (2014) Response of benthic macrofauna to an oil pollution: lessons from the "Prestige" oil spill on the rocky shore of Guéthary (south of the Bay of Biscay, France). Deep-Sea Res II 106:192–197

Chun-Li P, Grosskopf KR, Kibert CJ (1994) Construction waste management and recycling strategies in the United States In: Kibert CJ (ed) Proceedings of the first conference of CIB TG 16 on sustainable construction. Tampa, pp 689–696

Coles D, McDowell KMJ (2003) Food packaging technology. Blackwell Publishing Ltd, Oxford

Corcoran E, Nellemann C, Baker E, Bos R, Osborn D, Savelli H (2010) Sick Water? The central role of wastewater management in sustainable development. A rapid response assessment. United Nations Environment Programme, UN-HABITAT, GRID-Arendal. Norway. www.grida.no. ISBN: 978-82-7701-075-5

Curriero FC, Patz JA, Rose JB, Lele S (2001) The association between extreme precipitation and waterborne disease outbreaks in the United States. Am J Public Health 91:1194–1199

Denaro R, Cappello S, Yakimov MM (2010) Vegetable oil wastes. In: Timmis KN (ed) Handbook of hydrocarbon and lipid microbiology. Springer, Berlin/Heidelberg, pp 2393–2399

Ebbs SD, Kochian LV (1997) Toxicity of zinc and copper to Brassica species: implications for phytoremediation. J Environ Qual 26:776–781

Europa (2003) Transnational corporations and developing countries: big business poor peoples, The Courier ACP-EU. European topic centre on waste european environment agency. Available from: http://reports.eea.europa.eu/technical_report_2001_69/en/tech_rep_69.pdf

Fram MS, Belitz K (2011) Occurrence and concentrations of pharmaceutical compounds in groundwater used for public drinking-water supply in California. Sci Total Enviro 409:3409–3417

Freeman AM, Lamon EC, Stow CA (2009) Nutrient criteria for lakes, ponds, and reservoirs: a Beyesian TREED model approach. Ecol Model 220:630–639

Garnier J, Brion N, Callens J, Passy P, Deligne C, Billen G, Servais P (2013) Claire Billen Modeling historical changes in nutrient delivery and water quality of the Zenne River (1790s–2010): the role of land use, waterscape and urban wastewater management. J Mar Syst 128:62–76

Gericke WA (1995) Environmental aspects of ferrochrome production. In: Proceedings Ferroalloys 7th international congress (INFACON XII). Trondheim, pp 131–140

Graves GA, Wan Y, Fike DL (2004) Water quality characteristics of storm water from major land uses in South Florida. J AmWater Resour 40(6):1405–1419

Hakeem KR (2015) Crop production and global environmental issues. Springer International Publishing AG, Cham, 598 pp

Hakeem KR, Sabir M, Ozturk M, Mermut A (2014) Soil remediation and plants: prospects and challenges. Academic/Elsevier, New York, p 724

Hamad TA, Agll AA, Hamad YM, Sheffield JW (2014) Solid waste as renewable source of energy: current and future possibility in Libya. Case Stud Therm Engineer 4:144–152

Harris S, Cormican M, Cummins E (2012) Antimicrobial residues and antimicrobialresistant bacteria: impact on the microbial environment and risk to human health—a review. Hum Ecol Risk Assess 18:767–809

Heberer T (2002) Tracking persistent pharmaceutical residues from municipal sewage to drinking water. J Hydrol 266:175–189

Houtman CJ, Kroesbergen J, Teunissen KL, Hoek JPVD (2014) Human health risk assessment of the mixture of pharmaceuticals in Dutch drinking water and its sources based on frequent monitoring data. Sci Total Environ 496:54–62

Hudak PF (2012) Chloride/bromide ratios in leachate derived from farm-animal waste. Environ human consumption: an Irish case study. Hum Ecol Risk Assess 19:1264–1284

Hunt RG, Sellers VR, Frankalin WE, Nelson JM, Rathje WL, Hughes WW, Wilson DC (1990) Estimates of the volume of MSW and selected components in trash cans and land fills. Tucson, Ariz.: Report prepared by The Garbage Project and Franklins Assn. Ltd. for the Council for Solid Waste Solutions

ISSF (2011) ISSF: International stainless steel forum 2011. Stainless steel demand index. http://www.worldstainless.org/Statistics/Demand+index/. Accessd 20 Jan 2012. J African Earth Sci 116:182–189

Jeanton HC, Schemberg D, Nabaz MN, Huneau F, Bertrand G, Lavastre V, Coustumer PL (2014) Evaluation of pharmaceuticals in surface water: reliability of PECs compared to MECs. Environ Int 73:10–21

Khan AAH (2009) Studies on indoor fungi and their control, thesis, Department of Biotechnology, School of Life Sciences, Swami Ramanand Teerth Marathwada University, Nanded

Kibert CJ (2000) Deconstruction as an essential component of sustainable construction. In: Proceedings of the second Southern African conference on sustainable development in the built environment, pretoria, pp 1–5. Available from: http://www.sustainablesettlement.co.za/event/SSBE/Proceedings/kibert.pdf

Killi D, Anaulf R, Kavdir Y, Haworth M (2014) Assessing the impact of agro-industrial olive wastes in soil water retention: implications for remediation of degraded soils and water availability for plant growth. Inte Biodeter Biodegrad 94:48–56

Lee CW, Choi JM, Pak CH (1996) Micronutrient toxicity in seed geranium (Pelargonium × hortorum Bailey). J Am Soc Hort Sci 121:77–82

Lee K, Azetsu-Scott K, Cobanli SE, Dalziel J, Niven S, Wohlgeschlaffen G, Yeats P (2005) Overview of potential impacts of produced water in Atlantic Canada. In: Armworthy SL, Cranford PJ, Lee K (eds) Offshore oil and gas environmental effects monitoring: approaches and technologies. Batelle Press, Columbus, pp 319–342

Li X, Poon CS, Liu PS (2001) Heavy metal contamination of urban soils and street dusts in Hong Kong. Appl Geochem 16:1361–1368

Litchfield JH (1987) Microbiological and enzymatic treatments for utilizing agricultural and food processing wastes. Food Biotech 1:29–57

Manaf LA, Samah MAA, Zukki NM (2009) Municipal solid waste management in Malaysia: practice and challenges. Waste Manag 29:2902–2906

Martinez HR, Blasco LN (2012) Estimation of dietary intake and content of lead and cadmium in infant cereals marketed in Spain. Food Control 26:6–14

Martinez-Martinez S, Acosta JA, Fazcano A, Carmona DM, Zornoza R, Cerda C (2013) Assessment of the lead and zinc contents in natural soils and tailing ponds from the Cartagena- La union mining district, SE Spain. J Geochem Explor 124:166–175

McKinney ML, Schoch R, Yonavjak L (2013) Environmental Science: systems and solution. 5th (ed) Jones and Bartlett Learning Inc, Burlington

Misra SG, Mani D (1992) Metallic pollution. Efficient Offset Printers, New Delhi

Muhammad S, Shah MT, Khan S (2011) Health risk assessment of heavy metals and their source apportionment in drinking water of Kohistan region, Northern Pakistan. Microchem J 98:334–343

Murthy YR, Tripathy SK, Kumar CR (2011) Chrome ore beneficiation challenges & opportunities – a review. Miner Eng 24:375–380

Nasrullah Naz R, Bibi H, Iqbal M, Durrani MI (2006) Pollution load in industrial effluent and ground water of Gadoon Amazai Induatrial Estate (GAIE) Swabi, NWFP. J Agri Bio Sci 1:18–24

National Statistics Organization of Thailand (2007) Population and labor statistics. Available from: http://wwwweb.nso.go.th/eng/stat/stat.htm

Neff J, Lee K, DeBlois EM (2011) Produced water: overview of composition, fates, and effects. In: Lee K, Neff J (eds) Produced water. Springer, New York

Norwegian Oil and Gas (2013). Environmental report 2013. The Norwegian Oil and Gas Association. https://www.norskoljeoggass.no/en/Publica/Environmental-reports/Environmental-report-2013/

Nyenje PM, Foppen JW, Kulabako R, Muwanga A, Uhlenbrooka S (2013) Nutrient pollution in shallow aquifers underlying pit latrines and domestic solid. J Environ Manage 122:15–24

Obasohan EE, Agbonlahor DE, Obano EE (2010) Water pollution: a review of microbial quality and health concerns of water, sediment and fish in the aquatic system. Afr J Biotech 9:423–427

Obiora SC, Chukwu A, Davies TC (2016) Heavy metals and health risk assessment of arable soils and food crops around Pb-Zn mining localities in Enyigba, South Eastern Nigeria. J Afr Earth Sci 116:1182–1189

Official Home Page of WHO: Guidelines for Drinking Water Quality 1996. From http.www.who.int.org. Retrieved March 4 2011. (TarekA.Hamad n, Abdulhakim A. Agll, Yousif M. Hamad, John W. Sheffield (ENVIRONMENTAL SCIENCE & TECHNOLOGY/VOL. 36, NO. 6, 2002). Solid waste as renewable source of energy: current and future possibility in Libya T.A.Hamadetal./Case Studies in Thermal Engineering 4(2014):144–152

Onen SA, Kocak F, Kucuksezgin F (2012) Evaluation of spatial and temporal variations of inorganic nutrient species in the Eastern Aegean Sea waters. Mare Pollut Bull 64:2849–2856

Otis RJ (2007) Estimates of nitrogen loadings to groundwater from onsite wastewater systems in the Wekiva Study Area: Task 2 Report, June 2007. Prepared for Florida Department of Health. Florida Department of Health, Tallahassee

Pandey J, Pandey U (2009) Atmospheric deposition and heavy metal contamination in an organic farming system in a seasonally dry tropical region of India. J Sustain Agr 33:361–378

Paruch AM, Mæhlum T (2012) Specific features of Escherichia coli that distinguish it from coliform and thermotolerant coliform bacteria and define it as the most accurate indicator of faecal contamination in the environment. Ecol Indic 23:140–142

Perrin JL, Rais N, Chahinian N, Moulin P, Ijjaali M (2014) Water quality assessment of highly polluted rivers in a semi-arid Mediterranean zone Oued Fez and Sebou River (Morocco). J Hydrol 510:26–34

Petrovic M, Skrbic B, Zivancev J, Climent LF, Barcelo D (2014) Determination of 81 pharmaceutical drugs by high performance liquid chromatography coupled to mass spectrometry with hybrid triple quadrupole–linear ion trap in different types of water in Serbia. Sci Total Environ 468–469:415–428

Pruvot C, Douay F, Herve F, Waterlot C (2006) Heavy metals in soil, crops and grass as a source of human exposure in the former mining areas. J Soils Sediment 4:215–220

Rahman H, Sabreen S, Alam S, Kawai S (2005) Effects of nickel on growth and composition of metal micronutrients in barley plants grown in nutrient solution. J Plant Nutr 28:393–404

Raja G, Venkatesan P (2010) Assessment of groundwater pollution and its impact in and around Punnam area of Karur district, Tamil Nadu, India. J Chem 2:473–478

Rattan RK, Datta SP, Chhonkar PK, Suribabu K, Singh AK (2005) Long-term impact of irrigation with sewage effluents on heavy metal content in soils, crops and groundwater—a case study. Agric Ecosyst Environ 109:310–322

Roca-Perez L, Martínez C, Marcilla P, Boluda R (2009) Composting rice straw with sewage sludge and compost effects on the soil–plant system. Chemosphere 75:781–787

Scoccianti V, Crinelli R, Tirillini B, Mancinelli V, Speranza A (2006) Uptake and toxicity of Cr(III) in celery seedlings. Chemosphere 64:1695–1703

Sharma P, Dubey RS (2005) Lead toxicity in plants. Braz J Plant Physiol 17:35–52

Sharma DC, Sharma CP, Tripathi RD (2003) Phytotoxic lesions of chromium in maize. Chemosphere 51:63–68

Shinee E, Gombojav E, Nishimura A, Hamajima N, Ito K (2008) Healthcare waste management in the capital city of Mongolia. Waste Manag 28:435–444

Sial RA, Chaudhary MF, Abbas ST, Latif MI, Khan AG (2006) Quality of effluents from Hattar industrial estate. J Zhejiang Univ Sci 12:974–980

Singh S, Young LS, Shen FT, Young CC (2014) Impacts of industrial waste resources on maize (*Zea mays* L.) growth, yield, nutrients uptake and soil properties. Waste Manag 34:1877–1883

Skeffington RA, Shewry PR, Peterson PJ (1976) Chromium uptake and transport in barley seedlings (*Hordeum vulgare* L.). Planta 132:209–214

Sreenath KH, Krishna KRS, Santhanam K (1995) Enzymatic liquefaction of some varieties mango pulp. Lebensm Wiss u Technol 28:196–200

Stadtman ER, Oliver CN (1991) Metal-catalyzed oxidation of proteins. J Biol Chem 266:2005–2008

Stapleton CM, David K, Jackson GF, Wyer MD (2000) Estimated inorganic nutrient inputs to the coastal waters of jersey from catchment and waste water sources. Water Res 34:787–796

Suciu I, Cosma C, Todica M, Bolboaca SD, Jantschi L (2008) Analysis of soil heavy metal pollution and pattern in Central Transylvania. Int J Mol Sci 9:434–453

Sulaiman A, Tabatabaei M, Hassan MA, Shirai Y (2009a) The influence of higher sludge recycling rate on anaerobic treatment of palm oil mill effluent in a semi-commercial single stage digester for renewable energy. Am J Biochem Biotechnol 5:1–6

Sulaiman A, Hassan MA, Shirai Y, Abd-Aziz S, Tabatabaei M, Busu Z (2009b) The effect of mixing on methane production in a semi-commercial closed digester tank treating palm oil mill effluent. Aust J Basic Appl Sci 3:1577–1583

Suthar S (2009) Vermicomposting of vegetable-market solid waste using Eisenia fetida: impact of bulking material on earthworm growth and decomposition rate. Ecolog Eng 35:914–920

Suthar S (2010) Recycling of agro-industrial sludge through vermi technology. Ecol Eng 36:1028–1036

Tabatabaei M, Rahim RA, Abdullah N, Wright ADG, Shirai Y, Sulaiman A, Hassan MA (2010) Importance of the methanogenic archaea populations in anaerobic wastewater treatments. Process Biochem 45:1214–1225

Tsang CF, Barnichon JD, Birkholzer J, Li XL, Liu HH, Sillen X (2012) Coupled thermo-hydro-mechanical processes in the near field of a high level radioactive wasterepository in clay formations. Int J Rock Mech Min Sci 49:31–44

US-EPA (2014) http://www2.epa.gov/nutrientpollution/effects-environment

Vajpayee P, Tripathi RD, Rai UN, Ali MB, Singh SN (2000) Chromium (VI) accumulation reduces chlorophyll biosynthesis, nitrate reductase activity and protein content in *Nymphaea alba* L. Chemosphere 41:1075–1082

Vamerali T, Bandiera M, Mosca G (2010) Field crops for phytoremediation of metal-contaminated land. A review. Environ Chem Lett 8:1–17

Verlicchi P, Al Aukidy M, Galletti A, Petrovic M, Barceló D (2013) Hospital effluent: investigation of the concentrations and distribution of pharmaceuticals and environmental risk assessment. Sci Total Environ 430:109–111

Wang Z, Yang L (2015) Delinking indicators on regional industry development and carbon emissions: Beijing-Tianjin-Hebei economic band case. Ecol Indic 48:41–48

WHO (1996) Hepatitis B Vaccine-making global progress update, Expanded Programme on Immunization

World Nuclear Association (WNA) (2013) Plans for new reactors worldwide. http://www.world-nuclear.org/info/current-and-future-generation/plans-for-newreactors-worldwide/. Accessed 22 Mar 2014
Wu S, Zhou S, Xingong L (2011) Determining the anthropogenic contribution of heavy metal accumulations around a typical industrial town: Xushe, China. J Geochem Explor 110:92–97
Xia Y (1999) Study on the current situation, effects and response of the foreign investment in pollution intensive industries in China. Manag World 3:108–123
Yu L, Weetjens E, Sillen X, Vietor T, Li XL, Delage P, Labiouse V, Charlier R (2013) Consequences of the thermal transient on the evolution of the damaged zone around a repository for heat-emitting high-level radioactive waste in a clay formation: a performance assessment perspective. Rock Mech Rock Eng 47:3–19
Zhang M, Pu J (2011) Mineral materials as feasible amendments to stabilize heavy metals in polluted urban soils. J Environ Sci 23:607–615

Soil Amendments for Heavy Metal Immobilization Using Different Crops

Mahar Amanullah, Amjad Ali, Wang Ping, Wang Quan, Shen Feng, Altaf Hussain Lahori, Li Ronghua, Mukesh Kumar Awasthi, Zhang Zengqiang, and Münir Öztürk

Contents

M. Amanullah (✉)
College of Natural Resources & Environment, Northwest A & F University, Yangling, 712100 Shaanxi, China

Centre for Environmental Sciences, University of Sindh, 76080 Jamshoro, Pakistan
e-mail: aman.fsw@gmail.com

A. Ali • W. Ping • W. Quan • S. Feng • A.H. Lahori • L. Ronghua
M.K. Awasthi • Z. Zengqiang
College of Natural Resources & Environment, Northwest A & F University, Yangling, 712100 Shaanxi, China

M. Öztürk (✉)
Botany Department and Center for Environmental Studies, Ege University, Izmir, Turkey

Faculty of Forestry, Universiti Putra Malaysia, Selangor, Malaysia
e-mail: Munirozturk@gmail.com

K.R. Hakeem et al. (eds.), *Soil Science: Agricultural and Environmental Prospectives*, DOI 10.1007/978-3-319-34451-5_17

Abstract Industrial development has put enormous negative impacts on the natural environment. Metal refining units using pyrometallurgical processes have emitted excessive load of heavy metals. Soil is the prime target of such contaminants and polluted in over the vast area. Soil pollution poses a great threat to public globally. Heave metals (HMs) are persistent in nature and do not undergo chemical or biological breakdown, but remain in the environment for long periods since their emission. Immobilization is an in-situ remediation technique that uses cost effective soil amendments to reduce Pb and Cd availability in the contaminated soils. This chapter has focused on heavy metal pollution in the soil through considering: sources, types of soil contaminants, global overview of soil pollution, environmental health risks and immobilization options using various soil amendments. This chapter describes commonly used and emerging cost effective amendments for heavy metal immobilization. There is a dire need to develop procedures to determine immobilization efficacy that could be used to assess the in situ short and long-term environmental stability of metal immobilization.

Keywords Heavy metal pollution • Soil remediation • Immobilization • Pyrometallurgical processes

1 Introduction

Under the nexus of environmental quality and population growth, the contamination of land resources through urban sprawl and industrialization are continuously rising and leaving negative ecological footprints on our environment (Ozturk et al. 2002, 2004, 2011a; Wong and Li 2004). The contaminated soils are a problem of major worldwide concern due to human health risks as well as other ecological hazards. Such areas at the same time pose huge economic burden for restoration and reclamation issues (Semenzin et al. 2007; Ozturk et al. 2011a, b). Much work is being done in this direction lately in particular the distribution and mobility of heavy metals (HMs) in soils. As a matter of fact some of the heavy metals are essential for living beings in trace quantities, but most are hazardous at high concentrations. For example, cadmium and lead occur naturally in soils, but if their levels in soil increase these elements get accumulated in humans and lead to serious health problems (Ok et al. 2004; Ozturk et al. 2008). Unsustainable waste disposal practices have resulted in a significant pile up in a diverse range of toxic heavy metals. Food chain contaminated with toxic metals is closely affiliated to the source, dosage, rate and magnitude of plant metal uptake, the extent of absorption by animals as well as physical and chemical properties of soil (Adriano 2001; Ozturk 2010a, b; Ozturk et al. 2010; Lan et al. 2014). Historically human health affected by metal toxicity gained paramount importance as a result of serious poisoning at large scale in the late 1950s when a large number of human poisoning tragic cases were recorded around

Minamata Bay in Japan due to consumption of contaminated fish with high doses of toxic metals (Knopf and Konig 2010).

Many countries have recorded excessive heavy metal concentrations in their soils and significant negative effects on human health can be expected (Ashraf et al. 2010, 2015). Typical examples are China, India and Bangladesh, where an alarming number of population is reported to be at risk from toxic metals (Bhattacharya et al. 2012; Hasanuzzaman et al. 2014, 2015). Similarly cadmium is reported to be a serious concern in the pacific area, where grazing animals have shown high bioaccumulation and made it unfit for human usage. This has affected the export of meat products to overseas markets. Lead in soils is also accumulating in high quantities because of the unsustainable use of herbicides and fungicides in the vicinity of urban horticultural sites, this again is a matter of concern in some countries (Gucel et al. 2009a, b; Pietrzak and Uren 2011).

Toxic heavy metals do not experience chemical and microbial degradation and the total concentration of metals do not change due to their persistent nature for a long period of time after their entry in to the soil system. As a result, a myriad of researchers are focussing immensely on the development of soil remediation technologies as global masses are increasingly vocal about the animal and human health implications resulting from contaminated soils (Ashraf et al. 2010; Hakeem et al. 2015; Ozturk et al. 2015a, b). Soil remediation technologies are now attracting a special attention since conventional soil reclamation practices, i.e. land filling and excavation is often costly and environmentally not paying as compared to alternative options which are cost-effective and environmentally sustainable such as soil immobilization/solidification (Mulligan et al. 2001; Kumpiene et al. 2008).

In the process of soil remediation, we cannot get rid of heavy metals like organic pollutants, but the toxicity and mobility of trace metals can be impaired by triggering the important immobilization process i.e. adsorption, precipitation, complexation and redox reactions. To reduce bioavailability of toxic metals, soil amelioration is employed for diffuse distribution (Park et al. 2011a; Zhao et al. 2010). Biological and chemical stabilization of toxic metals using organic, i.e. biosolids and inorganic amendments i.e. phosphate compounds and lime are suitable options to minimize metal bioavailability. In an urban environment, more localized toxic heavy metals are found and for the metal stabilization in such circumstances, the process to be followed is chemical washingfollowed by phytoextraction for remediation (Ozturk et al. 2012; Sabir et al. 2015). For safe disposal of toxic metals from urban soil, phytoextraction process and the subsequent recovery are considered for commercial and research studies (Robinson et al. 2009). However, when phytoextraction is unable to remediate toxic heavy metals from contaminated sites due to some unavoidable reasons, then alternate options, i.e. in situ immobilization are considered as an important part of environmental management. Both organic and inorganic soil amendments are part of remediation techniques applied to manage contaminated soils (Ozturk et al. 2015a, b).

2 Sources of Heavy Metals in the Soil

Sources of HMs include mining, industrial processing, pesticides, chemical fertilizers and automobile exhausts in industrializing countries (Granero and Domingo 2002; Huang et al. 2007; Govil et al. 2008; Johri et al. 2010; Fang et al. 2014; Mohsenzadeh and Shahrokhi 2014). Among these sources, heavy metals contribute a major part as per their cytotoxicity, mutagenicity, and carcinogenic nature (Hamman 2004; Mahavi 2005). The elements like cadmium, lead, mercury, arsenic, chromium, copper (Cu), nickel (Ni), selenium (Se) and few others are more toxic to living organisms even at low levels (Aziz et al. 2015; Sabir et al. 2015). Soil is the major sink for heavy metals released into the environment by anthropogenic activities around the globe. The heavy metal stress is first faced by soil flora and fauna (Kirpichtchikova et al. 2006).

Heavy metals include around 53 elements and majority are hazardous as well as toxic for our environment. The reason for this is that they show persistence, are non-biodegradable and possess higher density (>5 g/cm^3). The introduction of HMs through food chain poses a lethal threat for living beings (US ATSDR. 2007; Miransari 2011; Bharti and Banerjee 2012; Bhattacharya et al. 2014; Zhou et al. 2014).

Cadmium (Cd) is considered as persistent, inorganic and toxic metal for human and biota even at a low concentration (Kirkham 2006; Wahid et al. 2008; Groppa et al. 2012). It is highly water soluble (Liang et al. 2014), can cause soil pollution and adverse effects on plant growth and development (DalCorso et al. 2010; Rafiq et al. 2014; Robson et al. 2014). by In general anthropogenic activities like, mining, industrialization and production of phosphate fertilizers are the main source for its release into the environment (Clemens 2006). Under normal natural conditions the soil cadmium content lies around 0.1–0.5 mg kg^{-1}, but there are reports of higher values reaching up to 150 mg kg^{-1} in areas associated with batteries, plastics and paint manufacturing, mining, electroplating, alloy preparation, fertilizers and composts (Gallego et al. 2012).

Lead (Pb) is readily absorbed in soil, contributed by atmospheric deposition, natural weathering processes and considered as a notorious environmental pollutant (Nagajyoti et al. 2010). It is added through mining, Pb ore smelters, fertilizers, pesticides (lead arsenate), pigments, batteries, fossil fuels, manure, sludge, electricity, heat production and is also released from the car exhausts as tetraethyl lead $(CH_3CH_2)_4Pb$), due to its use in automobiles as an antiknock in the countries using leaded gasoline (Cheng and Hu 2010; Li et al. 2013). The rapid economic growth around the globe also contributes to an increment in the Pb concentration in urban and industrial areas (Dermont et al. 2008). Annual lead level in air should not exceed 0.5 μgm^{-3} (WHO 2000, 2011). In the ambient air its range is reported as from 7.6x10^5 μgm^{-3} in remote areas such as Antarctica, to >10 μgm^{-3} near stationary sources such as smelters (USATSDR 2007). The lead concentration in air for some industrial European countries has been reported to show a range of 0.003 to 0.033 μgm^{-3} in 1990, but reached a level ranging between 0.003–0.010 μgm^{-3} in 2003

(UNEP 2008). The increase in the lead content in our environment actually started rising from the year of its use in 1923 in the form of tetraethyl as an anti-knock agent in fuel, leading to an increase in the load of lead in our atmosphere (Walraven et al. 2014). In some roadside soils the concentrations as high as 3000 $mgkg^{-1}$ have been reported (Bakirdere and Yaman 2008; Chen et al. 2010; Khan et al. 2011). The legislations in European countries enforced from 1970 onwards against the use of lead in petrol has helped in the reduction of lead levels upto safe limits in man countries (Von Storch et al. 2003; Pacyna et al. 2009). Chemical forms of lead depend on the source of lead. Like in atmosphere lead exist in the form of $PbSO_4$ and $PbCO_3$, it is released as $PbCl_2$, PbO, PbS from coal combustion and from oil combustion mainly in the form PbO as insoluble mineral particles (Kabata-Pendias and Mukherjee 2007). Lead particle size ranges between 0.1–1.0 μm depending on the source of emission. The particles in the atmosphere are easily deposited in the terrestrial and aquatic ecosystems by dry or wet deposition (Mishra and Acharya 2004).

The sources of mercury (Hg) are mainly chloroalkali productions, pharmaceutical (thermometers and medical drugs) and cosmetic preparations, electrical instruments, batteries, pulp and paper industries (Martínez-Juárez et al. 2012; Wu et al. 2012). It is the only metal which can be found in liquid state at room temperature, and is also found in the soil, water and air (de Lacerda et al. 2007), with no biological function (Cursino et al. 2003). Hg is a global pollutant released into the environment by natural (volcanic emissions) and anthropogenic activities as well. Its anthropogenic releases are associated with the use of pesticides and fungicides, forest fires, gold extraction, industries of electrolytic chlorine–soda, dental amalgams, vinyl chloride monomer (VCM) production units, fluorescent lamps and thermometers (Micaroni et al. 2000; Churchill et al. 2004). The atmospheric depositional flux of Hg has increased approximately three times since the industrial revolution (Biester et al. 2007). Recent estimations of global anthropogenic Hg emission are considered to be much higher than those coming from natural sources and range from 6.6 to 9.4×10^6 kg $year^{-1}$ (Lohman et al. 2008; Issaro et al. 2010). Mercury can be found in elemental (Hg^0), methyl mercury (MeHg) and ionic form (Hg^{+2}), and in the atmosphere it exists as (Hg^0), which is converted to MeHg by methanogenic bacteria (Bizily et al. 2000). MeHg is easily absorbed by plants (Ozturk 2010a, b). Ionic form (Hg^{+2}), found in the aquatic environment is of major concern due to its high toxicity to the living systems by entering into the food chain (Bizily et al. 2000; Boularbah et al. 2006).

3 Types of Soil Contaminants

3.1 Fertilizers and Pesticides

An extensive use of chemical fertilizers is practiced all around the world to feed the ever increasing human population. The major components of fertilizers are; nitrogen, phosphorus and potassium. Apart from these some heavy metals such as Pb and

Cd are also introduced into the environment as contaminants. Similarly, the use of pesticide, insecticide and fungicide sprays on crops and weeds also lead to the introduction of elements like zinc, chromium, copper, mercury, manganese, and lead in the soil surface (Luo et al. 2009). The examples of such pesticides are Bordeaux mixture (copper sulfate) and copper oxychloride (Ozturk 2010a, b). Lead arsenate is used in fruit orchards to control some insects (Wuana and Okieimen 2011).

3.2 Biosolids and Manures

The application of biosolids (livestock manures, composts, municipal sewage sludge) to the land also leads to an accumulation of arsenic, cadmium, chromium, copper, lead, mercury, nickel, selenium, molibednum and antimony among many other elements in the soil (Basta et al. 2005). Poultry, cattle and pig manures are commonly applied to crops and pastures for high yield as solids or slurries, these are rich in heavy metals and lead to food web through consumption of the agricultural products. The dumping of these wastes in open air not only leads to air as well as ground water pollution but also the volatilization of organic compounds like chlorofluorocarbons and leaching of heavy metals into underground layers.

3.3 Wastewater Irrigation

The waste water produced by the municipal and industrial activities is released without any proper treatment in most of the developing countries. The farming community is unaware of the health hazard of the waste water and uses it for the irrigation purpose in rural areas. It is estimated that 20 million ha of land is irrigated with waste water on global scale. The crops grown on such areas are then supplied to urban areas and lead to an introduction of heavy metals in to the food web. The most common heavy metals found in the waste water irrigation streams are silver, cadmium, boron, arsenic, barium, mercury, copper, nickel, iron, lead, manganese, and selenium (Farmaki and Thomaidis 2008). The prolonged use of such waste waters for irrigation purpose results in the bioaccumulation of metals which ends up with the soil pollution and destruction of soil microbial communities.

3.4 Mining, Milling Processes and Industrial Wastes

Mining, oil and gas exploration and metallurgical activities also contribute to heavy metal accumulation in soil. The mine tailings are discharged into natural depressions, which are generally rich in HMs like Pb and Zn (DeVolder et al. 2003).

The transition metals, lanthanoids, actinoids and metalloids, all have high density and are categorized as heavy metals. Their on-site disposal severely affects the growth and development of different plant organs. Similarly, petroleum gas leaks, leaded paint manufacturing, and leather processing industries top the list as a risk for living objects. Generally very high concentrations of lead and zinc are determined in the plants and soils adjacent to smelting works (El-Shahawi et al. 2014).

3.5 Airborne Sources of Fly Ash

Airborne sources for heavy metals include stack or duct emissions of air, gas, or vapors from storage areas or waste piles. These are generally released as particulates. Many metals can volatilize at high-temperatures forexample; arsenic, cadmium and lead. They get distributed by natural air currents and on condensation precipitate on large areas, which lead to soil and air pollution at the sites quite away from the source. The major source of lead is combustion of petrol, containing tetraethyl lead; zinc and cadmium are generally added to the soils adjacent to roads, the sources being tyres and lubricant oils. Luo et al. (2009) has reported in his work conducted in China that, 43–85% of pollutants like lead, arsenic, zinc, mercury and cadmium can be accepted as the major pollution contributors from the industry.

4 Global Overview of Soil Contamination

Approximately 1,400,000 land resources have been recorded to contain heavy metals in the Western Europe; of this total area, a very large number of sites are recorded as polluted (McGrath et al. 2001). There are 400,000 polluted sites in Germany, Uk, Denmark, Spain, Italy, Netherlands and Finland, whereas in Sweden, France, Hungary, Slovakia and Austria the number of contaminated land sites is reported to lie around 200,000. Nearly 10, 000 contaminated areas have been recorded from Greece and Poland and less than this number in Ireland and Portugal (Perez 2012).

Recently, the Ministry of Environmental Protection (MEP) and the Ministry of Land and Resources of the People's Republic of China have issued a joint report on the current status of soil contamination in China. According to the report, soils in some areas, especially those surrounding mining and industrial activities are seriously contaminated, while the quality of farmland soils is also of particular concern. The report is based on extensive surveys of soils conducted between 2005 and 2013, covering more than 70% of China's land area. According to the report approximately 16.1% of the farmland soils in the country are rich in heavy metals, and their level has exceeded the environmental quality standard set by the MEP; for agricultural soils the percentage of exceedance is reported as 19.4% (equivalent to nearly 26 million ha; assuming that that the area is proportional to the number of survey samples). Contamination by heavy metals and metalloids accounts for 82.4% of the soils, which are classified as being contaminated and the rest of the contamination comes from the organic contaminants. Among the heavy metals and metalloids, cadmium ranks as the first in the percentage of soil samples (7.0 %) exceeding the MEP limit (MEP and MRL, PRC 2015). One-sixth of the total agricultural lands in China is suffering from heavy metal pollution, and approximately 40% is reported to be disturbed by erosive activities and rapid deforestation (Liu et al. 2005; Hang Zhou et al. 2014). It The calculations have shown that cultivated agricultural lands irrigated with polluted water in China comes to about 7.3% of the total irrigated cropland. There is no control over the quantity of contaminated water (Luo et al. 2009). Similarly paddy fields in Korea are reported to suffer from pollution of heavy metals like cadmium, lead, and zinc amounting to 0.11 mg kg – 1. Likewise, concentration in Japanese paddy fields has been assessed to lie around 22.9 mg kg^{-1}, 75.9 mg kg^{-1} and 3.71 mg kg^{-1}.

Crude open dumping of sewage sludge and industrial effluents in the close proximity is regular practice in the countries like. Pakistan, India, Bangladesh and Sri Lanka. There are no sufficient number of treatment plants for the final disposal of hazardous hospital, municipal and industrial wastes, as a result crude open dumping in fresh water bodies is widespread (Lone et al. 2008). Approximately 600, 000 Brownfield sites in America show heavy metal pollution. The most of the soil and land contamination in America can be attributed to the crude disposal of waste from poultry farms. The number of chicken broilers has crossed 200 million in 2007, unchecked chicken waste disposal has resulted in an alarming increase in soil and water contamination in USA. In Pacific Islands (Cook Islands), about 9000 cubic meters of solid waste are dumped in the close proximity of land areas and fresh water bodies, resulting in massive pollution. If tourism keeps on increasing at the current pace, pollution is expected to increase 10-times in the coastal areas (Convard et al. 2005). The use of fertilizers, pesticides, herbicides, tourism and fossil fuel combustion in the Kiribati, Fiji, Nauru, Niue and Marshall Islands release an excessive quantity of solid wastes which are dumped unchecked into the oceans. On the other hand the polluted irrigation waters are continuously used for irrigating croplands (Convard et al. 2005).

5 Commonly Used Extraction Techniques

The heavy metals in the soil may exist in different forms and show different ways of bindings. In uncontaminated soils, HMs are mainly fixed to primary minerals and silicate making relatively immobile species but, in contaminated sites heavy metals are generally more mobile and fixed to other soil phases. In environmental science the assessment of the different ways of binding gives good information on the heavy metal availability together with their toxicity and mobility, as compared to their total concentration. Though an assessment of the different ways of binding is challenging and often impossible,various methods have been used for this purpose in soils, a large number based on contaminant desorption from the solid phase; whereas others focus on contaminant adsorption from solution to the solid phase. Leaching and desorption procedures are the most broadly accepted and applied among these approaches.

In soil science, the extraction procedures based on a single extraction are commonly used. These techniques are aimed to dissolve a phase in which heavy metal concentration correlates with the availability of the elements to plants. This approach is recognized for nutrients and macro elements and it is generally used in fertility science and quality of crops, for assessing deficiency or surplus quantity of the element in the soil, for forecasting essential element uptake. To a lesser extent, these are used for HMs. The usage of extraction techniques for contaminated soils is specifically focused to determine the mobility and potential bioavailability of contaminants.

Different extraction techniques for extractable HMs in soil have been developed and modified in recent years. There are two categories of extraction, i.e. the single extraction method, one extraction solution and on soil sample, whereas in the sequential extraction method, many extraction solutions are used step by step in the same sample. These two kinds of extraction are employed using not only different extraction protocols, but also different laboratory conditions.

Sequential extraction schemes are commonly used to determine the redistribution or partitioning of heavy metals in different chemical forms, i.e. adsorbed (exchangeable), soluble, organic occluded and precipitated (Table 1). With each successive step of the scheme, the bioavailability and solubility of HMs in the soil reduces, though the extraction methods differ (Basta and Gradwohl 2000; Zakir and Shikazono 2011). Specific chemical species measured through chemical extraction procedures have been successfully linked with plant metal uptake in assuming the plant availability of metals in soils (Abedin et al. 2012).

Lead has been associated with pathological changes of organs and damnification of the central nervous system, leading to decrements of intelligence quotients (IQ) in children. It is an abundant toxic metal responsible for soil, water as well as air pollution, and poses a serious threat to humans, animals and plant systems. In humans it is stored in soft tissues, bones and teeth (95 %), affecting nervous system and brain, skeletal, circulatory, enzymatic, endocrine, immune systems and delay in physical and mental development (Kasten-Jolly et al. 2012) (Zhang et al. 2012). It can also cause abdominal cramps, anemia, ataxia, stunted growth, infertility in both genders, high blood pressure, encephalopathy, hepatitis, nephritic syndrome and weakness of the joints if the levels exceed those recommended by WHO (2000, 2011) and the allowable standard of 0.15 mg/L.

Mercury particularly in its organic form, mainly exhibits neurotoxicity and teratogenicity (Jarup 2003; Oskarsson et al. 2004). It can easily enter cell membranes including the brain, where it can interfere with fetal development and cause autism in neonates (Camargo 2002; Shi et al. 2014). Hg has high tendency for binding to proteins and mainly affects the renal and nervous systems (Martínez-Juárez et al. 2012). The outbreak of Minamata disease was caused by consuming the fish rich in mercury leading to Hg poisoning. Higher intake of Hg in food can lead pulmonary

Table 1 Commonly used sequential extraction methods

Extraction techniques	References
BCR-3STEP sequential extraction	Guevara-Riba et al. (2004)
Modified BCR- 3STEP sequential extraction	Cuong and Obbard (2006)
Modified tessier extraction	Esslemont (2000)
Tessier extraction step	Wepener and Vermeulen (2005)
BCR-3STEP sequential extraction	Yuan et al. (2004)
4- Step sequential extraction	Pempkowiak et al. (1999)
3-Step sequential extraction	Usero et al. (1998)
BCR-Sequential step	Morillo et al. (2004)
BCRSEP optimized	Zhang et al. (2003)
BCRSEP optimized	Elass et al. (2004)
BCRSEP	Pueyo et al. (2003)
New SEP developed	Fuentes et al. (2004)
Na_2-EDTA 0.1 M	Mahavi et al. (2005)
Sposito's procedure	Soumia et al. (2005)
Kerstner–Forstner sequential extraction scheme	Katasonov et al. (2005)
H_2O, 0.1 N Ca $(NO_3)_2$ and 0.05 N EDTA	Garau et al. (2007)
Ahnstrom and parker extraction scheme	Silveira et al. (2006)
DTPA/TEA method	Contin et al. (2008)
Deionized water & 0.1 M Ca $(NO_3)_2$ extraction	Sang-Hwan et al. (2009)
DTPA, $CaCl_2$, EDTA & ammonium acetate extraction scheme	Tica et al. (2011)
USEPA TCLP method 1311	Rabindra et al. (2012)
Aqua regia extraction	Sik et al. (2011)

dysfunction (chest pain) and dyspnea (Bernard 2008; Martínez-Juárez et al. 2012; Mathialagan and Viraraghavan 2002). It is lethal if ingested in very minute quantity in any of the available form found in the environment (Shi et al. 2014). The permissible limit of Hg is 2 μg L^{-1} (2 ppb) in drinking water (US EPA 2009). The ingestion of methyl mercury is said to cause delay in the onset of walking, talking and diminish learning ability (US EPA 2004, 2009; Moreno et al. 2005).

The main anthropogenic activities, like waste product discharge, fertilizer applications, mining activities, paint and dye industries are the main sources responsible for the copper release into the environment. It is present in human body in very trace amount and is required for many enzymatic reactions like production of blood hemoglobin in our body. In plants Cu plays a vital role in seed production, disease resistance and regulation of water. Higher human intake of Cu can however, lead to central nervous system irritation, GI disorders, haemolysis, anemia, liver and kidney damage (Wuana and Okieimen 2011).

Arsenic is carcinogenic in nature at higher concentrations, but toxicity depends on the chemical form and solubility. It is available as arsenate and arsenite in nature; reported as the main contaminant originating from anthropogenic activities and ending up with the arsenic pollution (Rathinasabapathi et al. 2006). Human exposure to arsenic can lead to diarrhea, vomiting and abdominal pain, skin darkening, cardiac, circulatory and respiratory diseases. The cancer-causing ability of arsenic is due to mono and dimethyl-arsenous acids, which are genotoxic, leading to the inhibition of DNA repair (Butcher 2009; Scragg 2006).

Chromium can be found in two oxidation states (Cr^{IV} and Cr^{III}). Chromium (VI) can be reduced to Cr (III) by soil organic matter. Major Cr (VI) species include chromate ($CrO_4{}^{2-}$) and dichromate ($Cr_2\,O_7{}^{2-}$). Chromium (III) is the dominant form of Cr at acidic pH (<4), while leachability of Cr (VI) increases at alkaline pH. Cr (VI) is highly soluble in water and causes digestive tract and lung cancer, epigastric pain, nausea, vomiting, severe diarrhea, allergic dermatitis and hemorrhage (Sugasini et al. 2014; Scragg 2006).

Selenium in trace amount is essential as it is part of important enzymes and proteins. However, at higher levels it is lethal. It can incorporate into the amino acids and protein leading to a change in their activity by replacing sulphur (Hamilton 2004; Jarzynska and Falandysz 2011; Milovanović et al. 2014).

The dietary intake of zinc is 15 mg day^{-1}, beyond this level its consumption has toxic effects on humans (Lemire et al. 2008). Zinc, when consumed in large quantities, can cause respiratory and GI disorders, as well as damage to the heart, brain and kidneys (Schwartz et al. 2003). At higher concentrations Zn can cause nausea, epigastric pain, vomiting and diarrhea. Higher intake affects the nervous system and iron balance, leading to the problems with cholesterol metabolism (Mertens et al. 2007; Lemire et al. 2008). It may also increase the acidity of water. The fishes found in the waters polluted with zinc are able to biomagnify it up the food chain.

Nickel is an element that occurs in the environment only at very low levels and is essential in small doses, but it can be dangerous when the maximum tolerable amounts are exceeded. It is mostly used in metal plating industries, combustion of fossil fuels, and nickel mining and electroplating (Khodadoust et al. 2004). Generally released into the environment by power plants and incinerators, which precipitate on the soil surface. Ni mostly affect the soil microbial activities, but usually microorganisms develop resistance to Ni. Its biomagnifications in plants has not been investigated fully.

6 Effects on Plant and Soil Microbial Activities

HMs catalyze different enzymatic and redox reactions, carry electron and are the main component of DNA and RNA metabolism. However, some of these such as Cd and Pb are not necessary for plant growth, especially their high levels can adversely affect plant growth (Miransari 2011). Soils contaminated with Cd, Pb, Hg, As, Cu and Mn can lead to reduction in Urease (50.1 %), Acid phosphatase (47.4 %), Dehydrogenase (39.8 %) in soil (Akmal. et al. 2005; Lee et al. 2002; Oliveira and

Pampulha 2006). The existence of HMs shows deleterious effects on flora, fauna and lead to groundwater pollution through leaching. They reduce the quality of agricultural products and pose a serious threat for public health and biota (Khosravi et al. 2009; Mohsenzadeh and Shahrokhi 2014). The heavy metals can also severely reduce the biomass production, seed, essential oil and leaf area of major crops. Furthermore they can disturb the cell metabolism, cause leaf abscission and senescence, lower the photosynthesis and plant nutrition and hamper the production of enzymes in plants (Chibuike and Obiora 2014; Ghani 2010; Liu et al. 2012; Moradi et al. 2012).

The introduction of HMs in the soil by anthropogenic and natural processes, not only affect the human and plant health, but also the soil ecology. Soil microbial population contributes in nitrogen fixation, assimilation and degradation of organic residues, to release nutrients (fertility) for plant growth and development. An enrichment of HMs in the soil decreases the litter decomposition, nitrogen fixation and nitrogen transformation, decrease in carbon mineralization and fixation, disturb nutrient cycling and impair enzyme synthesis and activity in soil, sediments and water. Soil enzymatic activities are known as receptive and early signs of natural and anthropogenic disturbances. Molecular techniques can provide better results about the qualitative and quantitative measures of microbial diversity in undisturbed and contaminated soils. Soil microbial community is considered as the most sensitive bio-indicator of pollution effects on bioavailability and biogeochemical processes. Long term toxic effects have been reported on biologically mediated soil processes by certain HMs (Lee et al. 2002; Li et al. 2014). The pollutants can reduce the soil microbial biomass, enzyme activities and microbial community structure (Reid et al. 2000, 2001; Keurentjes et al. 2011).

However, the metal exposure also leads to tolerance in the microbial community (Friedlova 2010). Fungi and actinomycetes are more tolerant to heavy metal stresses in polluted soils than bacteria. These metal tolerant microbial populations can be used as potential agents in remediation of the environments contaminated with HMs. The resistant bacteria, which can cope with the higher metal concentration of Pb (2000 ppm & 160 ppm) and Cd (6 ppm), in the soil in lab experiment (Al Gaidi 2010; Sobolev and Begonia 2008). (Zarei et al. 2010) have been isolated. 1 out of 9 AMF from higher Pb and Zn contaminated environments, which can resist higher concentration of Pb and Zn. Such isolates of fungi can prove effective for future remediation processes.)

7 Conventional Approaches for HMs Removal

Every year the concentration of HMs is exceeding in the natural environment (Govindasamy et al. 2011). According to Meers et al. (2010) the atmospheric deposition of Zn, Pb and Cd has contaminated 700 km^2 in the Netherlands and the Champagne region in Belgium. Approximately 3.88×10^6 ha of the area in China is reported to have gotten polluted as a consequence of mining. It also includes an average area of 46,700 ha destroyed annually. These degraded areas, although covered by vegetation, but severe soil contamination is eventually leading to erosion (Xia 2004). Therefore, remediation of heavy metal polluted soils is dire need of the time to reduce the possible effects on ecological systems. Technical and financial implications have made soil remediation a difficult exercise (Barcelo and Poschenrieder 2003; Purkayastha et al. 2014). Over the year, several biological, physical and chemical approaches have been used to achieve this task. The traditional soil cleanup methods include; soil incineration, soil vitrification, landfill and excavation, soil flushing, soil washing, stabilization and electrokinetic solidification (Sheoran et al. 2011; Wuana and Okieimen 2011). Usually, the chemical and physical approaches confront with limitations, i.e. intensive labor, high cost, disturbance of indigenous soil microflora and irreversible changes in soil physicochemical properties. Hence, more intensive studies are required to introduce efficient, cost effective, environmentally sustainable approaches for detoxification of heavy metal contaminated soils. One such an innovative method is phytoremediation which is perceived as a solar energy driven eco-friendly solution to the soil contaminated with heavy metals.

8 Immobilization of HMs Using Soil Amendments

Many efforts have been made in developing remediation technologies to minimize or control, HMs contaminations in soil. Immobilization of toxic heavy metals can be achieved mainly through precipitation, complexation and adsorption reactions which result in the redistribution of metals from solution phase to a solid phase, thus reducing their transport and bioavailability in the soil (Bolan et al. 2003a; Porter et al. 2004).

8.1 Phosphate Compounds

Various studies have reported convincing evidences for remediative significance for both water-soluble (diammonium phosphate, DAP) and water-insoluble (apatite, also known as PR) P compounds to immobilize metals in soils, thus decreasing their bioavailability for plant uptake, as their transport becomes immobile (Bolan et al. 2003b, c). Phosphate compounds increase the immobilization of metals in soils by different processes, such as, direct metal absorption/substitution of phosphate compounds, precipitation of metals with phosphate solution as metal phosphates. An application of phosphate compounds to the soil can cause direct adsorption of metals through increased phosphate anion-induced metal adsorption and increased surface charge of these compounds depending upon the sources.

Precipitation as metal–phosphate has proven one of the main ways for the immobilization of HMs in soils (Bolan et al. 2014). These fairly stable metal–phosphate compounds have extremely low solubility over a wide pH range, which makes phosphate application an attractive technology for managing heavy metal contaminated soils, but phosphate is also a fossil source and high application rates cause eutrophication of water resources. In typical arable soils, precipitation of heavy metals is unlikely, but in modest metal contaminated soils, this process can play a major role in the immobilization of such metals.

Chrysochoou et al. (2007) have reported the potential of apatite to immobilize heavy metals in contaminated soils or solution by metal-phosphate precipitation. The precipitants are usually termed as chloropyromorphite or as hydroxypyromorphite. Two methods have been recommended for the reaction of dissolved Pb with apatite. Firstly, in apatite, Pb (II) can be substituted for Ca (II), thereby, through adsorption of Pb or by dissolution of hydroxyapatite (Ca_{10} $(PO_4)_6$ $(OH)_2$), (Ca, Pb) apatite could be formed followed by co-precipitation of mixed apatite. Secondly, followed by precipitation of pure hydroxypyromorphite (Pb_{10} $(PO_4)_6(OH)_2$), Pb^{2+} can react with apatite through hydroxyapatite dissolution.

A substitute process, which happens to be significant in temperate soils, involves metal-ligand complexation in solution and followed the reduction in cation charge, that most likely decreases adsorption. The formation of soluble Cd-Phosphate complexes reduces Cd (II) sorption in soils in the occurrence of phosphate. Free Cd (II) activity, rather than total dissolved Cd (II) value, is a regulatory factor in Cd (II) sorption. The efficacy of phosphate-induced metal immobilization can be improved by enhancing solubility of phosphate compounds in the soil. The combination of rock phosphate and phosphoric acid, potentially immobilizes Pb. To enhance heavy metal immobilization in soil, phosphate solubilizing bacteria have been used which gradually release phosphate from insoluble phosphate rock (Park et al. 2011a, c). There are some drawbacks related to the usage of phosphate compounds for metal immobilization; after the phosphate treatment, the leaching of phosphate must be considered. A mixture of soil with a phosphate addition in the molar ratio of H_3PO_4: hydroxyapatite of 0.75:1 is suggested to be optimum to reduce phosphorus leaching.

The second problem concerns the influence of bacteria on the stability of pyromorphite. It is assumed that the microbes enhance the dissolution of mineral P, promoting its transformation into pyromorphite. The controversy of the method stems from the need of introducing a living, extraneous strain of bacteria in an uncontrolled environment, although the long-term effect of such a treatment is unknown. Park et al. (2011a, b) showed that PSB (phosphate-solubilizing bacteria) can affect the stability of pyromorphite and that the effectiveness of the process depends on the availability of dissolved phosphates in solution. However, the interaction between microbes and minerals are complex and some aspects of the potential involvement of PSB in remediation treatments remain unclear. There is some evidence that various organic compounds, microbial metabolites and plant activity may increase the dissolution of pyromorphite and cause a secondary Pb release (Formina et al. 2004; Manecki and Maurice 2008; Debela et al. 2010, 2013; Topolska et al. 2013).

8.2 Lime Treatment

Lime is basically used to ameliorate soil acidity, but at the same time it is getting wide acceptance as a potential option among the scientific community to reduce metal poisoning in the soil. Liming materials are in a diverse range with a difference in their potential to neutralize the acidity. Over the years, liming has been employed as a regular traditional practice to decrease levels of heavy metals in the edible parts of agricultural crops. Liming increases sorption of metals by decreasing the H^+ concentration and enhancing negatively charged ions. Lime and red mud as the addition of alkaline material to enhance the concentration of the residual fraction of heavy metals in the contaminated soil. The precipitation of metals results due to an increase in pH by lime and red mud (Garau et al. 2007).

The competition between Ca (II) and metal ions and reduced mobility in soils by precipitation and adsorption result in reducing metal uptake by plants due to the effect of liming on the root surface. An in-situ field trial of heavy metal remediation in contaminated soil proved that by combing red mud and lime soil pH increases and metal bioavailability is reduced, thus resulting in the appearance of a vegetative cover in Pb and Cd contaminated soils (Gray et al. 2006).

Lime can be employed with the combination of other amendments to decrease metal availability is soil amendments. When lime is mixed with biosolids, it decreases electrical conductivity (EC) and enhances pH by precipitation of soluble ions (Fang et al. 2014). Lime also effectively increases pH when lime is added during composting, it decreases leaching and bioavailability of metals (Singh and Kalamdhad 2013).

8.3 Cement-Based Solidification/Stabilization

Soil contaminated with HMs usually need solidification/stabilization (s/s) prior to use, in order to lower the leaching rate and bioavailability. Cement is the most adaptable binder currently available for their immobilization. The selection of cements and operating parameters depends upon an understanding of the chemistry of the system (Chen et al. 2009). Cement-based stabilization/solidification technology is an attractive option for the management of soils polluted with toxic metals to facilitate land use and reduce the release of contaminants into the environment. The efficacy of cement-based solidification/stabilization can be improved by modifying cement phase compositions and controlling temperature, water/solid ratios, particle size, and other factors which affect setting and strength development and long-term durability of solidified waste forms. The potentially dangerous HMs may adversely affect the cementing matrix; pretreatment to render such substances harmless, e.g., by adsorbent adding, is necessary in some cases (Jang and Kim 2000).

Portland cement (15 %, w/w) was used for solidification/stabilization (S/S) in the HMs contaminated soils from the former industrial sites. Soils formed solid monoliths with cement. Concentrations of cadmium and lead in water extracts from S/S soils generally decreased. Their concentrations in the TCLP extracts from S/S soils were lower than from original soils (Voglara and Leˇstan 2010). Formulations of 15% (w/w) of ordinary Portland cement (OPC), calcium aluminate cement (CAC) and pozzolanic cement (PC) and additives: plasticizers cementol delta extra (PCDE) and cementol antikorodin (PCA), polypropylene fibers (PPF), polyoxyethylene-sorbitan monooleate (Tween 80) and aqueous acrylic polymer dispersion (Akrimal) were used for solidification/stabilization (S/S) of soils from an industrial Brownfield contaminated with up to 157, 32 mg kg−1 of cadmium and lead respectively. Based on the model calculation, the most efficient S/S formulation was CAC+Akrimal, which reduced soil leachability of cadmium and lead into deionized water below the limit of quantification and into TCLP solution by up to 55 and 185 times, respectively; and the mass transfer of elements from soil monoliths by up to 740, 746, 104,000, 4.7, 343 and 181-times respectively (Voglara and Leˇstan 2010). It appears that there are several attractive areas for further investigation in the field of cement-based solidification/stabilization of cadmium and lead in contaminated soils. For example, the phased development of cement pastes in the presence of cadmium and lead, the dissolution and precipitation of stoichiometries phases and of solid solutions, and surface phenomena controlling cadmium and lead immobilization.

8.4 Animal Manure and Biosolids

Animal manures and biosolids are the significant sources of organic composts. Biosolids are considered traditionally as one of the key sources of metal contamination in the soil. Recent developments in the industrial wastewater and sewage treatment technologies have successfully decreased metal contamination in biosolids. Moreover, alkaline materials are used for stabilizing metals in biosolids. Zeolites are also valuable as metal scavengers in the metal contaminated biosolids. Heavy metals are stabilized by natural zeolites which transform them in the exchangeable and carbonate to residual fractions (Zorpas et al. 2000).

Cattle, poultry and dairy are key sources of global animal manure byproducts. Majority of the manure products are less contaminated with metals except arsenic in poultry and, zinc and copper in swine manure respectively. However, recent developments in the manure as a byproduct treatment has brought a reduction in the bioavailability of metals. For example, treatment of poultry manure with alum [$Al_2(SO_4)_3$] has reduced the concentration of water soluble cadmium. Low metal contaminated manure byproducts can be employed in stabilizing metals in the soil. Various studies have assessed the role of biosolids in the HM contamination in the soil, only few studies have documented the advantageous impact of organic amendment as a sink of the stabilization of toxic metals in the soil (Brown et al. 2003; Li et al. 2000). Alkaline-stabilized biosolids are known as designer sludge with exceptional quality of low metal bioavailability, and can be employed as potential sinks for decreasing metal bioavailability in the sediments and soil. Complexation, redox reaction and adsorption are the immobilization processes accomplished through such amendments.

Metals make both soluble and insoluble complexes with organic constituents in the soil, the mechanisms specially depend on the type of the organic matter. The organic constituent of soil component has great affiliation with metal cations due to the existence of functional groups or legends which can form chelates with metals. The alcoholic, phenolic, carbonyl and carboxyl functional groups dissociate due to increasing pH, thus enhancing legand ions affinity towards metal cations. Through the prevention of sulfide oxidation/hydrolysis, the decrease in the metal concentrations can be credited to an increase in soil pH (Hartley et al. 2004). Likewise, compost amendment has increased the growth of *Lupinus albus* (white lupin) together with a decrease in the uptake of lead by reducing metal bioavailability in the soils (Castaldi et al. 2005). High fulvic and high compost humic acid concentrations are credited for the high metal binding capacities of compost (O'Dell et al. 2007; Perminova and Hatfield 2005).

8.5 Oxides of Metals

The oxides of iron, aluminium and manganese play an important role in metal geochemistry in soils. Atmospheric nature and highly active surface area make them potential for immobilization and sorption of diverse soil contaminants. Co-precipitation, forming inner-surface complexes and specific sorption result in a strong metal binding by metal oxides. Similarly synthesized industrial by-products and naturally occurring oxides have been worked upon for their potentiality to be employed for soil remediation. Recent developments in the application of metal oxides and its precursors for chemical immobilization of metals in contaminated soils have been investigated at length (Komarek et al. 2013).

Oxides of manganese (birnessite and a phyllomanganates group of minerals), oxyhydroxides (goethite, ferrihydrite, lepidocrocite, feroxyhite and akaganeite) and oxides of iron (magnetite, hematite and maghemite) mostly occur in soil. $AsO_4{}^{3-}$ and Pb (II) form inner sphere surface complexes with hydrous ferric oxide and lead and cadmium form mononuclear complexes on goethite and ferrihydrite surfaces (Knox et al. 2001). The surfaces of oxides of iron hydrous have a significant role to play in metal retention. The lead and cadmium have been immobilized with the amalgamation of iron (II) and (III) sulfates (Hartley et al. 2004). The useful re-use of iron oxides-based residue of drinking water treatment may be a beneficial soil amendment for various cations and anions. Oxides of manganese exist in soil, i.e. cryptomelane, todorokite, hausmannite and birnessite. Out of these birnessite exhibits the highest adsorption capacity on cadmium and lead among all. The sequence of greatest sorption capacity by birnessite was Pb (II)>Cd(II) and the maximum adsorption of metals on birnessite has been attributed to hydrolysis constants which show metal adsorption through manganese oxides present primarily in the form of hydroxylation cations (Feng et al. 2007). Oxides of manganese (IV) are well-known to precisely adsorb lead. Cryptomelane is representative of manganese oxides and has been employed to decrease lead bioavailability (Hettiarachchi et al. 2000). Fungi and bacteria enzymatically oxidize manganese (II), produce insoluble oxides of Mn (III, IV) and form biogenic manganese oxides with great sorption potential for metals including their oxidation. It takes place through complexation below or above, incorporation into the vacancies and at the structural vacancy sheet, edges by isomorphical substitution for lead by biogenic manganese oxide produced by bacteria (Miyata et al. 2007).

Nano particle sized oxides of metals for examples the oxides of manganese, ferric, titanium, aluminum, cerium and magnesium are referred to as potential adsorbents for metal remediation due to their high activities and surface area (Hua et al. 2012). Ammonium acetate fuel was used for the synthesis of nanosized alumina and is effective in the lead adsorption in soil, but very few studies deal with the use of metal oxides in soil environments (Rahmani et al. 2010).

8.6 *Biochar*

Recent development in scientific research has proved that biochar has the capacity to adsorb metals in contaminated soils and serves as green environmental sorbent. Various parameters are involved in assessing the main role of biochar for environmental protection. Feedstock type and pyrolysis condition are the important factors affecting the sorption capacity of biochars (Joseph et al. 2010; Kookana et al. 2011). Some studies have documented the reduction of cadmium and lead in contaminated soils amended with biochar. In view of this there is need for an assessment of the biochar efficiency in the immobilization of contaminants in multi-metal contaminated soils including the metal immobilization mechanism of biochar to examine the long-term efficacy. Hardwood-derived biochar is effective in immobilizing HMs in contaminated soils at an alkaline pH induced by biochar which then effectively decreases heavy metal solubility (Bees et al. 2010). However, biochar produced at low pyrolysis temperature regimes supports the heavy metal immobilization (Uchimiya et al. 2011). Remarkable heavy metal immobilization by biochar produced at low temperature resulted in the augmented release of available calcium, potassium and phosphorus in the soil environment. Biochar derived by dairy manure has a great quantity of available phosphorus which stabilizes lead by making insoluble hydroxypyromorphite ($Pb_5 (PO_4)_3(OH)$) in soil. The cotton seed hull-derived biochar produced at 350°C with high oxygen concentration results in high uptake of HMs based on the role of O-containing functional groups on biochar surfaces in metal binding (Uchimiya et al. 2011).

9 Conclusion and Future Prospects

Unlike organic contaminants, HMs do not undergo microbial or chemical degradation and persist for a long time after their introduction in to the environment. Their bioavailability plays a vital role in the remediation of contaminated soils. In this chapter, the remediation of heavy metal contaminated soils by manipulating their bioavailability using a range of soil amendments has been discussed. Immobilizing amendments such as precipitating agents and sorbent materials decrease the bioavailability of heavy metals. Immobilizing agents can be used to reduce the transfer of heavy metals to food chain via plant uptake and leaching to groundwater. One of the major limitations of the immobilization technique is the long-term stability of the immobilized heavy metals which need regular monitoring.

The main crux of the current review this chapter enlightens the facts related to the immobilization options with regard to engineering of the bioavailability of lead and cadmium heavy metals followed by the remediation of polluted soil. The future prospects in this connection are that more research is required related to the influence of immobilizing agents on the physiological processes of the plants, in particular photosynthetic activity, hydrolysis, uptake of nutrients by plants, tropisms, development of plant hormones including their functions, nastic movements, photomorphogenesis, circadian rhythms, photoperiodism, plant responses to environmental stress, seed dormancy, germination, osmotic pressure, turgor potential and stomatal function in relation to plant-water nexus. The other aspects which need to be studies are; chelate-assisted phytoremediation and the root–soil interface. The phytotoxicity research is needed to focus on the effects of heavy metal contaminated soils; amended with solidifying agents; on agricultural species under both filed and laboratory conditions as well as plant growth and the activity of enzymes required for photosynthesis, cell division and respiration, water absorption and transpiration, and chlorophyll, carotenoids, and adenosine triphosphate (ATP) synthesis. More in-situ studies are needed to validate the significance of diverse ranges of immobilizing soil amendments to restore contaminated soils. These field-based experiments are essential to assess the effects of soil amendments on co-contaminants. There is a dire need to develop procedures to determine immobilization efficacy that could be used to assess the in situ short and long-term environmental stability of metal immobilization. Additional in situ studies should be undertaken to determine phyto-poisoning, ecological receptors end points and human biomagnification to demonstrate the heavy metal risk management resulting from soil amendments.

References

Abedin J, Beckett P, Spiers G (2012) An evaluation of extractants for assessment of metal phytoavailability to guide reclamation practices in acidic soils capes in northern regions. Can J Soil Sci 92:253–268

Adriano DC (2001) Trace elements in terrestrial environments: biogeochemistry, bioavailability and risks of metals, 2nd edn. Springer, New York

Akmal M, Xu J, Li Z, Wang H, Yao H (2005) Effects of lead and cadmium nitrate on biomass and substrate utilization pattern of soil microbial communities. Chemosphere 60(4):508–514. doi:http://dx.doi.org/10.1016/j.chemosphere.2005.01.001

Al GA (2010) Effect of heavy metals on soil microbial processes and population. Egypt Acad J biolog Sci 2(2):9–14

Ashraf M, Ozturk M, Ahmad MSA (eds) (2010) Plant adaptation and phytoremediation. Science+Business Media, New York, 481 pp

Ashraf MY, Roohi M, Iqbal Z, Ashraf M, Ozturk M, Gücel S (2015) Cadmium (Cd) and Lead (Pb) induced inhibition in growth and alteration in some biochemical attributes and mineral accumulation in Mung bean [*Vigna radiata* (L.) Wilczek]. Communications in soil science and plant analysis-Taylor & Francis ISSN: 0010–3624 (Print) 1532–2416 (Online)

Aziz H, Sabir M, Ahmad HR, Aziz T, Zia-ur-Rehman M, Hakeem KR, Ozturk M (2015) Alleviating effect of calcium on nickel toxicity in rice. CLEAN-Soil Air Water 43(6):901–909

Bakirdere S, Yaman M (2008) Determination of lead, cadmium and copper in roadside soil and plants in Elazig, Turkey. Environ Monit Assess 136(1–3):401–410

Barcelo J, Poschenrieder C (2003) Phytoremediation: principles and perspectives. Contrib Sci 2:333–344

Basta NT, Gradwohl R (2000) Estimation of Cd, Pb, and Zn bioavailability in smelter contaminated soils by a sequential extraction procedure. J Soil Contam 9:149–164

Basta NT, Ryan JA, Chaney RL (2005) Trace element chemistry in residual-treated soil: key concepts and metal bioavailability. J Environ Qual 34(1):49–63

Bees L, Moreno-Jimenez E, Gomez-Eyles JL (2010) Effects of biochar and green waste compost amendments on mobility, bioavailability and toxicity of inorganic and organic contaminants in a multi-element polluted soil. Environ Pollut 158:2282–2287

Biester H, Bindler R, Martinez-Cortizas A, Engstrom DR (2007) Modeling the past atmospheric deposition of mercury using natural archives. Environ Sci Technol 41(14):4851–4860

Bernard A (2008) Cadmium & its adverse effects on human health. Indian J Med Res 128(4):557–564

Bharti S, Banerjee TK (2012) Phytoremediation of the coalmine effluent. Ecotoxicol Environ Saf 81(0):36–42. doi:http://dx.doi.org/10.1016/j.ecoenv.2012.04.009

Bhattacharya S, Gupta K, Debnath S, Ghosh UC, Chattopadhyay D, Mukhopadhyay A (2012) Arsenic bioaccumulation in rice and edible plants and subsequent transmission through food chain in Bengal basin: a review of the perspectives for environmental health. Toxicol Environ Chem 94:429–441

Bhattacharya S, Das A, Prashanthi K, Palaniswamy M, Angayarkanni J (2014) Mycoremediation of Benzo[a]pyrene by pleurotus ostreatus in the presence of heavy metals and mediators. 3 Biotech 4(2):205–211. doi:10.1007/s13205-013-0148-y

Bizily SP, Rugh CL, Meagher RB (2000) Phytodetoxification of hazardous organomercurials by genetically engineered plants. Nat Biotechnol 18(2):213–217

Bolan NS, Adriano DC, Naidu R (2003a) Role of phosphorus in (im) mobilization and bioavailability of heavy metals in the soil–plant system. Rev Environ Contam Toxicol 177:1–44

Bolan NS, Adriano DC, Duraisamy A, Mani P (2003b) Immobilization and phytoavailability of cadmium in variable charge soils. III. Effect of biosolid compost addition. Plan Soil 256:231–241

Bolan NS, Adriano DC, Mani P, Duraisamy A, Arulmozhiselvan S (2003c) Immobilization and phytoavailability of cadmium in variable charge soils: I. Effect of phosphate addition. Plant Soil 250:83–94

Bolan N, Kunhikrishnan A, Thangarajana R, Kumpiene J, Park J, Makino T, Kirkham MB, Scheckel K (2014) Remediation of heavy metal (loid)s contaminated soils – To mobilizeor to immobilize? J Hazard Mater 266:141–166

Brown S, Chaney RL, Hallfrisch JG, Xue Q (2003) Effect of biosolids processing on lead bioavailability in an urban soil. J Environ Qual 32:100–108

Butcher DJ (2009) Phytoremediation of arsenic: fundamental studies, practical applications, and future prospects. Appl Spectrosc Rev 44(6):534–551
Boularbah A, Schwartz C, Bitton G, Aboudrar W, Ouhammou A, Morel JL (2006) Heavy metal contamination from mining sites in South Morocco: 2. Assessment of metal accumulation and toxicity in plants. Chemosphere 63(5):811–817
Camargo JA (2002) Contribution of Spanish–American silver mines (1570–1820) to the present high mercury concentrations in the global environment: a review. Chemosphere 48(1):51–57
Castaldi P, Santona L, Melis P (2005) Heavy metal immobilization by chemical amendments in a polluted soil and influence on white lupin growth. Chemosphere 60:365–371
Chen S, Xu M, Ma Y, Yang J (2009) Evaluation of different phosphate amendments on availability of metals in contaminated soil. Ecotoxicol Environ Saf 6:278–285
Chen X, Xia X, ZhaoY PZ (2010) Heavy metal concentrations in roadside soils and correlation with urban traffic in Beijing, China. J Hazard Mater 181:640–646
Cheng H, Hu Y (2010) Lead (Pb) isotopic fingerprinting and its applications in lead pollution studies in China: a review. Environ Pollut 158(5):1134–1146
Chibuike GU, Obiora SC (2014) Heavy metal polluted soils: Effect on plants and bioremediation methods. Appl Environ Soil Sci 2014:1–12. doi:10.1155/2014/752708
Chrysochoou M, Dermatas D, Grubb DG (2007) Phosphate application to firing range soils for Pb immobilization: the unclear role of phosphate. J Hazard Mater 144:1–14
Churchill R, Meathrel C, Suter P (2004) A retrospective assessment of gold mining in the Reedy Creek sub-catchment, northeast Victoria, Australia: residual mercury contamination 100 years later. Environ Pollut 132(2):355–363
Clemens S (2006) Toxic metal accumulation, responses to exposure and mechanisms of tolerance in plants. Biochimie 88(11):1707–1719
Convard, Nancy SA, Tomlinson A, Welch C (2005) Strategies for preventing and mitigating land-based sources of pollution to trans-boundary water resources in the Pacific region, SPREP
Contin M, Mondini C, Leita L, Zaccheo P, Crippa L, Nobili MD (2008) Immobilization of soil toxic metals by repeated additions of Fe (II) sulphate solution. Geoderma 147:133–140
Cursino L, Mattos SVM, Silva NO, Chartone-Souza E, Nascimento AMA (2003) Measurement of volatilized mercury by a mini-system: a simple, reliable and reproducible technique. Bras Arch Biol Technol 46(4)
Cuong TD, Obbard JP (2006) Metal speciation in coastal marine sediments from Singapore from Singapore using a modified BCR- sequential extraction procedure. Appl Geochem 21:1335–1346
DalCorso G, Farinati S, Furini A (2010) Regulatory networks of cadmium stress in plants. Plant Signal Behav 5(6):663
Debela F, Arocena JM, Thring RW, Withcombe T (2010) Organic acid-induced release of lead from pyromorphite and its relevance to reclamation of Pb-contaminated soils. Chemosphere 80:450–456
Debela F, Arocena JM, Thring RW, Withcombe T (2013) Organic acids inhibit the formation of pyromorphite and Zn-phosphate in phosphorous amended Pb- and Zn-contaminated soil. J Environ Manag 116:156–162
DeVolder PS, Brown SL, Hesterberg D, Pandya K (2003) Metal bioavailability and speciation in a wetland tailings repository amended with biosolids compost, wood ash, and sulfate. J Environ Qual 32(3):851–864
de Lacerda LD, dos Santos AF, Marins RV (2007) Emissão de mercúrio para a atmosfera pela queima de gás natural no Brasil. Química Nova 30(2):366
Dermont G, Bergeron M, Mercier G, Richer-Laflèche M (2008) Metal-contaminated soils: remediation practices and treatment technologies. Pract Period Hazard Toxic Radioact Waste Manage 12(3):188–209
Elass K, Laachach A, Azzi M (2004) Three-stage sequential extraction procedure for metal partitioning in polluted soils and sediments. Ann Chim-Rome 94:325–332

El-Shahawi M, Bashammakh A, Orief M, Alsibaai A, Al-Harbi E (2014) Separation and determination of cadmium in water by foam column prior to inductively coupled plasma optical emission spectrometry. J Ind Eng Chem 20(1):308–314

Esslemont G (2000) Heavy metals in seawater, marine sediments and corals from the Townsville section, Great Barrier Reef Marine Park. Qld Mar Chem 71:215–231

Fang Y, Sun X, Yang W et al (2014) Concentrations and health risks of lead, cadmium, arsenic, and mercury in rice and edible mushrooms in China. Food Chem 147(0):147–151. doi:http://dx.doi.org/10.1016/j.foodchem.2013.09.116

Farmaki E, Thomaidis N (2008) Current status of the metal pollution of the environment of Greece- a review. Global nest Int J 10(3):366–375

Feng XH, Zhai LM, Tan WF, Liu F, He JZ (2007) Adsorption and redox reactions of heavy metals on synthesized Mn oxide minerals. Environ Pollut 147:366–373

Formina M, Alexander IJ, Hillier S, Gadd GM (2004) Zinc phosphate and pyromorphite solubilization by soil plant-symbiotic fungi. Geomicrob J 21:351–366

Friedlova M (2010) The influence of heavy metals on soil biological and chemical properties. Soil Water Res 5(1):21–27

Fuentes E, Pinochet H, Potin-Gautier M, Graegori ID (2004) Fractionation and redox speciation of antimony in agricultural soils by hydride generation- -atomic fluorescence spectrometry and stability of Sb (III) and Sb (V) during extraction with different extractant solutions. J AOAC Int 87:60–67

Gallego SM, Pena LB, Barcia RA et al (2012) Unravelling cadmium toxicity and tolerance in plants: insight into regulatory mechanisms. Environ Exp Bot 83:33–46

Garau G, Castaldi P, Santona L, Deiana P, Melis P (2007) Influence of red mud, zeolite and lime on heavy metal immobilization, culturable heterotrophic microbial populations and enzyme activities in a contaminated soil. Geoderma 142:47–57

Ghani A (2010) Toxic effects of heavy metals on plant growth and metal accumulation in maize (Zea maysL.). Iran J Toxicol 4(3):325–334

Govil PK, Sorlie JE, Murthy NN, Sujatha D, Reddy GL, Rudolph-Lund K (2008). Soil contamination of heavy metals in the Katedan Industrial Development Area, Hyderabad, India. Environmental Monitoring Assessment 140 ((1–3)):313–323

Govindasamy C, Arulpriya M, Ruban P, Francisca LJ, Ilayaraja A (2011) Concentration of heavy metals in seagrasses tissue of the Palk Strait, Bay of Bengal. Int J Environ Sci 2:145–153

Granero S, Doming JL (2002) Levels of metals in soils of Alcalá de Henares, Spain: human health risks. Environ Int 28(3):159–164. doi:http://dx.doi.org/10.1016/S0160-4120(02)00024-7

Gray C, Dunham S, Dennis P, Zhao F, McGrath S (2006) Field evaluation of in situ remediation of a heavy metal contaminated soil using lime and red-mud. Environ Pollut 142:530–539

Groppa M, Ianuzzo M, Rosales E, Vázquez S, Benavides M (2012) Cadmium modulates NADPH oxidase activity and expression in sunflower leaves. Biologia Plant 56(1):167–171

Gucel S, Ozturk M, Yucel E, Kadis C, Guvensen A (2009a) Studies on the trace metals in the soils and plants growing in the vicinity of copper mining area- Lefke, Northern Cyprus. Fresenius Environ Bull 18(3):360–368

Gucel S, Kocbas F, Ozturk M (2009b) Metal bioaccumulation by barley in Mesaoria Plain alongside the Nicosiafamagusta Highway, Northern Cyprus. Fresenius Environ Bull 18(11):2034–2039

Guevara-Riba A, Sahuquillo A, Rubio R, Rauret G (2004) Assessment of metal mobility in dredged harbour sediments from Barcelona. Spain Sci Total Environ 321:241–255

Hakeem KR, Sabir M, Ozturk M, Mermut A (2015) Soil remediation and plants: prospects and challenges. Academic Press, Elsevier, London, Waltham, 724 pp

Hamilton SJ (2004) Review of selenium toxicity in the aquatic food chain. Sci Total Environ 326(1–3):1–31. doi:http://dx.doi.org/10.1016/j.scitotenv.2004.01.019

Hamman S (2004) Bioremediation capabilities of white rot fungi. BI570 – review article. J Mol Biol 23(2):234–238

Hang Zhou XZ, Zeng M, Liao B-H, Liu L, Yang W-T, Wu Y-M, Qiu Q-Y, Wang Y-J (2014) Effects of combined amendments on heavy metal accumulation in rice (*Oryza sativa* L.) planted on contaminated paddy soil. Ecotoxicol Environ Saf 101:226–232

Hartley W, Edwards R, Lepp NW (2004) Arsenic and heavy metal mobility in iron oxide-amended contaminated soils as evaluated by short-and long-term leaching tests. Environ Pollut 131:495–504

Hasanuzzaman M, Nahar K, Alam Md M, Bhowmik PC, Hossain Md A, Rahman MM, Prasad MNV, Ozturk M, Fujita M (2014) Potential use of halophytes to remediate saline soils. Hindawi Publishing Corporation, BioMed Research International, Article ID 589341, pp:12

Hasanuzzaman M, Nahar K, Hakeem KR, Ozturk M, Fujita M (2015) Arsenic toxicity in plants and possible remediation. Chapter 16. In: Hakeem et al (eds) Soil remediation and plants: prospects and challenges, Academic Press, Elsevier, New York-Waltham, pp 433–501

Hettiarachchi GM, Pierzynski GM, Ransom MD (2000) In situ stabilization of soil Pb using phosphorus and manganese oxide. Environ Sci Technol 34:4614–4619

Hua M, Zhang S, Pan B, Zhang W, Lv L, Zhang Q (2012) Heavy metal removal from water/wastewater by nanosized metal oxides: a review. J Hazard Mater 211:317–331

Huang SS, Liao QL, Hua M (2007) Survey of heavy metal pollution and assessment of agricultural soil in Yangzhong district, Jiangsu Province, China. Chemosphere 67(11):2148–2155. doi:http://dx.doi.org/10.1016/j.chemosphere.2006.12.043

Issaro N, Besancon S, Bermond A (2010) Thermodynamic and kinetic study of the single extraction of mercury from soil using sodium-thiosulfate. Talanta 82(5):1659–1667

Jang A, Kim INS (2000) Technical Note: Solidification and stabilization of Pb, Zn, Cd and Cu in tailing waste using cement and fly ash. Min Eng 13:1659–1662

Jarup L (2003) Hazards of heavy metal contamination. Br Med Bull 68:167–182

Jarzynska G, Falandysz J (2011) Selenium and 17 other largely essential and toxic metals in muscle and organ meats of Red Deer (*Cervus elaphus*)--consequences to human health. Environ Int 37(5):882–888. doi:10.1016/j.envint.2011.02.017

Johri N, Jacquillet G, Unwin R (2010) Heavy metal poisoning: the effects of cadmium on the kidney. Biometals 23(5):783–792

Joseph S, Camps-Arbestain M, Lin Y, Munroe P, Chia C, Hook J, Singh B (2010) An investigation into the reactions of biochar in soil. Soil Res 48:501–515

Kabata-Pendias A, Mukherjee AB (2007) Trace elements from soil to human. Springer, Berlin

Kasten-Jolly J, Pabello N, Bolivar VJ, Lawrence DA (2012) Developmental lead effects on behavior and brain gene expression in male and female BALB/cAnNTac mice. Neurotoxicology 33(5):1005–1020

Katasonov ON, Fedotov PS, Karandashev VK, Spivakov BY (2005) Application of rotating coiled columns to the fractionation of soil particles and to the sequential extraction of heavy-metal species from silty, dusty, and sandy fractions. J Anal Chem 60:684–690

Keurentjes JJB, Angenent GC, Dicke M, van der Putten WH, de Ruiter PC, Struik PC et al (2011) Redefining plant systems biology: from cell to ecosystem. Trend Plant Sci 16:183–190

Khan MN, Wasim AA, Sarwar A, Rasheed MF (2011) Assessment of heavy metal toxicants in the roadside soil along the N-5, National Highway, Pakistan. Environ Monit Assess 182(1–4):587–595

Khosravi F, Savaghebi GH, Farah BH (2009) Effect of Potassium chloride on Cd uptake by colza in a polluted soil. Water Soil J 23:28–35

Khodadoust AP, Reddy KR, Maturi K (2004) Removal of nickel and phenanthrene from kaolin soil using different extractants. Environ Eng Sci 21(6):691–704

Kirkham M (2006) Cadmium in plants on polluted soils: Effects of soil factors, hyperaccumulation, and amendments. Geoderma 137(1):19–32

Kirpichtchikova TA, Manceau A, Spadini L, Panfili F, Marcus MA, Jacquet T (2006) Speciation and solubility of heavy metals in contaminated soil using X-ray microfluorescence, EXAFS spectroscopy, chemical extraction, and thermody namic modeling. Geochim Cosmochim Acta 70(9):2163–2190

Knopf B, Konig H (2010) Biomethylation of heavy metals in soil and terrestrial invertebrates: soil heavy metals. Springer, Berlin/Heidelberg, pp 315–328

Knox A, Seaman J, Mench M, Vangronsveld J (2001) Remediation of metal-and radionuclides-contaminated soils by in situ stabilization techniques. In: Iskandar IK (ed) Environmental restoration of metals-contaminated soil. Lewis Publishers, Boca Raton, pp 21–61

Komarek M, Vanˇek A, Ettler V (2013) Chemical stabilization of metals and arsenic in contaminated soils using oxides – a review. Environ Pollut 172:9–22

Kookana R, Sarmah A, Van Zwieten L, Krull E, Singh B (2011) Biochar application to soil: agronomic and environmental benefits and unintended consequences. Adv Agron 112:103–143

Kumpiene J, Lagerkvist A, Maurice C (2008) Stabilization of As, Cr, Cu, Pb and Zn in soil using amendments – A review. Waste Manag 28:215–225

Lan Z, Bi KS, Chen XH (2014) Ligustrazine attenuates elevated levels of indoxyl sulfate, kidney injury molecule-1 and clusterin in rats exposed to cadmium. Food Chem Toxicol 63:62–68

Lee I-S, Kim OK, Chang Y-Y, Bae B, Kim HH, Baek KH (2002) Heavy metal concentrations and enzyme activities in soil from a contaminated Korean shooting range. J Biosci Bioeng 94(5):406–411. doi:http://dx.doi.org/10.1016/S1389-1723(02)80217-1

Lemire J, Mailloux R, Appanna VD (2008) Zinc toxicity alters mitochondrial metabolism and leads to decreased ATP production in hepatocytes. J Appl Toxicol 28(2):175–182

Liang J, Shohag MJ, Yang X (2014) Role of sulfur assimilation pathway in cadmium hyperaccumulation by Sedum alfredii Hance. Ecotoxicol Environ Saf 100:159–165. doi:10.1016/j.ecoenv.2013.10.026

Li YM, Chaney RL, Siebielec G, Kerschner BA (2000) Response of four turf grass cultivars to limestone and biosolids-compost amendment of a zinc and cadmium contaminated soil at Palmerton. Pa J Environ Qual 29:1440–1447

Li Z, Feng X, Li G (2013) Distributions, sources and pollution status of 17 trace metal/metalloids in the street dust of a heavily industrialized city of Central China. Environ Pollut 182:408–416

Li C, Jiang W, Ma N (2014) Bioaccumulation of cadmium by growing *Zygosaccharomyces rouxii* and *Saccharomyces cerevisiae*. Bioresour Technol 155(0):116–121. doi:http://dx.doi.org/10.1016/j.biortech.2013.12.098

Liu XM, Wu QT, Banks MK (2005) Effect of simultaneous establishment of Sedum alfridii and Zea mays on heavy metal accumulation in plants. Int J Phytoremediation 7:43–53

Liu S, Yang C, Xie W, Xia C, Fan P (2012) The effects of cadmium on germination and seedling growth of suaeda salsa. Procedia Environmental Sciences 16(0):293–298. doi:http://dx.doi.org/10.1016/j.proenv.2012.10.041

Lohman K, Seigneur C, Gustin M, Lindberg S (2008) Sensitivity of the global atmospheric cycle of mercury to emissions. Appl Geochem 23(3):454–466

Lone MI, He ZL, Stoffella PJ, Xiao EY (2008) Phytoremediation of heavy metal polluted soils and water: progresses and perspectives. J Zhejiang Univ Sci B 9(3):210–220

Luo LMY, Zhang S, Wei D, Zhu YG (2009) Inventory of trace element inputs to agricultural soils in China. J Environ Manag 90(8):2524–2530

Mahavi AH, Mesdaghinia AR, Naghipoor D (2005) Comparison of heavy metals extractions efficiency in contaminated soils by various concentration of EDTA. Pak J Bio Sci 8:1081–1085

Manecki M, Maurice PA (2008) Siderophore promoted dissolution of pyromorphite. Soil Sci 173:821–830

Martínez-Juárez VM, Cárdenas-González JF, Torre-Bouscoulet ME, Acosta-Rodríguez I (2012) Biosorption of mercury (II) from aqueous solutions onto fungal biomass. Bioinorg Chem Appl 2012:156190. doi:10.1155/2012/156190

Mathialagan T, Viraraghavan T (2002) Adsorption of cadmium from aqueous solutions by perlite. J Hazard Mater 94(3(14)):291–303

McGrath SP, Zhao FJ, Lombi E (2001) Plant and rhizoshphere processes involved in phytoremediation of metal-contaminated soils. Plant Soil 232:207–214

Meers E, Slycken SV, Adriaensen K, Ruttens A, Vangronsveld J, Laing GD, Witters N, Thewys T, Tack FMG (2010) The use of bio-energy crops (*Zea mays*) for 'phytoremediation' of heavy metals on moderately contaminated soils: a field experiment. Chemosphere 78:35–41

Mertens J, Degryse F, Springael D, Smolders E (2007) Zinc toxicity to nitrification in soil and soilless culture can be predicted with the same biotic ligand model. Environ Sci Technol 41(8):2992–2997

Micaroni R, Bueno M, WdF JARDIM (2000) Compostos de mercúrio. Revisão de métodos de determinação, tratamento e descarte. Química Nova 23(4):487–495

Milovanović I, Brčeski I, Stajić M, Korać A, Vukojević J, Knežević A (2014) Potential of pleurotus ostreatus mycelium for selenium absorption. Scientific World Journal 2014:681834. doi:10.1155/2014/681834

Miransari M (2011) Hyperaccumulators, arbuscular mycorrhizal fungi and stress of heavy metals. Biotechnol Adv 29(6):645–653. doi:http://dx.doi.org/10.1016/j.biotechadv.2011.04.006

Mishra M, Acharya UR (2004) Protective action of vitamins on the spermatogenesis in lead-treated Swiss mice. J Trace Elem Med Biol 18(2):173–178

Miyata N, Tani Y, Sakata M, Iwahori K (2007) Microbial manganese oxide formation and interaction with toxic metal ions. J Biosci Bioeng 104:1–8

Mohsenzadeh F, Shahrokhi F (2014) Biological removing of Cadmium from contaminated media by fungal biomass of Trichoderma species. J Environ Health Sci Eng 12:102. doi:10.1186/2052-336X-12-102

Moradi SA, Moghaddam PR, Koocheki AR, Danesh S, Fotovat A (2012) Effect of cadmium and lead on quantitative and essential oil traits of peppermint (*Mentha piperita* L.). Not Sci Biol 4(4):101–109

Moreno FN, Anderson CW, Stewart RB, Robinson BH (2005) Mercury volatilisation and phytoextraction from base-metal mine tailings. Environ Pollut 136(2):341–352

Morillo J, Usero J, Gracia I (2004) Heavy metal distribution in marine sediments from the southwest coast of Spain. Chemosphere 55:431–442

Mulligan CN, Yong RN, Gibbs BF (2001) Remediation technologies for metal-contaminated soils and groundwater: an evaluation. Eng Geol 60:193–207

Nagajyoti PC, Lee KD, Sreekanth TVM (2010) Heavy metals, occurrence and toxicity for plants: a review. Environ Chem Lett 8: 199e216 8, 199e216

O'Dell R, Silk W, Green P, Claassen V (2007) Compost amendment of Cu–Zn mine spoil reduced toxic bioavailable heavy metal concentrations and promotes establishment and biomass production of Bromus carinatus (Hook and Arn.). Environ Pollut 148:115–124

Ok YS, Lee H, Jung J, Song H, Chung N, Lim S (2004) Chemical characterization and bioavailability of cadmium in artificially and naturally contaminated soils. Agri Chem Biotech 47:143–146

Oliveira A, Pampulha ME (2006) Effects of long-term heavy metal contamination on soil microbial characteristics. J Biosci Bioeng 102(3):157–161. doi:http://dx.doi.org/10.1263/jbb.102.157

Ozturk M (2010a) Natural plant systems in the renovation of wastewaters (Part I). Refresher course on "Trace elements in the environment: contamination cleanup to phytoproducts", June 11–17, Istanbul-Turkey

Ozturk M (2010b) Natural plant systems in the renovation of wastewaters (Part II). Refresher course on "trace elements in the environment: contamination cleanup to phytoproducts", June 11–17, Istanbul-Turkey.

Ozturk M, Celik A, Yarci C, Aksoy A, Feoli E (2002) An overview of plant diversity, land use and degradation in the Mediterranean region of Turkey. Environ Manag Health 13(5):442–449

Ozturk M, Ozcelik H, Sakcali MS, Guvensen A (2004) Land degradation problems in the euphrates basin. Turkey Environews 10(3):7–9

Ozturk M, Yucel E, Gucel S, Sakcali S, Aksoy A (2008) Plants as biomonitors of trace elements pollution in soil. In: Prasad MNV (ed) Trace elements: environmental contamination, nutritional benefits and health implications, Chapter 28, Wiley, USA, pp 723–744

Ozturk M, Sakcali S, Gucel S, Tombuloğlu H (2010) Boron and plants. In: Ashraf et al (ed) Plant adaptation & phytoremediation, Springer, Part 2, New York, USA, pp 275–311
Ozturk M, Mermut A, Celik A (2011a) Land degradation, urbanisation, land use & environment. NAM S. & T, Delhi, 445 pp
Ozturk M, Okmen M, Guvensen A, Celik A, Gucel S (2011b) Land degradation, urbanisation and biodiversity in the Gediz Basin-Turkiye. In: Ozturk et al (eds) Urbanisation, land use, land degradation and environment, NAM proceedings, Daya Publishing House, Delhi, pp 74–93
Ozturk M, Memon AR, Gucel S, Sakcali MS (2012) Brassicas in Turkey and their possible role in the phytoremediation of degraded habitats. In: Anjum NA (ed) The plant family Brassicaceae: contribution towards phytoremediation, vol 21, Environmental pollution book series. Springer, New York, pp 265–288
Ozturk M, Ashraf M, Aksoy A, Ahmad MSA (eds) (2015a) Phytoremediation for green energy. Springer Science+Business Media, New York, 191 pp
Ozturk M, Ashraf M, Aksoy A, Ahmad MSA (eds) (2015b) Plants, pollutants & remediation. Springer Science+Business Media, New York, 407 pp
Oskarsson A, Widell A, Olsson IM, Grawe KP (2004) Cadmium in food chain and health effects in sensitive population groups. BioMetals 17(5):531–534
Pacyna JM, Pacyna EG, Aas W (2009) Changes of emissions and atmospheric deposition of mercury, lead, and cadmium. Atmos Environ 43(1):117–127
Park JH, Bolan NS, Chung JW, Naidu R (2011a) Environmental monitoring of the role of phosphate compounds in enhancing immobilization and reducing bioavailability of lead in contaminated soils. J Environ Monit 13:2234–2242
Park JH, Choppala GK, Bolan NS, Chung JW, Chuasavathi T (2011b) Biochar reduces the bioavailability and phytotoxicity of heavy metals. Plant Soil 348:439–451
Park JH, Bolan NS, Megharaj M, Naidu R (2011c) Comparative value of phosphate sources on the immobilization of lead, and leaching of lead and phosphorus in lead contaminated soils. Sci Total Environ 409:853–860
Pempkowiak J, Sikora A, Biernacka E (1999) Speciation of heavy metals in marine sediments vs their bioaccumulation by mussels. Chemosphere 39:313–321
Perez J (2012) The soil remediation industry in Europe: the recent past and future perspectives. pp. 2–22.
Perminova IV, Hatfield K (2005) Remediation chemistry of humic substances: theory and implications for technology. In: Perminova IV, Hertkorn N, Hatfield K (eds) Use of humic substances to remediate polluted environments: from theory to practice, NATO science series: IV. Earth and environmental sciences, vol 5, Springer, Dordrecht, pp 3–36
Pietrzak U, Uren N (2011) Remedial options for copper-contaminated vineyard soils. Soil Res 49:44–55
Porter SK, Scheckel KG, Impellitteri CA, Ryan JA (2004) Toxic metals in the environment: thermodynamic considerations for possible immobilization strategies for Pb, Cd, As, and Hg. Crit Rev Environ Sci Technol 34:495–604
Pueyo M, Sastre J, Hernández E, Vidal M, López-Sánchez JF, Rauret G (2003) Prediction of trace element mobility in contaminated soils by sequential extraction. J Environ Qual 32:2054–2066
Purkayastha D, Mishra U, Biswas S (2014) A comprehensive review on Cd(II) removal from aqueous solution. J Water Process Eng 2(0):105–128 doi:http://dx.doi.org/10.1016/j.jwpe.2014.05.009
Rabindra FB, Sanghwa O, Sik SW (2012) Assessment of metal bioavailability in smelter-contaminated soil before and after lime amendment. Ecotoxicol Environ Saf 80:299–307
Rafiq MT, Aziz R, Yang X (2014) Cadmium phytoavailability to rice (*Oryza sativa* L.) grown in representative Chinese soils. A model to improve soil environmental quality guidelines for food safety. Ecotoxicol Environ Saf 103:101–107
Rahmani A, Mousavi HZ, Fazli M (2010) Effect of nanostructure alumina on adsorption of heavy metals. Desalination 253:94–100

Rathinasabapathi B, Ma LQ, Srivastava M (2006) Arsenic hyperaccumulating ferns and their application to phytoremediation of arsenic contaminated sites. In: Teixeira da Silva JA (ed) Floriculture, ornamental and plant biotechnology, vol 3, Advances and topical issues. Global Science Books, Takamatu, pp 305–311

Reid BJ, Jones KC, Semple KT (2000) Bioavailability of persistent organic pollutants in soils and sediments a perspective on mechanisms, consequences and assessment. Environ Pollut 108:103–112

Reid BJ, MacLeod CJA, Lee PH, Morriss AWJ, Stokes JD, Semple KTA (2001) Simple 14C-respirometric method for assessing microbial catabolic potential and contaminant bioavailability. FEMS Microbiol Lett 196:141–146

Robson T, Braungardt C, Rieuwerts J, Worsfold P (2014) Cadmium contamination of agricultural soils and crops resulting from sphalerite weathering. Environ Pollut 184:283–289

Robinson BH, Banuelos G, Conesa HM, Evangelou MWH, Schulin R (2009) The phytomanagement of trace elements in soil. Crit Rev Plant Sci 28:240–266

Sabir M, Waraich EA, Hakeem KR, Ozturk M, Ahmad HR, Shahid M (2015) Phytoremediation: mechanisms and adaptations. Chapter 4. In: Hakeem et al (eds) Soil remediation and plants: prospects and challenges. Academic Press, Elsevier, New York-USA, pp 85–105

Sang-Hwan L, Jin-Soo L, Jeong CY, Jeong-Gyu K (2009) In situ stabilization of cadmium, lead, and zinc-contaminated soil using various amendments. Chemosphere 77:1069–1075

Semenzin E, Critto A, Carlon C, Rutgers M, Marcomini A (2007) Development of a site-specific ecological risk assessment for contaminated sites: part II. A multicriteria based system for the selection of bioavailability assessment tools. Sci Total Environ 379:34–45

Scragg A (2006) Environmental biotechnology, 2nd edn. Oxford University Press, Oxford

Schwartz C, Echevarria G, Morel JL (2003) Phytoextraction of cadmium with Thlaspi caerulescens. Plant Soil 249:27–35

Sheoran V, Sheoran A, Poonia P (2011) Role of hyperaccumulators in phytoextraction of metals from contaminated mining sites: a review. Crit Rev Environ Sci Technol 41:168–214

Shi W-J, Menn F-M, Xu T et al (2014) C60 reduces the bioavailability of mercury in aqueous solutions. Chemosphere 95:324–328

Silveira ML, Alleoni LRF, O'Connor GA, Chang AC (2006) Heavy metal sequential extraction methods – A modification for tropical soils. Chemosphere 64:1929–1938

Singh J, Kalamdhad AS (2013) Effects of lime on bioavailability and leachability of heavy metals during agitated pile composting of water hyacinth. Bioresour Technol 138:148–155

Sobolev D, Begonia MF (2008) Effects of heavy metal contamination upon soil microbes: lead-induced changes in general and denitrifying microbial communities as evidenced by molecular markers. Int J Environ Res Public Health 5(5):450–456

Soumia A, Muhammad H, Georges M, Jean-Cluad R (2005) Sequential extraction of heavy metals during composting of sewage sludge. Chemosphere 59:801–810

Sugasini A, Rajagopal K, Banu N (2014) A study on biosorption potential of aspergillus sp. of tannery effluent. Adv Biosci Biotechnol 5:853–860

The National People's Congress of the People's Republic of China, The Ministry of Environmental Protection; The Ministry of Land and Resources and The National People's Congress of the People's Republic of China; joint report on the national soil contamination survey 2015: http://www.mep.gov.cn/gkml/hbb/qt/201404/t20140417_270670, http://www.mlr.gov.cn/xwdt/jrxw/201211/t20121101_1152706, http://www.npc.gov.cn/npc/zhibo/zzzb22/2011-10/25/content_1676505

Tica D, Udovic M, Lestan D (2011) Immobilization of potentially toxic metals using different soil amendments. Chemosphere 85:577–583

Topolska J, Borowicz P, Manecki M, Bajda T, Kaschabek S, Merkel BJ (2013) The effect of gluconic acid secretions by phosphatase solubilizing pseudomonas putida bacteria on dissolution of pyromorphite Pb5 (PO4)3Cl and Pb remediation. Ann Soc Geol Pol 83:343–351

Uchimiya M, Chang S, Klasson KT (2011) Screening biochars for heavy metal retention in soil: role of oxygen functional groups. J Hazard Mater 190:432–441

UNEP (United Nations Environment Programme) (2008) Interim review of scientific information on lead. Version of March 2008

Ure AM, Quevauviller P, Muntau H, Griepink B (1993) Int J Environ Anal Chem 51:135–151

US ATSDR (United States Agency for Toxic Substances and Disease Registry) (2007) Toxicological profile for lead. U.S. Department of Health and Human Services. 1–582

US EPA (U.S. Environmental Protection Agency) (2004) IRIS integrated risk information system, lead and compounds (inorganic) (CASRN 7439-92-1). Available from http://www.epa.gov/ncea/iris/subst/0277.htm#noncar

US EPA (U.S. Environmental Protection Agency) (2009) Guidance manual for the integrated exposure uptake biokinetic model for children. Publication number 9285.7-215-1, EPA 540-R-93-081, PB93-963510. February 1994. Version 1.1.

Usero J, Gamero M, Morrillo J, Gracia I (1998) Comparative study of the sequential extraction procedures for metals in marine sediments. Environ Int 24:487–496

Voglara GE, Leˇstan D (2010) Solidification/stabilisation of metals contaminated industrial soil from former Zn smelter in Celje, Slovenia, using cement as a hydraulic binder. J Hazard Mater 178:926–933

Von Storch H, Costa-Cabral M, Hagner C (2003) Four decades of gasoline lead emissions and control policies in Europe: a retrospective assessment. Sci Total Environ 311(1):151–176

Wahid A, Ghani A, Javed F (2008) Effect of cadmium on photosynthesis, nutrition and growth of mungbean. Agron Sustain Dev 28(2):273–280

Walraven N, Van OS BJ, Klaver GT, Middelburg JJ, Davies GR (2014) The lead (Pb) isotope signature, behaviour and fate of traffic-related lead pollution in roadside soils in The Netherlands. Sci Total Environ 472:888–900. doi:10.1016/j.scitotenv.2013.11.110

Wepener V, Vermeulen LA (2005) A note on the concentrations and bioavailability of selected metals in sediments of Richards Bay Harbour, South Africa. Water SA 31:589–595

WHO (2011) Evaluation of certain food additives and contaminants: seventy-third report of the Joint FAO/WHO Expert Committee on Food Additives. WHO technical report series 960 http://whqlibdoc.who.int/trs/WHOTRS960eng.pdf

WHO (2000) Safety evaluation of certain food additives and contaminants. WHO food additives series 44: Lead. Available from http://www.inchem.org/documents/jecfa/jecmono/v44jec12.htm

Wong CSC, Li XD (2004) Pb contamination and isotopic composition of urban soils in Hong. Kong Sci Total Environ 319:185–195

Wuana RA, Okieimen FE (2011) Heavy metals in contaminated soils: a review of sources, chemistry, risks and best available strategies for remediation. ISRN Ecol 2011:1–20

Wu P, Li C, Chen J, Zheng C, Hou X (2012) Determination of cadmium in biological samples: an update from 2006 to 2011. Appl Spectrosc Rev just-accep(just-accep) doi:10.1080/05704928.2012.665401

Xia HP (2004) Ecological rehabilitation and phytoremediation with four grasses in oil shale mined land. Chemosphere 54:345–353

Yuan CG, Shi JB, He B, Liu JF, Liang LN (2004) Speciation of heavy metals in marine sediments from the East China Sea by ICP-MS with sequential extraction. Environ Int 30:769–783

Zakir HM, Shikazono N (2011) Environmental mobility and geochemical partitioning of Fe, Mn, Co, Ni and Mo in sediments of an urban river. J Environ Chem Ecotoxicol 3:116–126

Zarei M, Hempel S, Wubet T (2010) Molecular diversity of arbuscular mycorrhizal fungi in relation to soil chemical properties and heavy metal contamination. Environ Pollut 158(8):2757–2765. doi:http://dx.doi.org/10.1016/j.envpol.2010.04.017

Zhang M, Wang M, Xuebao T (2003) Potential leachability of heavy metal in urban soils from Hangzhou City. Acta Pedologica Sinica 40:915–920

Zhang X, Yang L, Li Y, Li H, Wang W, Ye B (2012) Impacts of lead/zinc mining and smelting on the environment and human health in China. Environ Monit Assess 184(4):2261–2273

Zhao K, Liu X, Xu J, Selim HM (2010) Heavy metal contaminations in a soil–rice system: Identification of spatial dependence in relation to soil properties of paddy fields. J Hazard Mater 181(1–3):778–787. doi:http://dx.doi.org/10.1016/j.jhazmat.2010.05.081

Zhou H, Zhou X, Zeng M (2014) Effects of combined amendments on heavy metal accumulation in rice (*Oryza sativa* L.) planted on contaminated paddy soil. Ecotoxicol Environ Saf 101(0):226–232. doi: http://dx.doi.org/10.1016/j.ecoenv.2014.01.001

Zorpas A, Constantinides T, Vlyssides A, Haralambous I, Loizidou M (2000) Heavy metal uptake by natural zeolite and metals partitioning in sewage sludge compost. Bioresour Technol 72:113–119

Climate Change: Impacts on Carbon Sequestration, Biodiversity and Agriculture

Zulfiqar Ahmad and Shermeen Tahir

Contents

Z. Ahmad (✉)
Institute of Soil and Environmental Sciences, University of Agriculture Faisalabad, Faisalabad, Pakistan
e-mail: zulfiqar1409@gmail.com

S. Tahir
Nuclear Institute for Agriculture and Biology (NIAB), Jhang Road, Faisalabad, Pakistan

K.R. Hakeem et al. (eds.), *Soil Science: Agricultural and Environmental Prospectives*, DOI 10.1007/978-3-319-34451-5_18

Abstract Climate change is a wider term and encompasses every aspect of biotic and abiotic life. Climate plays a very basic and significant role in the biology of living things. As a result the key factor amongst many to determine life in specific region thousands of years ago is the fact that various climate cycles existed at that place at that times. The recent decades have witnessed drastic changes in climate because of rise in atmospheric carbondioxide (CO_2) and ozone (O_3) levels leading to increase in temperature, melting of glaciers and rise in sea level. The ultimate trends that CO_2 and climate will constitute in the future are unknown. However, the researchers have been raising questions about carbon sequestration, food security, and crop productivity in the field of agriculture and extinction of species in the field of biodiversity. The term carbon sequestration implies the ways and means through which atmospheric carbon is transferred into the long lived pools and storing it safely in a way that it may not be re-emitted into the atmosphere. Anthropogenic activities, over a period of time have raised serious concerns to sequester carbon and lower down its concentration in the atmosphere, hence leading to drastic climate changes. Since it is not possible to cover all aspects of climate change, in this review we have emphasized on green house and non-green house aspects of climate change and their potential of global warming, implications on carbon sequestration sustainability and agriculture.

Keywords Climate change • Soil carbon sequestration • Sustainable agriculture • Global warming

1 Introduction

There are several signs of life on the earth even at the places life does not exist today. These signs are in the form of chalk or fossils. It is also important to note that the signs of life of different species found today are different from the species found today. This difference is due to evolution; however, the reason for the presence of one type of species at one place and absence at another place is the climate. Climate plays a very basic and significant role in the biology of living things. As a result the key factor amongst many to determine life in specific region thousands of years ago is the fact that various climate cycles existed at that place at that times (Cowie 2013; Hakeem 2015).

It is quite convenient to use this biological fact to study the climate. These biological remains are the imprints of past climate. Moreover, biology has a very significant influence on climate i.e. transpiration of rainforests can influence the flow of water in a catchment area, it means it can transform climate otherwise it would have been in the absence of living organisms. Looking from another angle, all living organisms have a certain optimum temperature range for their growth and development and flourish well in that range. Similarly, water requirement of all organisms are different for completing their life cycles and availability of water is controlled by the environment. Taking into account the connection of water and temperature with life, it is very easy to determine the habitats of different species and their ecosystems.

From this discussion if we can understand the significance of climate then it is imperative to understand the climate change for prediction of the fate of different species in different ecosystems. It is also possible to use reverse in an applied sense to record the past presence of various species and take it as an indicator of climate change. This is the very basic interrelationship between climate and life and affects all living things and we tend to shun the fact that every continent except Antarctica has signs of extinct civilizations and settlements which used to be the flourishing centers of the world which no more exist fundamentally because of the climate change.

If we assume that living things are not subject to climate change, there is sufficient evidence to prove the alteration of climate by global societies in a way which has deep global, biological, and regional implications that will impart profound impact on living things and create interactions between the two.

2 Difference Between Weather and Climate

To understand the biology of climate it is imperative to know about the climate and weather and differentiate between the two. We listen about climate and weather quite often however; there is a lot of confusion over the two terminologies. Every one of us checks the weather forecast for every planning and climate change is the news of the day. In simple words, climate is the forecast and weather is the outcome. Weather is what we see outside on any day of the year. For example it may be 40° C and sunny or 20° C and cloudy (National Oceanic and Atmospheric Administration 2015). Climate is the average of that weather. For example one can expect heavy rains in the Northeast in July or it to be hot and humid in Southeast in July. The outcome is that the climate change is a long term phenomenon and weather is a change for short period of time.

There are many factors of climate and weather at a place however; major elements are temperature, pressure, wind, humidity and precipitation. The study and analysis of these factors provides the basis of weather and climate forecasting. These factors make the basis of climatology.

3 The Greenhouse Effect

The better understanding of the greenhouse effect may be obtained by considering that the Earth has no atmosphere. To understand this process we have to consider the information on the Moon surface. As we know the Moon surface is airless, its day time temperature is 117 °C and it drops to −173 °C at night rendering a median of −28 °C. During the day time its surface either absorbs the Sun light and warms the rocks or reflects off. On the other hand the Earth's surface average temperature is 15 °C, which is due to the fact that the Earth's surface keeps it low, otherwise it would be 43 °C. This is due to the greenhouse effect of the Earth. This warming effect is natural and existed always. This warming occurs because all the radiations coming from the Sun are not reflected by the Earth's surface. Some of the radiations are trapped on the surface of the Earth just like on the surface of the Moon rocks trap them however, trapping of the radiations is more on the Earth's surface because its atmosphere is more transparent for the high wavelength radiations and opaque to the short wavelength radiations. Contrary to this rocks on the Moon surface are not transparent and only their surface warms, not strata in deep.

The reason of the reflection of some of the light back to the atmosphere is that for some of the light the atmosphere becomes transparent. These are mostly long wavelength radiations and are reflected back to the atmosphere while short wavelength radiations are trapped and cause the temperature to rise. Its function is similar to the glass of a greenhouse which allows the higher wavelength light in and traps some of the lower wavelength radiations; therefore this phenomenon is called greenhouse effect. That is the reason why the components of the atmosphere which show such properties are called greenhouse gases.

4 Greenhouse Gases

Many gases known as greenhouse gases occur naturally at various concentrations. These include water vapors (H_2O), methane (CH_4) nitrous oxide (N_2O) and CO_2. Other most important included in this category are chloloroflorocarbons (CFCs) which are artificial and are used as coolants or foam blowing. Carbon dioxide is the main gas whose concentration is increasing day by day due to human activities.

Tyndall (1861) recognized greenhouse gases and narrated their possible implications if their concentration changes in the atmosphere (Hulme 2009). He also speculated the possibilities of no warming effect if concentration of these gases decreased in the atmosphere. Consequently the Earth may face another ice age. However; this phenomenon is reverted due to increasing concentration of greenhouse gases which cause the warming effect. The main reason behind current global warming is the difference between natural greenhouse effect and the human generated effect (Cowie 2013). Today the atmosphere is changing with ever increasing concentrations

Table 1 Summary of major greenhouse gases (1765–2015)

Greenhouse gases	1765	1990	2005	2011	2015
$CO_{2\,(ppm)}$	280	354	380	392	400
$CH_{4\,(ppb)}$	730	1720	1770	1800	1840
CFC-11 $_{(ppt)}$	0.0	258.45	250.56	238.11	232.23
CFC-12 $_{(ppt)}$	0.0	485.20	542.80	529.57	519.73
$N_2O_{(ppt)}$	270.47	308.07	319.10	324.32	328.11

The values of CFC-11, CFC-12 and N_2O are average of concentrations at Northern and Southern hemispheres. Cowie (2013); Leung et al. 2014; Bullister 2015. *ppt* parts per trillian, *ppm* parts per million. http://cdiac.ornl.gov/ftp/oceans/CFC_ATM_Hist/CFC_ATM_Hist_2015/NDP_095(2015).pdf

of CO_2 due to burning of fossil fuels. The following table illustrates the periodic increase of CO_2 concentration in the atmosphere (Table 1).

Since the Industrial Revolution the Earth has encountered various cycles of warmth although the warmness is not directly proportional to the increase of greenhouse gases but the fact is that the warming of the Earth has taken place. Today we know that with less concentration of greenhouse gases in the atmosphere the Earth cools and there is ice age; when the Earth was cool and the atmosphere carbon dioxide concentration was less.

Methane and CO_2 are part of global carbon cycle and nitrous oxide forms part of nitrogen cycle and all the three gases have natural and human origin. Nitrous oxide is released by the decomposition of organic matter by tropical forests, soils and oceans. Organic matter burning and fertilizers are among the anthropogenic activities that contribute towards N_2O in the atmosphere. Removal of N_2O from atmosphere is carried out by a process known as photosynthesis in the presence of sunlight which results into the release of N_2 and $O_{2.}$

Another important greenhouse gas that has not captured a lot of attention so far is water vapor. It is a powerful greenhouse gas and is playing a significant role in global warming. Atmosphere has sufficient concentrations of water vapors which absorb the infra-red radiations from the Sun. These water vapors are present everywhere in the atmosphere, even over the Sahara Desert however; the concentration is not constant throughout the atmospheric column (National Oceanic and Atmospheric Administration 2015; Cowie 2013). These vapors absorb radiation and cause rise in temperature. In a warmer world there will be more evaporation from the surface of the Oceans, hence will contribute more water vapors that will double the warming.

5 Global Warming Potential of Greenhouse Gases

Each greenhouse gas has different physico-chemical properties and has different warming potential to the atmosphere. Their properties are quantified for global warming potential (GWP) which is defined as a "comparative index for unit mass of a gas measured against the warming potential of a unit mass of carbon dioxide over a specific period of time". Since all greenhouse gases have different atmospheric residence time so GWP must be expressed in certain time period otherwise this will make no sense. If we take the example of methane, whose atmospheric residence time is 12 years and a quarter after 2 years which means that average life time of a molecule of methane would be 12 year[2]. On the other hand N_2O has an average residence time of over 100 years. If we compare the GWP of N_2O and CH_4 over a decade will give different warming value. It means GWP of the greenhouse gas will vary with the nature of the gas and the time of residence. Even the GWPs determined by IPCC vary in their different reports. There are several other agents that also contribute to the global warming, among these chloroflorocarbons (CFCs), hydrochloroflorocarbons (HCFCs) and hydrofluorocarbons (HFCs) are well known. Human made chemicals have low atmospheric concentrations and their contribution is less than a quarter of present warming (Cowie 2013).

6 Non-Greenhouse Influences of Climate

There are several non-greenhouse aspects which play role although not a major one in the climate changes. Milankovitch variation and changing sun light-reflecting properties of ice caps explain these processes. Although there is a clear understanding among the climate community that man-made addition of greenhouse gases are the main contributors towards the climate change yet non-greenhouse factors cannot be ignored altogether. One of these is solar output which has played a key role for billions of years on the Earth. The Sun being a main star in this sequence is getting warmer over a period of billions of years and has imparted considerable implications on the biosphere. However for a shorter time period of hundreds to thousands of years it is relatively stable. Even shorter time scale of the Sun has impact on the climate. This impact is not very significant and only causes a variation of 0.08 % in the irradiance which may have the potential to change global climate by 0.02–0.04 °C (Foukal et al. 2004).

Cowie (2013) has reported that the scientists have been taking space-born measurements of solar output and have been correlating it with sun sport activity since 1978. The process is not clear but the interaction exists. The solar component is another term which is five times in magnitude as compared to the solar variation reflected by sun sport activity. Moreover there exists a correlation between solar output and global temperature but it is not clear that this relationship is real or partial.

It can be concluded that there are several factors impacting the climate positively and/or negatively. Out of these factors some are strong and some are weak; for example greenhouse gases are strong factor and the variations in the sun's output over a thousand of years are weak factors. However their superimposition may help trigger larger changes in the concentrations of methane and CO_2 as the Earth moves between inter glacial and glacial modes. Therefore it must not be surprising to note that greenhouse gases are not the sole source of global warming. There are other non-greenhouse factors such as solar output, volcanic eruption, oceanic and atmospheric circulations and marine release of methane. Although these factors contribute towards the global warming yet they are not the main contributors. At the same time these cannot be neglected and they may be contributing to global warming in future especially the circulation changes.

7 Global Warming and Impacts on Future

The response of natural system and species to climate change is an intricate process, moreover certainly become a bit more complex. Some responses are quite opposite to the expectations. One thing is common and that is the fact that our planet is warming up. The data presented by different researchers (Cowie 2013; National Oceanic and Atmospheric Administration 2015; Hakeem 2015) shows that the temperature of the North of the northern hemisphere increased between 1601 and 1974, however in late 20th century the temperature returned to normal. The reason could be increased snowfall which could had delayed the snow melting and onset of spring greening. This change is not unexpected as already discussed that warmer planet would increase ocean evaporation and increased precipitation will occur in the form of rain or snow. Another reason could be other environmental factors which have impacted the growth of plants as species respond to a number of factors which may act positively or negatively. One of the most suitable examples to late 20th century warming is the Alaskan white Spruce (*Picea glauca*). Current warming has increased thermal growing period at high latitudes. During the year there are days when plants can grow and increase the primary biological activity and CO_2 sequestration is increased by photosynthesis; consequently water consumption is increased.

8 Soil Carbon Sequestration

The term carbon sequestration implies the ways and means through which atmospheric carbon is transferred into the long lived pools and storing it safely in a way that it may not be reemitted into the atmosphere. It means increase in Soil Organic Carbon (SOC) and soil inorganic carbon (SIC) through judicious practices and better management and use of land. The observed rates of SOC in the ecosystems depend on various factors which include soil texture, soil profile properties, and climate (Armstrong et al. 2003; Grace et al. 1995; West and Post 2002). Prolonged and sustainable management practices can sustain the rates of carbon sequestered over a period of time (Grace et al. 1995; West and Post 2002). All the systems which ensure the enhanced addition of biomass to the soil, cause minimum soil degradation, conserve soil and water, improve soil structure, enhance microbial activity and species diversity and enhance soil fertility contribute towards the SOC sequestration. Mulching, zero tillage, agroforestry, multiple cropping systems and integrated nutrient management in the form of manures, composts, and bio-solids also fall in the same category (Silver et al. 2000). The restorations of degraded soil with intact resilience capacity have more potential of SOC sequestration. The SIC sequestration is low and is accelerated by different biogenic processes and leaching of carbonates into ground water (Nordt et al. 2000; Levy 1984) especially in the soils irrigated with water having low concentrations of carbonates.

9 Impact of Global Warming on Carbon Sequestration

About 5 % of global carbon is stored by soils and vegetation and their contribution of carbon to the atmosphere is about 50 % (Barraclough et al. 2015). Their response to the climate is not clear yet. It is expected that the processes (loss by respiration and gain by plants residues) contributing towards the soil carbon concentration will accelerate with increase in temperature (Smith et al. 2008). Bond-Lamberty and Thomson (2010) have confirmed an increase in soil respiration over time with the increase in air temperature. The most important and a bit less certain is that these two opposing processes will show changed behavior under warm climate. If the losses of soil carbon increase through respiration more quickly as compared to carbon returns through plant debris, a positive feedback can be drawn on impact of climate change on carbon sequestration (Barraclough et al. 2015). The determination and calculation of changes in soil carbon pool over a period of time can provide an undeniable proof on the balance between gains and losses.

Determining small changes against a longer and different background is challenging because the separation of effects caused by land-use from those caused by climate change are very difficult. A major study that let the scientists to link these changes was conducted by National Soil Inventory of England and Wales (NSI) who have reported large decrease in soil carbon during periods between 1978–1983 and 1995–2005 (Bellamy et al. 2005). However Emmet et al. (2010) reported that the changes suggested by NSI study were unrelated to climate change and even other studies have reported no significant changes in soil carbon over times (Tomlinson and Milne 2006; Barraclough et al. 2015).

An alternative method to study the soil carbon concentration over time was adopted by Fantappie et al. (2011) in which they used 'space-for-time substitution' in which changes in carbon across climatic gradients at one time were used to predict carbon under a future climate. Similar models were used by Barraclough et al. (2015) in a combined form to derive regression models between soil carbon concentrations and mean annual temperature and rainfall for each of the 11 land uses reported by Bellamy et al. (2005). They also concluded that their findings were not consistent with Bellamy et al. (2005) altogether. The reported changes were more related with the changes in land-use and reduction in carbon returns from grazing animals. They estimated that only 9–22 % changes can be attributed to climate and indirect temperature related mechanisms could be responsible for these changes.

10 Carbon Sequestration and Sustainability

The world agriculture and degraded soil have 50–66 % sink capacity of carbon loss. Soil management, rainfall, farming systems, soil structure, texture, and temperature are the properties on which soil organic carbon sequestration depends (Lal 2004; Hakeem et al. 2014). The strategies recommended by scientists to increase soil carbon pool include zero tillage, cover crops, soil restoration, reforestation, manuring, soil fertility management, sludge application, water conservation, improved grazing, judicious harvesting and efficient irrigation. An additional yield of crops (up to 20 %) can be achieved by adding one ton of carbon to soil carbon pool. It will enhance food security and will offset the fossil fuel emission by 0.4–1.2 giga tons (Gt) of carbon per year which is equivalent to 5–15 % of the global fossil fuel emission (Sartaj et al. 2016).

The atmospheric and biotic carbon pools are 3.3 times and 4.5 times less than the soil organic carbon (SOC) pool. The SOC exists in the form of dynamic equilibrium between gains and losses. The conversion of SOC from organic to inorganic pool causes severe depletion in the soil. The depletion increases with increased degradation of soil. This depletion enhances when the output of carbon exceeds the input. Some soils have lost as much as 20–80 % of their carbon to the atmosphere. The decrease in SOC may degrade soil quality, productivity, soil ecology and water quality. The emission of the carbon to atmosphere increased with the passage of time and this was 270 ± 30 Gt from fossil fuels and 136 ± 55 Gt from terrestrial ecosystem between 1850 and 1998 (Lal 2004).

11 Soil Carbon Sequestration and Climate Change

The estimates of total carbon sequestration capacity of soil are variable, the range starts from 0.4 to 0.6 Gt C per year (Sauerbeck 2001) and ends at 0.6–1.2 Gt C per year (Lal 2003a, b) and the potential is finite in time and capacity. Until the alleviation of the fossil fuels take effect the C sequestration has bought us time (Sartaj et al. 2016). Some of the main issues related with carbon sequestration in agricultural ecosystem are discussed.

11.1 Agricultural Chemicals

Most of the practices used in agriculture involve C-based inputs. It takes 0.86 kg C kg^{-1} N, 0.17 kg C kg^{-1} P_2O_5, 0.12 kg C kg^{-1} K_2O, 0.36 kg C kg^{-1} lime, 4.7 kg C kg^{-1} of herbicides, 5.2 kg C kg^{-1} of fungicides, 4.9 kg C kg^{-1} of insecticides (Lal 2004) and 150 kg C ha^{-1} for pumping. Similarly, for enhancing use efficiency and mineralizing losses, the wise use of C-based inputs is necessary, nevertheless, the inputs are needed for carbon sequestration but they are mandatory for food production and ensuring sustainability of soil and water.

11.2 Essential Nutrients

Carbon is one of the major essential nutrients and plants cannot complete their life cycle without it (Lal 2004). For the sequestration of 1 Gt of carbon in the soil it would require 80 Mt of N, 20 Mt of P and 15 Mt of K as compared to global fertilizer use which was 136 Mt in 2000 (IFDC 2000). However, there are other sources of carbon sequestration i.e. biological nitrogen fixation, recycling from subsoil aerial decomposition, use of bio solids and crop residues. Lal (2004) has reported that one ton of cereal residue can contribute 12–20 kg of N, 1–4 kg of P, 7–30 kg of K, 4–8 kg of Ca and 2–4 kg Mg. If the residues produced are incorporated into soil (3 Gt) instead of removal for fuel and other uses, would sequester C and improve soil quality. The same crop residues may be used for obtaining biofuel [ethanol or energy (H_2)]. This residue may be used to sequester carbon for the production of biofuel which has enormous economic value.

11.3 Soil Degradation and Deposition

Most of the erosional processes (wind or water) preferentially remove SOC from the sediments and is redistributed over the landscape or re-deposited in the depression sites and aquatic ecosystems. During the processes the C may be buried but most of it is emitted into the atmosphere in the form of CO_2 through mineralization or CH_4 through methanogenesis. The intensity of buried C may vary from 0.4–0.6 Gt C per year as compared to re-emitted which is 0.8–1.2 Gt C per year (Lal 2003b). For sustainable agricultural productivity and land use, effective soil erosion control is essential.

11.4 Better Farming Practices

The annual nutrient losses in sub Saharan Africa due to low inputs is estimated to be 40 kg of NPK ha^{-1} in cultivated land (Sanchez 2002). The effect of SOC mining is similar to that of fossil fuel combustion. Therefore, the better farm management practices must increase SOC pool and enhance fertility which would ultimately lead to increased crop yield and soil quality.

11.5 Social Aspects

For trading carbon credits, its commodification is very important. In Europe carbon trading markets were established in 2002 (Johnson and Heinen 2004). Currently, the low prices of SOC may be increased with regulations and emission caps. In the European Union countries for the concept of SOC credits to make a routine matter as a part of the solutions to mitigate climate change the ability exists to measure short term changes in SOC (Lal et al. 2001), but the need of the hour is to base price of soil C on off site and on site societal benefits otherwise the undervaluing of soil C may lead to tragedy of the commons (Lal 2004).

11.6 Water and Carbon Cycle

Resources of fresh renewable water are decreasing very rapidly. The data shows 56 % increase in cereal production between 1997 and 2050 on ever shrinking land area with less available water (Lal 2004; Rosegrant and Cline 2003). Thus it is very important to develop a link between hydrologic and carbon cycles. This may be achieved by conservation of water resources, improving agronomic yield of the crop and carbon sequestration. The stocks of SOC in rain-fed region can be increased by water conservation, increasing water use efficiency, judicious water harvesting and farming systems. In dry land SOC stocks may be enhanced by zero tillage farming to manage drought: a real win-win situation (Lal et al. 2004).

11.7 Global Warming and Soil Carbon Sequestration

Global warming is a global issue and century-scale problem and soil C sequestration is a separate but related issue having its own benefits of improving water quality, increasing soil productivity, restoring ecosystems and degraded soils. Keeping fossil fuel emission on aside achievable SOC potential provides multiple societal benefits. Moreover, soil carbon sequestration plays a bridging role among three global issues i.e. climate change, desertification and biodiversity.

11.8 Greenhouse Gases

Soil capacity to oxidise CH_4 is increased by enhancing SOC pools. This is a special case under zero tillage farming system (Six et al. 2002). It may enhance emission of N_2O (Smith et al. 2001). These fluxes of CH_4 and N_2O can alter the CO_2 mitigation potential of soil management practices and must be taken care of along with SOC sequestration.

11.9 Tropical Soils

Tropical soils are severely degraded and depleted and have high carbon sink capacity and low rate of carbon sequestration. Since these soils have low crop yields so they need more attention for improvement of soil quality than the soils of high latitude. This improvement is more challenging because of poorly developed infra structures weak institutions and resource-poor agricultural systems. Such lands needs soil restorative policies for the mitigation of soil degrading trends.

11.10 Permanence

Soil carbon sequestration is environment friendly, cost effective, natural process and sequestered carbon remains in soils until recommended management practices like zero till farming and restorative land use are followed. The soil sink capacity and the permanence are related to many physical factors like clay content, mineralogy, temperature regimes, structural stability and ability to form and retain stable micro aggregates.

12 Food Security and Soil Carbon Sequestration

China, Central and South Asia, the Andrian region, the Caribbean and the acidic savannahs of South America (SSA) are the hot spots of global soil degradation and priority area for soil restoration and carbon sequestration. In these regions complete crop residue removal for fuel and fodder is a general norm and depletion of SOC pools from the root vicinity has affected soil productivity very badly. This is simply the revengeful attitude of the poor farmers that they have passed on their sufferings to the soil. They cultivate marginally fit soils with limited resources and get marginal output and live in poverty. The SOC is a source of soil fertility and production in the farming systems of SSA whose contribution is only 2.5 % of the chemical fertilizers consumed and 2 % of the world's arable and irrigated land area. Both of these factors are essential for SOC sequestration. In extremely degraded soils the benefit of recommended management practices cannot be realized in true spirit. An optimum concentration of SOC stock in the soil is needed to decrease the risks of erosion, hold water and nutrients improve tilth and soil structure and be a source of energy for living microorganism in the soil. The SOC acts as a bio-membrane that filters pollutants, provides oxygen in the coastal ecosystems, decreases sediments load in the rivers, degrades contaminants and plays significant role as a sink for atmospheric CO_2 and CH_4. The crop productivity increases by addition of fertilizers along with incorporation of crop residues in SSA (Pieri 1986) in the form of mulch (Yamoah et al. 2002).

Even in high-input commercial crops, increase in SOC may enhance crop yield (Bauer and Black 1994) especially in severely degraded soils (Johnston 1986) where it has been depleted. It is reported that an increase of 1 ton of SOC in cereal wheat, yields by 27 kg ha^{-1} in America (Bauer and Black 1994) and by 40 kg ha^{-1} in semi-arid parts of Argentina (Diaz-Zorita et al. 2002), 17 kg ha^{-1} of maize in Thailand (Petchawee and Chaitep 1995) 6 kg ha^{-1} of wheat and 3 kg ha^{-1} of maize in alluvial soils of northern India (Kanchikerimath and Singh 2001). For sustainable high yield, high SOC is necessary which improves nutrient and water use efficiency and microbial activity in the soil. For tropical soils the critical value of SOC is 1.1 % (Aune and Lal 1997) and increasing its concentration from a low level of 0.1–0.2 % to the critical value is a major challenge for tropical ecosystems. Nevertheless, a sharp decline in the SOC stock in SSA or anywhere in the world must be revised in order to ensure advance food security. In Kenya, a study was carried out for 18 years, the results revealed that the yield of maize and bean was 1.4 ton ha^{-1} per year in the control (without external inputs) and 6 tons ha^{-1} per year was recorded when stover was retained in the soil and fertilizer and manures were applied. The SOC pool recorded in the same field up to 15 cm depth were 2.3–6 tons ha^{-1} and 28.7 tons ha^{-1}, respectively (Kapkiyai et al. 1999). This sort of management is needed at global level to ensure food security. It is need of the hour that vicious circle of decreasing productivity, declining SOC pools and lower per hectare yield must be broken by improving soil health through SOC sequestration. This will help to free humanity from perpetuating poverty, hunger, substandard living and malnutrition.

13 Climate Change and Biodiversity

These days the researchers have focused their research on the predictions of responses of biodiversity to climate change (Dawson et al. 2011; McMahon et al. 2011; Beaumont et al. 2011; Salamin et al. 2010; Pereira et al. 2010; Bellard et al. 2012; Hakeem 2015). The response of biodiversity to climate change plays significant role in planning and alteration of the thinking of scientists to stop the biological changes due to climate change and decrease the climate change impacts on biodiversity (Parmesan et al. 2011). Currently the research suggests that there is relatively little evidence of the extinctions of species due to the impact of climate change. However, it is predicted that natural habitats of different species could be destroyed due to climate change over the next few decades (Leadley et al. 2010). However, to get the clear picture of future of the biodiversity is difficult because of the variability and the multiplicity of approaches. Therefore, it is extremely needed to review our present understanding of the effects of climate change on the future of biodiversity.

14 Impact of Climate Change on Biodiversity

Bellard et al. (2012) and Parmesan (2006) have reviewed that various components of climate change could affect the biodiversity. Climate change can decrease genetic diversity of populations due to quick migration and directional selection which may in turn impact changes in ecosystem resilience and functioning (Botkin et al. 2007; Meyers and Bull 2002). However, the focus of recent research is on organizational levels and only a little work has been done on the genetic effects of climate change and a very few number of species have been explored.

Beyond this, the web of interactions at community level is modified by various effects on population (Gilman et al. 2010; Walther 2010). In fact the species affected by climate change also affect the other species which rely on them. A study carried out in 2004 on interspecific systems of pollinators and parasites showed that 6300 species are facing the threat of extinctions because of the extinction of those species on which these species depend (Koh et al. 2004). Furthermore, for many species the initial impact of climate change may be attributed through effects on synchrony with species, habitat and food requirements (Kiers et al. 2010). The climate change has induced phonological shifts in pollinating insects and flowering plants which have caused a mismatch between pollinating insect population and flowering plants which has caused extinction of both pollinators and plant populations (Rafferty and Rafferty and Ives 2010). Other changes of interspecific relationships like with predator prey, parasites/host or competitors or mutualists have also changed the functioning of ecosystems and structure of community (Yang and Rudolf 2010).

Climate change may induce variation in plant communities at the higher degrees of biodiversity which are expected to affect biome integrity. It has been forecasted by the Millennium Ecosystem Assessment (MES) that 5–20 % shifts for Earth's terrestrial ecosystems will take place, in particular in the cool conifer forests, savannahs and boreal forests, tundra and shrub land (Sala et al. 2005). Special concerns have been raised for the 'tipping points' where irreversible shifts in biomass may take place in the ecosystems thresholds (Bellard et al. 2012).

Lapola et al. (2009)) have reported that the potential future biomes distribution in tropical regions of South America may change and tropical savannah may replace Amazonian rainforests. Similarly alpine and boreal forests at higher altitudes and latitudes may expand northwards towards more height at the expense of low stature alpine and tundra (Alo and Wang 2008). Rise in temperature and decrease in rainfall may case drying of lakes in Africa (Campbell et al. 2009). Oceans may become more acidic and warm which may destroy tropical coral reefs at large scale (Hoegh-Guldberg et al. 2007). The impact of climate change could be more severe for genetic and specific diversity for ecosystem in services, whose extreme form of decrease of fitness may be extinction of species. To cope with these adverse effects the biodiversity may adopt various mechanisms.

15 Response of Biodiversity to Climate Change

Because of climate change in a given region species may no longer be adapted to the environmental conditions and may fall outside of their climate niche. For survival of the species, populations or individuals must develop adaptive measures of various types which may be provided by two categories i.e. mechanisms and responses (Bellard et al. 2012).

15.1 Mechanisms

There are two mechanisms i.e. genetic and plastic. In genetic mechanisms species may adapt genetically to changing environment by mutation or selection and in plastic mechanisms species are provided with short term responses in behavior, whether or not the species will be able to adopt the mechanisms to cope with changing climate (Lavergne et al. 2010). Whether the adaptations will be motivated by micro-evolution (genetic adoption) or plasticity (short term response) as described by Salamin et al. (2010) and Charmantier et al. (2008). These adaptive means may involve transpacific diversity in physiological, morphological or behavioral traits which may have expression on various time scales within the spatial range of population (Botkin et al. 2007; Chevin et al. 2010). As reported in birds and mammals, the empirical evidence shows that plastic response is more than the genetic contribution (Hoffman and Sgro 2011). On the other side the evolution is very fast (Lavergne et al. 2010) in the case of introduced species and the phenotypic changes have increased the invasive potential (Phillips 2009). Recent work on evolutionary rescue also suggests that rapid evolution by mutation and selection could help species to adopt very severe and rapid environmental changes (Bell and Gonzales 2009).

15.2 Responses

There are three types of responses i.e. spatial, temporal and self. The first two are easily observed and well reputed responses to climate change and self-corresponds to changes which are less obvious physiological and behavioral responses which enable species to adapt to environmental changes in the same space and time paradigms.

In spatial responses species follow specific conditions that they track during adapting period. They generally take place through dispersion however, are not limited to this. Migration to different habitats at micro-habitat or local level can also take place. More than 1000 species have been reported to undergo latitudinal and altitudinal range shifts. This is more common in the case of species with greater dispersal capacities like insects, marine invertebrates and birds (Parmesan 2006) which has lend to the decrease range especially on mountain top and polar species (Ferero-Medina et al. 2010). In this case individuals may shift populations to maintain equilibrium in the environment they are adapted to, at the same time they may not be adapted to the other abiotic conditions such as novel biotic interactions or photoperiod (Visser 2008) in such cases micro-evolution or genetic mechanism may be required for persistence.

The individual species also respond to a climate change through shifts in time period on seasonal or daily basis. These are cyclic variations which take place over a period of time during one year. The best example in this case is temperature that varies on daily or yearly basis. The 20th century global warming has caused changes in the seasons of flowering, fruiting in plants and seasonal migration in birds (Parmesan 2006; Charmantier et al. 2008). The data shows that for last 50 years the phenological events in a wide range of species have been shifted to 5.1 days earlier per decade (Parmesan 2006). These variations help those species to maintain synchrony with changing abiotic factors. However, there could be some disruption due to increased asynchrony in insect-plant or predator prey system which could be a cause of species extinction. The third response of the species to climate change could be by adapting themselves to new conditions and do not track new optional environment in space and time. Unlike spatial, temporal changes are known *in situ* changes since they take place within the species. These may be in the form of physiological alterations that lead to environmental adaptations or changes in behavior of their food, energy and activity. Such changes are not very obvious like changes in time and space and have been reported during the 20th century climate change. For example in many ecosystems changes in growth, locomotion, reproduction and sex determination are temperature sensitive (Tewksbury et al. 2008). This is not true in all the cases like for many plastic phenotypic traits, in extreme climate change should reach a physiological limit. For instance metabolic rate and body size cannot increase or decrease indefinitely under prolonged climate change (Chevin et al. 2010). For such cases, to cope with climate change, strong genotypic selection is required because their spatial and temporal frames does not change hence limit the alterations of interspecific relationship.

Failure of species in any of the mechanisms or responses will lead their population to face extinction on local or global scale. Since there are so many responses for species to adapt to the climate change, therefore only a few taxa of them went extinct due to climate change during past century (Botkin et al. 2007). This is enough to dilute the temper catastrophic predictions about the possible effect of climate change on biodiversity. However, many populations responded inadequately to counter the quickly changing climate, moreover unlike the past the populations of living organisms have to manage to cope with additional threatening factors which may affect them, in synergy with climate change. Since today the world is facing undeniable facts of biodiversity crisis, the number of endangered species has been increasing with time. Some of the facts are narrated below:

16 Climate Change Impact on Agriculture

Agriculture is an important sector of world's economy. It provides as much of our food through crops, livestock and sea food. Its contribution towards economy is in trillion dollars. Livestock, agriculture and fisheries are dependent on specific temperature. It is very difficult to understand the overall impact of climate change on food supply. Increase in CO_2 and temperature can affect some crops positively in some cases, however, its effects on soil health, nutritional level and water availability may also be considered (EPA 2015). Rainfall frequency, floods and droughts could pose serious challenges for the farming communities. At the same time hot weather temperature may alter the habitats ranges and productivity of many fish and shell fish species and could destroy ecosystems. If seen holistically, changes in climate could create more problems for growing crops, raising animals and fish in places as was done in the past. The effects of climate change must be considered along with other allied factors which may affect agricultural practices and technology (Hakeem 2015).

Crop growth in the world has significant effects on the food supply of the world populations especially in US. According to an estimate about 30% of the wheat, corn and rice produced in US are exported in the global market (US Census Bureau 2011). Variations in the CO_2 concentrations, temperature and rainfall could affect the yield of these crops adversely. For example hot/warm weather may cause quick growth of crops and could reduce yield because the crops grow faster under warmer temperature, however, the time required for grain development would be decreased and yields would be low (USGCRP 2009).

17 Impact of Climate Change on Fisheries

Fish, across the world face many types of stresses which include water pollution, over fishing, heating of oceans. Climate change is worsening these changes and could lead to significant impacts. For instance the range of many fish species may change as several species of fish has particular range of temperature at which they can survive. One of the examples is cod fish found in the North Atlantic thrives best at temperature below 54 °F (EPA 2015). And their reproduction is even reduced when sea bottom temperature is above 47 °F. During the current century this temperature is expected to increase both thresholds. Several marine species are expected to move to colder areas lakes and streams or move to North world in the ocean which may lead to a new competition with new species over food and resources. In warm water some disease may affect the species more than in cold water because in warm water these are more prevalent, as in case of lobsters in New England. Similarly, variation in temperature and season could affect the migration and reproduction periods (CCSP 2008a, b). Many aspects of aquatic life are controlled by seasons for example the warm water in North West has affected the life cycles of salmon and it has become more vulnerable to disease (CCSP 2008a, b), which has caused a large decline in salmon population (Field et al. 2007).

In addition, the increase in CO_2 concentration is causing the acidification of oceans which are affecting shell fish by weakening shells made up of calcium. The acidity may destroy the structures of fish and shell fish ecosystem upon which they rely.

18 Influence of Climate Change on Crop Productivity

For the last several decades, the trends of climate changes in the world's agricultural zones have been very quick and obvious changes in the CO_2 and ozone (O_3) concentrations have been recorded. The actual changes that will occur due to rise in CO_2 concentrations and their influence on climate have raised questions about the security of food. One of them is whether the overall productivity of world's crops will be affected or not. It is estimated that for the next few decades, the global crop yields will increase by 1.8 % due to increase in CO_2 trends (Lobell and Gourdgi 2012) and rise in temperature will decrease yield by 1.5 % per decade. The main factors that will contribute towards this decline include higher O_3 and greater precipitation.

The global food security will be shaped by many factors which include rate of human population, disease, dietary preferences, income growth and distribution, demand for water and land resources for non-agricultural uses, carbon sequestration and rate of improvements in agricultural productivity. The crop yield factor has a special significance which is define as metric tons of grains produced per hectare of land. Sources of agricultural growth including level of funding for research and development, variation in soil fertility and quality, economics and supply of fertilizers, CO_2 and O_3 concentrations in atmosphere and changes in rainfall and temperature are of multi-faced nature. This information focusses on variations in CO_2 and O_3 levels in agricultural regions and their impacts on crops production. This will give us insight on the part of a full story on crop production which will lead us to full story about the future of global food security. For instance this information has no clues about different ways that global change can affect world's food security via different pathways other than agricultural productivity i.e. rate of income growth or influences on human disease occurrences.

19 Climate Changing Trends in the World's Cropping Areas

The data on observed trends, over the past several decades show that air temperature has been increasing in the major cereal cropping areas in the world. Lobell and Gourdji (2012) have reported linear trends in minimum and maximum temperature from 1980 to 2011. Roughly the average trends for maximum and minimum temperature were 0.3 °C and 0.2 °C per decade respectively.

CO_2 concentration in the atmosphere has increased from 278 μL L^{-1} in 1750 to 390 μL L^{-1} in 2000 (Global Carbon Project 2011). Increase in global average tropo-spherc O_3 lcvcl is from 15 nL L^{-1} to 35 nL L^{-1} from prcindustrial cra to thc prcscnt. Ever increasing pollution can raise this concentration to 100 nL L^{-1} (Wilkinson et al. 2012) which could be damaging to the crops (Oltmans et al. 2006). Solar dimming has also been observed from 1950 to 1980 which is associated with increased pollution and aerosol load (Wild 2012).

The projected trends show that the major factor of global warming will be rise in temperature in the agricultural regions. The data shows average model projected rates of global warming from 2040 to 2060 will be similar to those observed from 1950 to 1980 (Lobell and Gourdji 2012) per decade. However there is no concrete evidence to establish whether minimum temperature will rise faster or slower than maximum temperature (Lobell et al. 2007).

This shows that expected rate of global warming is consistent with the past which may be significantly lower or higher for any one or two decades such as global mean temperature (the average of ocean and land) does not rise for one decade due to 1998 El Naino. Unlikely it is quite possible that we could record 10 years trend of as high as 1 °C in the global mean temperature which will be as much as 2 °C in major agriculture areas of the world because the ocean warms slowly than the land (Easterling and Wehner 2009).

CO_2 concentration is expected to increase in next century because 80 % reduction in its emission is required just to stabilize the current atmospheric levels (Meehl et al. 2007). Up to 2050, 25 μL L^{-1} increases in CO_2 concentration per decade is expected which will raise the overall level to 500 nL L^{-1} by that time (IPCC 2001). In developing countries O_3 precursors emission is expected to raise however, its prediction is difficult due to uncertainty in emission pathways and air pollution control (Cape 2008).

20 Response of Crops to Climate Change

There are primary mechanisms which have effects on agriculture among these are: increasing temperature, severe hydrological cycles, increasing CO_2 concentration in the atmosphere and increase of tropospheric O_3 levels. The mechanisms through which these factors affect crop physiology are discussed below.

Yield of crops is affected by temperature through five ways. First, it enhances growth and development and reduces crop duration, which ultimately leads to reduction in yield (Stone 2001). Second, rates of photosynthesis, respiration and grain filling are affected by temperature without any distinction of C4 or C3 plants (Crafts-Brandner and Salvucci 2002). High temperature during day or night can affect photosynthesis, however, warming during the night increases rate of respiration at the cost of any benefit to photosynthesis. Third, temperature raises the vapor pressure deficit (VPD) between air and leaf, which leads to reduced water use efficiency in the form of more water loss per unit of carbon gain (Ray et al. 2002) and plants close their stomata, reducing photosynthesis and increasing heat related impacts. Fourth, high temperature can damage plant cells directly, reduces spring and autumn frost risk which would lead to frost-free growing season. Contrary to this warming during the critical reproductive periods may induce heat stress, leading to sterility, reduction in yields and risk of crop failure (Teixeira et al. 2013). Fifth, high temperature along with elevated CO_2 in the atmosphere can favor the growth and survival of many pests, insects and diseases in agricultural crops (Ziska et al. 2011).

Increased agricultural droughts will cause water stress in crops which will be harmful especially during the reproductive periods of cereal crops (Hatfield et al. 2011). Alterations in the timing of the rainy season may compel the farmers to shift sowing times or more intense rains will result into flooding and water logging and damage crop production (Lobell and Gourdji 2012). Unlike temperature increase in atmosphere CO_2 levels has some positive effects on crops like fertilization effect of CO_2 in C_3 crops by alleviating photosynthesis pathways. It also increases water use efficiency by decreasing stomatal conductance in C3 and C4 plants (Ainsworth and Long 2005). Fifteen percent increase in yields in C3 plants is expected by raised CO_2 concentration. However, it is also expected that CO_2 fertilization will decrease nutritional quality of crops through decreased nitrate assimilation and lower protein content in harvestable yield (Taub et al. 2008).

Tropospheric O_3 is formed when air pollutants like methane, carbon monoxide and nitrogen oxides react with hydroxyl radicals and causes oxidative damage to photosynthetic machinery in the plants (Wilkinson et al. 2012). These pollutants are found in abundance in the agricultural regions across the globe (Van Dingenen et al. 2009). There are possibilities of interactive effect of CO_2 and O_3 which may reduce the damage caused by O_3 through reduced stomatal conductance. It will reduce damage caused by O_3 uptake by maintaining biomass production (McKee et al. 2000). This has raised a concern about the development of new crop varieties in cereals such as increased stomatal conductance has been induced by breeders to support the fact that higher respiration fluxes are related to increased photosynthetic rate and ultimately to yield (Reynolds et al. 1994) whereas higher stomatal conductance means more uptake of O_3 and vulnerability to sterility and reduced yield (Biswas et al. 2008).

The facts discussed here are not always conclusive because they may vary from region to region and cannot be applied across the world to estimate the response of crop production to changing global climate.

21 Climate Change and Future Strategies for Agricultural Crops Production

With the same or less available land and water resources, 56 % increase in cereal production is estimated by 2050 to feed the population (Lobell et al. 2012). Natural calamities like devastating rains and droughts are predicted to increase (Beddington et al. 2012; Hakeem 2015). Warming trends are expected to decrease global yield of agricultural crops by 1.5 % per decade.

The scientists have been working to revitalize sustainable increase in yield with fewer resources and several frameworks like ecological intensification, evergreen revolution and sustainable intensification have been suggested in the past (Cassman 1999a, b; Swaminathan 2000; Fan et al. 2012). Here a question arises that how can we achieve the objective of increased yield while having several constraints (land and water availability, climate change and environmental degradation).

For this, emphasis must be given to the challenge of applying good governance in modification of suboptimal crop and soil management with the prevailing knowledge of agricultural technologies and introducing advances in crop productivity. Two strategies will help to achieve these goals i) management of integrated crop-soil system which will deal with the existing limitations in the crop cultivars ii) development of new high yielding cultivars which may utilizes less water and nutrients and are more resistant to stresses like drought, pest attack, disease, waterlogging etc. (Fan et al. 2012).

22 Judicious Use and Improvement of Existing Resources and Technologies

Due to CO_2 fertilization the crop yield has increased over the past so many years however, the degradation of existing land and water resources and non-judicious crop management practices are very common. The available evidence shows that there is a huge gap between the total crop potential yield and the average farm yield at the farmers field (Fan et al. 2012). There are several factors which are responsible for this which include no to limited access to technologies, marketing problems and low profitability and poor crop and soil management (Fan et al. 2009). Across the globe, several cost effective and easy to use technologies have been developed and their use at the farmers scale must be emphasized, which can increase yield of grains by 9.2–14.6 % and can improve nitrogen productivity by 10.5–18.5 %. Split use of nitrogen, and changes in transplanting patterns have enhanced yield of rice up to 22 % in China (Fan et al. 2009). Similarly, water saving practices like alternate wetting and drying, irrigation for rice can increase rice yield (Davies et al. 2010). Other techniques for example mulching, deficit watering for upland crops and alternate furrow irrigation in maize have also been reported to increase yield (Yang and Zhang et al. 2010; Wang et al. 2009b). Decrease in emission of greenhouse gases can be achieved by adopting nitrogen management practices i.e. N_2O and CO_2 (Huang and Tang 2010).

For the adoption of new technologies, it must be ensured that all farmers have access and purchasing power; for this purpose economic incentives can play important role. Farming subsidies may be beneficial to motivate farmers to adopt new technologies and suitable management practices.

23 Innovations in Crop Production

For ensuring food security greater improvement and innovation in crop production must be carried out by developing a multidisciplinary approach including the joint ventures of plant scientists, soil scientist, agronomists, social scientist, agro-ecologists, plant breeders and microbiologists. This approach will help to understand coupling mechanisms that exist between climate and crops, soil and plant ecology and plant biology and various rhizospheric components and their management (Yang and Zhang 2010). For this purpose three points must be emphasized i) integration of soil fertility and nutrients management with intensive cultivation systems ii) utilization of different nutrient resources must be integrated with supply to the crop needs iii) take all possible measures to maintain soil fertility and quality (Zhang et al. 2011).

Genetic improvement in the crop cultivars with improved yield potential through conventional and genetic engineering will be critical for future food security (Foulkes et al. 2010). Yield potential has been defined as the yield of a crop under optimum growing conditions (Evans 1996). When crop reaches to 80 % of its potential it becomes very difficult to improve it on sustainable basis through conventional practices. It suggests that at this stage the improvement of a crop will depend on the improvement of yield potential. Here we need to breed cultivars which have high yield potential, or resource efficient and resistant to biotic and abiotic stresses (Morison et al. 2008). Conventional breeding techniques must be combined with advanced breeding methods such as genetic engineering and marker based selection. This will help more specific selection of required germplasm among multiple traits and breeding cycle will be fast. This technology will help to achieve the challenge of identification of the suitable genes needed for breeding, their incorporation in to elite genotypes and evaluation in the field trials, adopting new genetically modified crops and increasing consumer's acceptance (Zhang et al. 2007).

References

Ainsworth EA, Long SP (2005) What have we learned from 15 years of free air CO_2 enrichment (FACE)? A meta-analytic review of the responses of photosynthesis, canopy properties and plant production to rising CO_2. New Phytol 165:351–371

Alo CA, Wang GL (2008) Potential future changes of the terrestrial ecosystem based on climate projections by eight general circulation models. J Geophy Res Biogeosci 113:16

Armstrong RD, Millar G, Halpin NV, Reid DJ, Standley J (2003) Using zero tillage, fertilizers and legume rotations to maintain productivity and soil fertility in opportunity cropping systems on a shallow Vertosol. Aust J Exp Agric 43:141–153

Aune JA, Lal R (1997) Agricultural productivity in the tropics and critical limits of properties of oxisols, ultisols, and alfisols. Trop Agric 74:96–103

Barraclough D, Smith P, Worrall F, Black HIJ, Bhogal A (2015) Is there an impact of climate change on soil carbon contents in England and Wales? Eur J Soil Sci 66:451–462

Bauer A, Black AL (1994) Quantification of the effect of soil organic matter content on soil productivity. Soil Sci Soc Am J 58:185–193

Beaumont LJ, Pitman A, Perkins S, Zimmermann NE, Yoccoz NG, Thuiller W (2011) Impacts of climate change on the world's most exceptional eco-regions. Proc Natl Acad Sci U S A 108:2306–2311

Beddington JR, Asaduzzaman M, Clark ME, Fernández Bremauntz A, Guillou MD, Howlett DJB, Jahn MM, Lin E, Mamo T, Negra C, Nobre CA, Scholes RJ, Van Bo N, Wakhungu J (2012) What next for agriculture after Durban? Science 335:289–290

Bell G, Gonzalez A (2009) Evolutionary rescue can prevent extinction following environmental change. Ecol Lett 12:942–948

Bellamy PH, Loveland PJ, Bradley RI, Lark RM, Kirk GJD (2005) Carbon losses from all soils across England and Wales 1978–2003. Nature 437:245–248

Bellard C, Bertelsmeier C, Leadley P, Thuiller W, Courchamp F (2012) Impacts of climate change on the future of biodiversity. Ecol Lett 15:365–377

Biswas DK, Xu H, Li YG, Sun JZ, Wang XZ, Han XG, Jiang GM (2008) Genotypic differences in leaf biochemical, physiological and growth responses to ozone in 20 winter wheat cultivars released over the past. Glob Change Biol 14:46–59

Bond-Lamberty B, Thomson A (2010) Temperature-associated increases in the global soil respiration record. Nature 464:579–582

Botkin DB, Saxe H, Araujo MB, Betts R, Bradshaw RHW, Cedhagen T, Chesson P, Dawson TP, Etterson JR, Faith DR, Ferrier S, Guisan A, Hansen AS, Hilbert DW, Loehle C, Margules C, New M, Sobel MJ, Stockwell DRB (2007) Forecasting the effects of global warming on biodiversity. Bioscience 57:227–236

Bullister JL (2015) Atmospheric histories (1765–2015) for CFC-11, CFC-12, CFC-113, CCl4, SF6 and N_2O. US department of energy. Available from: http://cdiac.ornl.gov/ftp/oceans/CFC_ATM_Hist/CFC_ATM_Hist_2015/NDP_095(2015).pdf. 20 Sept 2015

Campbell A, Kapos V, Scharlemann JPW, Bubb P, Chenery A, Coad L, Dickson B, Doswald N, Khan MSI, Kershaw F, Rashid M (2009) Review of the literature on the links between biodiversity and climate change: impacts, adaptation and mitigation. In: Diversity SotCoB (ed) CBD technical series n°42. Secretariat of the Convention on Biological Diversity, Montreal, p 124

Cape JN (2008) Surface ozone concentrations and ecosystem health: past trends and a guide to future projections. Sci Total Environ 400:257–269

Cassman K (1999a) Ecological intensification of cereal production systems: yield potential, soil quality, and precision agriculture. Proc Natl Acad Sci U S A 96:5952–5959

Cassman K (1999b) Ecological intensification of cereal production systems: yield potential, soil quality, and precision agriculture. Proc Natl Acad Sci USA 96:5952–5959

CCSP (2008a) Analyses of the effects of global change on human health and welfare and human systems. A report by the U.S. climate change science program and the subcommittee on global change research. In: Gamble JL (eds) Ebi KL, Sussman FG, Wilbanks TJ (authors) U.S. Environmental Protection Agency, Washington, DC

CCSP (2008b) Preliminary review of adaptation options for climate-sensitive ecosystems and resources. A report by the U.S. climate change science program and the subcommittee on global change research. Julius SH, West JS (eds) Baron JS, Griffith B, Joyce LA, Kareiva P, Keller BD, Palmer MA, Peterson CH, Scott JM (authors). U.S. Environmental Protection Agency, Washington, DC

Charmantier A, McCleery RH, Cole LR, Perrins C, Kruuk LEB, Sheldon BC (2008) Adaptive phenotypic plasticity in response to climate change in a wild bird population. Science 320:800–803

Chevin LM, Lande R, Mace GM (2010) Adaptation, plasticity and extinction in a changing environment: towards a predictive theory. PLoS Biol 8:e1000357

Cowie J (2013) Climate change: biological and human aspects. 2nd edn. Cambridge University Press, Cambridge, pp 4–256

Crafts-Brandner SJ, Salvucci ME (2002) Sensitivity of photosynthesis in a C4 plant, maize, to heat stress. Plant Physiol 129:1773–1780
Davies WJ, Zhang J, Yang J, Dodd IC (2010) Novel crop science to improve yield and resource use efficiency in water-limited agriculture. J Agric Sci 149:123–131
Dawson TP, Jackson ST, House JI, Prentice IC, Mace GM (2011) Beyond predictions: biodiversity conservation in a changing climate. Science 332:53–58
Díaz-Zorita MN, Duarte GA, Grove JH (2002) A review of no-till systems and soil management for sustainable crop production in the sub-humid and semiarid Pampas of Argentina. Soil Tillage Res 65:1–18
Easterling DR, Wehner MF (2009) Is the climate warming or cooling? Geophys Res Lett L08706
Emmet BA, Reynolds B, Chamberlain PM, Rowe E, Spurgeon D, Brittain SA, Frogbrook Z, Hughes S, Lawlor AJ, Poskitt J, Potter E, Robinson DA, Scott A, Wood C, Woods C (2010) Countryside survey: soils report from 2007. Technical report no 9/07, NERC/Centre for Ecology and Hydrology, Wallingford
EPA (2015) Climate impacts on agriculture and food supply. Available from: http://www.epa.gov/climatechange/impacts/agriculture.html. 15 Sept 2015
Evans L (1996) Crop evolution, adaptation and yield. Cambridge University Press, Cambridge
Fan M, Lu S, Jiang R, Liu X, Zhang F (2009) Triangular transplanting pattern and split nitrogen fertilizer application increase rice yield and nitrogen fertilizer recovery. Agron J 101:1421–1425
Fan M, Shen J, Yuan L, Jiang R, Chen X, Davies WJ, Zhang F (2012) Improving crop productivity and resource use efficiency to ensure food security and environmental quality in China. J Exp Bot 63:13–24
Fantappie M, L'Abate G, Costanini EAC (2011) The influence of climate change on the soil organic carbon content in Italy from 1961 to 2008. Geomorphol 135:343–352
Field CB, Mortsch LD, Brklacich M, Forbes DL, Kovacs P, Patz JA, Running SW, Scott MJ (2007) North America. In: Climate change. 2007: impacts, adaptation and vulnerability. Contribution of working group II to the fourth assessment report of the Intergovernmental Panel on Climate Change. In: Parry ML, Canziani OF, Palutikof JP, van der Linden PJ, Hanson CE (eds) Cambridge University Press, Cambridge
Forero-Medina G, Joppa L, Pimm SL (2010) Constraints to species elevational range shifts as climate changes. Conserv Biol 25:163–171
Foukal P, North G, Wigley T (2004) A stellar view on solar variations and climate. Science 306:68–69
Foulkes MJ, Slafer GA, Davies WJ, Berry PM, Sylvester- Bradley R, Martre P, Calderini DF, Reynolds MP (2010) Raising yield potential in wheat: optimizing partitioning to grain yield while maintaining lodging resistance. J Exp Bot 62:469–486
Gilman SE, Urban MC, Tewksbury J, Gilchrist GW, Holt RD (2010) A framework for community interactions under climate change. Trend Ecol Evol 25:325–331
Global Carbon Project (2011) Carbon budget and trends 2010. Available from: http://www.globalcarbonproject.org/carbonbudget. Sept 12 2015
Grace PR, Oades JM, Keith H, Hancock TW (1995) Trends in wheat yields and soil organic carbon in the permanent rotation trial at the waite agricultural research institute, South Australia. Aust J Exp Agric 35:857–864
Hakeem KR (2015) Crop production and global environmental issues. Springer International Publishing AG, Cham, p 598
Hakeem KR, Sabir M, Ozturk M, Mermut A (2014) Soil remediation and plants: prospects and challenges. Academic/Elsevier, New York, p 724
Hatfield JL, Boote KJ, Kimball BA, Ziska LH, Izaurralde RC, Ort D, Thomson AM, Wolfe D (2011) Climate impacts on agriculture: implications for crop production. Agron J 103:351–370
Hoegh-Guldberg O, Mumby PJ, Hooten AJ, Steneck RS, Greenfield P, Gomez E, Harvell CD, Sale PF, Edwards AJ, Caldeira K, Knowlton N, Eakin CM, Iglesias-Prieto R, Muthiga N, Bradbury

RH, Dubi A, Hatziolos ME (2007) Coral reefs under rapid climate change and ocean acidification. Science 318:1737–1742

Hoffman AA, Sgro CM (2011) Climate change and evolutionary adaptation. Nature 470:479–485

Huang Y, Tang Y (2010) An estimate of greenhouse gas (N_2O and CO_2) mitigation potential under various scenarios of nitrogen use efficiency in Chinese croplands. Glob Change Biol 16:2958–2970

Hulme M (2009) On the origin of 'the greenhouse effect': John Tyndall's 1859 interrogation of nature. Weather 64:121–123

IFDC (2000) International fertilizer development Centre. World fertilizer consumption, Muscle Shoals

IPCC (2001) Climate change 2001: the scientific basis. Intergovernmental Panel on Climate Change. Cambridge University Press, Cambridge

Johnson E, Heinen R (2004) Carbon trading: time for industry involvement. Environ Intl 30:279–288

Johnston AE (1986) Soil organic matter, effects on soils and crops. Soil Use Manag 2:97–105

Kanchikerimath M, Singh D (2001) Soil organic matter and biological properties after 26 years of maize–wheat–cowpea cropping as affected by manure and fertilization in a Cambisol in semi-arid region of India. Agri Ecosys Environ 86:155–162

Kapkiyai JJ, Karanja NK, Qureshi JN, Smithson PC, Woomer PL (1999) Soil organic matter and nutrient dynamics in a Kenyan nitisol under long-term fertilizer and organic input management. Soil Biol Biochem 31:1773–1782

Kiers ET, Palmer TM, Ives AR, Bruno JF, Bronstein JL (2010) Mutualisms in a changing world: an evolutionary perspective. Ecol Lett 13:1459–1474

Koh LP, Dunn RR, Sodhi NS, Colwell RK, Proctor HC, Smith VS (2004) Species co-extinctions and the biodiversity crisis. Science 305:1632–1634

Lal R (2003a) Global potential of soil carbon sequestration to mitigate the greenhouse effect. Crit Rev Plant Sci 22:151–184

Lal R (2003b) Soil erosion and the global carbon budget. Environ Intl 29:437–450

Lal R (2004) Soil carbon sequestration impacts on global climate change and food security. Science 304:1623–1627

Lal R, Kimble JM, Follett RF (2001) Methodological challenges toward balance soil C pools and fluxes. In: Lal R, Kimble JM, Follett RF, Stewart BA (eds) Assessment methods for soil carbon. Lewis Publishers, Boca Raton

Lapola DM, Oyama MD, Nobre CA (2009) Exploring the range of climate biome projections for tropical South America: the role of CO_2 fertilization and seasonality. Global Biogeochem Cy 23:1–16

Lavergne S, Mouquet N, Thuiller W, Ronce O (2010) Biodiversity and climate change: integrating evolutionary and ecological responses of species and communities. Ann Rev Ecol Evol Syst 41:321–350

Leadley P, Pereira HM, Alkemade R, Fernandez-Manjarres JF, Proenca V, Scharlemann JPW, Walpole MJ (2010) Biodiversity scenarios: projections of 21st century change in biodiversity and associated ecosystem services. In: Secretariat of the convention on biological diversity (ed. Diversity SotCoB). Technical series no. 50. Published by the Secretariat of the Convention on Biological Diversity, Montreal, pp 1–132

Leung DYC, Caramanna G, Maroto-Valer M (2014) An overview of current status of carbon dioxide capture and storage technologies. Renew Sust Rev 39:426–443

Levy R (ed) (1984) Chemistry of irrigated soils. Van Nostrand-Reinhold, New York, pp 182–229

Lobell DB, Gourdji SM (2012) The influence of climate change on global crop productivity. Plant Physiol 160:1686–1697

Lobell DB, Bonfils C, Duffy PB (2007) Climate change uncertainty for daily minimum and maximum temperatures: a model inter-comparison. Geophys Res Lett 34:L05715

McKee IF, Mulholland BJ, Craigon J, Black CR, Long SP (2000) Elevated concentrations of atmospheric CO_2 protect against and compensate for O_3 damage to photosynthetic tissues of field-grown wheat. New Phytol 146:427–435
McMahon SM, Harrison SP, Armbruster WS, Patrick BPJ, Beale CM, Edwards ME, Kattge J, Midgley G, Morin X, Prentice IC (2011) Improving assessment and modelling of climate change impacts on global terrestrial biodiversity. Trend Ecol Evol 26:249–259
Meehl GA, Stocker TF, Collins WD, Friedlingstein P, Gaye AT, Gregory JM, Kitoh A, Knutti R, Murphy JM, Noda A, Raper SCB, Watterson IG, Weaver AJ, Zhao ZC (2007) Global climate projections. In: Solomon S, Qin D, Manning M, Chen Z, Marquis M, Averyt KB, Tignor M, Miller HL (eds) Climate change 2007: the physical science basis. Cambridge University Press, Cambridge, pp 747–845
Meyers LA, Bull JJ (2002) Fighting change with change: adaptive variation in an uncertain world. Trend Ecol Evol 17:551–557
Morison JIL, Baker NR, Mullineaux PM, Davies WJ (2008) Improving water use in crop production. Philos Trans R Soc Spec Issue Sust Agric 363:639–658
National Oceanic and Atmospheric Administration (2015) Climate monitoring, greenhouse gases. Available from: https://www.ncdc.noaa.gov/monitoring references/faq/greenhouse-gases.php. 15 Sept 2015
Nordt LC, Wilding LP, Drees LR (2000) Pedogenic carbonate transformation in leaching soil sysems: implications for the global C cycle. In: Lal R, Kimble JM, Eswaran H, Stewart BA (eds) Global climate change and pedogenic carbonate. Lewis Publishers, Boca Raton, pp 43–64
Oltmans SJ, Lefohn AS, Harris JM, Galbally I, Scheel HE, Bodeker G, Brunke E, Claude H, Tarasick D, Johnson BJ, Simmonds P, Shadwick D, Anlauf K, Hayden K, Schmidlin F, Fujimoto T, Akagi K, Meyer C, Nichol S, Davies J, Redondas A, Cuevas E (2006) Long-term changes in tropospheric ozone. Atm Environ 40:3156–3173
Parmesan C (2006) Ecological and evolutionary responses to recent climate change. Ecol Evol 37:637–669
Parmesan C, Duarte CM, Poloczanska E, Richardson AJ, Singer MC (2011) Overstretching attribution. Nat Clim Change 1:2–4
Pereira HM, Leadley PW, Proenca V, Alkemade R, Scharlemann JPW, Fernandez-Manjarres JF, Araújo MB (2010) Scenarios for global biodiversity in the 21st century. Science 330:1496–1501
Petchawee S, Chaitep W (1995) Organic matter management for sustainable agriculture. In: LeFroy RDB, Blaci GJ, Craswell ET (eds) Organic matter management in upland systems in Thailand. ICIAR, Canberra, pp 21–26
Phillips BL (2009) The evolution of growth rates on an expanding range edge. Biol Lett 5:802–804
Pieri C (1986) Fertilisation des cultures vivrie'res et fertilite' des sols en agriculture paysanne subsaharienne. Agronomie Tropicale 41:1–20
Rafferty NE, Ives AR (2010) Effects of experimental shifts in flowering phenology on plant–pollinator interactions. Ecol Lett 14:69–74
Ray JD, Gesch RW, Sinclair TR, Hartwell AL (2002) The effect of vapor pressure deficit on maize transpiration response to a drying soil. Plant Soil 239:113–121
Reynolds MP, Balota M, Delgado M, Amani I, Fischer RA (1994) Physiological and morphological traits associated with spring wheat yield under hot, irrigated conditions. Aust J Plant Physiol 21:717–730
Rosegrant MW, Cline SA (2003) Global food security: challenges and policies. Science 302:1917–1919
Sala OE, Detlef van Vuuren D, Pereira HM, Lodge D, Alder J, Cumming G, Dobson A, Wolters V, Xenopoulos MA (2005) Chapter 10: biodiversity across scenarios. In: Ecosystems and human well-being, vol. 2: scenarios. Island Press, Washington, DC, pp 375–408
Salamin N, Wüest RO, Lavergne S, Thuiller W, Pearman PB (2010) Assessing rapid evolution in a changing environment. Tren Ecol Evol 25:692–698

Sanchez PA (2002) Soil fertility and hunger in Africa. Science 295:2019–2020
Sartaj AW, Sofi MN, Chand S, Hakeem KR (2016) Soil carbon sequestration: as a climate change adaptation and mitigation strategy – An overview. Int J Plant Anim Environ Sci 6(1):227–232
Sauerbeck DR (2001) CO_2 emissions and C sequestration by agriculture – perspectives and limitations. Nutr Cycl Agroecosys 60:253–266
Silver WL, Ostertag R, Lugo AE (2000) The potential for carbon sequestration through reforestation of abandoned tropical agricultural and pasture lands. Restor Ecol 8:394–407
Six J, Feller C, Denef K, Ogle S, Moraes JC, Albrecht A (2002) Soil organic matter, biota and aggregation in temperate and tropical soils – effects of no-tillage. Agronomie 22:755–775
Smith P, Goulding K, Smith K, Powlson D, Smith J, Falloon P, Coleman K (2001) Enhancing the carbon sink in European agricultural soils: including trace gas fluxes in estimates of carbon mitigation potential. Nutr Cycl Agroecosys 60:237–252
Smith P, Fang C, Dawson JJC, Moncreiff JB (2008) Impact of global warming on soil carbon. Adv Agron 97:1–43
Stone P (2001) The effects of heat stress on cereal yield and quality. In: Basra AS (ed) Crop responses and adaptations to temperature stress. Food Products Press, Binghamton, pp 243–291
Swaminathan M (2000) An evergreen revolution. Biol Inst Biol 47:85–89
Taub DR, Miller B, Allen H (2008) Effects of elevated CO_2 on the protein concentration of food crops: a meta-analysis. Glob Change Biol 14:565–575
Teixeira EI, Fischer G, van Velthuizen H, Walter C, Ewert F (2013) Global hot-spots of heat stress on agricultural crops due to climate change. Agri Forest Meteor 170:206–215
Tewksbury JJ, Huey RB, Deutsch CA (2008) Ecology – putting the heat on tropical animals. Science 320:1296–1297
Tomlinson RW, Milne RM (2006) Soil carbon stocks and land cover in Northern Ireland from 1939 to 2000. Appl Geogr 26:18–39
Tyndall J (1861) On the absorption and radiation of heat by gases and vapours. Philos Mag 22:169–285
U.S. Census Bureau (2011) The 2011 statistical abstract: international statistics. U.S. Census Bureau, Washington, DC USA http://www.census.gov/library/publications/2010/compendia/statab/130ed.html
USGCRP (2009) Global climate change impacts in the United States. In: Karl TR, Melillo JM, Peterson TC (eds) United States global change research program. Cambridge University Press, New York
Van Dingenen R, Raes F, Krol MC, Emberson L, Cofala J (2009) The global impact of O_3 on agricultural crop yields under current and future air quality legislation. Atmos Environ 43:604–618
Visser ME (2008) Keeping up with a warming world; assessing the rate of adaptation to climate change. Proc R Soc B-Biol Sci 275:649–659
Walther GR (2010) Community and ecosystem responses to recent climate change. Phil Trans Soc B-Biol Sci 365:2019–2024
Wang Y, Xie Z, Malhi S, Vera C, Wang J (2009) Effects of rainfall harvesting and mulching technologies on water use efficiency and crop yield in the semi-arid Loess Plateau, China. Agric Water Manag 96:374–382
West TO, Post WM (2002) Soil organic carbon sequestration rates by tillage and crop rotation: a global data analysis. Soil Sci Soc Amer J 66:1930–1940
Wild M (2012) Enlightening global dimming and brightening. Bull Am Meteorol Soc 93:27–37
Wilkinson S, Mills G, Illidge R, Davies WJ (2012) How is ozone pollution reducing our food supply? J Exp Bot 63:527–536
Yamoah CF, Bationo A, Shapiro B, Koala S (2002) Trend and stability analyses of millet yields treated with fertilizer and crop residues in the Sahel. Field Crops Res 75:53–62
Yang LH, Rudolf VHW (2010) Phenology, ontogeny and the effects of climate change on the timing of species interactions. Ecol Lett 13:1–10

Yang J, Zhang J (2010) Crop management techniques to enhance harvest index in rice. J Exp Bot 61:3177–3189

Zhang F, Cui Z, Wang J, Li C, Chen X (2007) Current status of soil and plant nutrient management in China and improvement strategies. Chinese Bull Bot 24:687–694

Zhang F, Shen J, Zhang J, Zuo Y, Li L, Chen X (2010) Rhizosphere processes and management for improving nutrient use efficiency and crop productivity: implications for China. Adv Agron 107:1–32

Zhang F, Cui Z, Fan M, Zhang W, Chen X, Jiang R (2011) Integrated soil–crop systems management: reducing environmental risk while increasing crop productivity and improving nutrient use efficiency in China. J Environ Qual 40:1051–1057

Ziska LH, Blumenthal DM, Runion GB, Hunt ER, Diaz-Soltero H (2011) Invasive species and climate change: an agronomic perspective. Climatic Change 105:13–42

Index

K.R. Hakeem et al. (eds.), *Soil Science: Agricultural and Environmental Prospectives*, DOI 10.1007/978-3-319-34451-5

MIX
Papier aus verantwortungsvollen Quellen
Paper from responsible sources
FSC® C105338

If you have any concerns about our products,
you can contact us on
ProductSafety@springernature.com

In case Publisher is established outside the EU,
the EU authorized representative is:
Springer Nature Customer Service Center GmbH
Europaplatz 3, 69115 Heidelberg, Germany

Printed by Libri Plureos GmbH
in Hamburg, Germany